I0065920

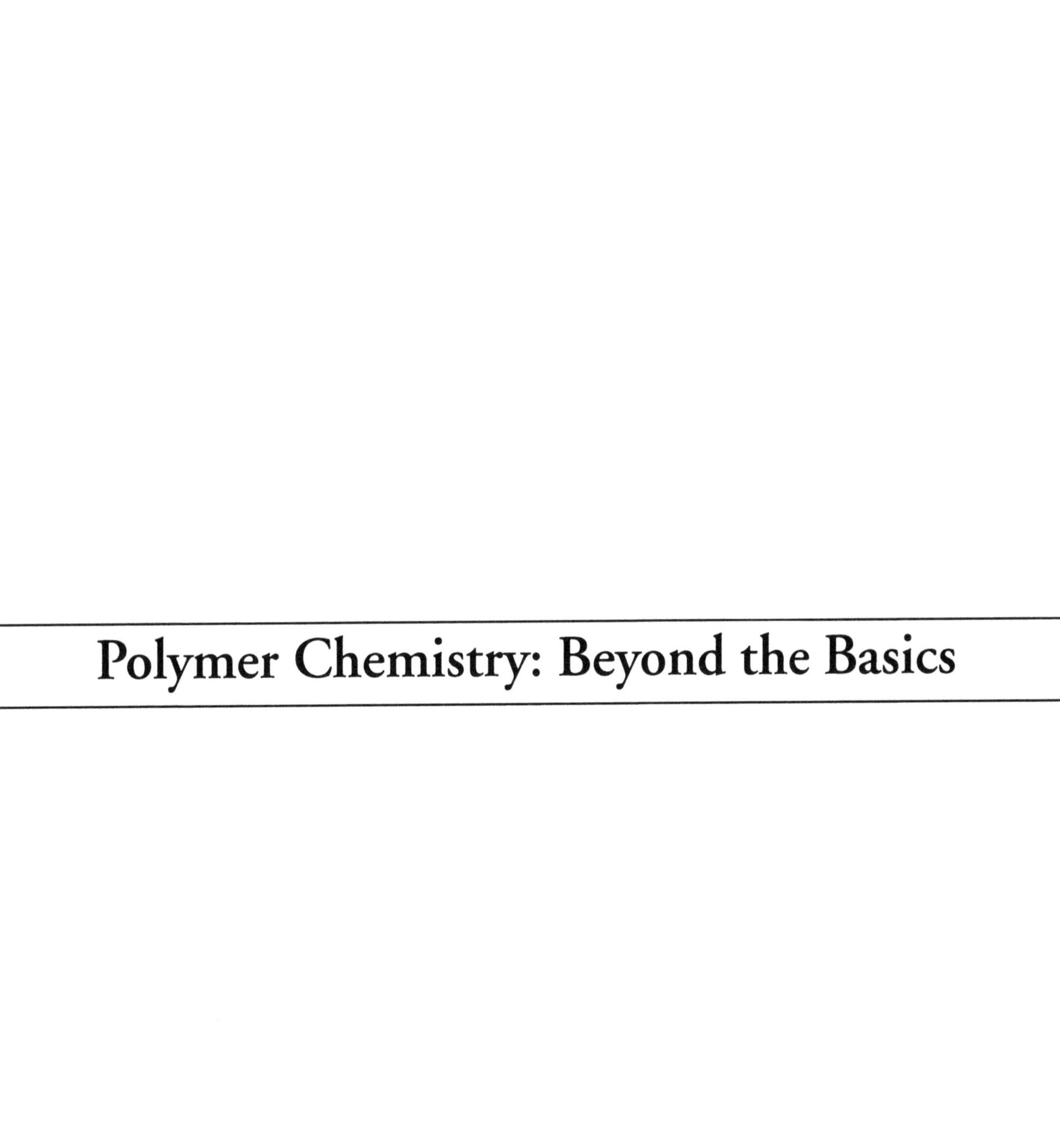

Polymer Chemistry: Beyond the Basics

Polymer Chemistry: Beyond the Basics

Editor: Mick Reece

New York

Published by NY Research Press
118-35 Queens Blvd., Suite 400,
Forest Hills, NY 11375, USA
www.nyresearchpress.com

Polymer Chemistry: Beyond the Basics
Edited by Mick Reece

International Standard Book Number: 978-1-63238-588-8 (Hardback)

Cataloging-in-Publication Data

Polymer chemistry : beyond the basics / edited by Mick Reece.
p. cm.
Includes bibliographical references and index.
ISBN 978-1-63238-588-8
1. Polymers. 2. Polymerization. I. Reece, Mick.
QD381 .P65 2018
547.7--dc23

Contents

Preface

This book aims to highlight the current researches and provides a platform to further the scope of innovations in this area. This book is a product of the combined efforts of many researchers and scientists, after going through thorough studies and analysis from different parts of the world. The objective of this book is to provide the readers with the latest information of the field.

Polymer chemistry is a sub-discipline of chemistry that deals with properties, structures and chemical synthesis of polymers. It helps in the creation and analysis of polymers with specific physical and chemical properties. Synthetic polymers have a wide range of industrial applications such as in electronics, construction, packaging, agriculture, biotechnology, etc. There has been rapid progress in this field and its applications are finding their way across multiple industries. Scientists and students actively engaged in this field will find this book full of crucial and unexplored concepts.

I would like to express my sincere thanks to the authors for their dedicated efforts in the completion of this book. I acknowledge the efforts of the publisher for providing constant support. Lastly, I would like to thank my family for their support in all academic endeavors.

Editor

1

The Growth and Survival of *Mycobacterium smegmatis* Is Enhanced by Co-Metabolism of Atmospheric H_2

Chris Greening[1], Silas G. Villas-Bôas[2], Jennifer R. Robson[1], Michael Berney[1,3], Gregory M. Cook[1]*

1 University of Otago, Department of Microbiology and Immunology, Dunedin, New Zealand, **2** University of Auckland, The Centre for Microbial Innovation, Auckland, New Zealand, **3** Albert Einstein College of Medicine, Department of Microbiology and Immunology, Bronx, New York, United States of America

Abstract

The soil bacterium *Mycobacterium smegmatis* is able to scavenge the trace concentrations of H_2 present in the atmosphere, but the physiological function and importance of this activity is not understood. We have shown that atmospheric H_2 oxidation in this organism depends on two phylogenetically and kinetically distinct high-affinity hydrogenases, Hyd1 (MSMEG_2262-2263) and Hyd2 (MSMEG_2720-2719). In this study, we explored the effect of deleting Hyd2 on cellular physiology by comparing the viability, energetics, transcriptomes, and metabolomes of wild-type vs. Δ*hyd2* cells. The long-term survival of the Δ*hyd2* mutant was significantly reduced compared to the wild-type. The mutant additionally grew less efficiently in a range of conditions, most notably during metabolism of short-chain fatty acids; there was a twofold reduction in growth rate and growth yield of the Δ*hyd2* strain when acetate served as the sole carbon source. Hyd1 compensated for loss of Hyd2 when cells were grown in a high H_2 atmosphere. Analysis of cellular parameters showed that Hyd2 was not necessary to generate the membrane potential, maintain intracellular pH homeostasis, or sustain redox balance. However, microarray analysis indicated that Δ*hyd2* cells were starved for reductant and compensated by rewiring central metabolism; transcripts encoding proteins responsible for oxidative decarboxylation pathways, the urea cycle, and ABC transporter-mediated import were significantly more abundant in the Δ*hyd2* mutant. Metabolome profiling consistently revealed an increase in intracellular amino acids in the Δ*hyd2* mutant. We propose that atmospheric H_2 oxidation has two major roles in mycobacterial cells: to generate reductant during mixotrophic growth and to sustain the respiratory chain during dormancy.

Editor: Riccardo Manganelli, University of Padova, Medical School, Italy

Funding: This work and MB were financially supported by a Marsden Grant from the Royal Society of New Zealand and NIH grant AI26170. CG was supported by an Otago Postgraduate Scholarship from the University of Otago. GMC is supported by a James Cook Fellowship from the Royal Society of New Zealand. The funders had no role in study design, data collection and analysis, decision to publish, or preparation of the manuscript.

Competing Interests: The authors have declared that no competing interests exist.

* Email: gregory.cook@otago.ac.nz

Introduction

In recent years, it has emerged that a number of soil *Actinobacteria* of the genera *Mycobacterium*, *Streptomyces*, and *Rhodococcus* oxidise the trace concentrations of H_2 found in the lower atmosphere [1,2,3,4]. In addition to being biogeochemically important [5], scavenging of tropospheric H_2 is physiologically unusual; all other characterised hydrogen-oxidising organisms are only capable of recycling the high concentrations of H_2 evolved through other biological processes or geothermal activity [6]. The purpose and importance of hydrogen scavenging in the physiology of *Actinobacteria* nevertheless remains to be understood. It is also to be determined whether this process influences the composition of microorganisms in soil ecosystems.

Work in our laboratory has resolved the determinants of hydrogen scavenging. The soil bacterium *Mycobacterium smegmatis* catalyses atmospheric H_2 oxidation using two high-affinity, membrane-associated, oxygen-dependent [NiFe]-hydrogenases [3]. Both of these enzymes are expressed during exponential growth, though their expression and activity is significantly higher during the transition to stationary phase due to carbon-limitation. The fast-acting Group 2a [NiFe]-hydrogenase Hyd1 (MSMEG_2262-2263) is responsible for the majority of whole-cell H_2 oxidation. In contrast, the Group 5 [NiFe]-hydrogenase Hyd2 (MSMEG_2720-2719) is a much slower-acting enzyme in whole-cells [7,3]. Despite its low activity, Hyd2 has been shown to be important for the growth of *M. smegmatis* [8]. Furthermore, orthologs of this enzyme are more widely distributed among sequenced *Actinobacteria* and are apparently responsible for the tropopheric H_2 uptake of streptomycetes and rhodococci [9,4]. It should also be noted that *M. smegmatis* also encodes a further hydrogenase, Hyd3; this enzyme is only expressed during oxygen-limitation, where we propose it serves to couple the reoxidation of NAD(P)H to the evolution of hydrogen [7,8].

In this work, we provide insight into the physiological role of hydrogen scavenging by observing the effect of deleting Hyd2 throughout exponential growth, upon entry into stationary phase, and during long-term survival. Using a combinatorial approach, we show that hydrogen scavenging is required for the efficient metabolism of certain carbon sources and infer that atmospheric H_2 is a source of reductant for mycobacterial metabolism.

Materials and Methods

Bacterial strains and growth conditions

All bacterial strains used in this study are listed in **Table S1**. *Mycobacterium smegmatis* mc^{2}155 [10] and derived mutants [7,8] were maintained on LB agar plates supplemented with 0.05% (w/v) Tyloxapol (Sigma-Aldrich). For broth culture, *M. smegmatis* was grown in Hartmans de Bont (HdB) minimal medium [11] supplemented with the stated carbon sources, 0.05% Tyloxapol, and 10 µM $NiSO_4$. Cultures were incubated at 37°C with agitation (200 rpm) in 30 mL medium in 125 mL aerated conical flasks. Culture volumes were upscaled to 500 mL in 2.5 L flasks for transcriptome analysis and 100 mL in 500 mL flasks for metabolome analysis. Cells were inoculated to an initial optical density of 0.005. Optical densities to assess growth were measured at 600 nm (OD_{600}) in a Jenway 6300 spectrometer. Cultures were diluted in 0.85% saline to bring the OD_{600} below 0.5 when measured in cuvettes of 1 cm light path length. To count colony forming units (CFU mL^{-1}), each culture was serially diluted in phosphate-buffered saline (PBS) (pH 7.0) and spotted on to agar plates [12]. A markerless deletion of the Hyd2 large subunit (MSMEG_2719) was complemented with a pOLYG vector containing the *hyd2* operon (MSMEG_2720-2718) in order to minimise disruption to hydrogenase maturation and folding [8]. β-galactosidase assays and amperometric hydrogen measurements were performed as previously described [7].

Challenge experiments

For acid challenge experiments, the strains were grown on HdB media at pH 7.0 to $OD_{600} = 1.0$. They were subsequently pelleted (7,000×g, 10 min, RT), washed in 100 mM citrate/phosphate buffer (pH 7.0), and resuspended in 100 mM citrate/phosphate buffer (pH 3.0 or pH 5.0). All buffer preparations contained 22 mM glycerol, 0.05% Tween80, and trace metals. Following acid challenge, the survival of cells was measured by measuring colony forming units (CFU mL^{-1}). The minimum inhibitory concentrations (MICs) of pH 5.0-challenged cells to the protonophore carbonyl cyanide *m*-chlorophenylhydrazone (CCCP) was determined using serial dilutions as previously described [13].

Measurement of internal pH and membrane potential

Internal pH and membrane potential was measured in *M. smegmatis* grown on HdB minimal medium at 2 h following the induction of stationary phase. Internal pH was calculated by determining the partitioning of a radioactive probe between intracellular and extracellular fractions. Cultures of 1 mL were incubated with 11 µM [^{14}C] benzoate (10–25 mCi mmol^{1}) (pH 7.5) (37°C, 10 min) and centrifuged through silicone oil (BDH Laboratory Supplies) (16,000×g, 5 min, RT). A 20 µl sample of the supernatant was removed. The tubes were otherwise frozen (−80°C, 60 min) and the cell pellets were removed with dog nail clippers. Samples of the supernatant (extracellular fraction) and pellet (intracellular fraction) were dissolved in scintillation fluid (Amersham). The relative concentrations of [^{14}C] benzoate in each sample was measured using a LKB Wallac 1214 Rackbeta liquid scintillation counter (Perkin Elmer Life Sciences). The internal pH was calculated from the uptake of [^{14}C] benzoate using the Henderson-Hasselbalch equation as previously described [14]. Membrane potential was measured by a equivalent method by detemining the partioning of 5 µM [^{3}H] methyltriphenylphosponium iodide ([^{3}H]TPP^{+}) (30–60 Ci mmol^{-1}). The membrane potential was calculated from the uptake of [^{3}H]TPP^{+} using the Nernst equation [15].

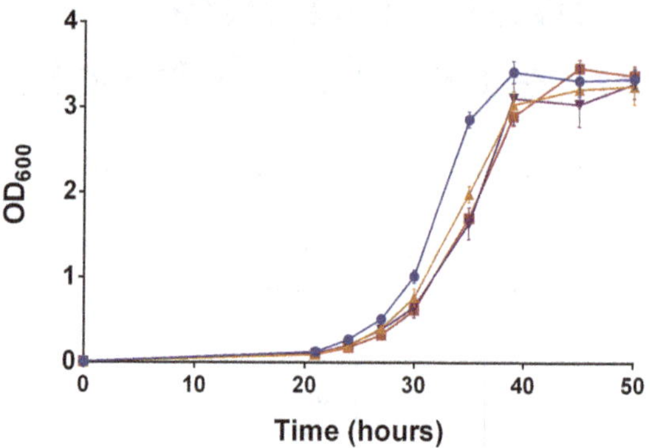

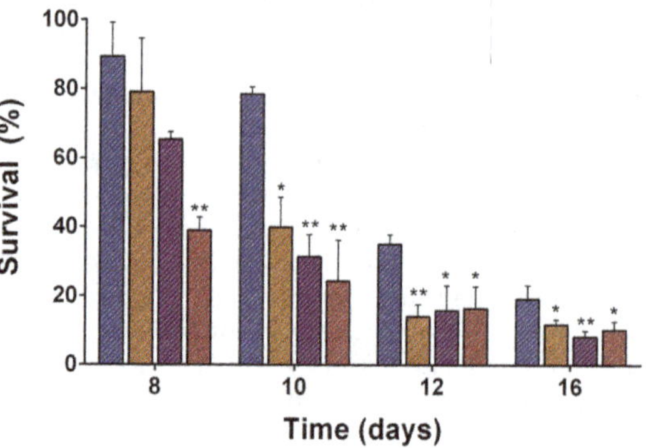

Figure 1. The growth and survival of *Mycobacterium smegmatis* in the presence and absences of hydrogenases. (A) Growth of *M. smegmatis* mc^{2}155 and *hyd* mutants into carbon-limitation. (B) Survival of *M. smegmatis* mc^{2}155 and *hyd* mutants during carbon-limitation. The strains were grown in aerated conical flasks on HdB minimal medium supplemented with 22 mM glycerol. Growth is shown in OD_{600}. Survival is shown in percentage colony forming units relative to day four. Legend: Blue circles/bars = Wild-type; Red squares/bars = Δ*hyd*123; Orange point-up triangles/bars = Δ*hyd*1; and Purple point-down triangles/bars = Δ*hyd*2. Error bars show standard deviations from biological triplicates. * = $p<0.05$, ** = $p<0.01$, *** = $p<0.001$ difference relative to wild-type bars (Student's T-test, unpaired, two-tailed).

Measurement of [NAD$^+$]/[NADH] ratios

1 mL cultures were centrifuged (150,000×g, 3 min, RT) and resuspended in either 0.3 mL 0.2 M HCl (for NAD$^+$ extraction) or 0.3 mL 0.2 M NaOH (for NADH extraction). The cultures were heated (55°C, 10 min), cooled (0°C, 5 min), and neutralized with either 0.3 mL 0.1 M NaOH (for NAD$^+$ extraction) or 0.3 mL 0.1 M HCl (for NADH extraction). After centrifugation (150,000×g, 3 min RT), the supernatants were collected. 200 µl supernatant was transferred into cuvettes containing 50 µl 1 M bicine (pH 8.0), 50 µl 40 mM EDTA (pH 8.0), 50 µl 4.2 mM 3-[4,5-dimethylthiazol-2-yl]-2,5-diphenyltetrazolium bromide, and 50 µl 16 mM phenazine ethosulfate. NAD$^+$ and NADH concentrations were measured by addition of 50 µl ethanol and 5 U yeast alcohol dehydrogenase II [16]. The rate of reduction of 3-[4,5-dimethylthiazol-2-yl]-2,5-diphenyltetrazolium bromide was measured photometrically at 570 nm and was proportional to the concentration of standards of each cofactor.

RNA extraction

M. smegmatis mc^{2}155 and Δ*hyd*2 cells were grown synchronously in aerated conical flasks. At 1 hour following entry into stationary phase due to carbon-limitation, the cells were harvested for RNA extraction and microarray analysis. 500 mL of each

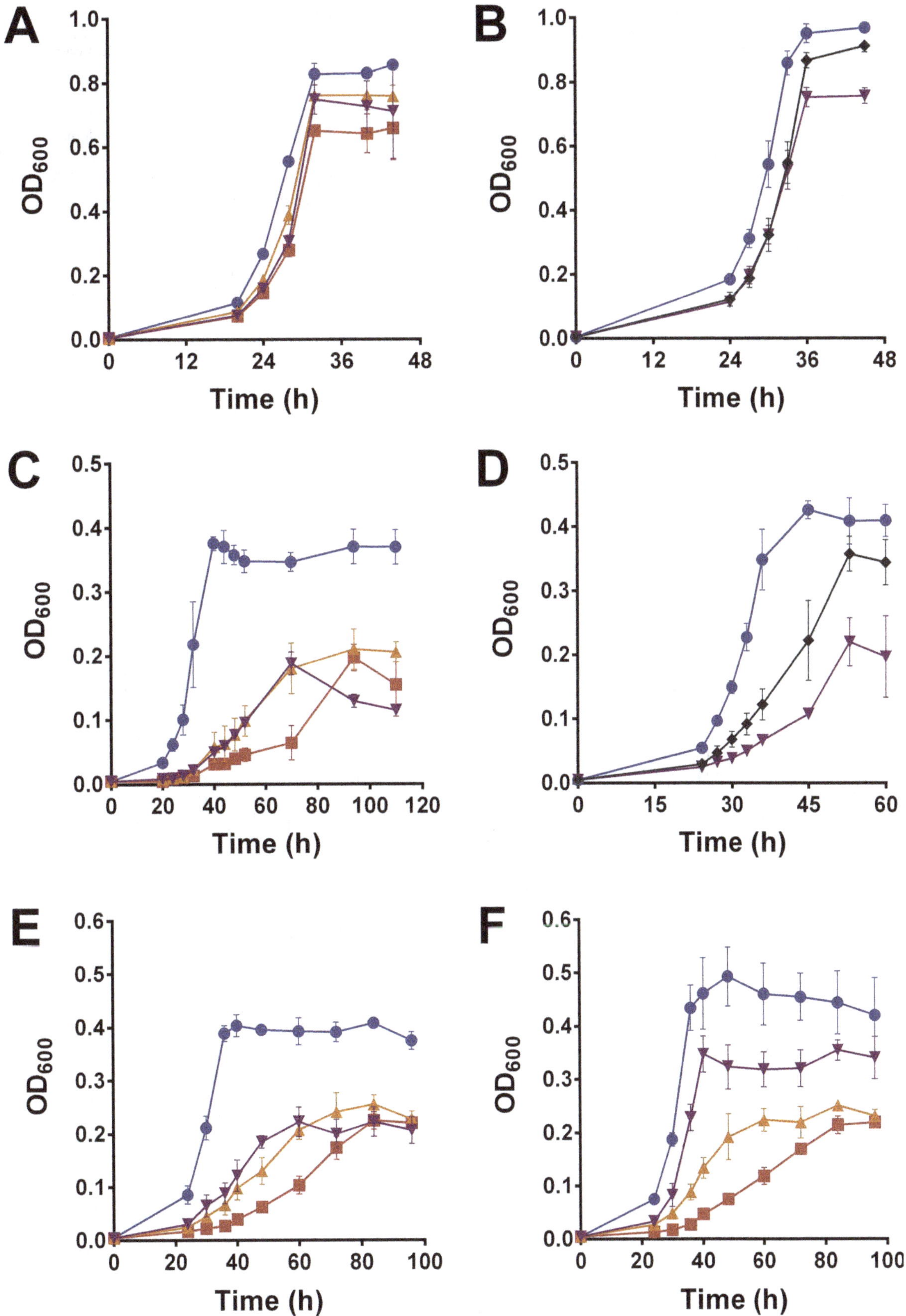

Figure 2. Observation, complementation, and recovery of mutant growth phenotypes. Strains were grown on HdB minimal medium. (A) Growth on 5.5 mM glycerol. (B) Complementation on 5.5 mM glycerol in the presence of 50 μg mL^{-1} hygromycin. (C) Growth on 12.5 mM acetate. (D) Partial complementation on 12.5 mM acetate in the presence of 50 μg mL^{-1} hygromycin. (E) Growth on 12.5 mM acetate in serum vials injected

with 10% pure N_2. (F) Growth on 12.5 mM acetate in serum vials injected with 10% pure H_2. Legend: Blue circles = Wild-type (or wild-type with empty pOLYG vector for complementation); Red squares = Δ*hyd*123; Orange point-up triangles = Δ*hyd*1; Purple point-down triangles = Δ*hyd*2 (or Δ*hyd*2 with empty pOLYG vector for complementation); and Grey diamonds = Δ*hyd*2 with pOLYG vector expressing MSMEG_2720-2719. Error bars show standard deviations from biological triplicates.

culture were mixed with 1000 mL cold glycerol saline (3:2 v/v) (-20°C), centrifuged (27000×g, 20 min, −20°C), and resuspended in glycerol saline (1:1 v/v) (−20°C). Cell lysis was achieved by three cycles of bead-beating in a Mini-Beadbeater (Biospec) at 5,000 rpm for 30 sec. Total RNA was extracted using TRIzol reagent (Invitrogen) according to the manufacturer's instructions. DNA was removed from the RNA preparation by treatment with 2 U RNase-free DNase using the TURBO DNA-*free* kit (Ambion), according to the manufacturer's instructions. The concentration and purity of the RNA was determined using a NanoDrop ND-1000 spectrophotometer, and its integrity was confirmed on a 1.2% agarose gel.

Microarray analysis

Transcriptome analysis employed glass slide DNA microarrays provided by the Pathogen Functional Genomics Research Center (PFGRC), which is funded by the National Institute of Allergy and Infectious Diseases. The arrays represented every open reading frame of the genome of *M. smegmatis* mc^2155 with 7,736 unique 70-mers spotted in triplicate. Samples for microarray analysis were prepared and hybridised based on standard operating protocols (SOP) M007 and M008 from The Institute of Genomic Research (TIGR) [17]. 5 µg extracted total RNA was reverse-transcribed and aminoallyl (aa)-labelled using 3 µg random primers (Invitrogen), SuperScript III reverse trancriptase (Invitrogen), and a 25 mM aa-dUTP labelling mix (2:3 aa-dUTP to dTTP) (Sigma-Aldrich). The synthesised cDNA was labelled with cyanine-3 (Cy3) or cyanine-5 (Cy5) fluorescent dyes (GE Healthcare BioSciences) for 2 h. After measurement of the concentration of cDNA and incorporated dyes (NanoDrop ND-1000 spectrophotometer), the labelled probes were mixed in equal ratios according to instructions in SOP M007. Prior to microarray hybridisation, the microarray slides were blocked, washed, and dried as described in SOP M008. The slides were immediately hybridised with the prepared samples and incubated overnight. After hybridisation, slides were washed with progressively more stringent buffers and dried as per SOP M008. Slides were immediately scanned using an Axon GenePix4000B microarray scanner (Molecular Devices) and analysed with the TM4 suite programs Spotfinder, MIDAS, and MeV as previously described [18]. Gene expression ratio (fold change from Δ*hyd*2 vs. wild-type) was calculated from the normalised signal intensities. Microarrays were hybridised using RNA from each of the four biological replicates. Cy3 and Cy5 dye swaps were employed between replicates.

Quantitative RT-PCR

cDNA was synthesized from 1 µg of RNA for each sample with the SuperScript III Reverse Transcriptase Kit (Invitrogen). After cDNA synthesis, quantitative RT-PCR was performed using Platinum SYBR Green qPCR SuperMix-UDG with ROX (Invitrogen) according to the manufacturer's instructions. Primers (Integrated DNA Technologies) for 10 genes (**Table S2**) were designed with the publicly available Primer3 software. Primer pairs were optimised to ensure efficient amplification. The real-time PCR reactions were conducted in ABI Prism 7500 (Applied Biosystems). Relative gene expression was determined from calculated threshold cycle (C_T) values that were normalised to the gene *sigA* (MSMEG_2758) as an internal normalisation standard.

Metabolome analysis

M. smegmatis mc^2155 and Δ*hyd*2 cells were grown synchronously in aerated conical flasks. At 1 hour following entry into stationary phase due to carbon-limitation, samples were collected. To prepare samples for extracellular metabolite analysis, 15 mL of culture were centrifuged (27000×g, 10 min, RT) and the supernatant was stored at −20°C. To prepare samples for intracellular metabolite analysis, 15 mL of each culture were quenched with 30 mL cold glycerol saline (3:2 v/v) (−20°C) [19], centrifuged (27000×g, 20 min, −20°C), and resuspended in 1 mL glycerol saline (1:1 v/v) (−20°C) for storage. The samples were recentrifuged prior to metabolite extraction and the pellets were submitted to metabolite extraction. Before extraction, 20 µl of the internal standard (10 mM L-alanine-d_4) was added to each intracellular and extracellular sample. The metabolites were extracted and derivatised as described in existing protocols [20]. The intracellular and extracellular metabolites were analysed using a gas chromatograph (GC-7890) coupled to a mass spectrometer (MSD5975) (Agilent Technologies) with a quadrupole mass selective detector (EI) operated at 70 eV. The results

Table 1. Energetic parameters of wild-type and Δ*hyd*2 cells two hours following the induction of stationary phase.

Carbon Source	5.5 mM Glycerol		12.5 mM Acetate	
Strain	**WT**	**Δ*hyd*2**	**WT**	**Δ*hyd*2**
Growth Yield (OD_{600})	0.86±0.01	0.73±0.04	0.37±0.03	0.19±0.03
Growth Rate (h^{-1})	0.19±0.04	0.17±0.01	0.17±0.01	0.07±0.01
External pH	6.3±0.1	6.5±0.1	7.5±0.1	7.5±0.1
Internal pH	7.3±0.2	7.1±0.1	7.1±0.1	7.2±0.2
Membrane Potential (mV)	−166±7	−163±5	−153±4	−155±9
NAD^+ Concentration (µM)	1.6±0.4	1.6±0.1	0.90±0.14	0.95±0.12
NADH Concentration (µM)	3.6±0.5	2.6±0.5	0.55±0.14	0.53±0.12

Error margins represent standard deviations from three biological replicates.

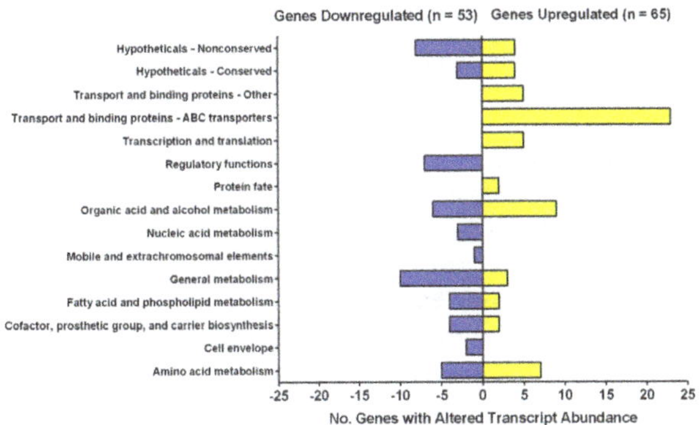

Figure 3. Genes with significant changes in expression in Δ*hyd2* vs. wild-type cells. Both strains were grown synchronously on HdB minimal medium supplemented with 22 mM glycerol. RNA was extracted at 2 hours following induction of stationary phase. The genes were classified as significantly upregulated if expression ratio >2.0, *p* value≤0.05. The genes were classified as significantly downregulated if expression ratio <0.5, *p* value≤0.05. The number of genes affected are listed by functional category. The full list of genes in each category is shown in **Table S4**, **Table S5**, and **Dataset S1**.

were processed using R package Metab [21], and samples were analysed using the mass spectral deconvolution and identification system [20]. The final concentration of metabolites was determined using the GC peak intensity of methyl chloroformate derivatives. Compounds considered false-positives were eliminated, and the intensity of each metabolite was normalised relative to the intensity of the internal standard. Samples of five technical replicates were collected and analysed from each of the three biological replicates.

Results

Hydrogen scavenging enhances survival of *M. smegmatis*

To determine the physiological importance of H_2 utilization, the growth and survival of *Mycobacterium smegmatis* mc[2]155 and hydrogenase mutants was measured in a range of conditions. All strains were grown in aerated conical flasks on HdB minimal salts medium supplemented with different carbon sources and the non-metabolisable detergent Tyloxapol. During growth on 22 mM glycerol (pH 7.0), there was no significant difference in the specific growth rates (~0.25 h^{-1}) or final growth yields (~10^8 CFU mL^{-1}) of the strain lacking Hyd1 (Δ*hyd*1), Hyd2 (Δ*hyd*2), and all three hydrogenases (Δ*hyd*123) (**Figure 1A**). However, the long-term survival of these strains was compromised following the onset of carbon-limitation. Each strain lost viability at a significantly and reproducibly faster rate than the wild-type. At eight days and at all sampling points thereafter, the Δ*hyd*123 strain produced at least 40% fewer colony forming units compared to the wild-type. For example, after 12 days, 3.6×10^7 CFU mL^{-1} were counted for the wild-type strain compared to 1.7×10^7 CFU mL^{-1} for the Δ*hyd*123 strain The long-term survival of the Δ*hyd*1 and Δ*hyd*2 strains was also reduced (**Figure 1B**). Thus, atmospheric H_2 is among the substrates that *M. smegmatis* employs to maintain viability when deprived of organic carbon sources.

M. smegmatis grows optimally by co-metabolising organic carbon sources and atmospheric H_2

Major growth phenotypes were observed when the hydrogenase mutants were grown at lower carbon concentrations or oxidised carbon sources. When the concentrations of glycerol was reduced

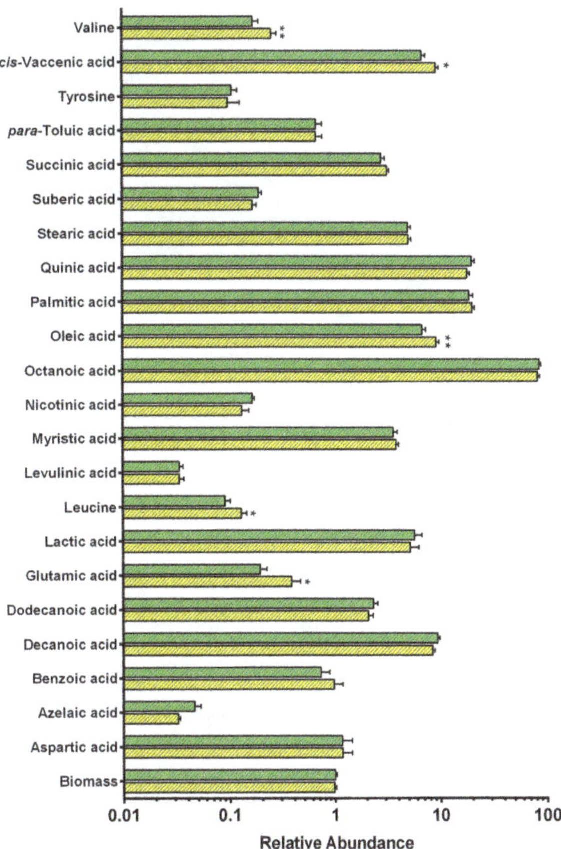

Figure 4. Profiles of intracellular metabolites in Δ*hyd2* vs. wild-type cells. Metabolites were detected by gas chromatography-mass spectrometry (GC/MS). The values show the relative abundance of each metabolite detected in the samples (arbitrary units) on a logarithmic scale. Legend: Green = Wild-type, Yellow = Δ*hyd*2. Means were calculated from three biological replicates and five technical replicates for each strain. *p* values were determined using a Student's T-test. * = $p<0.05$, ** = $p<0.01$ difference relative to wild-type bars.

from 22 mM to 5.5 mM, the Δ*hyd*123 strain grew to a final OD_{600} approximately 25% lower than the wild-type (**Figure 2A**). Growth of the mutant strain was also defective on a range of other carbon sources. There was a twofold reduction in the growth yield of the Δ*hyd*123 strain compared to the wild-type when the short-chain fatty acid acetate was available as the sole carbon source. Furthermore, the specific growth rate of the Δ*hyd*123 strain (0.03 h^{-1}) was fivefold lower than the wild- type (0.17 h^{-1}) (**Figure 2C**). When the phenotypes were traced to the single mutants, it was revealed that the growth of the Δ*hyd*1 and Δ*hyd*2 strains were intermediate to those of the wild-type and Δ*hyd*123 on 5.5 mM glycerol (**Figure 2A**) and 12.5 mM acetate (**Figure 2C**). These phenotypes indicate that these enzymes have overlapping roles during growth and survival, but are not redundant. It was possible to complement the Δ*hyd*2 phenotypes by expressing the genes of the operon MSMEG_2720-2718 from the Hsp60 promoter-driven hygromycin-resistant shuttle vector pOLYG [22] (**Figure 2B&D**); a reduction of growth rate was still observed in the complementation strain, likely due to inefficient hydrogenase maturation or differences in gene copy number, but the final growth yield was similar to the wild-type strain.

During growth on 12.5 mM acetate as the sole carbon source, it was also possible to partially recover the phenotype of the Δ*hyd*2 strain, but not the Δ*hyd*1 or Δ*hyd*123 strains, by growing the cells

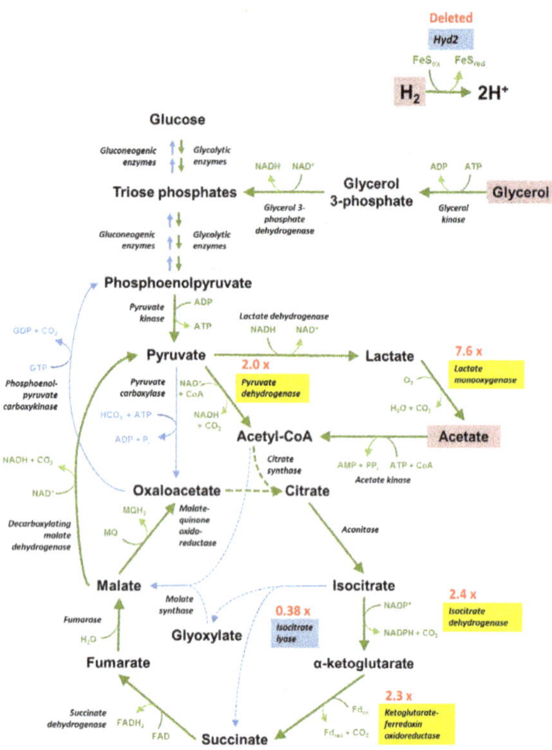

Figure 5. Altered balance of catabolic/anabolic carbon metabolism in Δ*hyd2* cells. Cells lacking Hyd2 compensate for the loss of electrons derived from H_2 by increasing oxidation of organic carbon sources. There is an increased flux though the tricarboxylic acid cycle due to upregulation of enzymes involved in oxidative decarboxylation (e.g. ketoglutarate-ferredoxin oxidoreductase) (highlighted in yellow) and downregulation of those involved in anaplerosis (i.e. isocitrate lyase) (highlighted in blue). We model that loss of CO_2 through oxidative decarboxylation reactions is principally responsible for the decreased biomass of Δ*hyd2* cells. Oxidative pathways are depicted with green arrows, whereas reductive pathways are represented with blue arrows. The red text shows the expression ratios of the significantly upregulated or downregulated genes in Δ*hyd2* vs. wild-type cells.

in the presence of 10% H_2 (**Figure 2E&F**). The Δ*hyd2* strain grew at a similar rate to the wild-type ($\mu_{wt} = 0.14\ h^{-1}$; $\mu_{\Delta hyd2} = 0.13\ h^{-1}$; $\mu_{\Delta hyd1} = 0.065\ h^{-1}$) in this condition; these cells also attained a higher final growth yield ($OD_{600\ wt} = 0.45$; $OD_{600\ \Delta hyd2} = 0.35$; $OD_{600\ \Delta hyd1} = 0.22$) on 10% H_2 than when ambient H_2 was absent ($OD_{600\ wt} = 0.40$; $OD_{600\ \Delta hyd2} = 0.21$; $OD_{600\ \Delta hyd1} = 0.22$) (**Figure 2F**). No differences were observed when equivalent volumes of exogeneous N_2 were introduced (**Figure 2E**). We model that Hyd1 ($V_{max\ (app)} = 12$ nmol g dw^{-1} min^{-1} [7]) can compensate for the loss of Hyd2 by oxidising the majority of this exogenous H_2. However, the converse compensation does not occur because the activity of the Hyd2 ($V_{max\ (app)} = 2.5$ nmol g dw^{-1} min^{-1} [7]) strain is too low to consume exogenous H_2 at rates meaningful for cell growth.

We tested whether the type of carbon source had an influence on hydrogenase expression or activity. Following growth on 5.5 mM glycerol compared to 12.5 mM acetate, the three hydrogenases were expressed at similar levels (**Figure S2A–C**) and whole-cells oxidised hydrogen at equivalent rates (**Figure S2D**). It was therefore clear that differences in hydrogenase activity were not accountable for the differences in the growth phenotypes. The Δ*hyd2* strain grew to proportionately higher yields when stationary-phase cells were spiked with 12.5 mM acetate. During growth on 50 mM acetate, the Δ*hyd2* strain also remained significantly impaired ($\mu = 0.08\ h^{-1}$; $OD_{600\ final} = 1.2$) compared to the wild-type ($\mu = 0.15\ h^{-1}$; $OD_{600\ final} = 1.8$). Repeated subculturing of acetate-grown cells also caused no improvement of growth rates or yields. Thus, the mutant cells consumed all the acetate provided in media, but less efficiently coupled its oxidation to growth.

Δ*hyd2* cells maintain redox balance, membrane potential, and pH gradients

It has been shown that both oxidative and evolving hydrogenases can be important for acid tolerance in *Enterobacteriaceae* [23,24,25]. We therefore tested the hypothesis that Hyd2 is required for intracellular pH homeostasis in *M. smegmatis*. Consistently, the hydrogenase deletions were significantly impaired compared to the wild-type when the external pH was lowered below pH 6.5 (**Figure S1A&B**); at pH 5.5, there was a significant difference in the specific growth rates ($\mu_{wt} = 0.12\ h^{-1}$; $\mu_{\Delta hyd123} = 0.05\ h^{-1}$) and final growth yields ($OD_{600\ wt} = 0.24$; $OD_{600\ \Delta hyd123} = 0.11\ h^{-1}$) of the strains (**Figure S1C**). However, the percentage survival, intracellular pH, and protonophore susceptibility of exponentially-grown cells challenged at pH 5.0 or pH 3.0 was similar between the strains (**Table S3**). It is therefore clear that the pH-dependent phenotypes do not reflect a defect in pH homeostasis and instead may be a secondary consequence of reduced electron input and altered metabolic flux in cells unable to scavenge H_2.

Based on previous studies on the physiological roles of uptake hydrogenases [6], we predicted that the probable physiological role of Hyd2 was therefore to generate a proton-motive force through the respiratory chain and/or reduce coenzymes required for reductive processes. To distinguish these possibilities, we measured the membrane potential, pH gradient, and NAD^+/NADH ratios of wild-type and Δ*hyd2* strains following growth on 5.5 mM glycerol or 12.5 mM acetate. However, all three parameters were again similar between the wild-type and mutant strains and were within normal ranges (**Table 1**). We hypothesised that the energetic parameters of the Δ*hyd2* strain were maintained at normal levels due to compensation of any deficiencies by organic electron acceptors.

Organic electron donors compensate for loss of H_2 oxidation

We performed a microarray analysis to confirm whether organic electron donors compensate for loss of tropospheric H_2 oxidation. The transcriptome of Δ*hyd2* versus wild-type cells was compared following growth on HdB minimal medium supplemented with 22 mM glycerol and 0.05% Tyloxapol. These conditions were selected because, while Hyd2 is expressed and active [7,3], its deletion did not induce a significant growth phenotype aside from an extended lag phase (**Figure 1A**); cells therefore fully compensate for the loss of the expressed hydrogenases. In this condition, we determined that 65 genes were significantly upregulated (ratio >2.0; $p \leq 0.05$) (**Table S4**) and 53 genes were significantly downregulated (ratio <0.5; $p \leq 0.05$) (**Table S5**) in the Δ*hyd2* strain compared to the wild-type (**Figure 3; Dataset S1**). The genes affected were distinct to those previously observed to be affected by slow growth or oxygen-limitation [8]. We performed quantitative RT-PCR to confirm the quality of the microarray data; the expression ratios of the selected genes correlated well with the microarray results (**Figure S3**).

The majority of the transcriptional changes involved genes implicated in substrate transport and binding, organic acid and alcohol metabolism, and amino acid metabolism (**Figure 3**). There were particularly extensive changes in central intermediary metabolism. MSMEG_3706, a bifunctional enzyme that catalyses key reactions in the glyoxylate shunt (isocitrate lyase) and methylcitrate cycle (methylcitrate lyase) [26], was significantly downregulated; this enzyme is usually upregulated during slow growth of *M. smegmatis* and *M. bovis BCG* [8]. In compensation, several predicted glycolytic and tricarboxylic acid cycle enzymes were upregulated, i.e. pyruvate dehydrogenase, isocitrate dehydrogenase, ketoglutarate-ferredoxin oxidoreductase [27], and lactate 2-monooxygenase [28]. All of these upregulated enzymes catalyse oxidative decarboxylation reactions that yield reduced cofactors concomitant with the loss of CO_2. To compensate for downregulation of the methylisocitrate cycle [29], the strain also increased expression of the enzymes of the methylmalonyl-CoA pathway (propionyl-CoA carboxylase, methylmalonyl-CoA mutase) that converts propionate to succinate in an ATP-dependent manner [30]. These changes suggest that *M. smegmatis* compensates for loss of hydrogen oxidation by re-routing carbon flux from anabolic to catabolic pathways.

In amino acid metabolism, the operon encoding the determinants of the urea cycle was upregulated (MSMEG_3769-3773). These ATP-consuming enzymes convert the carbon components of amino acids into the tricarboxylic acid intermediate fumarate, while removing excess nitrogen as urea. Transcripts encoding the predicted NAD-dependent glutamate synthase MSMEG_6458-6459 were also significantly more abundant. We also observed that the putative operons encoding six ABC transporters were upregulated, including those predicted to transport trehalose, methionine, branched-chain amino acids, and alkane sulfonates. Some of these compounds may be scavenged from the cell envelope; it has previously been observed that trehalose is produced by mycobacteria as a byproduct of mycolic acid cell envelope biosynthesis, and the recycling of this compound by a homologous ABC transporter is essential for virulence in *Mycobacterium tuberculosis* [31].

We observed no significant changes in the expression of any hydrogenase-related genes in this condition, including Hyd1 (MSMEG_2263 ratio = 1.5) or Hyd3 (MSMEG_3928 ratio = 0.9), suggesting that other hydrogenases do not compensate for loss of Hyd2 in this condition.

Δ*hyd2* cells have an altered intracellular metabolome

The intracellular and extracellular metabolomes of Δ*hyd2* and wild-type cells were determined by gas chromatography-mass spectrometry (GC/MS) under the same conditions as the microarray. The metabolome profile was generally similar between the strains, but there were statistically significant changes in the relative abundance of several amino acids and fatty acids (**Figure 4**; **Table S6**). These changes were consistent with the transcriptome data (**Table S4**; **Table S5**). The metabolome profile substantiates the finding that NAD-dependent glutamate synthase is upregulated in Δ*hyd2* cells. There was a twofold increase in the concentration of glutamate in this strain. Consistent with the import or recycling of branched-chain amino acids, there was also 30–40% more leucine and valine inside Δ*hyd2* cells (**Table S6**). This is in line with the mutant strains harnessing amino acids as electron donors in the absence of Hyd2.

Discussion

In conclusion, it is clear that hydrogen scavenging enhances the growth and survival of *Mycobacterium smegmatis* under a range of conditions. Single and double markerless deletions of the hydrogen-scavenging enzymes Hyd1 or Hyd2 grew to lower yields than the wild-type strain. Mutant strains were defective when cultured on minimal medium at low carbon concentrations, acidic pH, and, most significantly, on short-chain fatty acids. Reduced growth yields of the Δ*hyd2* strain have also been observed during growth on rich media, e.g. LBT (lysogeny broth supplemented with Tween80) [8]. All defects were observed when strains were grown in flasks aerated with ambient air, i.e. when H_2 is available at trace concentrations. In high H_2 atmospheres, the rapidly-oxidising hydrogenase Hyd1 could compensate for the loss of Hyd2, but the Δ*hyd1* strain remained defective. *M. smegmatis* therefore grows optimally through mixotrophic metabolism of available carbon sources and a ubiqitious hydrogen supply. H_2 scavenging is therefore likely to be a general feature of *M. smegmatis* growth and survival under physiological conditions, and may significantly influence the competitiveness of the bacterium in soil ecosystems.

Despite considerable investigation, it has proven challenging to resolve why hydrogen metabolism is so important during mixotrophic growth. The membrane potentials, pH gradients, and NAD^+/NADH ratios of the mutant and wild-type strains were similar even in phenotype-inducing conditions. This is likely to be due to extensive compensation for the loss of hydrogen scavenging by organic carbon sources. Transcriptome and metabolome analysis inferred that the Δ*hyd2* strain increased oxidation of organic carbon sources through central intermediary pathways. There was a pronounced increase in the transcript levels of the oxidative decarboxylation reactions of the tricarboxylic acid cycle, coupled with a downregulation of the glyoxylate shunt. Increased flux through oxidative pathways would increase the production of NADH, NADPH, and reduced ferredoxin with concomitant loss of CO_2 (**Figure 5**). Mutant cells also upregulated several ABC importers for several potential organic electron donors, e.g. trehalose, methionine, and branched-chain amino acids, that may be scavenged from the cell envelope. In addition, metabolomics analysis showed that the amino acids glutamate, valine, and leucine were more abundant in the cells of the Δ*hyd2* mutant. Transcriptomics indicate that they may be ultimately broken down into fumarate and urea through an upregulated urea cycle. Thus, amino acids may also serve as electron donors to compensate for loss of hydrogen scavenging.

To reconcile the available evidence, we propose that tropospheric H_2 principally serves as a source of reductant in carbon metabolism during exponential growth of *M. smegmatis*. The electrons yielded from its oxidation may provide the reduced compounds required for efficient carbon metabolism. This process enables cells to efficiently balance anabolic and catabolic processes to maximise yields during mixotrophic growth. Reductant is likely to be especially important during growth on acetate. This compound is more oxidised than glycerol and its metabolism results in higher NAD^+/NADH ratios (**Table 1**). In addition, acetate molecules cannot be simultaneously used for catabolic and anabolic reactions due to loss of CO_2. Increased flux through catabolic reactions in the Δ*hyd2* strain on acetate would therefore cause reduced biomass via loss of carbon as CO_2. Membrane-association may be an advantage in this case because it enables the hydrogenases to more efficiently bind the extracellular H_2 that diffuses into cells.

We have previously postulated that Hyd1 and Hyd2 enzymes directly couple the oxidation of H_2 to the reduction of O_2 via the electron transport chain [7]. The membrane association and oxygen-dependence of hydrogenase activity indicates it is physically and functionally linked to the respiratory chain [3,7]. However, it seems improbable that oxidation of nanomolar concentrations of H_2 could significantly influence proton-motive generation during exponential growth on millimolar concentrations of carbon sources; our phenotypic and transcriptome studies are more consistent with hydrogenases harnessing electrons for reductive cellular processes. It nevertheless remains conceivable that aerobic hydrogen respiration may be responsible for the enhanced long-term survival of wild-type cells compared to *Δhyd*123 cells during carbon-limitation. Hydrogenases are expressed at higher levels [8] and oxidise tropospheric H_2 more rapidly [7] in this condition. Tropospheric H_2 oxidation may therefore serve as a significant generator of proton-motive force when organic carbon supplies are exhausted; H_2 is a dependable fuel source given it is present at a constant, albeit trace, concentration throughout the troposphere. Expression and activity profiling suggests that Group 5 [NiFe]-hydrogenases have equivalent roles during the sporulation of streptomycetes and the adaptation of rhodococci to carbon-limitation [2,4].

The processes of using hydrogenases to generate reductant and generate proton-motive force need not be mutually exclusive. The NADH generated by the Group 3d [NiFe]-hydrogenase of *R. eutropha*, for example, can be simultaneously oxidised in the respiratory chain and used as reductant in the Calvin cycle [6,32,33]. Tropospheric H_2 oxidation may also be coupled to the reduction of a multifunctional redox carrier in *M. smegmatis*. Identification and characterisation of the electron acceptors of Hyd1 and Hyd2 is clearly a priority in order to elucidate the cellular processes where these enzymes contribute.

Supporting Information

Figure S1 Importance of hydrogen metabolism for growth at acidic pH. All strains were grown on HdB medium supplemented with 22 mM glycerol. pH was adjusted with concentrated HCl. (A) Final growth yield of strains grown at a range of pHs. (B) Specific growth rate of strains grown at a range of pHs. (C) Full growth curve of strains at pH 5.5. Legend: Blue circles = wild-type, Red squares = *Δhyd*123; Orange point-up triangles = *Δhyd*1; Purple point-down triangles = *Δhyd*2. Error bars show standard deviations from biological triplicates.

Figure S2 Expression and activity of hydrogenases on glycerol and acetate. (A) *hyd1*, (B) *hyd2*, and (C) *hyd3* expression measured with promoter-*lacZ* reporter plasmids. *M. smegmatis* mc^{2}155 harbouring either pJEM*hyd1-lacZ*, pJEM*hyd2-lacZ*, or pJEM*hyd3-lacZ* were grown in HdB minimal medium supplemented with either 5.5 mM glycerol or 12.5 mM acetate. Samples for β-galactosidase activity assays were withdrawn in mid-exponential (yellow bars) and early stationary (green bars) phase. (D) Hydrogenase activity of whole-cells grown in HdB minimal medium supplemented with 5.5 mM glycerol or 12.5 mM acetate during early stationary phase. Activity was measured amperometrically. Positive values indicate net H_2 evolution. Negative values indicate net H_2 consumption. Error bars show standard deviations from three biological replicates.

Figure S3 Validation of microarray data by quantitative RT-PCR. Microarray data were validated by comparing the gene expression changes of selected genes with that of qRT-PCR using the same RNA samples. Genes were chosen that were downregulated (MSMEG_1203, MSMEG_3706), upregulated (MSMEG_3194, MSMEG_3249, MSMEG_3962, MSMEG_3769, MSMEG_5059), or unchanged (MSMEG_4640) in the microarray. All bars show the expression ratio of genes in *Δhyd*2 vs. wild-type strains. Yellow bars show microarray data. Green bars show qRT-PCR data. Error bars represent standard deviations from four biological replicates. * = $p<0.05$, ** = $p<0.01$ difference relative to zero (Student's T-test).

Table S1 Bacterial strains and plasmids used in this work.

Table S2 qRT-PCR primers used in this study. The forward and reverse primers for each gene (MSMEG_XXXX) targeted is listed.

Table S3 Intracellular pH homeostasis of *M. smegmatis* mc^{2}155 following acid challenge. Percentage survival, internal pH, and protonophore susceptibility of wild-type and *hyd* mutants is shown following acid exposure. Cultures were grown on HdB supplemented with 22 mM glycerol to OD 1.0. Cells were subsequently challenged in 100 mM citrate/phosphate buffer at pH 5.0 or pH 3.0. Error margins show standard deviations from three biologically independent replicates.

Table S4 Genes significantly upregulated in *Δhyd2* vs. wild-type microarrays. Means and *p* values are calculated from four microarrays. The genes were classified as significantly upregulated if expression ratio >2.0, p value≤0.05. Less stringent criteria was sometimes used when genes were operonic with other upregulated genes, or when *p* values were perturbed by one clearly anomalous replicate. Asterisks are placed next to genes that did not meet the strict criteria, but are still very likely to be upregulated in the *Δhyd*2 strain.

Table S5 Genes significantly downregulated in *Δhyd2* vs. wild-type microarray. The mean gene expression ratio was calculated from the normalised signal intensities of four microarrays. The genes were classified as significantly downregulated if expression ratio <0.5 and p value≤0.05 (Student's T test). Less stringent criteria was sometimes used when genes were operonic with other downregulated genes, or when *p* values were perturbed by one clearly anomalous replicate. Asterisks are placed next to such genes that did not meet the strict criteria, but are still very likely to be downregulated in the *Δhyd*2 strain.

Table S6 List of intracellular and extracellular metabolites in *Δhyd2* vs. wild-type cells. Metabolites were detected by gas chromatography-mass spectrometry (GC/MS). Values show the relative abundance of each metabolite detected in the samples (arbitrary units). Means were calculated from three biological replicates and five technical replicates for each strain and are shown to two significant figures. *p* values were determined using a Student's T-test. Changes were classified as significant when $p\leq 0.05$ (Student's T-test).

Dataset S1 All genes with significant changes in expression ratios comparing *Δhyd2* to WT. See excel file.

Acknowledgments

We thank Dr. Yuri Zubenko, Sergey Tumanov, Margarita Markovskaya, and Elizabeth McKenzie of the Villas-Bôas Laboratory for technical assistance with the metabolomics analysis. Dr. Htin Aung, Marion

Weimar, and Kiel Hards of the Cook Laboratory are thanked for providing expert technical advice and support.

Author Contributions

Conceived and designed the experiments: CG SGVB JRR MB GMC. Performed the experiments: CG. Analyzed the data: CG SGVB MB GMC. Contributed to the writing of the manuscript: CG MB GMC.

References

1. Constant P, Poissant L, Villemur R (2008) Isolation of *Streptomyces* sp. PCB87, the first microorganism demonstrating high-affinity uptake of tropospheric H_2. ISME J 2: 1066–1076.
2. Constant P, Chowdhury SP, Pratscher J, Conrad R (2010) Streptomycetes contributing to atmospheric molecular hydrogen soil uptake are widespread and encode a putative high-affinity [NiFe]-hydrogenase. Environ Microbiol 12: 821–829.
3. Greening C, Berney B, Hards K, Cook GM, Conrad R (2014) A soil actinobacterium scavenges atmospheric H_2 using two high-affinity, oxygen-dependent [NiFe]-hydrogenases. Proc Natl Acad Sci USA 111: 4257–4261.
4. Meredith LK, Rao D, Bosak T, Klepac-Ceraj V, Tada KR, et al. (2013) Consumption of atmospheric H_2 during the life cycle of soil-dwelling actinobacteria. Environ Microbiol Rep 6: 226–238.
5. Ehhalt DH, Rohrer F (2009) The tropospheric cycle of H_2: a critical review. Tellus B 61: 500–535.
6. Schwartz E, Fritsch J, Friedrich B (2013) H_2-metabolizing prokaryotes. In: Rosenberg E, DeLong EF, Lory S, Stackebrandt E, Thompson F, editors. The Prokaryotes: Prokaryotic Physiology and Biochemistry 4th Edition. Springer. 119–199.
7. Berney M., Greening G, Hards K, Collins D, Cook GM (2014) Three different [NiFe]-hydrogenases confer metabolic flexibility in the obligate aerobe *Mycobacterium smegmatis*. Environ Microbiol 16: 318–330.
8. Berney M, Cook GM (2010). Unique flexibility in energy metabolism allows mycobacteria to combat starvation and hypoxia. PLoS One 5: e8614.
9. Constant P, Chowdhury SP, Hesse L, Pratscher J, Conrad R (2011) Genome data mining and soil survey for the novel Group 5 [NiFe]-hydrogenase to explore the diversity and ecological importance of presumptive high-affinity H_2-oxidizing bacteria. Appl Environ Microbiol 77: 6027–6035.
10. Snapper SB, Melton RE, Mustafa S, Kieser T, Jacobs WR Jr. (1990) Isolation and characterization of efficient plasmid transformation mutants of *Mycobacterium smegmatis*. Mol Microbiol 4: 1911–1919.
11. Berney M, Weimar MR, Heikal A, Cook GM (2012) Regulation of proline metabolism in mycobacteria and its role in carbon metabolism under hypoxia. Mol Microbiol 84: 664–681.
12. Miles AA, Misra SS, Irwin JO (1938) The estimation of the bactericidal power of the blood. J Hyg 38: 732–749.
13. Tran SL, Rao M, Simmers C, Gebhard S, Olsson K, et al. (2005) Mutants of *Mycobacterium smegmatis* unable to grow at acidic pH in the presence in the presence of the protonophore carbonyl cyanide *m*-chlorophenylhydrazone. Microbiol 151: 665–672.
14. Riebeling V, Thauer RK, Jungermann K (1975) The internal-alkaline pH gradient, sensitive to uncoupler and ATPase inhibitor, in growing *Clostridium pasteurianum*. Eur J Biochem 55: 445–453.
15. Rao M, Streur TL, Aldwell FE, Cook GM (2001) Intracellular pH regulation *Mycobacterium smegmatis* and *Mycobacterium bovis* BCG. Microbiol 147: 1017–1024.
16. Leonardo MR, Dailly Y, Clark DP (1996) Role of NAD in regulating the *adhE* gene of *Escherichia coli*. J Bacteriol 178: 6013–6018.
17. Hegde P, Qi R, Abernathy K, Gay C, Dharap S, et al. (2000) A concise guide to cDNA microarray analysis. Biotechniques 29: 548–556.
18. Hümpel A, Gebhard S, Cook GM, Berney M (2010) The SigF regulon in *Mycobacterium smegmatis* reveals roles in adaptation to stationary phase, heat, and oxidative stress. J Bacteriol 192: 2491–2502.
19. Villas-Bôas SG, Bruheim P (2007) Cold glycerol-saline: The promising quenching solution for accurate intracellular metabolite analysis of microbial cells. Anal Biochem 370: 87–97.
20. Smart KF, Aggio RBM, Van Houtte JR, Villas-Bôas SG (2010) Analytical platform for metabolome analysis of microbial cells using methyl chloroformate derivatization following by gas chromatography-mass spectrometry. Nature Prot 5: 1709–1729.
21. Aggio RB, Ruggiero K, Villas-Bôas SG (2010) Pathway activity profiling (PAPi). From the metabolite profile to the metabolic pathway activity. Bioinformatics 26: 2969–2976.
22. Garbe TR, Barathi J, Barnini S, Zhang Y, Abou-Zei C, et al. (1994) Transformation of mycobacterial species using hygromycin as selectable marker. Microbiol 140: 133–138.
23. McNorton MM, Maier RJ (2012) Roles of H_2 uptake hydrogenases in *Shigella flexneri* acid tolerance. Microbiol 158: 2204–2212.
24. Noguchi K, Riggins DP, Eldahan KC, Kitko RD, Slonczewski JL (2010) Hydrogenase-3 contributes to anaerobic acid resistance of *Escherichia coli*. PLoS One 5: e10132.
25. Zbell AL, Maier SE, Maier RJ (2008) *Salmonella enterica* serovar Typhimurium NiFe uptake-type hydrogenases are differentially expressed *in vivo*. Infect Immun 76: 4445–4454.
26. Gould TA, Van De Langemheen H, Muñoz-Elías EJ, McKinney JD, Sacchettini JC (2006) Dual role of isocitrate lyase 1 in the glyoxylate and methylcitrate cycles in *Mycobacterium tuberculosis*. Mol Microbiol 61: 940–947.
27. Baughn AD, Garforth SJ, Vilchèze C, Jacobs WR Jr. (2009) An anaerobic-type α-ketoglutarate ferredoxin oxidoreductase completes the oxidative tricarboxylic acid cycle of *Mycobacterium tuberculosis*. PLoS Pathog 5: e1000662.
28. Giegel DA, Williams CH Jr., Massey V (2000) L-lactate 2-monooxygenase from *Mycobacterium smegmatis*. J Biol Chem 265: 6626–6632.
29. Muñoz-Elías EJ, Upton AM, Cherian J, McKinney JD (2006) Role of the methylcitrate cycle in *Mycobacterium tuberculosis* metabolism, intracellular growth, and virulence. Mol Microbiol 60: 1109–1122.
30. Savvi S, Warner DF, Kana BD, McKinney JD, Mizrahi V, et al. (2008) Functional characterization of a vitamin B_{12}-dependent methylmalonyl pathway in *Mycobacterium tuberculosis*: implications for propionate metabolism during growth on fatty acids. J Bacteriol 190: 3886–3895.
31. Kalscheuer R, Weinrick B, Veeraraghavan U, Besra GS, Jacobs WR Jr. (2010) Trehalose-recycling ABC transport LpqY-SugA-SugB-SugC is essential for virulence of *Mycobacterium tuberculosis*. Proc Acad Natl Sci USA 107: 21761–21766.
32. Schneider K, Schlegel HG (1976) Purification and properties of soluble hydrogenase from *Alcaligenes eutrophus* H16. Biochim Biophys Acta 452: 66–80.
33. Schwartz E, Voigt B, Zühlke D, Pohlmann A, Lenz O, et al. (2009) A proteomic view of the facultatively chemolithoautotrophic lifestyle of *Ralstonia eutropha* H16. Proteomics 9: 5132–5142.

2

Photodynamic Antimicrobial Polymers for Infection Control

Colin P. McCoy[1]*, Edward J. O'Neil[2], John F. Cowley[1], Louise Carson[1], Áine T. De Baróid[1], Greg T. Gdowski[1], Sean P. Gorman[1], David S. Jones[1]

1 Queen's University Belfast, School of Pharmacy, Belfast, United Kingdom, **2** Blue Highway, Inc., Center for Science & Technology, Syracuse University, Syracuse, New York, United States of America

Abstract

Hospital-acquired infections pose both a major risk to patient wellbeing and an economic burden on global healthcare systems, with the problem compounded by the emergence of multidrug resistant and biocide tolerant bacterial pathogens. Many inanimate surfaces can act as a reservoir for infection, and adequate disinfection is difficult to achieve and requires direct intervention. In this study we demonstrate the preparation and performance of materials with inherent photodynamic, surface-active, persistent antimicrobial properties through the incorporation of photosensitizers into high density poly(ethylene) (HDPE) using hot-melt extrusion, which require no external intervention except a source of visible light. Our aim is to prevent bacterial adherence to these surfaces and eliminate them as reservoirs of nosocomial pathogens, thus presenting a valuable advance in infection control. A two-layer system with one layer comprising photosensitizer-incorporated HDPE, and one layer comprising HDPE alone is also described to demonstrate the versatility of our approach. The photosensitizer-incorporated materials are capable of reducing the adherence of viable bacteria by up to 3.62 Log colony forming units (CFU) per square centimeter of material surface for methicillin resistant *Staphylococcus aureus* (MRSA), and by up to 1.51 Log CFU/cm^2 for *Escherichia coli*. Potential applications for the technology are in antimicrobial coatings for, or materials comprising objects, such as tubing, collection bags, handrails, finger-plates on hospital doors, or medical equipment found in the healthcare setting.

Editor: Michael Hamblin, MGH, MMS, United States of America

Funding: This work was supported by funding from the Department for Employment and Learning (Northern Ireland), and Blue Highway, Inc. Blue Highway, Inc. assisted in the study design, data analysis, preparation of the manuscript and decision to publish. The Department of Employment and Learning, Northern Ireland, had no role in study design, data collection and analysis, decision to publish, or preparation of the manuscript.

Competing Interests: CPM, SPG and DSJ received consultancy payments from Blue Highway, Inc. during the period the research was carried out. EO and GTG are affiliated with Blue Highway, Inc. All other authors have declared that no competing interests exist. No products are currently in development and no products have been marketed as a result of these affiliations.

* Email: c.mccoy@qub.ac.uk

Introduction

Hospital-acquired (nosocomial) infections pose a global healthcare concern. It has been estimated that 1 in 10 patients will acquire an infection after admission to a healthcare institution [1]. Such infection presents a serious risk to the morbidity and mortality of the most vulnerable individuals, who are being cared for in the very environment where recuperation and improvement in health and wellbeing is intended. The financial burden to healthcare systems is also alarming; a decade ago the annual economic cost of nosocomial infection in the US was determined to be in the region of $6.7 billion [1]. The costs of implementing strategies to prevent nosocomial infection is likely much less than the value of resources consumed in treatment of these infections once they occur [2]. A major concern in the treatment of nosocomial infection is the emergence of bacterial pathogens displaying resistance to a broad range of antibacterial chemotherapeutic drugs [3]. While the discovery of antibiotics has proved one of the most important advances in healthcare in the 20th century, their widespread use is a double-edged sword. There has been a profound effect on the selective adaptation of bacteria, with multi-drug resistant strains (MDR) emerging at an alarming rate, and threatening the end of the "antibiotic era" [4,5]. Antibiotic resistant bacteria pose a serious problem in hospitals; strains of methicillin resistant *Staphylococcus aureus* (MRSA) appear well adapted to the healthcare environment and have spread internationally (epidemic MRSA, EMRSA) [6]. With the discovery of the next generation of new and efficacious antibiotics lagging behind the emergence of MDR bacteria, the importance of hygiene and disinfection practices in healthcare institutions requires particular emphasis; such interventions may prevent cross-colonization of patients due to contamination of inanimate objects, such as handrails, bedding and medical equipment, or even the skin of healthcare workers and patients, acting as a reservoir for nosocomial infection [7]. In addition to MRSA, there is evidence to suggest that other pathogens may be transmitted by means of environmental reservoirs, including viral pathogens (influenza virus, norovirus, hepatitis, coronavirus), and problematic Gram-negative pathogens (*Escherichia coli*, *Clostridium difficile*, *Pseudomonas aeruginosa*, vancomycin-resistant enterococci) [7]. The present mainstay employed in controlling hospital infection is cleansing using biocides (antiseptics and disinfectants), such as quaternary ammonium compounds (QACs), halogen-releasing agents, and phenolics. The activity of such biocide agents depends on several factors, most notably concentration, period of contact, pH, temperature, numbers and nature of microorganisms to be inactivated [8]. There is also concern that intensive exposure of nosocomial pathogens to biocides may allow for the selection of biocide resistant/tolerant bacterial strains [8–12]. Unlike antibiotics that act at a specific cellular target or interfere with a defined metabolic process, biocides act in a non-specific manner at a variety of cellular targets, such as the bacterial outer membrane or cell wall, the cytoplasmic membrane, proteins, genetic material, and other cytosolic components [13]. Despite this, MRSA strains that show resistance to antiseptics and disinfectants have been isolated from clinical samples, with resistance due to the presence of genes encoding for energy-dependent drug efflux mechanisms, and these genes also confer cross-resistance to a diverse range of antimicrobial drugs [10]. In addition to acquired genetic resistance, bacteria in the biofilm mode of growth are inherently tolerant to inactivation using biocides [14]. Bacterial biofilms are defined as a sessile community of bacteria, characterized by cells that are irreversibly attached to a substratum or interface, or to each other, and are imbedded in a matrix of extracellular polymeric substances (EPS) that they have produced, and exhibit an altered phenotype with respect to growth rate and gene transcription [15]. The biofilm provides an environment where antimicrobial penetration is hindered, genetic exchange and resistance transfer are facilitated, and a change in physiological state, such as stationary phase dormant zones, are a significant factor in the resistance to antibacterial challenge [16,17]. Numerous studies have demonstrated the difficulties in biofilm eradication using biocides commonly employed for cleansing purposes in the hospital setting [18–21]. There is therefore a logical interest in the development of antibacterial surfaces, serving to reduce microbial bioburden on these materials. By preventing the interaction with and adherence of bacterial cells to a surface, the initial stages of biofilm formation are disrupted, effectively removing the foundation which the bacterial biofilm requires. Antimicrobial polymeric coatings, fabrics, and paints are examples of approaches that have attracted interest to date [22–24].

This current study investigates the use of photosensitizer incorporation into polymers, with this approach intended to impart an antimicrobial and/or anti-adherent property to the material surface. Photosensitizers such as porphyrins and phenothiazines have been used clinically in photodynamic therapy (PDT) of malignancies [25–27], and have potential application in photodynamic antimicrobial chemotherapy (PACT), with photodynamic inactivation of MDR bacteria proving equally as effective as antibiotic-susceptible strains [28–31]. The mechanism of action of photodynamic therapy relies on the fact that photosensitizers are capable of reacting in the presence of visible light to produce cytotoxic effects. Phototoxic effects are initiated when, on absorption of an appropriate wavelength of light, the photosensitizer molecule is excited to the higher energy triplet state. This energy can be dissipated in one of two ways, *via* electron-transfer from the photosensitizer to a substrate, producing radical ions, which can react with oxygen, forming cytotoxic molecules such as superoxide, hydroxyl and lipid-derived radicals, or *via* direct energy transfer to oxygen, to produce the higher energy state singlet oxygen, which is highly reactive and can oxidize biological molecules such as proteins, nucleic acids and lipids, resulting in cytotoxicity [29,31,33].

Previously, we demonstrated how incorporation of photosensitizers into hydrogels can generate singlet oxygen on a biomaterial surface, with intended application in the design of infection-resistant medical devices [34,35,37]. Here, we generalise this concept, and demonstrate the facile production of a model two-layer poly(ethylene) (PE) film system, with one layer comprising sensitizer-incorporated PE, and the other a backing layer with the mechanical properties desired for the end application. Such a photodynamic, infection-resistant material may find broad application as coatings or covers for various inanimate objects commonly found in a hospital environment such as handrails, high-tech medical equipment (in particular touch-screens of IT devices), or materials for the manufacture of difficult-to-clean polymer surfaces such as coiled telephone cables or keypads. Incorporation of photosensitizer into PE by a hot-melt extrusion process, mechanical performance of sensitizer-incorporated polymer, leaching behaviour of sensitizer from the material, and antimicrobial properties of the material against a methicillin resistant strain of *S. aureus* and the Gram negative pathogen, *Escherichia coli*, upon light illumination are detailed. The results illustrate the viability of such materials as an effective general means for creating antimicrobial surfaces with the potential to control the spread of bacterial pathogens.

Materials and Methods

Materials

High-density poly(ethylene) (HDPE) was obtained from Q-Chem Ltd, Doha, Qatar (Marlex HHM TR-144, BN. 11100464). Low-denisty poly(ethylene) (LDPE) was obtained from Lydonell-BAsel Industries, Rotterdam, The Netherlands. Photosensitizers, cationic 5,10,15,20-Tetrakis(1-methyl-4-pyridinio)porphyrin tetra(p-toluenesulfonate) >97% (TMPyP) was obtained from Tri-PorTech GmbH, Lübeck, Germany. Neutral 5,10,15,20-Tetraphenyl-21H,23H-porphine 97% (TPP), Toluidine blue O (Tolonium chloride) 97% (TBO), and Methylene blue ≥82% (MB) were obtained from Sigma-Aldrich, Gillingham, UK, and were used without further purification.

Extrusion of poly(ethylene) and sensitizer-incorporated poly(ethylene) materials

Extrusion was performed using a Dr Collin ZK 25 co-rotating twin screw extruder, with paired general purpose screws containing mixing section (25 mm dia; L/D ratio 36), and equipped with Dr Collin 250 mm slot die (coat hanger formation). Extrusion was controlled via a Dr Collin ECS-30 system and sheets were collected using a Dr Collin CR 136–350 chill-roll unit with three roll stack. Extruded materials were prepared in 1 kg batches containing pure PE or mixtures of either TMPyP, TPP, TBO, or MB and HDPE at sensitizer concentrations of 0.05% (0.50 g) and 0.40% (4.00 g) (w/w). The required weights of HDPE and sensitizer were mixed together until a consistent and even coating of HDPE with sensitizer was achieved. The extruder was pre-heated to 230°C along the screw and at the die. Pure HDPE was extruded initially in order to ensure the correct conditions had been obtained. Melt temperature was set at 222°C, screw speed at 60 rpm and pressure at the die was measured to be 114 bar. The chill roll was maintained at 110°C and roller speed was 1.2 metres per minute. Once a film of consistent quality and thickness was obtained, the hopper was emptied of remaining PE and was filled with the required sensitizer-containing PE mixture. Initial extrudate was discarded due to uneven mixing, resulting from

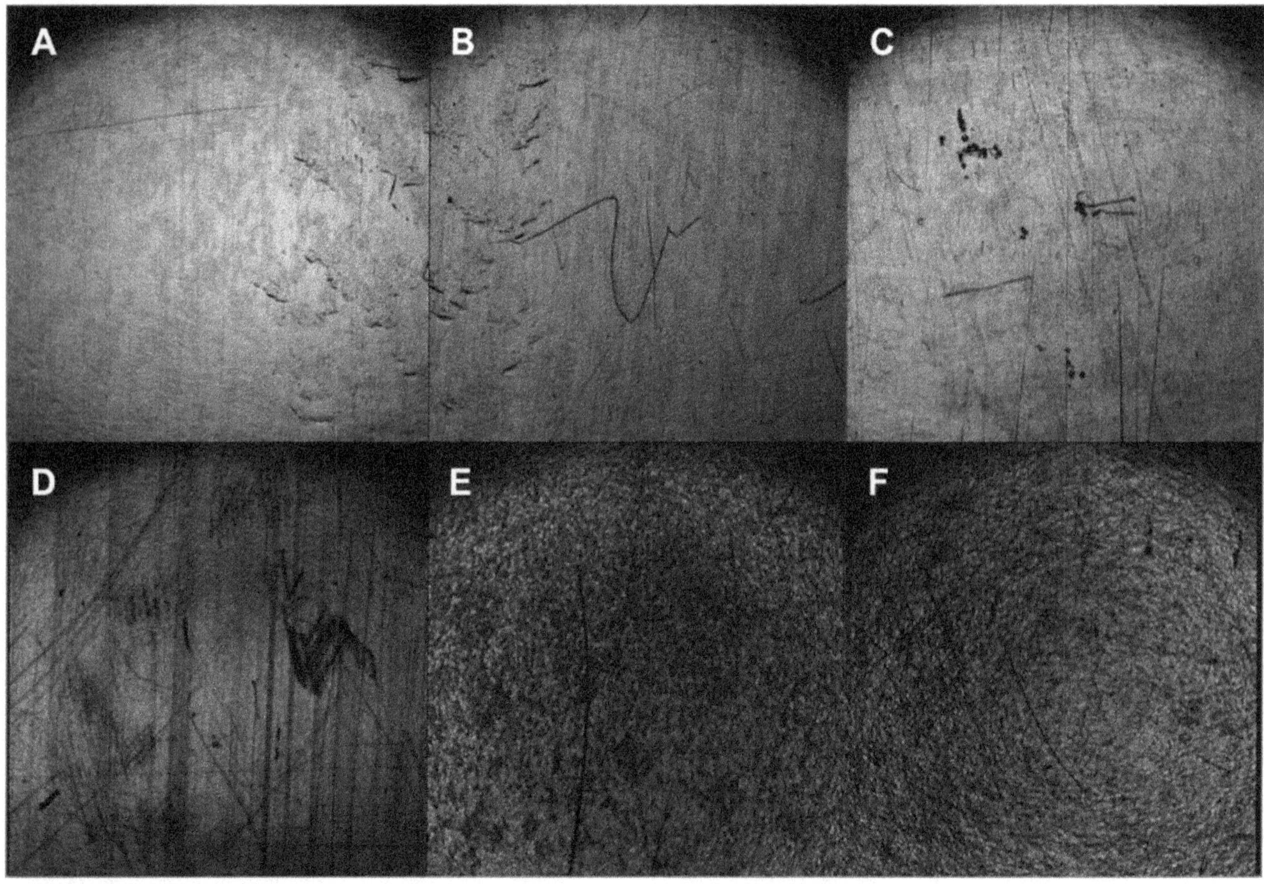

Figure 1. CLSM reflctance image of (A) HDPE control, (B) 0.40% TPP-HDPE, (C) 0.40% TMPyP-HDPE surfaces, (D) HDPE control, (E) 0.40% MB-HDPE, (F) 0.04% TBO-HDPE. All images represent an area measuring 3.75 mm x 3.75 mm.

the extrusion of the remainder of the pure PE resident in the screw, with collection beginning once pigmentation was uniform. Similarly, extrudate at the end of the batch was discarded as reduction in the available mass of the mixture resulted in reduced pressure, affecting the homogeneity of the film. The residence time of the sensitizer-PE mixture within the extruder was approximately 5 minutes.

Production of twin layer sheets by platen press

Platen pressing was achieved using a Dr Collin P 200 P platen press, capable of maintaining a maximum temperature of 300°C and maximum pressure of 250 bar, with an effective operating area of 196×196 mm^2. Sections of extruded sheets, one pure HDPE and a second sensitizer-incorporated HDPE, were cut to approximately 190×190 mm^2 and placed one on top of the other inside a PTFE envelope. The envelope was placed on a tray and set in the platen press, pre-heated to 150°C, after which a five stage automated program was initiated, with temperature and pressure not exceeding 150°C and 70 bar respectively.

Characterization of sensitizer-incorporated PE by UV-visible spectroscopy and confocal laser scanning microscopy

Confocal laser scanning microscopy was performed using a Leica DM RE upright microscope in conjunction with a Leica TCS SP2 system, and images were analyzed using Leica LAS AF imaging software. All images captured were 2048×2048 pixels, with a line average of 16 and were the average of 16 individual scans. The microscope pinhole was set at a 5.89 airy (600 μm) diameter. Reflectance images were recorded by setting the excitation beam wavelength to 488 nm and detecting emission in the range 480–500 nm. Fluorescence images required setting a suitable excitation wavelength and emission detection dependent on the photosensitizer under examination. For TPP, wavelengths used were 514 nm (excitation) and 600 nm –800 nm (emission), and for TMPyP wavelengths used were 514 nm (excitation) and 600 nm –720 nm (emission), while the emission wavelengths of MB and TBO lay outside the detection range of our instrumentation. Optical microscopy was therefore used to examine the distribution and homogeneity of MB and TBO within the polymer.

As a possible end application of these materials is antimicrobial covers for touch-screen devices, it is important to determine the transparency and optical clarity of the photosensitizer-incorporated films. UV-visible spectroscopy was performed using a Perkin Elmer Lambda 650 UV-visible spectrophotometer, to determine the optical transmittance of the materials in the visible region between 390–750 nm. This was achieved by attaching samples to the wall of a quartz cuvette, scanning this region of the electromagnetic spectrum, and determining mean transmission across this range of wavelengths. This analysis is used to determine the percentage of visible light that is transmitted through the photosensitizer-incorporated materials.

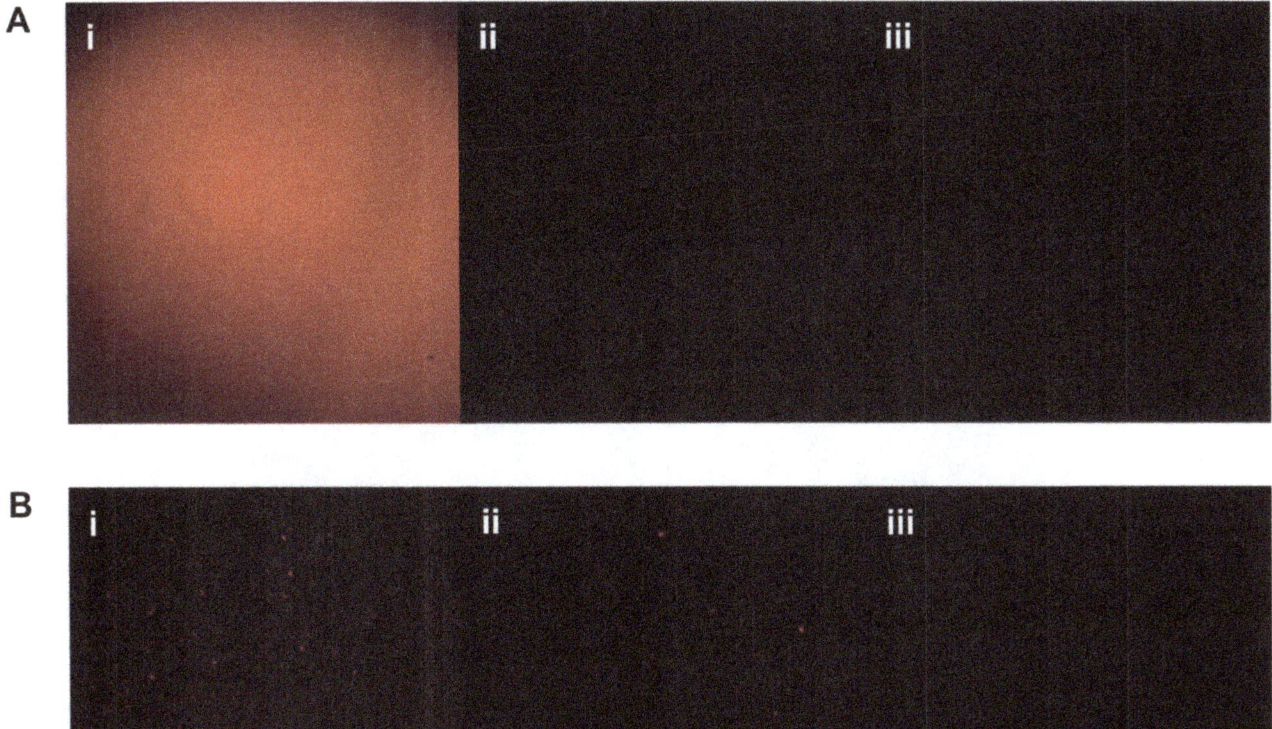

Figure 2. CLSM fluorescence micrographs of (A) TPP-HDPE at (i) 0.40%, (ii) 0.05%, (iii) 0% TPP (control) and of (B) TMPyP-HDPE at (i) 0.40%, (ii) 0.05%, (iii) 0% TMPyP (control). All images represent an area measuring 1.5 mm x 1.5 mm.

Characterization of mechanical performance of sensitizer-incorporated materials

Mechanical analysis of films was performed on dumb-bell shaped samples (length 30 mm, thickness of narrow portion 2 mm) cut using a Ray-Ran hand-operated cutting press, and was tested using a Stable Micro Systems TA.XT plus texture analyzer, fitted with a 40 kg load cell. The dimensions of the narrow portion of the dumb-bell (width and thickness) were measured using a digital micrometer, and the samples were secured between the mobile upper and static lower clamps of the texture analyzer. The distance separating the upper and lower clamps was used to measure the gauge length of the sample. In order to test the sample, the upper clamp was raised at a speed of 50 $mm.min^{-1}$, until sample fracture occurred. From the resultant stress–strain relationship, the mechanical properties of the samples (yield point, ultimate tensile strength (UTS), Young's modulus and percentage elongation) were calculated. A minimum of five replicates of each sample were performed and the effect of incorporation of varying concentrations of both photosensitizers on the mechanical properties of PE was assessed for statistical significance using a one-way ANOVA, with post-hoc comparisons made using Tukey's HSD test. Significance was denoted by a value of $p<0.05$.

Characterization of leaching behaviour

Photosensitizer-incorporated samples (30×20 mm) were cut and pressed firmly 1000 times with one clean, washed finger. Adherent photosensitizer was washed from the finger by immersing in 10 mL deionised water for 30 seconds. Solutions were analyzed using a Perkin Elmer Lambda 650 UV-visible spectrophotometer; the concentration of photosensitizer in solution was determined via the calculated molar extinction coefficient, (ε) at λ_{max}, of solutions of known concentrations. Calculated ε values are as follows: TBO 39,750 mol^{-1} dm^{-3} cm^{-1} (λ_{max} 610 nm); TPP 406,750 mol^{-1} dm^{-3} cm^{-1} (λ_{max} 420 nm); MB 74,028 mol^{-1} dm^{-3} cm^{-1} (λ_{max} 664 nm); TMPyP 193,000 mol^{-1} dm^{-3} cm^{-1} (λ_{max} 446 nm). Using the same sample, the materials were pressed a further 1000 times, repeating the procedure as before, and continuing to touch samples in increments of 1000, up to 10,000 times. Analyses were carried out in triplicate.

Photosensitizer-incorporated samples (75×50 mm) were cut and wiped 50 times with a medical wipe moistened in a 1% (w/v) solution of non-ionic surfactant, Tween 20, as a model for typical hospital surface cleaning products. Leaching of the highly-colored photosensitizer by this surface cleansing procedure was assessed by examination of the degree of staining of the medical wipe. Using the same sample, the materials were wiped a further 50 times, repeating the procedure as before, and continuing to clean samples in increments of 50, up to 500 times. Analyses were repeated in triplicate.

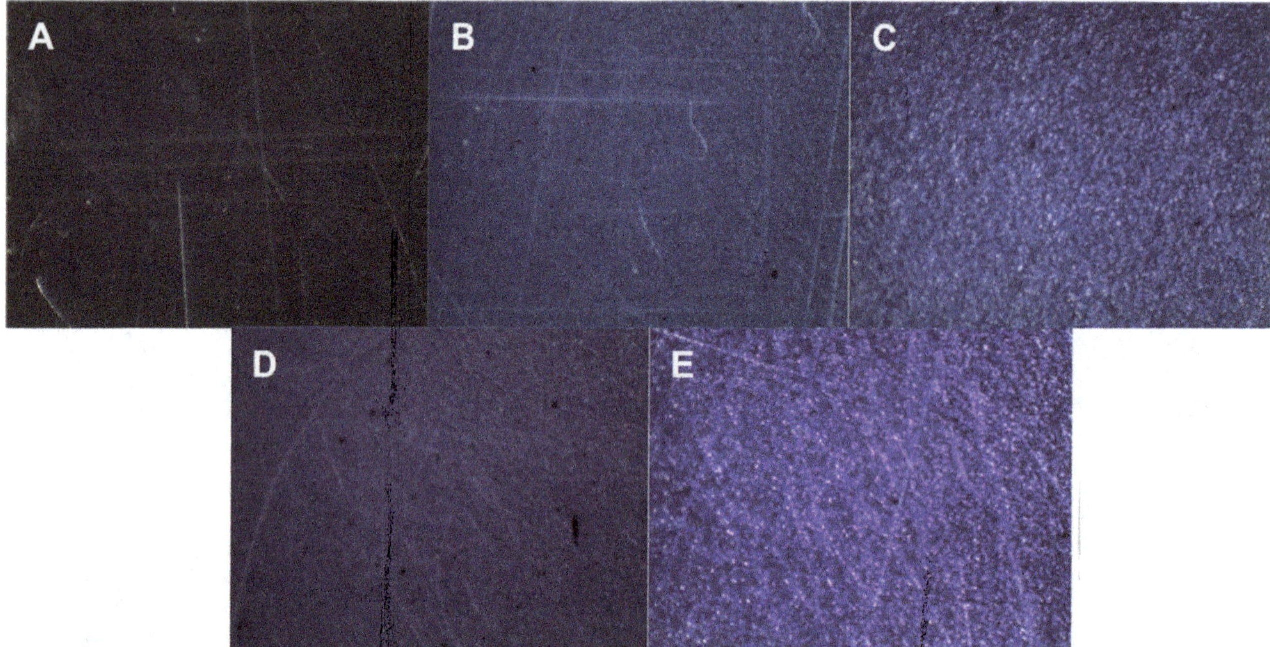

Figure 3. Optical microscope images of (A) PE, (B) 0.05% MB-PE, (C) 0.40% MB-PE, (D) 0.05% TBO-PE, and (E) 0.40% TBO-PE. All images represent an area measuring 1.0 mm x 1.0 mm.

Characterization of antimicrobial behaviour

Bacterial adherence to materials was tested using methicillin-resistant *Staphylococcus aureus* (MRSA) ATCC 33591; bacteria were grown aerobically at 37°C in Müller–Hinton Broth (MHB) for 18 hours. During log phase of growth, the broth culture was centrifuged at 3000 rpm for 12 minutes, the cell pellet re-suspended in phosphate buffered saline (PBS). The suspension was diluted, such that the optical density was 0.3 at 540 nm (approx. 2.0×10^8 cfu/mL). The suspension was serially diluted in PBS, to form the final inoculum (approx. 4.0×10^5 cfu/mL). Sensitizer-loaded samples and controls (blank HDPE without sensitizer) were cut to 15×10 mm. Drops of bacterial inoculum (60 µL – approx. 1.6×10^4 cfu/cm^2) were placed into a sterile petri dish, after which each drop was covered with a single sensitizer-loaded or control sample. Petri dishes containing inoculated samples were inverted, in order to facilitate light irradiation of the polymer surface in contact with the bacteria, and placed under two 230W halogen bulbs situated 24 cm from dishes, providing 1340 µW light across the sample surface, for 2 hours. Samples are held in place against the base of the inverted petri dish by adhesive capillary force.

Post-illumination, samples were removed from petri dishes and placed in 10 mL PBS, inverting constantly for 30 seconds, to remove non-adherent bacteria; samples were then transferred to individual bijoux bottles containing 2 mL Quarter Strength Ringers Solution (QSRS). Samples were sonicated for 10 minutes in order to remove and suspend adherent bacteria, using a Branson 3510 ultrasonic cleaner providing a fixed puissance of 42 KHz. The number of surviving microorganisms were determined by spread plating on Müller–Hinton Agar (MHA) plates. Plates were allowed to dry, inverted and incubated at 37°C, 50% RH for 18–24 hours. Testing was performed using five replicates of each sensitizer-loaded material (at both high and low concentrations), along with corresponding polyethylene control. Lead materials identified as effective against MRSA were then carried forward for further testing with Gram-negative *Escherichia coli* NCTC 8196. These lead materials (MB 0.4%, TBO 0.4%, and TMPyP 0.4%) were tested against *E. coli* as per methodology described above. Statistical significance was determined against dark HDPE control and was carried out using a two-tailed student's t-test, $p<0.05$.

Results

Distribution of sensitizer in materials and optical transmittance of sensitizer incorporated films

HDPE sheets containing varying concentrations of sensitizers were subject to examination by confocal laser microscopy. Samples were first viewed in reflectance mode, in order to provide images of the material surfaces. No distinct visual alterations to the polymer surface were seen upon addition of TPP to the HDPE mixture, however, inspection of TMPyP incorporated HDPE appears to show some aggregates at the material surface; there also appears to be some modification to the polymer surface upon addition of the higher concentrations of MB and TBO, likely due to the presence of aggregates at the material surface, compared with the relatively smooth surface of the pure HDPE, as illustrated in Figure 1.

The use of fluorescence imaging allowed visualization of the sensitizer throughout the materials. At both 0.40% and 0.05% TPP loading, an even fluorescence signal was observed on the surfaces of the samples, indication that the photosensitizer has been well mixed with the polymer and has produced a largely homogenous surface, as displayed in Figure 2. However, fluorescence imaging of the surface of TMPyP incorporated HDPE revealed incomplete mixing of the photosensitizer with the material, at both the 0.05% and 0.40% concentrations (Figure 2).

As fluorescence imaging of materials containing TBO and MB was not possible, optical microscope images were collected and revealed a relatively even colouring with minimal alteration in the surface compared to HDPE; however, the images show small

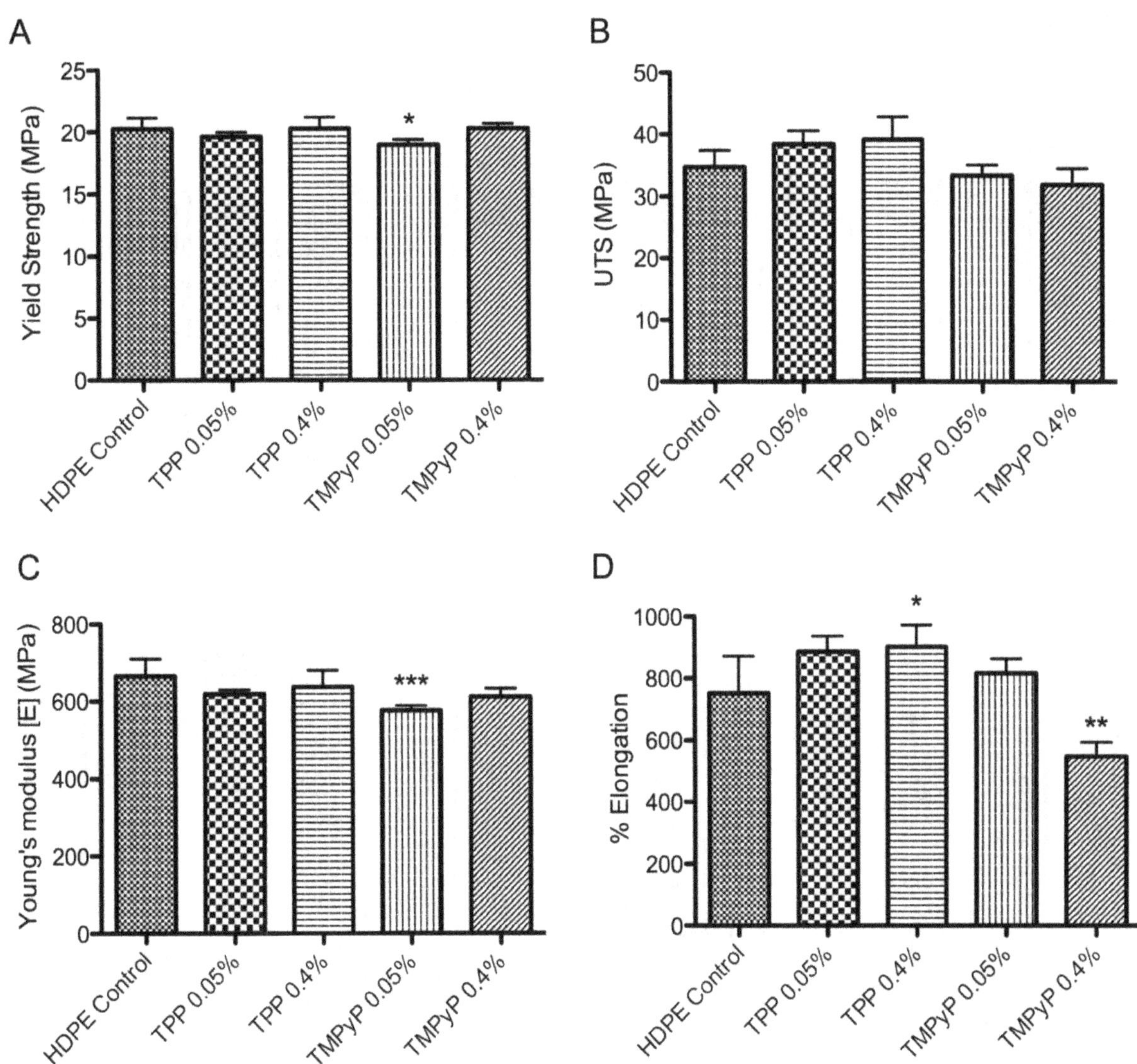

Figure 4. Effect of incorporation of TPP or TMPyP on extruded HDPE sheets (at 0.05 or 0.40% w/w) on: (A) yield strength, (B) ultimate tensile strength, (C) Young's Modulus and (D) percentage elongation at break. * indicates level of significance from HDPE control, graph displays means+standard deviation.

darkened spots indicating non-homogeneity and incomplete mixing of the sensitizers (Figure 3).

UV-visible spectroscopic analysis of photosensitizer-loaded samples was performed to determine the transmittance of light through the sample, in order to give a measure of optical clarity and transparency. The mean transmittance of pure polyethylene film, across the visible range (390 nm–750 nm), was 68%. Compared with this, the transmittance of low concentration porphyrin-incorporated materials (0.05% TPP and TMPyP) was approximately 49%, in both cases. Samples containing 0.4% TMPyP exhibited transmittance of 37%, while 0.4% TPP reduced transmittance to 33%. For phenothiazine incorporated films, the transmittance of low concentration materials (0.05% TBO and MB) was approximately 46% and 44%, respectively. Samples containing 0.40% MB exhibited transmittance of 28%, while 0.40% TBO reduced transmittance to 14%.

Mechanical performance

Mechanical testing of prepared HDPE samples, containing either TPP, TMPyP, MB or TBO at concentrations of 0.05 or 0.40% (w/w), was performed and the results compared to the mechanical properties of pure HDPE. Using the data collected, four measures of the mechanical performance of the materials were calculated – yield strength, ultimate tensile strength (UTS), Young's Modulus, and percentage elongation, which are shown in Figure 4 (porphyrin-based sensitizers, TPP and TMPyP) and Figure 5 (phenothiazine-based sensitizers, TBO and MB).

These results indicate that incorporation of TPP at either concentration causes no significant effect on the yield strength of PE (20.3±0.90 MPa vs. 19.6±0.36 (0.05%) and 20.3±0.92 (0.40%)). Incorporation of TMPyP at concentrations of 0.05% causes a statistically significant reduction in the yield strength of the material (19.0±0.41 MPa); while HDPE with a TMPyP-loading of 0.40% produces no significant difference in yield strength from that of PE (20.3±0.35 MPa). TBO at either concentration causes a significant reduction on the yield strength

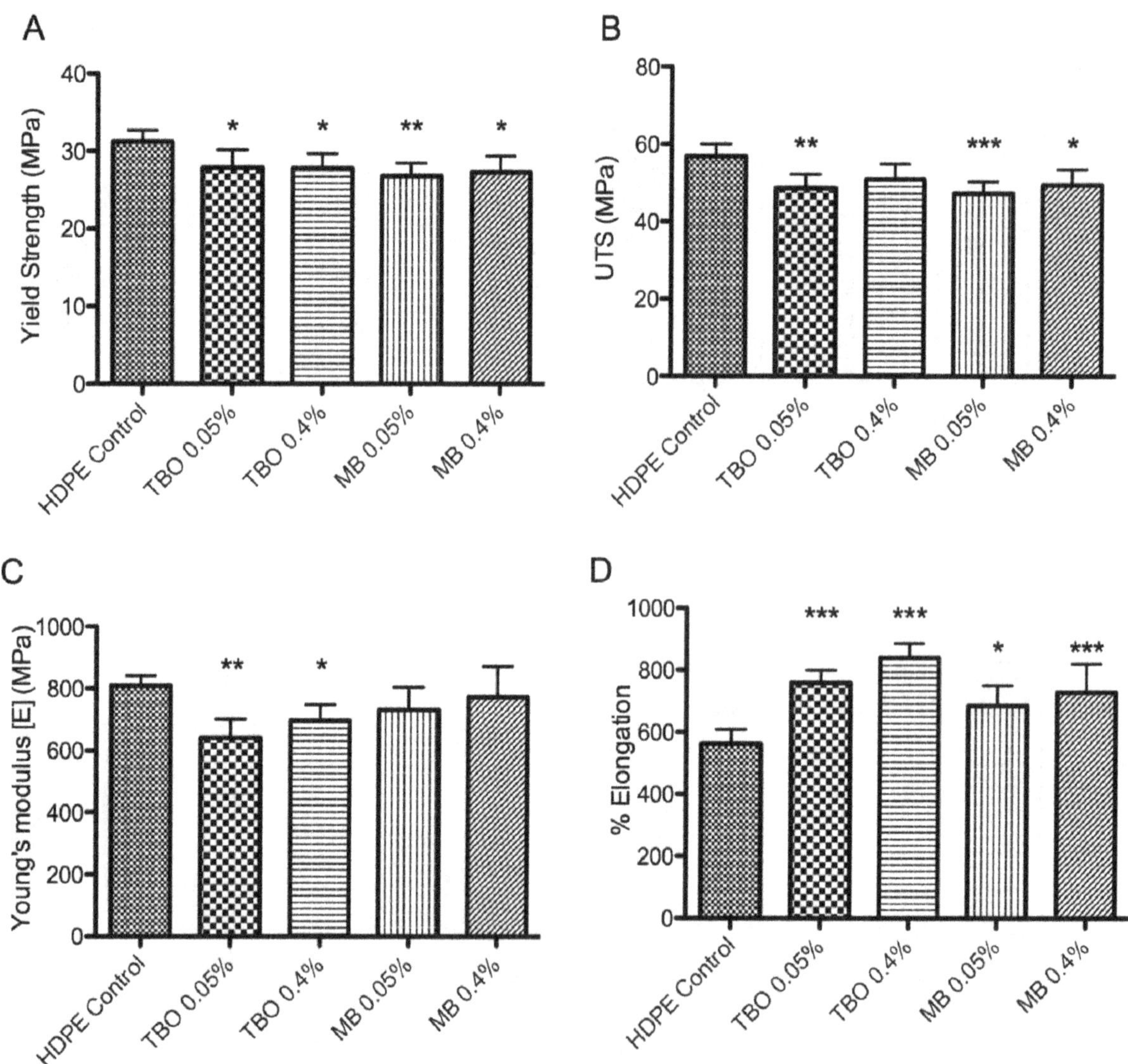

Figure 5. Effect of incorporation of TBO or MB on extruded HDPE sheets (at 0.05 or 0.40% w/w) on (A) yield strength, (B) ultimate tensile strength, (C) Young's Modulus and (D) percentage elongation at break. * indicates level of significance from HDPE control, graph displays means+standard deviation.

of PE (31.1±1.47 MPa vs. 27.9±2.27 (0.05%) and 27.8±1.94 (0.40%)). Similarly, incorporation of MB at concentrations of 0.05% and 0.40% also causes a significant reduction in the yield strength of the material (26.8±1.5 and 27.3±2.18, respectively).

Evidence suggests that TPP may have a strengthening effect on HDPE, successively increasing UTS as concentration is increased from 0.05% to 0.40% (34.8±2.7 MPa vs. 38.5±2.1 (0.05%) and 39.2±3.7 (0.40%)). Conversely, results suggest the TMPyP reduces UTS of HDPE, with increasing concentration (33.4±1.7 (0.05%) and 31.9±2.6 (0.40%)). However, statistical analysis finds no significant difference between photosensitizer-incorporated HDPE and standard HDPE sheets; similar analysis indicates TPP-loaded PE sheets (both concentrations) exhibit significantly greater UTS than the TMPyP counterparts. TBO and MB both affect ultimate tensile strength. Incorporation of MB at either concentration resulted in significant reduction in ultimate tensile strength (56.9±3.22 MPa vs. 47.2±3.21 (0.05%) and 49.5±4.00 (0.40%)). In a similar manner, the presence of a 0.05% loading of TBO results in a significant reduction in ultimate tensile strength; while, although not significant, 0.40% TBO also reduced tensile strength (48.6±3.73 (0.05%) and 51.0±3.9 (0.40%)).

Addition of TPP produces no significant change in the Young's modulus from that of PE (665±45 MPa (PE) vs. 619±10 (0.05%) and 637±42 (0.40%)); however inclusion of 0.05% TMPyP in PE sheets results in significant reduction in Young's modulus (576±13 MPa), indicating increased elasticity and reduced stiffness of these samples. Higher concentrations of TMPyP (0.40%) produce no significant change compared with PE (611±21 MPa). Addition of 0.05% TBO produces significant alteration of the Young's modulus of PE (808±34 MPa vs. 641±60); addition of 0.40% TBO also results in significant reduction (695±53 MPa), however the reduction observed is not as great and suggests that on increasing TBO loading further, materials may return to similar levels to that of PE. Inclusion of MB in PE sheets results in a similar pattern to that of TBO, however reductions in Young's moduli (731±75 MPa (0.05%) and 772±95 (0.40%)) are not considered statistically significant. It should be noted that standard deviation of Young's moduli of photosensitizer-incorporated PE

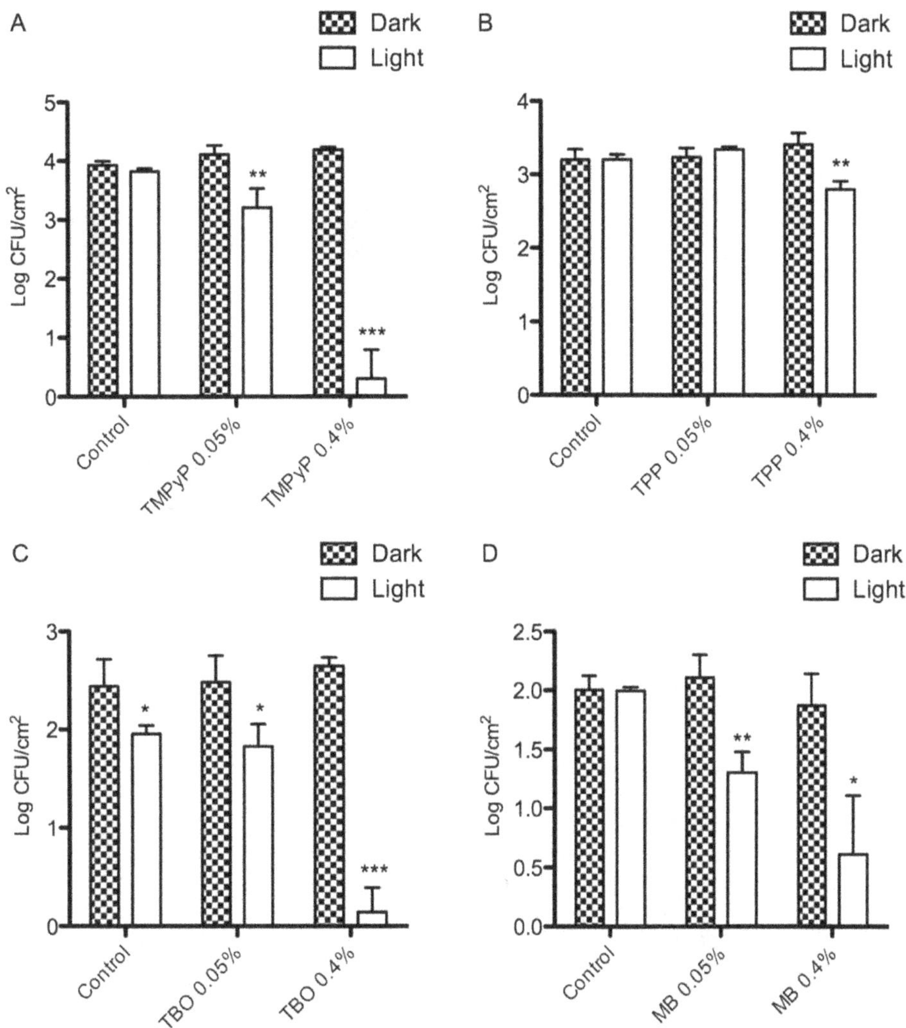

Figure 6. Reduction in adherence of viable MRSA on the surfaces of materials loaded with (A) TMPyP, (B) TPP, (C) TBO and (D) MB. Log CFU/cm^2 adhered bacteria were enumerated on control materials (HDPE without photosensitizer), and test samples in both dark and light-irradiated conditions. * indicates level of significance from HDPE dark control, graph displays means+standard deviation.

samples are greater than that observed for PE suggesting that inclusion of photosensitizers has resulted in greater variability with regard to this mechanical property.

Use of TPP, initially, produces moderate, but statistically insignificant, increase in elongation from that of PE (751±120% vs. 885±51 (0.05%)); however increasing concentration to 0.40% significantly increases elongation (901±72%). Similarly, 0.05% TMPyP results in no significant changes in elongation (816±46%); however at concentrations of 0.40%, TMPyP significantly reduces the percentage elongation of the sample (548±46%). Use of TBO produces significant increases in elongation, with increasing concentration, from that of PE (563±48% vs. 756±42 (0.05%) and 840±47% (0.40%)). Similarly, MB inclusion also results in significant increases in elongation (684±68% (0.05%) and 728±94% (0.40%)).

Leaching behaviour

Touch testing was performed on samples containing TPP, TMPyP, TBO, and MB, all at a concentration of 0.4%. It was observed that after 10,000 touches of the TMPyP, MB and TBO samples, no photosensitizer had leached from the material. A small amount of leaching was observed from TPP-loaded HDPE, equating to 0.0045% of the total TPP in the sample tested.

Leaching of photosensitizer from the material during surface cleansing was assessed by wiping the material surface with a medical wipe moistened in 1% (w/v) Tween 20. Slight staining of the medical wipe was evident in the case of TMPyP, TPP and MB, although the amount could not be quantified due to the difficulty in extracting the photosensitizer from the wipe into solution for UV-visible spectroscopic analysis. For TBO surface cleansing did not result in the staining of the medical wipe, indicating that for this photosensitizer, leaching was negligible or failed to occur.

The experimental design used in these leaching tests reflects the environment the materials would experience in their intended end application. These results indicate that negligible leaching would occur during normal skin contact or wipe cleaning with detergent.

Characterization of antimicrobial behaviour

Antimicrobial adherence was performed by inoculating samples with MRSA and *E.coli* and determining the number of viable bacteria after two hours, in either light or dark conditions. Only lead materials that demonstrated good antibacterial activity

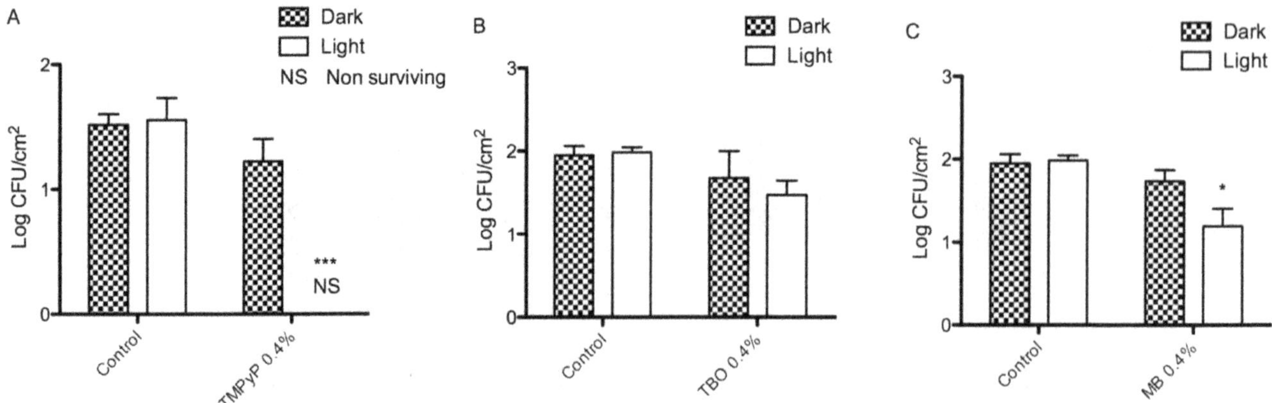

Figure 7. Reduction in adherence of viable *E. coli* on the surfaces of materials loaded with (A) TMPyP, (B) TPP, (C) TBO and (D) MB. Log CFU/cm^2 adhered bacteria were enumerated on control materials (HDPE without photosensitizer), and test samples in both dark and light-irradiated conditions. * indicates level of significance from HDPE dark control, graph displays means+standard deviation.

against Gram-positive MRSA were taken forward to testing against Gram-negative *E.coli*. Gram-negative bacteria are widely considered to be more tolerant to photoinactivation than their Gram-positive counterparts [29,36]. Reduction in bacterial adherence on light irradiation, when compared to identical material samples inoculated in the dark, and additional controls of non-photosensitizer incorporated materials in both light and dark conditions, are shown in Figures 6 and 7.

Figure 6 illustrates the results obtained against MRSA. It is clear that the greatest log-cycle reduction in colony-forming units (CFU) per square centimetre material are observed for those samples containing the higher (0.4%) concentration of photosensitizer. HDPE containing TPP 0.4% displayed the least antimicrobial activity, with a 0.40 log-cycle reduction in MRSA adherence when compared to dark HDPE control. MB 0.4%, TBO 0.4% and TMPyP 0.4% showed a 1.39, 2.30, and 3.62 log-cycle reduction against MRSA adherence respectively.

As shown in Figure 7, *E.coli* were inherently less adherent than MRSA on all materials, including the HDPE controls without photosensitizer. HDPE containing 0.4% TMPyP displayed excellent anti-bacterial and anti-adherent characteristics, with no viable organisms detected after light irradiation, equating to a 1.51 log-cycle reduction in adhered *E.coli,* when compared to HDPE dark control. MB 0.4% and TBO 0.4% did not perform as favourably, resulting in a 0.75, and a 0.47 log-cycle reduction in adherence, respectively.

Discussion

Several photosensitizing compounds, namely TMPyP, TPP (phorphyrin-based), TBO and MB (phenothiazine dyes), were successfully incorporated into high density poly(ethylene) (HDPE) films by a hot-melt extrusion process, and incorporated into a twin-layer polymer sheet by platen press. Our aim is to develop photodynamic antimicrobial surfaces that may be used in a healthcare environment to prevent the spread of nosocomial infection. The antimicrobial surface is the result of the generation of reactive oxygen species (ROS) due to illumination of photosensitizer molecules present in the polymer film, as demonstrated by our earlier work on related materials [28]. These ROS react indiscriminately with components of the bacterial cell, such as nucleic acids, proteins and lipids, eventually causing cell death [29,31–33].

The hot-melt extrusion process was successful in incorporating photosensitizer into HDPE films, although the homogeneity of photosensitizer distribution within the film appears to vary depending on the photosensitizer used. CLSM fluorescence microscopy revealed favourable homogenous distribution of TPP within the extruded polymer, and this is likely due to the neutral hydrophobic properties of TPP allowing for complete miscibility with the poly(ethylene). The other sensitizers used, TMPyP, TBO and MB, all show some evidence of photosensitizer aggregation and incomplete mixing, with the phenothiazines TBO and MB to a lesser extent than TMPyP. It is desirable to maintain a smooth material surface, as an increase is surface roughness or irregularity may promote adhesion of bacteria. Figure 1 shows that at the higher photosensitizer concentrations employed (0.40%), a smooth, regular surface was maintained in films incorporating the porphyrin-based TPP and TMPyP while TBO and MB did display evidence of surface modification, however the effects of this on the ability of bacteria to adhere to the material surface will only be evident through microbiological testing.

Characterization of mechanical properties of the prepared films by determination of yield point, ultimate tensile strength, Young's modulus and percentage elongation measurements reveal that incorporation of TPP and TMPyP have a minimal effect on the mechanical characteristics of the HDPE films, while incorporation of TBO or MB have a slight negative effect, suggesting that these sensitizers may adversely effect the quality of the material. However, if TBO or MB were incorporated into the twin layer system as the minor layer, the overall effect is likely negligible.

UV-visible spectroscopy demonstrated that incorporation of photosensitizer into the HDPE films has a negative effect on the optical clarity of the materials given that mean transmission over the visible range (390 nm –750 nm) is reduced in comparison to control HDPE, especially at the higher concentration of photosensitizer used. The implications of this observation depend on the intended purpose of the material. For example, a cover for a touch-screen computer would require the material to have good optical transmittance. It should be noted, however, that in this study the thickness of the photosensitizer-incorporated material is approximately 200 μm and with a different manufacturing process, such as blown-film extrusion, this may be reduced to 25 μm, and hence it is expected that optical transmittance of the photosensitizer incorporated films would improve dramatically.

The study of leaching behaviour from materials *via* touch-testing shows that TMPyP, TBO and MB are not removed from the material surface by this mechanism, while minimal amounts of TPP are lost after 10,000 touches. This favourable leaching behaviour is important should these materials be used as covers for equipment such as touch-screen covers, keypads or handrails.

Antimicrobial testing showed that the most effective photosensitizer for incorporating into antimicrobial HDPE films was TMPyP at the high concentration of 0.40%. This material displayed a 3.61 log-cycle reduction in viable MRSA, and 1.51 log-cycle reduction in *E. coli* after two hours light irradiation at 1340 $\mu W/cm^2$ per sample, using halogen bulbs as a light source. The ability of the TMPyP-incorporated HDPE in achieving a greater than one log-cycle reduction of a Gram-negative bacterium is encouraging, since the Gram-negative outer membrane functions as a barrier, protecting the cell and rendering it more resilient to photodynamic inactivation. It is the polycationic nature of the TMPyP molecule that imparts an advantage over the neutral TPP photosensitizer, as the cation is necessary for interaction and disruption of the outer membrane, so making inactivation with ROS possible [36].

The white light generated by the halogen bulb light source is inexpensive and representative of lighting conditions found in typical artificially-illuminated indoor environments, meaning that photosensitizer incorporated materials would not require any special conditions under which to produce their antimicrobial effect, as long as the environment in which they are used is adequately illuminated. Alternatively, for specific decontamination purposes, such as during deep-cleaning of hospital operating theatres, specialized lighting could be installed to emit at higher intensity, and at a wavelength specific for the excitation of the photosensitizer employed, for efficient generation of ROS and the resulting bactericidal effects.

This study, aimed at controlling the microbial bioburden and infectious reservoir on inanimate, everyday objects, builds upon our previous work in the development of photo-activated, antimicrobial polymers for the prevention of medical device associated infection [34,35,37], thereby expanding scope of photo-active materials for potential infection control applications in a healthcare setting. Other published studies have documented the incorporation of photosensitizing compounds into various materials with the aim of generating anti-microbial or anti-infective surfaces. For example, Krouit *et. al.* (2008) incorporated a cationic porphyrin into a cellulose-based material [38]. However, the microbiological assays used in their work make it difficult to conclude if the reduction in viable bacteria was due to contact of sessile bacteria with the material surface, or release and diffusion of porphyrin into the nutrient agar used, in a similar fashion to antibiotic testing 'zone of inhibition' assays. Conversely the methodology used in our present study is specific against assessing the photo-inactivation of sessile, surface immobilised, adherent bacteria.

Our photosensitizer incorporated HDPE films display similar antibacterial efficacy as a photoactive cotton fabric, developed by Ringot *et. al.* (2011). This group also demonstrated the inherent difficulty of photo-inactivation of Gram-negative species in comparison to Gram-positive. The cotton-based fabrics documented in their study showed up to 5 log-cycle reductions in viable *S. aureus*, but only an approximate 1 log-cycle reduction in *E.*coli [39]. These figures are comparable to those we have obtained in our present work.

While we have focused on the antibacterial activity of our developed materials, future work shall investigate their ability to resist colonization with other pathogens, such as yeasts. Alvarez *et. al.* (2012) have reported successful inactivation of *Candida albicans* using polysilsesquioxane films doped with porphyrin [40].

Conclusions

This study demonstrates a general method to manufacture light-activated antimicrobial surfaces through the incorporation of photosensitizers into polymer films, specifically exemplified using high density poly(ethylene) showing high levels of antimicrobial behaviour, even to resistant strains such as MRSA and Gram-negative organisms in the presence of visible light. The use of such materials in the healthcare setting as coatings or coverings of inanimate objects such as keypads, touchscreens or handrails, may remove these surfaces as reservoirs of nosocomial pathogens. This may assist in the prevention of hospital-acquired infections, which are currently detrimental to the morbidity and mortality of vulnerable patients, and also have a significant financial impact on the resources of modern healthcare systems and institutions.

Author Contributions

Conceived and designed the experiments: CPM EJO JFC GTG SPG DSJ. Performed the experiments: CPM JFC LC. Analyzed the data: CPM LC ATDB SPG DSJ. Contributed to the writing of the manuscript: CPM EJO JFC LC ATDB GTG SPG DSJ.

References

1. Graves N (2004) Economics and preventing hospital-acquired infection. Emerging Infectious Diseases 10: 561–566.
2. Plowman R, Graves N, Griffin MA, Roberts JA, Swan AV, et al. (2001) The rate and cost of hospital-acquired infections occurring in patients admitted to selected specialties of a district general hospital in England and the national burden imposed. The Journal of Hospital Infection 47: 198–209.
3. Yates RR (1999) New intervention strategies for reducing antibiotic resistance. Chest 115: 24S–27S.
4. Hancock REW (2007) The end of an era? Nature Reviews Drug Discovery 6: 28.
5. Nikaido H (2009) Multidrug resistance in bacteria. Annual Review of Biochemistry 78: 119–146.
6. Enright MC, Robinson DA, Randle G, Feil EJ, Hajo G, et al. (2002) The evolutionary history of methicillin-resistant *Staphylococcus aureus* (MRSA). Proceedings of the National Academy of Sciences 99: 7687–7692.
7. Hota B (2004) Contamination, disinfection, and cross-colonization: Are hospital surfaces reservoirs for nosocomial infection? Clinical Infectious Diseases 39: 1182–1189.
8. Russell AD (2002) Bacterial resistance to disinfectants. British Journal of Infection Control 3: 22–24.
9. McCay PH, Ocampo-Sosa AA, Fleming GTA (2010) Effect of subinhibitory concentrations of benzalkonium chloride on the competitiveness of *Pseudomonas aeruginosa* grown in continuous culture. Microbiology 156: 30–38.
10. Noguchi N, Hase M, Kitta M, Sasatsu M, Deguchi K, et al. (1999) Antiseptic susceptibility and distribution of antiseptic-resistance genes in methicillin-resistant *Staphylococcus aureus*. FEMS Microbiology Letters 172: 247–253.
11. McBain AJ, Gilbert P (2001) Biocide tolerance and the harbingers of doom. International Biodeterioration & Biodegradation 47: 55–61.
12. Block C, Furman M (2002) Association between intensity of chlorhexidine use and micro-organisms of reduced susceptibility in a hospital environment. Journal of Hospital Infection 51: 201–206.
13. McDonnell G, Russel D (1999) Antiseptics and disinfectants: Activity, action, and resistance. Clinical Microbiology Reviews 12: 147–149.
14. Johnson JR, Kuskowski MA, Wilt TJ (2006) Systematic review: Antimicrobial urinary catheters to prevent catheter-associated urinary tract infection in hospitalized patients. Annals of Internal Medicine 144: 116–126.
15. Donlan RM, Costerton JW (2002) Biofilms: Survival mechanisms of clinically relevant microorganisms. Clinical Microbiology Reviews 15: 167–193.
16. Hamill TM, Gilmore BF, Jones DS, Gorman SP (2007) Strategies for the development of the urinary catheter. Expert Review of Medical Devices 4: 215–225.
17. Hall-Stoodley L, Costerton JW, Stoodley P (2004) Bacterial biofilms: From the natural environment to infectious diseases. Nature Reviews Microbiology 2: 95–108.

18. Knobloch JK, Horstkotte MA, Rohde H, Kaulfers P, Mack D (2002) Alcoholic ingredients in skin disinfectants increase biofilm expression of *Staphylococcus epidermidis*. Journal of Antimicrobial Chemotherapy 49: 683–687.
19. Bardouniotis E, Huddleston W, Ceri H, Olson ME (2001) Characterization of biofilm growth and biocide susceptibility testing of mycobacterium phlei using the MBEC (TM) assay system. FEMS Microbiol Lett 203: 263–267.
20. Rajamohan G, Srinivasan VB, Gebreyes WA (2009) Biocide-tolerant multidrug-resistant *Acinetobacter baumannii* clinical strains are associated with higher biofilm formation. Journal of Hospital Infection 73: 287–289.
21. Epstein AK, Pokroya B, Seminarab A, Aizenberg J (2011) Bacterial biofilm shows persistent resistance to liquid wetting and gas penetration. Proceedings of the National Academy of Sciences 108: 995–1000.
22. Fulmer PA, Wynne JH (2011) Development of broad-spectrum antimicrobial latex paint surfaces employing active amphiphilic compounds. ACS Applied Materials and Interfaces 3: 2878–2884.
23. O'Hanlon SJ, Enright MC (2009) A novel bactericidal fabric coating with potent *in vitro* activity against meticillin-resistant *Staphylococcus aureus* (MRSA). International Journal of Antimicrobial Agents 33: 427–431.
24. Pasquier N, Keul H, Heine E, Moeller M (2007) From multifunctionalized poly(ethylene imine)s toward antimicrobial coatings. Biomacromolecules 8: 2874–2882.
25. Wilson BC, Patterson MS (2008) The physics, biophysics and technology of photodynamic therapy. Physics in Medicine and Biology 53: R61–R109.
26. Kennedy JC, Pottier RH (1992) New trends in photobiology: Endogenous protoporphyrin IX, a clinically useful photosensitizer for photodynamic therapy. Journal of Photochemistry and Photobiology B: Biology 14: 275–292.
27. Mathai S, Smith TA, Ghiggino KP (2007) Singlet oxygen quantum yields of potential porphyrin-based photosensitizers for photodynamic therapy. Photochemical and Photobiological Sciences 6: 995–1002.
28. Craig RA, McCoy CP, Gorman SP (2010) Use of a sacrificial probe for singlet oxygen in photodynamic biomaterials. Journal of Pharmacy and Pharmacology 62: 1359–1360.
29. Hamblin MR, Hasan T (2004) Photodynamic therapy: A new antimicrobial approach to infectious disease? Photochemical and Photobiological Sciences 3: 436–450.
30. Ryskova L, Buchta V, Slezak R (2010) Photodynamic antimicrobial therapy. Central European Journal of Biology 5: 400–4006.
31. Wainwright M (1998) Photodynamic antimicrobial chemotherapy (PACT). Journal of Antimicrobial Chemotherapy 42: 13–28.
32. Wainwright M (2004) Photoantimicrobials - a PACT against resistance and infection. Drugs of the Future 29: 85.
33. Maisch T, Bosl C, Szeimies RM, Love B, Abels C (2007) Determination of the antibacterial efficacy of a new porphyrin-based photosensitizer against MRSA ex vivo. Photochemical and Photobiological Sciences 6: 545–551.
34. Brady C, Bell SEJ, Parsons C, Gorman SP, Jones DS, et al. (2007) Novel porphyrin-incorporated hydrogels for photoactive intraocular lens biomaterials. Journal of Physical Chemistry B 111: 527–534.
35. Parsons C, McCoy CP, Gorman SP, Jones DS, Bell SEJ, et al. (2009) Anti-infective photodynamic biomaterials for the prevention of intraocular lens-associated infectious endophthalmitis. Biomaterials 30: 597–602.
36. Hamblin MR, O'Donnell DA., Murthy N, Rajagopalan K, Michaud N, et al. (2002) Polycationic photosensitizer conjugates: effects of chain length and Gram classification on the photodynamic inactivation of bacteria. Journal of Antimicrobial Chemotherapy 49: 941–951.
37. McCoy CP, Craig RA, McGlinchey SM, Carson L, Jones DS, et al. (2012) Surface localisation of photosensitisers on intraocular lens biomaterials for prevention of infectious endophthalmitis and retinal protection. Biomaterials 33: 7952–7958.
38. Krouit M, Granet R, Krausz P (2008) New photoantimicrobial films composed of porphyrinated lipophilic cellulose esters. Bioorganic & Medicinal Chemistry Letters 16: 1651–1656.
39. Ringot C, Sol V, Barrière M, Saad N, Bressollier P, et al. (2011) Triazinyl porphyrin-based photoactive cotton fabrics: preparation, characterization, and antibacterial activity. Biomacromolecules 12: 1716–1723.
40. Alvarez MG, Gómez ML, Mora SJ, Milanesio ME, Durantini EN (2012) Photodynamic inactivation of candida albicans using bridged polysilsesquioxane films doped with porphyrin. Bioorganic & Medicinal Chemistry 20: 4032–4039.

3

Catalytic Profile of *Arabidopsis* Peroxidases, AtPrx-2, 25 and 71, Contributing to Stem Lignification

Jun Shigeto, Mariko Nagano, Koki Fujita, Yuji Tsutsumi*

Faculty of Agriculture, Kyushu University, Fukuoka, Japan

Abstract

Lignins are aromatic heteropolymers that arise from oxidative coupling of lignin precursors, including lignin monomers (*p*-coumaryl, coniferyl, and sinapyl alcohols), oligomers, and polymers. Whereas plant peroxidases have been shown to catalyze oxidative coupling of monolignols, the oxidation activity of well-studied plant peroxidases, such as horseradish peroxidase C (HRP-C) and AtPrx53, are quite low for sinapyl alcohol. This characteristic difference has led to controversy regarding the oxidation mechanism of sinapyl alcohol and lignin oligomers and polymers by plant peroxidases. The present study explored the oxidation activities of three plant peroxidases, AtPrx2, AtPrx25, and AtPrx71, which have been already shown to be involved in lignification in the *Arabidopsis* stem. Recombinant proteins of these peroxidases (rAtPrxs) were produced in *Escherichia coli* as inclusion bodies and successfully refolded to yield their active forms. rAtPrx2, rAtPrx25, and rAtPrx71 were found to oxidize two syringyl compounds (2,6-dimethoxyphenol and syringaldazine), which were employed here as model monolignol compounds, with higher specific activities than HRP-C and rAtPrx53. Interestingly, rAtPrx2 and rAtPrx71 oxidized syringyl compounds more efficiently than guaiacol. Moreover, assays with ferrocytochrome *c* as a substrate showed that AtPrx2, AtPrx25, and AtPrx71 possessed the ability to oxidize large molecules. This characteristic may originate in a protein radical. These results suggest that the plant peroxidases responsible for lignin polymerization are able to directly oxidize all lignin precursors.

Editor: Keqiang Wu, National Taiwan University, Taiwan

Funding: This work was supported by Japan Society for the Promotion of Science (JSPS) KAKENHI (B) Grant Number 23380102 (Y.T.) and JSPS KAKENHI Young Scientists (B) Grant Number 25850123 (J.S.). The funders had no role in study design, data collection and analysis, decision to publish, or preparation of the manuscript.

Competing Interests: The authors have declared that no competing interests exist.

* Email: y-tsutsu@agr.kyushu-u.ac.jp

Introduction

Lignin is a main component of vascular plant cell walls and possesses a complex and irregular structure. In angiosperms, lignins consist mainly of two monolignols, coniferyl (4-hydroxy-3-methoxycinnamyl) and sinapyl (3,5-dimethoxy-4-hydroxycinnamyl) alcohols, which polymerize through at least five different linkage types and result in 4-hydroxy-3-methoxyphenyl (guaiacyl, G) and 3,5-dimethoxy-4-hydroxyphenyl (syringyl, S) units, respectively. Monolignols are supplied to the cell wall and polymerized to fill, together with hemicellulose, the spaces between cellulose microfibrils; this polymerization proceeds through oxidative coupling catalyzed by plant peroxidases [1]. Based on the "End-wise" polymerization process, monolignol radicals can be coupled to a growing lignin polymer to produce a lignin macromolecule [2].

Plant peroxidases, which include large numbers of isoforms, participate in a broad range of physiological processes besides lignification, including suberin formation, phytoalexins synthesis, metabolism of reactive oxygen and nitrogen species, and programmed cell death [3]. To date, there is limited information available regarding the role of individual isoforms. Their contribution to lignification have been evaluated in several studies that have demonstrated that the up- or down-regulation of a target peroxidase gene is an effective strategy. For example, overexpression of a basic peroxidase in tomato leads to an increase in lignin content [4], and suppression of PrxA3a in aspen decreases lignin content [5]. Transgenic tobacco suppressed TP60 causes great decreases (up to 50%) in lignin content [6] and xylem with both fibers and vessels having thin cell walls [7]. Studies designed to identify plant peroxidases that contribute to lignification have also employed other approaches, such as enzyme purification using the enzyme's oxidation abilities toward monolignols and lignin polymers as an index. It has been reported that some plant peroxidases could oxidize sinapyl alcohol so far [8]. However, only cationic cell wall-bound peroxidase (CWPO-C), a peroxidase isozyme from *Populus alba* L. (poplar) cell wall, has been verified to serve oxidation of monolignols and lignin polymer [9,10]. Previously, this research group has focused on seven *Arabidopsis* plant peroxidases selected using amino acid similarities to CWPO-C as the probe and found that AtPrx2 or AtPrx25 deficiency led both decreased total lignin content and altered lignin structure, including cell wall thinning in the stem. In addition, AtPrx71 deficiency led an altered stem lignin structure, although the lignin content is not decreased [11]. These results provided *in vivo* evidence that AtPrx-2, 25, and 71 are involved in *Arabidopsis* stem lignification.

On the other hand, the catalytic mechanism of lignin polymerization by plant peroxidases, including the above three peroxidases, toward monolignols and growing lignin polymers is still being discussed. Because of the lack of oxidation activities toward lignin polymers and sinapyl alcohol, well-studied plant peroxidases, such as horseradish peroxidase C (HRP-C) and AtPrx53, are not matched as lignin polymerization enzymes. CWPO-C's unique oxidation ability does qualify as such an enzyme, and its activity provided by two protein surface tyrosine residues (Tyr74 and Tyr177) that can form a radical which is then available as an oxidation active site [12,13]. The biochemical characterization of CWPO-C *in vitro* has clarified that it can catalyze lignin polymerization without suffering steric hindrance owing to the substrate molecular size by conducting a one-electron oxidation reaction on monolignols and lignin oligomers and polymers on the protein surface [12]. Although CWPO-C's substrate oxidation system allows explanation of lignin polymerization catalyzed by plant peroxidases, further studies are required to reveal CWPO-C's physiological function. AtPrx-2, 25, and 71 are attractive for characterization of their oxidation activities because they have high amino acid similarity to CWPO-C and have already been shown to be responsible for lignification. AtPrx2 conserves its Tyr78 corresponding to catalytic Tyr74 of CWPO-C and shares 44% amino acid identity with CWPO-C. AtPrx25, with 64% amino acid identity with CWPO-C, is the only peroxidase that conserves its Tyr177 in *Arabidopsis*. And AtPrx71 possesses the highest amino acid sequence identity (68%) with CWPO-C. It would, therefore, be interesting to verify whether their oxidation activities are as effective for lignin polymerization as is CWPO-C. This study focused on three lignification related peroxidases, AtPrx-2, 25 and 71, in *Arabidopsis*, and the ability of their recombinant proteins to oxidize monolignols model compounds and large molecule were characterized *in vitro*. The relevance of their oxidation activity is discussed with respect to the common lignin polymerization mechanism catalyzed by peroxidases in vascular plants.

Materials and Methods

Plasmid construction and production of rAtPrx inclusion bodies

A cDNA library constructed from whole tissues of *Arabidopsis* (ecotype Columbia) was used as a template for PCR amplification of the targeted genes with KOD-Plus-DNA polymerase (Toyobo Co., Ltd., Osaka, Japan). Gene-specific primers containing *Bam*HI and *Sac*I restriction enzyme sites (Table S1) were used in PCR to amplify the coding sequence for the predicted mature proteins of AtPrx-2, 25, 53, and 71. Subsequent protein construction employing these primers eliminated from the precursor AtPrx proteins the N-terminal amino acids that would constitute the signal-peptide. The amplified fragment was ligated into pBluscript II KS (+) (Stratagene Corp., La Jolla, CA, USA) using restriction site *Bam*HI/*Sac*I and sequenced to ensure that no mutation was introduced during the subcloning process. Every cDNA was then transferred as a *Bam*HI/*Sac*I fragment into pQE-30-Xa to yield pQE-30-Xa-AtPrx. *Escherichia coli* (*E. coli*) M15 (pREP4) cells were transformed with pQE-30-Xa-AtPrx and grown in LB medium containing ampicillin and kanamycin (each at 50 $\mu g \cdot mL^{-1}$) at 37°C and with vigorous shaking. When the $A_{600\ nm}$ reached 0.2, ITPG was added to a final concentration of 0.4 mM. After further growth for 4 h, bacterial pellets were pelleted by centrifugation at 10 000×g for 10 min and hydrophobic proteins containing inclusion bodies prepared with Bugbuster Protein Extraction Reagent (Novagen, EMD Biosciences, Inc., Darmstadt, Germany) containing benzonase nuclease and lysozyme (25 units$\cdot mL^{-1}$ and 200 $\mu g \cdot mL^{-1}$, respectively) according to the manufacturer's protocol. The final pellets were stored at −80°C until the day of the refolding experiments.

Production and purification of recombinant AtPrx and CWPO-C

For each recombinant protein, the composition of the refolding mixture and refolding time were optimized in preliminary experiments (Table 1). After refolding, each refolding mixture of rAtPrx-2, 53, and 71 was dialyzed against 50 mM HEPES-NaOH (pH 7.0), 20 mM Tris-HCl (pH 8.5), and 50 mM MES-NaOH (pH 5.7), respectively. Then, insoluble material was sedimented by centrifugation at 14 000×g for 20 min and the supernatant filtered using cellulose acetate filters with 0.45 µm pore size (Advantec MFS, Inc., Dublin, CA, USA). Renatured rAtPrx-2 and 71 were purified using a cation-exchange chromatography column (Mono Q; GE Healthcare Bio-Sciences UK Ltd., Little Chalfont, UK) connected to a fast protein liquid chromatography system (Amersham Pharmacia Biotech, Inc., Piscataway, NJ, USA) with a linear gradient of 0–1.0 M NaCl in HEPES-NaOH and 20 mM Tris-HCl (pH 7.0 and 8.5, respectively). Renatured rAtPrx53 was purified using an anion-exchange chromatography column (Mono S; GE Healthcare Bio-Sciences UK Ltd.) with a linear gradient of 0–0.1 M NaCl in 20 mM Tris-HCl (pH 8.5). Finally, for renatured rAtPrx25 purification, NaCl was added into a refolding mixture of rAtPrx25 to a final 2.0 M concentration. After centrifugation at 14 000×g for 20 min, the supernatant was loaded onto a hydrophobic interaction chromatography column (HiPrep Butyl FF 16/10; GE Healthcare Bio-Sciences UK Ltd.) and eluted with a NaCl gradient of 2.0–0 M in buffer containing 50 mM Tris-HCl (pH 7.5) and 100 mM $CaCl_2$. The resulting enzyme fractions with the highest specific activity were pooled and concentrated using Pierce Protein Concentrators (20 ml/20K MWCO; Thermo Fisher Scientific Inc., Rockford, IL, USA). The concentrated sample (2 ml) was applied onto a HiLoad 16/60 Superdex 75 column (GE Healthcare Bio-Sciences UK Ltd.), and eluted with buffer containing 50 mM Tris-HCl (pH 7.5) and 100 mM $CaCl_2$ and 200 mM NaCl at a 0.25 ml$\cdot min^{-1}$ flow rate. Fractions containing renatured rAtPrx with the highest specific activity were used for peroxidase assays. The production and purification procedures of rCWPO-C were the same as previously described [13].

Peroxidase assay

The specific activity of rAtPrx toward guaiacol, 2,6-DMP, and syringaldazine were measured using a previously described method [13] with slight modification of the substrate concentrations. The optimal substrate concentration for maximum reaction velocity was determined beforehand and employed in measuring specific activities. The assay mixture contained 50 mM Tris-HCl (pH 7.5), a reducing substrate (5–30 mM guaiacol, 20–100 mM 2,6-DMP, and 20–30 µM syringaldazine), and various concentrations of recombinant proteins and was preincubated at 30°C. In oxidation assays of rAtPrx25 activity, including the oxidation assay of Cc^{2+} described below, unexpectedly high calcium chloride was required to avoid deactivation, and 100 mM $CaCl_2$ was added to the assay mixtures. The assay reaction was initiated by H_2O_2 addition to a final 1 mM concentration. Cc^{2+} was prepared by reducing Cc^{3+} with sodium hydrosulfite and the sodium hydrosulfite removed using a PD-10 column (GE Healthcare Bio-Sciences UK Ltd.). In a Cc^{2+} oxidation assay, the reaction mixture (500 µl) contained 30 µM Cc^{2+}, various concentrations of recombinant proteins in the native or heat-denatured state (0–0.4 µM), and 0.1 mM H_2O_2 in 50 mM Tris-HCl buffer (pH 7.5) and preincubated at 30°C. The reaction was then carried out at 30°C. The progress of Cc^{2+} oxidation was monitored by spectral changes, evaluated in terms of decreased absorbance at 550 nm indicating oxidation of Cc^{2+} to Cc^{3+}.

Homology modeling

The three-dimensional structures of AtPrx-2, 25, and 71 were predicted with the homology-modeling server SWISS-MODEL (Swiss Institute of Bioinformatics, Lausanne, Switzerland; http://swissmodel.expasy.org/) [14]. The soybean peroxidase crystal structure [Protein Data Bank (PDB) entry 1FHF] was employed as the most appropriate template for AtPrx2, as it shares the highest amino acid sequence identity with soybean peroxidase among the known peroxidase crystal structures. In the same way, the ATP A2 crystal structure [PDB entry 1PA2] was selected as the most appropriate template for AtPrx-25 and 71. The obtained structures were visualized and analyzed with PyMol (an Open Source molecular viewing engine, DeLano Scientific LLC).

Results

Refolding and purification of recombinant proteins

Peroxidase activity was investigated using recombinant protein of each AtPrx, termed rAtPrx-2, 25, and 71, overexpressed in *E. coli*. Recombinant AtPrx53 was also produced as a comparison enzyme. As each recombinant protein from *E. coli* was recovered as an inclusion body lacking enzymatic activity, *in vitro*-refolding was applied to obtain the active renatured form. On the basis of known plant and fungal recombinant peroxidase refolding conditions [13,15,16], the compositions of refolding mixtures for rAtPrx-2, 25, and 53 were optimized by adjusting the concentrations of denaturizing agent (urea or guanidine), calcium chloride, hemin, oxidizing agent (GSSG or cystine), reducing agent (GSH or cystein), and protein as well as the pH and incubation time (Fig. S1). The refolding mixture for rAtPrx71 possessed sufficiently high peroxidase activity under the same refolding conditions as used for rCWPO-C such that further optimization was not needed. In the case of rAtPrx25, urea as the denaturizing agent resulted in misfolded-protein and cysteine/cysteine was found to be a better agent than GSSG/GSH for reconstructing the disulfide bond. Table 1 shows refolding conditions of inclusion body proteins after the optimization. The optimized refolding mixture of each recombinant protein, except for rAtPrx25, was purified by ion exchange chromatography on a MonoS or MonoQ column following dialysis in the appropriate buffer (see Materials and Methods). rAtPrx25 was purified by hydrophobic interaction chromatography on a HiPrep Butyl FF 16/10 column and gel filtration chromatography on a HiLoad 16/60 Superdex 75 column. After purification, the specific activities of rAtPrx-2, 25, 53, and 71 were increased to 108, 293, 47, and 2.8-fold, respectively, and the Reinheitszahl values (RZ) determined by the absorbance ratio A_{400}/A_{280} were 1.8, 1.8, 2.8, and 2.7, respectively (Table 2). Purified rAtPrx fractions each presented a single band on CBB staining after SDS–PAGE (Fig. 1, inset), and their absorption spectrum exhibited a single Soret peak and two visible peaks (Fig. 1). An abundance of misfolded-protein resulted in an abnormally broad Soret peak (Fig. S2). Thus, as all purified enzyme preparations exhibited a spectrum specific to peroxidase, it was concluded that these recombinant proteins were effectively renatured.

Table 1. Refolding conditions of inclusion body proteins expressed in *E. coli.*

	Denaturizing agent (M)	pH	$CaCl_2$ (mM)	Hemin (μM)	Oxidizing agent (mM)	Reducing agent (mM)	Protein conc. (mg/ml)	Time (h)	Glycerol (%)
rAtPrx2	Urea, 1.75	11	80	5	GSSG, 0.35	GSH, 0.11	0.1	96	5
rAtPrx25	Guanidine, 0.25	9.5	160	10	Cystine, 0.50	Cystein, 0.50	0.2	24	5
rAtPrx53	Urea, 2.75	9.5	5	5	GSSG, 0.70	GSH, 0.21	0.3	72	5
rAtPrx71	Urea, 3.25	9.5	40	10	GSSG, 0.70	GSH, 0.21	0.2	48	5

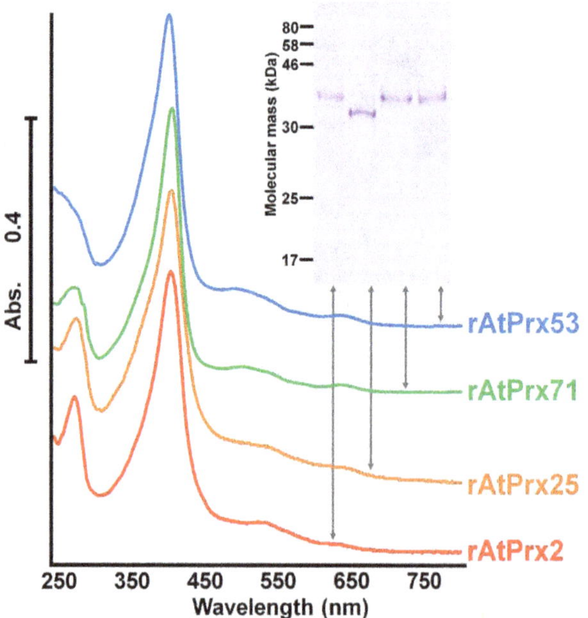

Figure 1. Absorption spectra of purified recombinant AtPrx2, AtPrx25, AtPrx53, and AtPrx71. Spectra from 250 to 800 nm of 10 μM proteins measured by UV-visible spectrometry and inset shows CBB staining of purified proteins (400 ng) after SDS–PAGE (12% gel) with molecular mass markers given in kDa on left.

Oxidation activity of recombinant proteins for monolignol model compounds

Because two methoxy groups of the syringyl unit cause severe steric hindrance compared with the guaiacyl unit, known plant peroxidases have difficulty oxidizing syringyl compounds, such as sinapyl alcohol. The oxidation activities of the recombinant enzymes were evaluated using three substrates, guaiacol, 2,6-dimethoxyphenol (2,6-DMP), and syringaldazine, because the oxidation product formation from these substrates could be monitored spectrophotometrically. Guaiacol, a representative of guaiacyl compounds, was used as a model for coniferyl alcohol, and 2,6-DMP and syringaldazine as models for sinapyl alcohol. The optimal substrate concentrations for maximal reaction velocity were determined in preliminary experiments. Table 3 shows each recombinant protein's specific activity for these substrates. rAtPrx53 showed higher specific activity for guaiacol than rAtPrx-2, 25, 53, and rCWPO-C, and the relative oxidation ratio for guaiacol/2,6-DMP/syringaldazine determined to be 1/0.12/0.026. This activity characteristic was quite similar to that of HRP-C. In contrast, rAtPrx-2, 25, and 71 showed higher specific activity for both syringyl compounds than rAtPrx53 and HRP-C. In particular, rAtPrx-2 and 71 preferred syringyl more than guaiacyl compounds as substrates, and the relative oxidation ratios were determined to be 1/2.1/15 and 1/3/3.5, respectively. Those properties were analogous to those of CWPO-C.

Oxidation of ferrocytochrome *c* by rAtPrx proteins

The oxidation ability of these recombinant enzymes for large substrates was evaluated using ferrocytochrome *c* (Cc^{2+}). Both Cc^{2+} and lignin polymer have too large molecular size to enter the plant peroxidase heme pocket. When Cc^{2+} is converted to Cc^{3+} by a single-electron oxidation, the absorption maximum at 550 nm decreases according to the degree of oxidation (Fig. 2A), and thus, Cc^{2+} oxidation is easily monitored as decreases in 550 nm absorption. The reaction was started by adding H_2O_2 to a reaction tube containing Cc^{2+} and the recombinant protein at different concentrations. No significant absorbance decreases were observed in reactions containing 0.1 and 0.2 μM rAtPrx53, which indicated that Cc^{2+} could not enter to the peroxidase heme pocket (Fig 2B). In contrast, 0.05, 0.1, and 0.2 μM rCWPO-C converted

Table 2. Purification of recombinant *Arabidopsis* peroxidases after refolding *in vitro*.

	Protein (mg)	Total activity (μmol/min)	Specific activity (μmol/min/mg protein)	Yield (%)	Purification (fold)	RZ (A400/A280)
rAtPrx2						
(1) Refolding mixture	5.0	59.6	11.9	100	1.0	0.59
(2) Dialysis	3.2	62.1	19.4	104	1.6	0.51
(3) Mono S	0.013	16.2	1291	27.2	108	1.8
rAtPrx25						
(1) Refolding mixture	30	27.6	0.920	100	1.0	0.81
(2) HiPrep Butyl FF	0.64	19.2	30.1	69.7	33	0.28
(3) Superdex75	0.031	8.4	270	30.3	293	1.8
rAtPrx53						
(1) Refolding mixture	14	96.6	7.07	100	1.0	0.61
(2) Dialysis	12	94.3	7.79	97.7	1.1	0.80
(3) Mono Q	0.058	19.4	333	20.1	47	2.8
rAtPrx71						
(1) Refolding mixture	0.65	218	337	100	1.0	0.60
(2) Dialysis	0.39	152	389	69.8	1.2	0.90
(3) Mono S	0.064	61.0	953	28.0	2.8	2.7

rAtPrx2 and rAtPrx71 activities determined by monitoring oxidation product of syringaldazine at 530 nm. rAtPrx25 and rAtPrx53 activities determined by monitoring oxidation product of guaiacol at 470 nm.

Table 3. Oxidation of guaiacol, 2,6-DMP, and syringaldazine by purified recombinant proteins and HRP-C.

Peroxidase	Purification process	Guaiacol				2,6-DMP				Syringaldazine			
rAtPrx2	Mono S	86	±	1.8	(1.0)	176	±	8.4	(2.1)	1291	±	84	(15)
rAtPrx25	Superdex 75	270	±	7.1	(1.0)	93	±	0.54	(0.34)	276	±	1.1	(1.0)
rAtPrx71	Mono S	270	±	4.3	(1.0)	805	±	19	(3.0)	953	±	3.5	(3.5)
rAtPrx53	Mono Q	333	±	0.45	(1.0)	39	±	0.80	(0.12)	8.6	±	0.20	(0.026)
rCWPO-C*	HiPrep Butyl FF16/10	120	±	2.4	(1.0)	480	±	29	(4.0)	1077	±	25	(8.9)
HRP-C*	Commercial Product	372	±	5.1	(1.0)	50	±	1.8	(0.14)	114	±	2.2	(0.31)

Oxidation activity (mean ±standard deviation, $n = 3$) expressed as μmol·min^{-1} per mg protein; relative activities in parentheses; relative activity of each protein determined by comparison with guaiacol activity, set to 1.0.
* Shigeto et al. 2012 [13].

Cc^{2+} to Cc^{3+} and the observed oxidation velocities was clearly dependent on the protein concentration (Fig. 2C). rAtPrx-2, 25, and 71 also converted Cc^{2+} to Cc^{3+} (Fig. 2D, E, and F) at velocities dependent on protein concentration. The oxidation activities of rAtPrx-2, 25, and 71 for Cc^{2+} were estimated to be one-third, half, and equal to that of rCWPO-C, respectively.

Prediction of the oxidation mechanism

The oxidation abilities of rAtPrx-2, 25, and 71 toward Cc^{2+} suggested that the substrate oxidation site existed on the proteins' surfaces, similar to Try74 and Tyr177 in CWPO-C. It has been reported that tryptophan can act as a redox active residue, similar to tyrosine, in fungal enzymes [17–19] and the Y74W CWPO-C variant [20]. Exposed tyrosine and tryptophan on protein surfaces can thus be considered as unique substrate oxidation sites and, specifically, the tyrosine and tryptophan residues present in these three peroxidases but not present in AtPrx53 and HRP-C were considered preferential candidates for oxidation sites (Fig. 3A, underlined residues). Following homology modeling of AtPrx-2, 25, and 71 by SWISS-MODEL was conducted to narrow the possible oxidation sites exposed on the proteins' surfaces. Tyr78 and Trp117 of AtPrx2 (Fig. 3B), Tyr177 and Trp246 of AtPrx25 (Fig. 3C), and Trp232 and Trp254 of AtPrx71 were on the shortlist for putative oxidation sites (Fig. 3D). Tyr78 of AtPrx2 and Tyr177 of AtPrx25 corresponded to the catalytic Try74 and Tyr177 in CWPO-C, respectively. Although AtPrx71 did not possess a similar tyrosine residue, a unique tryptophan residue at position 232 was located near the heme.

Discussion

Oxidation of both growing lignin polymers and monolignols is indispensable for lignin biosynthesis. The molecular sizes of the lignin oligomers or polymers are too large to enter the heme-pocket active site of plant peroxidases [21]. Furthermore, as shown in Table S2, only a few peroxidases in the literatures can oxidize S unit. These facts posed a question for the progress of the lignin polymerization reaction mechanism. Based on modeling studies, AtPrx53 was concluded to have Pro139 overlapping the substrate-binding site such that a methoxy group in sinapyl alcohol hinders substrate docking [22]. Instead, it is suggested that a coniferyl alcohol radical acts as an intermediate in the oxidation of synapyl alcohol and lignin polymer [23]. On the other hand, a distinct characteristic of lignifying xylem peroxidases is their ability to oxidize syringyl compounds [24], and radical transfer from coniferyl alcohol to polymeric lignols occurred only slightly [10]. These observations imply that the lignin polymerization mechanism could not be explained sufficiently by a radical transfer system *via* a mediator. In the last ten-odd years, three plant peroxidases, ZePrx from *Zinnia elegans* [25], BPX1 from *Betula pendula* [26], and CWPO-C from *Populus alba* L. [9], have been reported to have higher oxidation activities toward sinapyl than coniferyl alcohol. Notably, CWPO-C exhibits higher oxidation activity for sinapyl alcohol by approximately ten times that of HRP-C, is able to oxidize Cc^{2+}, and synthesizes lignin polymer *in vitro* [10]. However, the oxidation ability of plant peroxidase catalyzing lignin polymerization *in vivo* remains unclear.

In this study, the oxidation activities of three genus *Arabidopsis* plant peroxidases, AtPrx-2, 25, and 71, formerly confirmed to be involved in lignification, were evaluated using monolignol model compounds and Cc^{2+}, the latter representing an oxidative property of the lignin polymer. As expected, rAtPrx53 and HRP-C exhibited substantially lower specific activities for the two syringyl compounds than for guaiacol and also did not oxidize Cc^{2+}. In

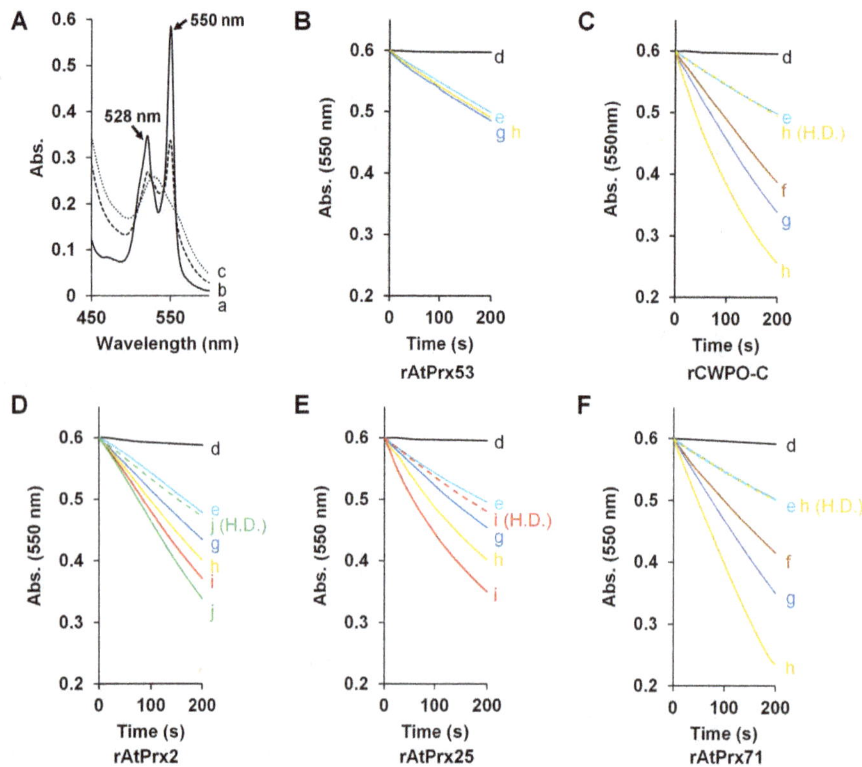

Figure 2. Spectrophotometric demonstration of Cc^{2+} oxidation. A: Spectral changes of Cc^{2+} during incubation with 0.1 μg rCWPO-C. Optical spectra between 450 and 600 nm recorded for Cc^{2+} in 50 mM Tris-HCl before (trace a) and after incubation with 0.1 mM H_2O_2 and 0.1 μM rCWPO-C for 180 sec (trace b) and 300 sec (trace c). B–F: Time course of Cc^{2+} oxidation by recombinant peroxidases. Absorbance monitoring started immediately after H_2O_2 addition; spontaneous Cc^{2+} oxidation monitored without H_2O_2 addition (trace d); protein concentrations: trace e, 0 μM; trace f, 0.05 μM; trace g, 0.1 μM; trace h, 0.2 μM; trace i, 0.3 μM; and trace j, 0.4 μM; complementary trace (H.D), absorption change with heat-denatured protein at given concentration.

contrast, rAtPrx-2, 25, and 71 readily oxidized the same two syringyl compounds as well as Cc^{2+}; the summarized oxidation results are depicted in a cobweb chart (Fig. 4). The oxidation abilities for both syringyl compounds and Cc^{2+} were conserved in all three AtPrx enzymes. This observation reflected that AtPrx-2, 25, and 71 can oxidize these substrates without the steric hindrance that restricts substrates from entering their heme pocket active site. In other words, these enzymes have oxidation site(s) exposed or available on their protein surfaces, and it is highly likely that their oxidation activities toward Cc^{2+} as well as their activity toward the monolignol model compounds were dependent on these site(s). rAtPrx-2 and 71 showed higher oxidation activities for two syringyl compounds than for guaiacol, a property shared with CWPO-C which cannot oxidize guaiacol or syringyl compounds in its heme pocket [13]. This suggested that guaiacol and syringyl compound oxidation of rAtPrx-2 and 71 took place largely on the protein surfaces for the following two reasons. First, 2,6-DMP and syringaldazine were more easily oxidized than guaiacol because of their redox potential under steric hindrance-free conditions. The methoxy group's electron-donating property has been reported to reduce the redox potential of phenolic compounds, and guaiacol has a higher redox potential than 2,6-DMP and syringaldazine [27]. This situation is true for sinapyl alcohol, which has a lower redox potential than coniferyl alcohol [28]. And second, AtPrx-2 and 71 would not have large enough heme pockets to easily accept syringyl compounds. Their heme pocket entrances, estimated by homology modeling, are not larger than those of AtPrx53 and HRP-C (Fig. S3) and Pro139 is conserved as well, as it has been in AtPrx53 (Fig. 3A). AtPrx25 possessed intermediate oxidation characteristics between those of AtPrx53 and CWPO-C, and did not possess a sufficiently large heme pocket entrance, and conserves the mentioned proline. Considering the oxidation activity and predicted 3D-structure of AtPrx25, most of its oxidation activity for guaiacol might have originated from the heme pocket, and the oxidation activity on the protein surface toward monolignol model compounds would be lower than in CWPO-C, AtPrx-2, and 71. An oxidation reaction caused by an exposed protein radical will depend on the redox potential and not follow the presumed enzyme kinetics of a two-step reaction (step one, the substrate binds to the enzyme and step two, the substrate is converted to product and released). Therefore, to explain these enzymes' oxidation activities for monolignol model compounds, specific activity (μmol·min^{-1} per mg protein) was employed instead of steady-state kinetic constant values (K_m, k_{cat}, and turnover rate).

Poplar CWPO-C is the only peroxidase that has been shown to have exposed catalytic sites in plant peroxidases [13]. CWPO-C uses exposed Tyr74 and Tyr177 as the oxidation site, and conserved tyrosines, Tyr78 of AtPrx2 and Tyr177 of AtPrx25, were considered here as the most likely oxidation sites. Although AtPrx71 did not have an exposed tyrosine, unique Trp232 and Trp254 exposed on the protein surface might play a similar role. Exposed catalytic sites have been identified in some fungal peroxidases, including versatile peroxidase with catalytic Trp164 [18,19], *Phanerochate chrysosporium* lignin peroxidase with catalytic Trp171 [17], and *Trametes cervina* lignin peroxidase

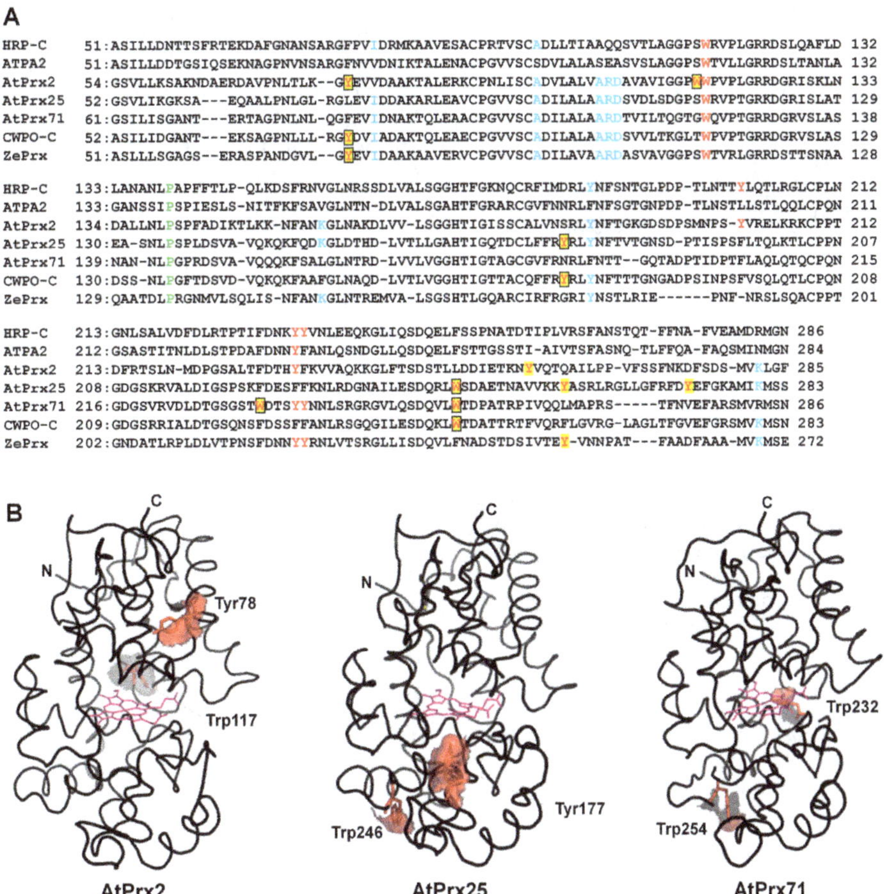

Figure 3. Primary and predicted 3-D structure of AtPrx2, AtPrx25, and AtPrx71. A: Amino acid alignment of plant peroxidases, HRP-C, AtPrx-53, 2, 25, 71, and ZePrx; conserved Pro139 in AtPrx53 in green structural motifs common to S peroxidases previously proposed (Ros Barceló et al. 2007, Novo-Uzal et al. 2013) in blue; tyrosine and tryptophan residues, presented in three peroxidases, AtPrx-2, 25, and 71, but not in AtPrx53 and HRP-C highlighted in yellow; possible oxidation sites enclosed in box. B: Predicted 3-D structures of AtPrx-2, 25, and 71 by homology-modeling performed with SWISS-MODEL using PDB entry (http://www.pdb.org/pdb/home/home.do) 1FHF (for AtPrx2) and 1PA2 structure (for AtPrx-25 and 71) as template; N and C-terminus of protein molecules labeled; and possible oxidation sites on protein surface in red.

with catalytic Tyr181 [29]. Long-range electron transfer (LRET) pathways from the exposed catalytic site to the buried heme cofactor were suggested by crystallographic studies of these peroxidases [17,29,30]. There may be a LRET pathway in CWPO-C, AtPrx-2, 25 and 71 similar to versatile peroxidase and lignin peroxidases, though crystallographic study is necessary to discuss detailed LRET pathway in CWPO-C, AtPrx-2, 25 and 71.

Plant peroxidases able to oxidize guaiacyl moieties but unable to oxidize efficiently syringyl moieties are called G peroxidase. Unlike G peroxidase, plant peroxidases able to oxidize both guaiacyl and syringyl moieties have been called S peroxidase [8] so far. It was reported that CWPO-C has not only a property of S peroxidase, but also has the oxidation ability for large molecule [10]. Therefore, CWPO-C should be included in new category of "all-round peroxidase". From the similar point of view, AtPrx-2, 25, and 71 are classified to the all-round peroxidase. The structural motifs of S peroxidases, which are amino acid sequences conserved among S peroxidases but not in G peroxidases, were proposed by Ros Barceló et al. (2007) [8] and Novo-Uzal et al. (2013) [31]. AtPrx-2, 25, and 71 conserve, indeed, all or part of these motifs (Fig. 3A). However, HRP-C, a representative G peroxidase, conserves part of these motifs and most AtPrxs (68 out of 73) share at least two of these motifs, indicating that it is difficult to search for S peroxidases using only these motifs. When it is presumed that S peroxidases oxidize syringyl compounds by a protein surface oxidation site, as in CWPO-C, it becomes a discriminating criterion regarding the peroxidase possessing exposed tyrosine and tryptophan, which form the substrate oxidation site, to distinguish G from S peroxidases. ZePrx, a typical S peroxidase, has Tyr73 that corresponds to Tyr74 in CWPO-C, and thus ZePrx might be likely to use the tyrosine as an oxidation site and able to oxidize not only syringyl compounds but larger substrates as well. It means that ZePrx may also be all-round peroxidase. Recently, AtPrx72, a homolog of ZePrx, has also been shown to be involved in *Arabidopsis* stem lignification by analysis of *Arabidopsis* knockout mutants [32]. The predicted mature AtPrx72 protein shows some exposed tyrosines. Such oxidation characteristics in plant peroxidases involved in lignification with *in vivo* evidence, including AtPrx72, are of interest and will provide additional knowledge regarding the mechanism of plant peroxidase-mediated lignin polymerization.

In conclusion, recombinant proteins of three plant peroxidases, AtPrx-2, 25, and 71, responsible for lignin polymerization in the *Arabidopsis* stem were produced, and their oxidation characteristics were examined. The results indicated that these peroxidases oxidized the large molecules required for direct oxidation of

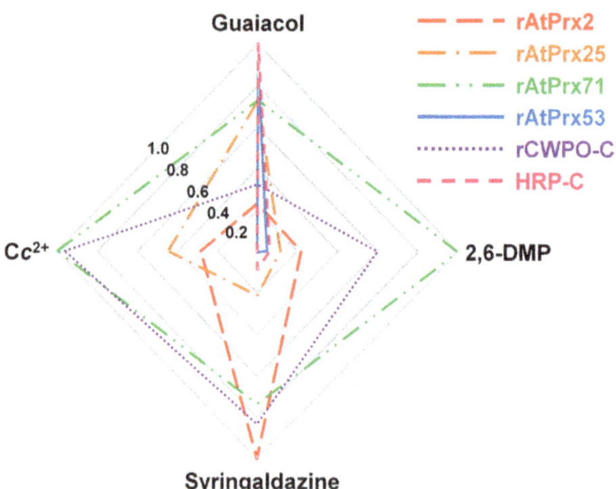

Figure 4. A cobweb chart illustration of the relative oxidation activities of recombinant peroxidases and native HRP-C. Peroxidase activities listed in Table 3 converted to relative values calculated with the highest activity among six peroxidases set to 1.0; Cc^{2+} oxidation activity estimated by decreased Cc^{2+} concentration listed in Figure 3, except for HRP-C; and Cc^{2+} oxidation by HRP-C previously reported (Sasaki et al. 2004).

sinapyl alcohol and lignin polymers, unlike most known plant peroxidases. This characteristic clearly explained the lignin polymerization mechanism. Taking their oxidation activities and estimated structure into consideration, it was suggested that AtPrx-2, 25, and 71 have oxidation site(s) on their protein surfaces.

Supporting Information

Figure S1 Parameter optimization for the *in vitro* refolding of recombinant AtPrx proteins. Denaturizing agent concentration (A), hemin concentration (B), pH (C), $CaCl_2$ concentration (D), oxidizing agent concentration (E), Reducing agent concentration (F), Protein concentration (G), and incubation time (H) were systematically varied, and shown as relative activity of maximal yields of each. The basic conditions were: 48 h incubation at 4°C in 50 mM Tris–HCl buffer (pH 9.5) containing 3.25 M urea, 10 μM hemin, 40 mM $CaCl_2$, 0.7 mM GSSG, 0.21 mM GSH, 0.2 mg/ml protein. Refolding efficiency was estimated by guaicol oxidation.

Figure S2 A normal and abnormal spectra of purified recombinant AtPrx25. To renature inactive recombinant AtPrx25 protein, a final concentration of 1 M urea or 0.25 M guanidine was used as denaturizing agent in refolding mixture; after purification (see Materials and methods), fraction with highest specific activity collected; and absorption spectrum measured by UV-visible spectrometry.

Figure S3 The entrance to the heme pocket of AtPrx proteins, HRP-C, and CWPO-C. Structure of AtPrx53 and HRP-C, and predicted structure of AtPrx2, 25, and 71 were as described in Materials and methods and Fig. 3 legend; predicted CWPO-C structure prepared as previously described [13]; and intensity of red deepness, level of hydrophobicity.

Table S1 Sequence of primers used to amplify cDNA sequences.

Table S2 Oxidation property of plant peroxidases for G and S units.

Author Contributions

Conceived and designed the experiments: JS YT. Performed the experiments: JS MN. Analyzed the data: JS MN KF YT. Contributed to the writing of the manuscript: JS YT.

References

1. Boerjan W, Ralph J, Baucher M (2003) Lignin biosynthesis. Annu Rev Plant Biol 54: 519–546.
2. Sarkanen KV (1971) Precursors and their polymerization. In: Sarkanen KV and Ludwig CH (eds) Lignins, Occurrence, Formation, Structure and Reactions. Wiley-Interscience, New York, pp.95–163.
3. Almagro L, Gómez Ros LV, Belchi-Navarro S, Bru R, Ros Barceló A, et al. (2009) Class III peroxidases in plant defence reactions. J Exp Bot 60: 377–390.
4. El Mansouri I, Mercado JA, Santiago-Dómenech N, Pliego-Alfaro F, Valpuesta V, et al. (1999) Biochemical and phenotypical characterization of transgenic tomato plants overexpressing a basic peroxidase. Physiol Plant 106: 355–362.
5. Li Y, Kajita S, Kawai S, Katayama Y, Morohoshi N (2003) Down-regulation of an anionic peroxidase in transgenic aspen and its effect on lignin characteristics. J Plant Res 116: 175–182.
6. Blee KA, Choi JW, O'Connell AP, Schuch W, Lewis NG, et al. (2003) A lignin-specific peroxidase in tobacco whose antisense suppression leads to vascular tissue modification. Phytochemistry 64: 163–176.
7. Kavousi B, Daudi A, Cook CM, Joseleau JP, Ruel K, et al. (2010) Consequences of antisense down-regulation of a lignification-specific peroxidase on leaf and vascular tissue in tobacco lines demonstrating enhanced enzymic saccharification. Phytochemistry 71: 531–542.
8. Ros Barceló AR, Ros LV, Carrasco AE (2007) Looking for syringyl peroxidases. Trends Plant Sci 12: 486–491.
9. Aoyama W, Sasaki S, Matsumura S, Hirai H, Tsutsumi Y, et al. (2002) Sinapyl alcohol-specific peroxidase isoenzyme catalyzes the formation of the dehydrogenative polymer from sinapyl alcohol. J Wood Sci 48: 497–504.
10. Sasaki S, Nishida T, Tsutsumi Y, Kondo R (2004) Lignin dehydrogenative polymerization mechanism: a poplar cell wall peroxidase directly oxidizes polymer lignin and produces in vitro dehydrogenative polymer rich in beta-*O*-4 linkage. FEBS Lett 562: 197–201.
11. Shigeto J, Kiyonaga Y, Fujita K, Kondo R, Tsutsumi Y (2013) Putative Cationic Cell-Wall-Bound Peroxidase Homologues in *Arabidopsis*, AtPrx2, AtPrx25, and AtPrx71, Are Involved in Lignification. J Agric Food Chem 61: 3781–3788.
12. Sasaki S, Nonaka D, Wariishi H, Tsutsumi Y, Kondo R (2008) Role of Tyr residues on the protein surface of cationic cell-wall-peroxidase (CWPO-C) from poplar: potential oxidation sites for oxidative polymerization of lignin. Phytochemistry 69: 348–355.
13. Shigeto J, Itoh Y, Tsutsumi Y, Kondo R (2012) Identification of Tyr74 and Tyr177 as substrate oxidation sites in cationic cell wall-bound peroxidase from *Populus alba* L. FEBS J 279: 348–357.
14. Arnold K, Bordoli L, Kopp J, Schwede T (2006) The SWISS-MODEL workspace: a web-based environment for protein structure homology modelling. Bioinformatics 22: 195–201.
15. Smith AT, Santama N, Dacey S, Edwards M, Bray RC, et al. (1990) Expression of a synthetic gene for horseradish peroxidase C in *Escherichia coli* and folding and activation of the recombinant enzyme with Ca^{2+} and heme. J Biol Chem 265: 13335–13343.
16. Miki Y, Morales M, Ruiz-Dueñas FJ, Martínez MJ, Wariishi H, et al. (2009) *Escherichia coli* expression and *in vitro* activation of a unique ligninolytic peroxidase that has a catalytic tyrosine residue. Protein Expr Puri 68: 208–214.
17. Doyle WA, Blodig W, Veitch NC, Piontek K, Smith AT (1998) Two substrate interaction sites in lignin peroxidase revealed by site-directed mutagenesis. Biochemistry 37: 15097–15105.
18. Ruiz-Dueñas FJ, Morales M, Mate MJ, Romero A, Martínez MJ, et al. (2008) Site-directed mutagenesis of the catalytic tryptophan environment in *Pleurotus eryngii* versatile peroxidase. Biochemistry 47: 1685–1695.
19. Ruiz-Dueñas FJ, Morales M, García E, Miki Y, Martínez MJ, et al. (2009) Substrate oxidation sites in versatile peroxidase and other basidiomycete peroxidases. J Exp Bot 60: 441–452.
20. Pham LT, Kim SJ, Ahn US, Choi JW, Song BK, et al. (2013) Extension of Polyphenolics by CWPO-C Peroxidase Mutant Containing Radical-Robust Surface Active Site, Appl Biochem Biotechnol 172: 792–805.
21. Nielsen KL, Indiani C, Henriksen A, Feis A, Becucci M, et al. (2001) Differential activity and structure of highly similar peroxidases. Spectroscopic, crystallo-

graphic, and enzymatic analyses of lignifying *Arabidopsis thaliana* peroxidase A2 and horseradish peroxidase A2. Biochemistry 40: 11013–11021.

22. Østergaard L, Teilum K, Mirza O, Mattsson O, Petersen M, et al. (2000) *Arabidopsis* ATP A2 peroxidase. Expression and high-resolution structure of a plant peroxidase with implications for lignifications. Plant Mol Biol 44: 231–243.
23. Takahama U, Oniki T, Shimokawa H (1996) A possible mechanism for the oxidation of sinapyl alcohol by peroxidase-dependent reactions in the apoplast: enhancement of the oxidation by hydroxycinnamic acids and components of the apoplast. Plant Cell Physiol 37: 499–504.
24. Harkin JM, Obst JR (1973) Lignification in trees: indication of exclusive peroxidase participation. Science 180: 296–308.
25. Gabaldón C, López-Serrano M, Pedreño MA, Ros Barceló A (2005) Cloning and molecular characterization of the basic peroxidase isoenzyme from *Zinnia elegans*, an enzyme involved in lignin biosynthesis. Plant Physiol 139: 1138–1154.
26. Marjamaa K, Kukkola E, Lundell T, Karhunen P, Saranpää P, et al. (2006) Monolignol oxidation by xylem peroxidase isoforms of Norway spruce (*Picea abies*) and silver birch (*Betula pendula*). Tree Physiol 26: 605–611.
27. Xu F, Shin W, Brown SH, Wahleithner JA, Sundaram UM, et al. (1996) A study of a series of recombinant fungal laccases and bilirubin oxidase that exhibit significant differences in redox potential, substrate specificity, and stability. Biochim Biophys Acta 1292: 303–311.
28. Kobayashi T, Taguchi H, Shigematsu M, Tanahashi M (2005) Substituent effects of 3,5-disubstituted *p*-coumaryl alcohols on their oxidation using horseradish peroxidase-H_2O_2 as the oxidant. J Wood Sci 51: 607–614.
29. Miki Y, Calviño FR, Pogni R, Giansanti S, Ruiz-Dueñas FJ, et al. (2011) Crystallographic, kinetic, and spectroscopic study of the first ligninolytic peroxidase presenting a catalytic tyrosine. J Biol Chem 286: 15525–15534.
30. Pérez-Boada M1, Ruiz-Dueñas FJ, Pogni R, Basosi R, Choinowski T, et al. (2005) Versatile peroxidase oxidation of high redox potential aromatic compounds: site-directed mutagenesis, spectroscopic and crystallographic investigation of three long-range electron transfer pathways. J Mol Biol 354(2): 385–402.
31. Novo-Uzal E, Fernández-Pérez F, Herrero J, Gutiérrez J, Gómez-Ros LV, et al. (2013) From Zinnia to Arabidopsis: approaching the involvement of peroxidases in lignification. J Exp Bot 64: 3499–3518.
32. Herrero J, Fernández-Pérez F, Yebra T, Novo-Uzal E, Pomar F, et al. (2013) Bioinformatic and functional characterization of the basic peroxidase 72 from Arabidopsis thaliana involved in lignin biosynthesis. Planta 237: 1599–1612.

4

Cytotoxicity of Polyaniline Nanomaterial on Rat Celiac Macrophages *In Vitro*

Yu-Sang Li[9], Bei-Fan Chen[9], Xiao-Jun Li, Wei Kevin Zhang, He-Bin Tang*

Department of Pharmacology, College of Pharmacy, South-Central University for Nationalities, Wuhan, PR China

Abstract

Polyaniline nanomaterial (nPANI) is getting popular in many industrial fields due to its conductivity and stability. The fate and effect of nPANI in the environment is of paramount importance towards its technological applications. In this work, the cytotoxicity of nPANI, which was prepared by rapid surface polymerization, was studied on rat celiac macrophages. Cell viability of macrophages treated with various concentrations of nPANI and different periods ranging from 24 to 72 hours was tested by a MTT assay. Damages of nPANI to structures of macrophages were evaluated according to the exposure level of cellular reactive oxygen species (ROS) and change of mitochondrial membrane potential (MMP). We observed no significant effects of nPANI on the survival, ROS level and MMP loss of macrophages at concentrations up to 1 μg/ml. However, higher dose of nPANI (10 μg/ml or above) induced cell death, changes of ROS level and MMP. In addition, an increase in the expression level of caspase-3 protein and its activated form was detected in a Western blot assay under the high dose exposure of nPANI. All together, our experimental results suggest that the hazardous potential of nPANI on macrophages is time- and dose-dependent and high dose of nPANI can induce cell apoptosis through caspase-3 mediated pathway.

Editor: Myon-Hee Lee, East Carolina University, United States of America

Funding: This study was supported by the National Natural Science Foundation of China (81373842, 81101538), and Natural Science Foundation of China Hubei (2013CFB451). The funders had no role in study design, data collection and analysis, decision to publish, or preparation of the manuscript.

Competing Interests: The authors have declared that no competing interests exist.

* Email: hbtang2006@mail.scuec.edu.cn

9 These authors contributed equally to this work.

Introduction

Nano-conductive-polymer materials develop rapidly in recent years. They have been widely applied in chemical, electronic, aerospace and medical field thanks to the combined excellent characteristics of both nanomaterial and conductive polymer. Along with the increasing production and use of nanomaterials, there has been a growing concern about the hazardous of inevitable unintended human exposure [1–3]. These nanometer-sized materials can access, mostly through the lungs, gastrointestinal tract, skin, injection and implantation, into human body posing significantly serious health threats [4–6].

As one of the most famous conductive polymer, nano-structured PANI (nPANI) was rapidly developed in industries and clinical fields with a lot of advantages, including availability of its raw materials, easy preparation, maneuverability of particle size and morphology, good electrical conductivity, environmental stability, biocompatibility and the reversible doping-dedoping process [7–10]. Recently, limited researches [11–14] had been carried out on the biocompatibility and cytotoxicity of PANI and its composites, but there was no definite answer whether the use of nPANI is safe or not. Therefore, the uncertainty of its toxicity is still a big problem for wide range of bioapplications [15].

Macrophages are important cells of the immune system that are responsive to cell debris and pathogens. Its applicability in immuno-toxicology has been established for cytotoxicity of exogenous chemicals [16]. In the present study, nPANI was prepared in the form of nano fibers by the rapid surface polymerization process [17–19]. Its cytotoxicity and toxic mechanism on rat celiac macrophages were examined by cell death, cellular reactive oxygen species (ROS) level, loss of mitochondrial membrane potential (MMP) and apoptosis-associated caspase activation.

Materials and Methods

Preparation of nPANI

Typically, nPANI was prepared with a surface polymerization method, by using ammonium persulfate (APS; Sinopharm, China) as the oxidant and hydrochloric acid as the doping agent. The procedures were described in details in our previous report [20]. After polymerization, the prepared neutral nPANI solution was then further dialyzed by saline to remove hypertoxic oligomer. The well cleaned nPANI was stocked in the form of neutral aqueous suspension (15 mg/ml) prior to use.

Characterization of nPANI

The morphology of nPANI was observed with a scanning electron microscope (SEM, Hitachi S-4800, Japan). The UV-Vis absorption spectra of nPANI suspensions were measured on a

Varian Cary 50 UV-Vis spectrophotometer (Varian Cary 50, Agilent). The Fourier infrared spectra of nPANI were recorded on a Bruker VERTEX 70 infrared spectrometer in a transmission mode.

Animal care

The care and use of animals for this study were performed according to the Guide for Animal Experimentation, South-Central University for Nationalities and the Committee of Research Facilities for Laboratory Animal Sciences, South-Central University for Nationalities, China. The protocols were approved by the Committee on the Ethics of Animal Experiments of the South-Central University for Nationalities, China (Permit Number: 2011-SCUEC-AEC-002). All efforts were made to minimize suffering.

Cell culture

The isolated primary macrophages from adult Wistar rat abdominal cavity (6–9 weeks of age) were maintained in Dulbecco's-modified eagle's medium (DMEM, Gibco, USA) containing 10% (v/v) fetal bovine serum (Gibco, USA), 1% penicillin-streptomycin solution, and suspended to a density of 6.0×10^5 cells/ml. Cells were seeded in a 96-well plates with 200 μl per well. Eight wells were supplied for each exposing nPANI concentration, three of them were used as background correction, and the rest were experimental group. Cells were allowed to attach for 12 hours onto the 96-well plate under 5% CO_2 at 37°C. Then, they were exposed with regular growth medium containing nPANI at different concentrations ranging from 0.1 to 100 μg/ml or pure water as a negative control unless otherwise instructed.

Cell morphology and viability assays

Cell viability was measured by a MTT assay. After being exposed to different concentrations of nPANI for 24 to 72 hours, cells were imaged by an inverted phase contrast microscope (Nikon Eclipse Ti-S, Japan). Then cells were rinsed with PBS and recovered by incubating with fresh culture medium without serum for an hour. After that, 5 mg/ml MTT (3-(4,5-dimethylthiazol-2-yl)-2,5-diphenyltetrazolium bromide, Sigma, USA) was added directly to the medium. After adding MTT, all samples were incubated for 4 hours at 37°C. Medium was then removed from cells, and DMSO was added to the wells to dissolve the formazan crystals. The absorbance was measured at 490 nm after oscillation of 15 minutes. All of the experiments were performed in triplicates. The results were expressed as percentages of the control (without nPANI).

Reactive oxygen species (ROS) assay

The generation of superoxide radical and hydrogen peroxide were detected by 2,7-dichlorodihydrofluorescein diacetate (DCF-DA) staining. For ROS assay, 6.0×10^5 cells per well were cultured in 35 mm dishes, and incubated with nPANI at different concentrations for 6 hours. Then ROS detection kit (Beyotime, China) was used for the assay. ROSup (50 μg/ml) was used as the positive control. Fluorescent intensity was detected by Enzyme-labeled meter (TECAN infinite M200, Switzerland) at an excitation wavelength of 488 nm and an emission wavelength of 525 nm.

Mitochondrial membrane potential assay

JC-1 fluorescent probe was used to determine the mitochondrial membrane potential change. Cells in this assay were cultured as the former. Cells treated with carbonyl cyanide 3-chlorophenylhydrazone (CCCP; 0.01 mM) were considered as the positive control. All the processes were conducted following the test kit (Beyotime, China). Fluorescent intensity of JC-1 monomer was detected by Enzyme-labeled meter (TECAN infinite M200, Switzerland) at an excitation wavelength of 490 nm and an emission wavelength of 530 nm, while that of JC-1 polymer was detected at an excitation wavelength of 525 nm and an emission wavelength of 590 nm. In this study, the ratio of JC-1 monomers and polymers was used as a representation of MMP (mitochondrial membrane potential).

Western blot assay

After incubating with control, 10 and 100 μg/ml nPANI for 24 hours, cells were lysed with protein lysis buffer (Beyotime, China). The concentration of total protein was detected by Lowry method. Equal amounts of protein were fractionated by 10% SDS gels and transferred to polyvinylidene difluoride membranes (Millipore Corporation, USA). After blocked with 5% nonfat dry milk, the membranes were incubated with an anti-caspase3 antibody (1:1000 dilution, Boster, China) or anti-actin antibody (1:1000 dilution, Boster, China) at 4°C overnight. After removing primary antibodies, the membranes were washed 3 times for 5 minutes by TBST (Tris-Buffered Saline, 0.1% Tween-20) solution and followed by exposure of secondary antibodies (1:5000 dilution, goat anti-rabbit or goat anti-mouse, Boster, China) for 2 hours at 4°C. Finally, after wash, bands on the membranes were visualized by developer and fixing solution [21]. The protein bands were quantified using the ImageJ software (NIH).

Statistical analysis

One-way analysis of variance (ANOVA) was used as the statistical test. Results shown in the figures are expressed as means ± standard error of mean (SEM). A value of $P<0.05$ was considered statistically significant.

Results

Characterization of nPANI

The nPANI was prepared by oxidative polymerization of aniline, resulting in the production of a mixture containing polymer, aniline oligomer, diphenylamine and benzidine. The polymerization was initiated by adding APS into aniline solution. Once started, the solution turned from initial colorless to weak green, and finally to dark blue. After the polymerization, the nPANI product was cleaned by dialysis, and the suspension was neutralized with diluted ammonia water, resulted in a pure neutral water solution.

Scanning electron microscopy revealed the morphology of polymerized nPANI, as shown in Fig. 1A. Polymers presented a typical nanofiber structure, being in consistent with the previous report [20]. Fig. 1B showed the UV-vis absorption spectra of nPANI aqueous solutions at different pHs. At acidic pH, nPANI is in its acid doped state. The absorptions of acid doped PANI at pH 2 were located at 350 and 750 nm, which are attributed to π-π* and π-polaron transitions, respectively. This suggested that the nPANI material at pH 2 had a higher level of proton doping [22]. When the suspension was adjusted to neutral or alkaline, the aforementioned bands were blue shifted to 330 and 580 nm at pH 7, and 330 and 550 nm at pH 12, respectively, representing the dedoping of nPANI. The state of nPANI changed from emeraldine salt to emeraldine base during such a dedoping process, which is consistent with the observation for conventional PANI reported by Wan and Yang [23].

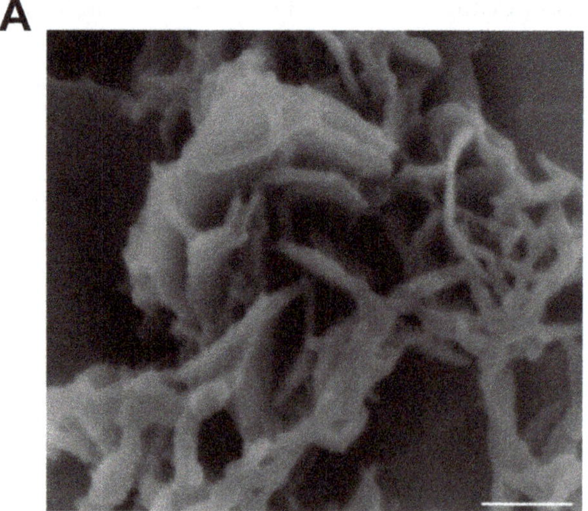

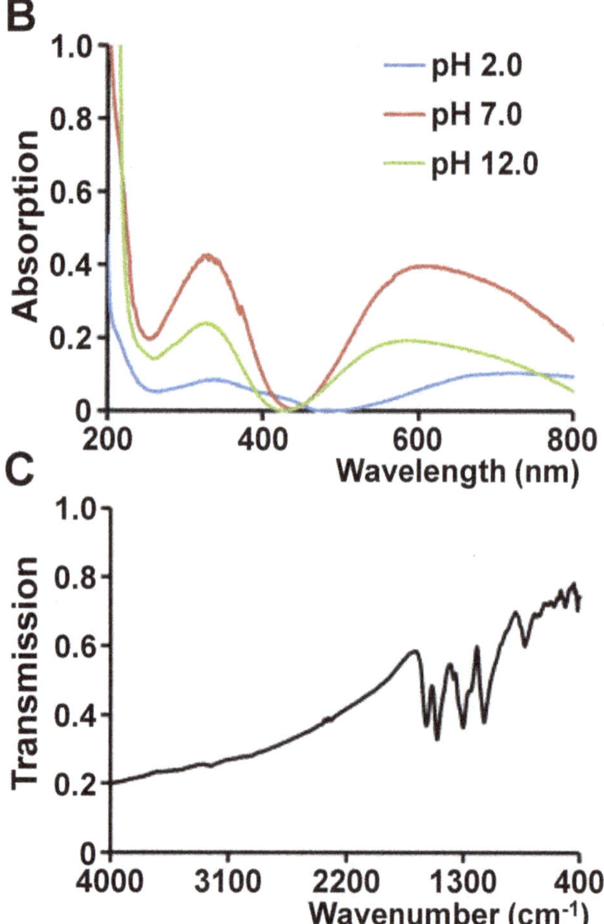

Figure 1. Characterization of nPANI. (A) The scanning electron microscope image of nPANI. Scale bar is 500 nm. (B) UV-vis absorption spectra of nPANI aqueous solutions at pH 2.0, pH 7.0, and pH 12.0. Ultra DI-water was used as a background solution. (C) FTIR spectrum of nPANI. KBr was used as a background.

Fig. 1C showed the FTIR spectrum of nPANI, where the characteristic bands of PANI were observed [24,25]. For example, the C = C stretching vibration of quinoid, C = C stretching vibration of benzene rings, C-N stretching of secondary aromatic ring, and out-of-plane bending vibrations of C-H occurred at 1580, 1498, 1303 and 828 cm^{-1}, respectively.

Cell damage and proliferation

To investigate the cytotoxicity of nPANI, rat celiac macrophages were selected. Cell death, loss of mitochondrial membrane potential (MMP), cellular reactive oxygen species (ROS) level and apoptosis-associated caspase activation were quantified in the present study.

As we know, morphological features played a leading role in the description of cell death. In Fig. 2A, the control macrophages adherent to the culture dish firmly with clear and smooth contour. The density and morphological features of cells exposed in 0.1 and 1 μg/ml nPANI for 24 hours showed no significant difference from the control group. However, in groups treated with 10 and 100 μg/ml nPANI, the density of cells was less and the cells were more transparent with vaguer rims compared with the control group. Moreover, vacuoles of different sizes could be visualized in these cells, and we also noticed tiny blue particles in the vacuoles.

Next, we quantified the cytotoxicity of nPANI on the macrophages, using concentrations of 0.1–100 μg/ml and treatment time of 24–72 hours. As shown in Fig. 2B, the cytotoxicity of nPANI on macrophages was dose- and time-dependent. When the concentration of nPANI was less than 1 μg/ml, cell viability remained to be 100% ($P>0.05$) in 72 hours. However, when the concentration went beyond 10 μg/ml, cell apoptosis existed in 24 hours after the incubation of nPANI ($P<0.05$, $P<0.001$, respectively). With the exposure time extended, more cells were dying in these two groups. In one word, the cytotoxicity of nPANI is dose- and time-dependent.

ROS generation

Environmental stress, such as UV exposure or heat, can increase ROS levels dramatically due to significant damage to cell structures. Therefore, to characterize the cytotoxicity of nPANI, ROS production of cells was also monitored. In Fig. 2C, hardly any change of ROS production was detected in low concentration (less than 1 μg/ml) treated group, while in the ROSup (positive control) treated group a huge increase was seen ($p<0.001$). Consistent with the cell viability assay, nPANI with concentration more than 10 μg/ml could promote intracellular active oxygen level and hence induce oxidative stress reaction and cell death.

Mitochondrial depolarization

Mitochondrial depolarization impairs the efficacy of the electron transport chain leading to massive ROS production. In order to further characterize the cytotoxicity of nPANI, we measured mitochondrial membrane potential (MMP) of cells. As shown in Fig. 2D, After 6-hour incubation, the ratio of JC-1 monomers to polymers in macrophages exposed in 0.1 and 1 μg/ml nPANI was not significant compare with the control group, suggesting that low concentration of nPANI could not alter MMP level of macrophages. However, 10 and 100 μg/ml nPANI could significantly change the mitochondrial membrane potential of macrophages.

Increased caspase-3 expression

Since high dose nPANI could induce cell apoptosis, we examined the expression and activation of caspase-3, an apoptosis related protein, under such condition. In Fig. 3, high dose nPANI (10, 100 μg/ml) induced more expression of caspase-3 protein in macrophages 24 hours. In addition, activated caspase-3 could be

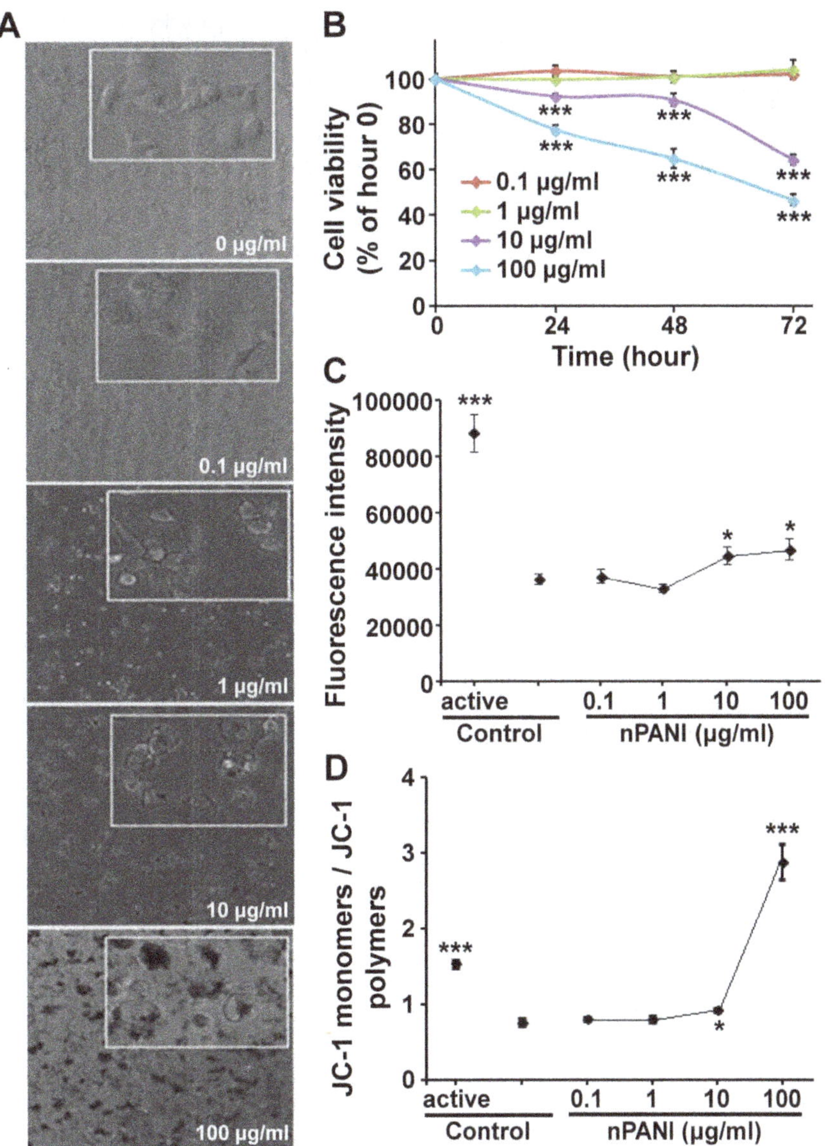

Figure 2. Cytotoxicity of nPANI. (A) Morphology of macrophages under phase-contrast microscope (20× objective, inset: 40×) treated with nPANI at different concentrations. (B) Cell viability of macrophages treated with nPANI at different concentrations for different time periods. (C) ROS production by macrophages after being incubated with nPANI at different concentrations for 6 hours. (D) Effect of 6 hours incubation of nPANI on the mitochondrial membrane potential of macrophages. All data were expressed by mean ± S.E.M. of six measurements and each experiment was performed in triplicate. * and *** denote $p<0.05$ and 0.001 versus the corresponding control.

observed, suggesting that caspase-3 was activated to different degrees in both treated groups. Therefore we speculated that caspase-3 might play a regulatory role in nPANI induced apoptosis of macrophages.

Discussion

In the present study, we have explored the toxicity of nPANI at different concentrations (0.1~100 μg/ml) on macrophages. The results showed that when concentration of nPANI was higher than 10 μg/ml, macrophage's survival was significantly threatened, and the harm was dose- and time-dependent. In general, expose of low concentration of nPANI did not show serious toxicity on macrophage in our research, which is consistent with the results of Yslas et al [13]. However, the upper limit of safety threshold of nPANI towards macrophages (less than 10 μg/ml) is lower than that towards lung fibroblast cells (25 μg/ml) [14], suggesting a necessity of studying cytotoxicity of nPANI towards different cell types.

It has been demonstrated that biological effects (such as eukaryotic cell toxicity, anti bacteria effect, and light toxicity of Lolium perenne) of nanomaterials (all kinds of nanometer metal oxide and carbon nanomaterials) are mediated by ROS [26]. Since nanomaterials are usually conductors or semi-conductors, their electric charge would interfere with electronic transduction of cells and thus increase the intracellular level of ROS and affect the permeability of cell membrane. As a result, these nanomaterials would enter cells, accumulate to form a vicious spiral, and destroy cell's osmotic balance [11]. Finally, the normal function of cells is absolutely restricted, resulting in cell lysis or apoptosis. As shown in Fig. 2C, we have shown that oxidative stress responses in

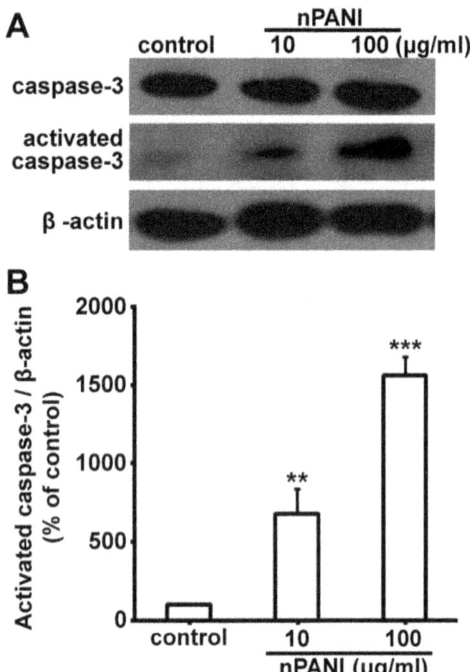

Figure 3. The effects of nPANI on the expressions of caspase-3 and activated caspase-3 expressions in macrophages. (A) Representative Western blots of the caspase-3 and β-actin expressions. (B) The relative levels were analyzed by determining the ratio of the activated caspase-3/β-actin. The values are the means ±S.E.M. of three replicates. ** and *** denote $p<0.01$ and 0.001 versus the control, respectively.

macrophages were elevated when the nPANI concentration go beyond 10 μg/ml.

In order to maintain a certain level of membrane potential, the permeability of mitochondrial membrane to outer electrolyte and foreign material is usually limited. From the early observations, we could clearly see that the macrophages swallow extracellular polyaniline nanoparticles. Microscopic examination also showed that when the nPANI concentration was more than 10 μg/ml, granular nPANI was visible in cells. This led to a speculation that nPANI could contact mitochondria directly after entering into cytoplasm and made the mitochondrial membrane porous, resulting in an increased permeability and unbalanced electrolytes inside the mitochondria. Therefore, the mitochondrial membrane potential would be lost and hence destroy the normal mitochondrial function, and cut off the respiratory chain. As a result, there would be peroxide accumulation in mitochondria and cytosol and cause oxidative stress reaction [27,28]. In addition, too-low mitochondrial membrane potential would bring about DNA fragmentation, eventually leading to cell death [29]. In accordance with this notion, the mitochondrial membrane potential showed decrease when the concentration of nPANI exceeded 10 μg/ml.

Caspase-3 is an important member of the caspase family. It is expressed widely in different tissue and cell types, and plays a crucial role in the final steps of cell apoptosis [30]. Here we have shown that the expression of active form of Caspase-3 was increased under the treatment of 10 and 100 μg/ml nPANI in a dose-dependent manner, further confirming the involvement of cell apoptosis process in nPANI induced cytotoxicity.

Conclusion

In this report, we have prepared nanometer-sized organic polyaniline, and this nPANI appeared to cause cytotoxicity on rat celiac macrophages in a time- and dose-dependent manner. We also demonstrated that the cytotoxicity of nPANI was triggered by generation of oxidative stress and change of intracellular mitochondrial membrane potential. In addition, we have shown that the pro-apoptotic protein caspase-3 took part in the nPANI-induced cell apoptosis. All together, we conclude that the safety upper limit for nPANI is less than 10 μg/ml.

Author Contributions

Conceived and designed the experiments: YSL HBT. Performed the experiments: BFC YSL HBT. Analyzed the data: XJL WKZ YSL HBT. Contributed to the writing of the manuscript: YSL HBT.

References

1. Service RF (2004) Nanotoxicology. Nanotechnology grows up. Science 304: 1732–1734.
2. Nel A, Xia T, Madler L, Li N (2006) Toxic potential of materials at the nanolevel. Science 311: 622–627.
3. Fischer HC, Chan WC (2007) Nanotoxicity: the growing need for in vivo study. Curr Opin Biotechnol 18: 565–571.
4. Pang C, Selck H, Misra SK, Berhanu D, Dybowska A, et al. (2012) Effects of sediment-associated copper to the deposit-feeding snail, Potamopyrgus antipodarum: a comparison of Cu added in aqueous form or as nano- and micro-CuO particles. Aquat Toxicol 106–107: 114–122.
5. Patil G, Khan MI, Patel DK, Sultana S, Prasad R, et al. (2012) Evaluation of cytotoxic, oxidative stress, proinflammatory and genotoxic responses of micro- and nano-particles of dolomite on human lung epithelial cells A(549). Environ Toxicol Pharmacol 34: 436–445.
6. Radad K, Al-Shraim M, Moldzio R, Rausch WD (2012) Recent advances in benefits and hazards of engineered nanoparticles. Environ Toxicol Pharmacol 34: 661–672.
7. Kawasumi M (2004) The discovery of polymer-clay hybrids. Journal of Polymer Science Part A: Polymer Chemistry 42: 819–824.
8. Kim HS, Hobbs HL, Wang L, Rutten MJ, Wamser CC (2009) Biocompatible composites of polyaniline nanofibers and collagen. Synthetic Metals 159: 1313–1318.
9. Kurian M, Dasgupta A, Galvin ME, Ziegler CR, Beyer FL (2006) A Novel Route to Inducing Disorder in Model Polymer-Layered Silicate Nanocomposites. Macromolecules 39: 1864–1871.
10. Yang C, Du J, Peng Q, Qiao R, Chen W, et al. (2009) Polyaniline/Fe3O4 nanoparticle composite: synthesis and reaction mechanism. J Phys Chem B 113: 5052–5058.
11. Humpolicek P, Kasparkova V, Saha P, Stejskal J (2012) Biocompatibility of polyaniline. Synthetic Metals 162: 722–727.
12. Khan JA, Qasim M, Singh BR, Khan W, Das D, et al. (2014) Polyaniline/CoFe2O4 nanocomposite inhibits the growth of Candida albicans 077 by ROS production. Comptes Rendus Chimie 17: 91–102.
13. Yslas EI, Ibarra LE, Peralta DO, Barbero CA, Rivarola VA, et al. (2012) Polyaniline nanofibers: acute toxicity and teratogenic effect on Rhinella arenarum embryos. Chemosphere 87: 1374–1380.
14. Oh WK, Kim S, Kwon O, Jang J (2011) Shape-dependent cytotoxicity of polyaniline nanomaterials in human fibroblast cells. J Nanosci Nanotechnol 11: 4254–4260.
15. Villalba P, Ram MK, Gomez H, Bhethanabotla V, Helms MN, et al. (2012) Cellular and in vitro toxicity of nanodiamond-polyaniline composites in mammalian and bacterial cell. Materials Science and Engineering: C 32: 594–598.
16. Ross M, Matthews A, Mangum L (2014) Chemical Atherogenesis: Role of Endogenous and Exogenous Poisons in Disease Development. Toxics 2: 17–34.
17. Ayad M, El-Hefnawy G, Zaghlol S (2013) Facile synthesis of polyaniline nanoparticles; its adsorption behavior. Chemical Engineering Journal 217: 460–465.
18. Huang YF, Lin CW (2012) Facile synthesis and morphology control of graphene oxide/polyaniline nanocomposites via in-situ polymerization process. Polymer 53: 2574–2582.
19. Li Y, Gong J, He G, Deng Y (2011) Synthesis of polyaniline nanotubes using Mn2O3 nanofibers as oxidant and their ammonia sensing properties. Synthetic Metals 161: 56–61.

20. Li J, Tang H, Zhang A, Shen X, Zhu L (2007) A New Strategy for the Synthesis of Polyaniline Nanostructures: From Nanofibers to Nanowires. Macromolecular Rapid Communications 28: 740–745.
21. Lewis CW, Taylor RG, Kubara PM, Marshall K, Meijer L, et al. (2013) A western blot assay to measure cyclin dependent kinase activity in cells or in vitro without the use of radioisotopes. FEBS Lett 587: 3089–3095.
22. Singh K, Ohlan A, Pham VH, Balasubramaniyan R, Varshney S, et al. (2013) Nanostructured graphene/Fe(3)O(4) incorporated polyaniline as a high performance shield against electromagnetic pollution. Nanoscale 5: 2411–2420.
23. Wan M, Yang J (1995) Mechanism of proton doping in polyaniline. Journal of Applied Polymer Science 55: 399–405.
24. Chiang JC, MacDiarmid AG (1986) 'Polyaniline': Protonic acid doping of the emeraldine form to the metallic regime. Synthetic Metals 13: 193–205.
25. Macdiarmid AG, Chiang JC, Richter AF, Epstein AJ (1987) Polyaniline: a new concept in conducting polymers. Synthetic Metals 18: 285–290.
26. Artetxe U, García-Plazaola JI, Hernández A, Becerril JM (2002) Low light grown duckweed plants are more protected against the toxicity induced by Zn and Cd. Plant Physiology and Biochemistry 40: 859–863.
27. Jones CF, Grainger DW (2009) In vitro assessments of nanomaterial toxicity. Adv Drug Deliv Rev 61: 438–456.
28. Kroll A, Dierker C, Rommel C, Hahn D, Wohlleben W, et al. (2011) Cytotoxicity screening of 23 engineered nanomaterials using a test matrix of ten cell lines and three different assays. Part Fibre Toxicol 8: 9.
29. Chiu WH, Luo SJ, Chen CL, Cheng JH, Hsieh CY, et al. (2012) Vinca alkaloids cause aberrant ROS-mediated JNK activation, Mcl-1 downregulation, DNA damage, mitochondrial dysfunction, and apoptosis in lung adenocarcinoma cells. Biochem Pharmacol 83: 1159–1171.
30. Alnemri ES, Livingston DJ, Nicholson DW, Salvesen G, Thornberry NA, et al. (1996) Human ICE/CED-3 protease nomenclature. Cell 87: 171.

5

Relationships between Degree of Polymerization and Antioxidant Activities: A Study on Proanthocyanidins from the Leaves of a Medicinal Mangrove Plant *Ceriops tagal*

Hai-Chao Zhou[1,2,3]*, Nora Fung-yee Tam[2,3]*, Yi-Ming Lin[1], Zhen-Hua Ding[1], Wei-Ming Chai[1], Shu-Dong Wei[1]

1 Key Laboratory of the Ministry of Education for Coastal and Wetland Ecosystems, Xiamen University, Xiamen, China, **2** Department of Biology and Chemistry, City University of Hong Kong, Hong Kong SAR, China, **3** Futian-CityU Mangrove R&D Centre, City University of Hong Kong Shenzhen Research Institute, Shenzhen, China

Abstract

Tannins from the leaves of a medicinal mangrove plant, *Ceriops tagal*, were purified and fractionated on Sephadex LH-20 columns. ^{13}C nuclear magnetic resonance (^{13}C-NMR), reversed/normal high performance liquid chromatography electrospray ionization mass spectrometry (HPLC-ESI MS) and matrix-assisted laser desorption/ionization time-of-flight mass spectrometry (MALDT-TOF MS) analysis showed that the tannins were predominantly B-type procyanidins with minor A-type linkages, galloyl and glucosyl substitutions, and a degree of polymerization (DP) up to 33. Thirteen subfractions of the procyanidins were successfully obtained by a modified fractionation method, and their antioxidant activities were investigated using 2,2-diphenyl-1-picrylhydrazyl (DPPH) scavenging capacity and ferric reducing antioxidant power (FRAP) method. All these subfractions exhibited potent antioxidant activities, and eleven of them showed significantly different mean DP (mDP) ranging from 1.43 ± 0.04 to 31.77 ± 1.15. Regression analysis demonstrated that antioxidant activities were positively correlative with mDP when around mDP <10, while dropped and then remained at a level similar to mDP $= 5$ with around 95 $\mu g\ ml^{-1}$ for DPPH scavenging activity and 4 mmol AAE g^{-1} for FRAP value.

Editor: Silvia Mazzuca, Università della Calabria, Italy

Funding: This present work was supported by the National Natural Science Foundation of China (URL: http://isisn.nsfc.gov.cn/egrantweb/): no. 41306084 (Hai-Chao Zhou's funding), no. 41176090 (Zhen-Hua Ding's funding), and no. 31070522 (Yi-Ming Lin's funding). The funders had no role in study design, data collection and analysis, decision to publish, or preparation of the manuscript.

Competing Interests: The authors have declared that no competing interests exist.

* Email: zhouhc2013@gmail.com (HCZ); bhntam@cityu.edu.hk (NFYT)

Introduction

Ceriops tagal, a mangrove plant species in the Family of Rhizophoraceae, is one of the medicinal plants in East and Southeast Asia and is often used as a tradition herb to treat diseases such as hemorrhages and malignant ulcers [1,2]. The phytochemistry of *C. tagal* has been reported, however, the work mainly focused on the isolation and identification of terpenoid compounds in root, with little attention on leaf tannins [3,4]. The high level of tannins in the leaves of Rhizophoraceae is known to deter feeding by herbivores [5], but leaf tannins also show a diversity of other biological activities [6]. The unexplored tannins could be novel potential resources of bioactive compounds in mangrove plants. So far, the chemical properties of *C. tagal* tannins have not yet been determined and the structure-activity relationships of tannins are still not clear.

Tannins comprised as much as 20–40% dry weight in the leaf and bark of mangrove plants [7,8]. Compared with hydrolysable tannins, condensed tannins (proanthocyanidins) are more abundant. They are commonly found in mangrove plants [9,10,11,12,13,14] and are also the main component of the polyphenols in our diet [15]. Because of their antioxidant activities and other potentially health-promoting qualities, proanthocyanidins have attracted more and more research interests in recent years [16,17,18,19]. Proanthocyanidins are oligomers and polymers of flavan-3-ol that are bound together with B-type and A-type linkages [15]. The chemistry and biological features of proanthocyanidins largely depend on their structure, particularly the molecular weight that is also expressed as degree of polymerization (DP) [20,21]. Ariga et al. [22] found that the ability to scavenge free radicals was proportional to DP for simple flavonoid oligomers. Hagerman et al. [23] proved that tannins with highly polymerized and many hydroxyl groups are more potent antioxidants than the simple phenolics. Recently, some findings reported that the increasing DP may enhance the antioxidant power of condensed tannins [24,25]. However, those previous works either limitedly detected the simple oligomers or directly studied the bulk mixture of polymers. The relationships between DP and antioxidant activity of proanthocyanidins, therefore, remained largely unknown.

The present study, therefore, aims to (I) achieve a complete structural characterization of tannins from the leaf of a mangrove plant (*C. tagal*) by a combination of analytical techniques, including ^{13}C-NMR, MALDI-TOF MS, thiolysis degradation, and reversed/normal-phase HPLC-ESI MS; and (II) establish an efficient fractionation method to obtain a series of proanthocyanidins with different DP; and (III) explore the possible relationships between DP and antioxidant activity of condensed tannins.

Materials and Methods

2.1. Chemicals and Materials

Water used in this experiment was purified on a Millipore Milli-Q apparatus. HPLC grade dichloromethane, acetonitrile (CH_3CN), methanol, trifluoroacetic acid (TFA), acetic acid, and all analytical grade solvents (acetone, methanol, n-Butanol etc.) were obtained from Sinopharm (Sinopharm, Shanghai, China). Sephadex LH-20, Folin-Ciocalteu reagents, acetone-d_6, deuteroxide (D_2O), Amberlite IRP-64 cation-exchange resin, cesium chloride (CsCl), 2,5-dihydroxybenzoic acid (DHB), benzylmercaptan, 2,2-diphenyl-1-picrylhydrazyl (DPPH), 2,4,6-tripyridyl-S-triazine (TPTZ), ascorbic acid (AA), and all HPLC standards were purchased from Sigma (St. Louis, MO, USA). The mature leaves (the third pair with fully expanded and dark green) of a mangrove plant *C. tagal* (Rhizophoraceae *Ceriops*), were collected from a mangrove forest in the Dongzhai harbor (19°56′N, 110°34′E), Hainan, China. The leaves were immediately freeze-dried, ground, and stored at -20°C prior to analysis.

2.2. Extraction, Purification and Fractionation of Mangrove Tannins

The extraction procedure was conducted according to the method of Zhou et al. [25] The crude tannins extract (Fc) obtained was applied to a Sephadex LH-20 column (50×1.5 cm i.d.). The procedure of purification and fractionation of F_C was shown in Figure 1. Purified tannins (Fp) and nine subfractions (F1 to F9) were evaporated under vacuum to remove organic solvents followed by freeze-dried. The last two subfractions (F8 and F9) were re-loaded on a Sephadex LH-20 column (50×1.0 cm i.d.) to yield seven more subfractions as shown below.

2.3. Determination of Total Phenolics and Extractable Condensed Tannins

The procedure described by Zhou et al. [25] to estimate the total phenolics and extractable condensed tannins of the entire leaf of *C. tagal* was used, and the concentrations were determined by Folin-Ciocalteu [26] and butanol-HCl method [27], respectively. The content of the total phenolics (TP) and extractable condensed tannins (ECT) were calculated using the purified tannins (Fp) obtained from the purification of crude tannins extract (Fc) as the standard. Both of them were expressed as mg Fp equivalents per gram dry weight of leaves.

2.4. ^{13}C-NMR Analysis

The obtained purified tannins from *C. tagal* leaves were dissolved in acetone-d_6/D_2O. A Varian Metcury-600 spectrometer (Palo Alto, CA, USA) was employed in the 150 MHz mode [10].

2.5. Reversed-phase HPLC-ESI MS Analysis Followed by Thiolysis

The modified method described by Zhou et al. [9] was carried out to characterize Fp and obtain the mDP for each subfraction. Condensed tannins were thiolysis degraded by benzylmercaptan, and then the degradation products were analyzed on an Agilent 1200 system (Agilent, Palo Alto, CA, USA) interfaced to a QTRAP 3200 (Applied Biosystems, Foster, USA) with a 250 mm×4.6 mm i.d. 5.0 μm Hypersil ODS column (Elite, Dalian, China).

2.6. Normal-phase HPLC-ESI MS Analysis

Normal-phase HPLC-ESI-MS analysis was conducted according to the method of Hellstrom et al. [28] with minor modifications. The HPLC-MS equipment consisted of an Agilent

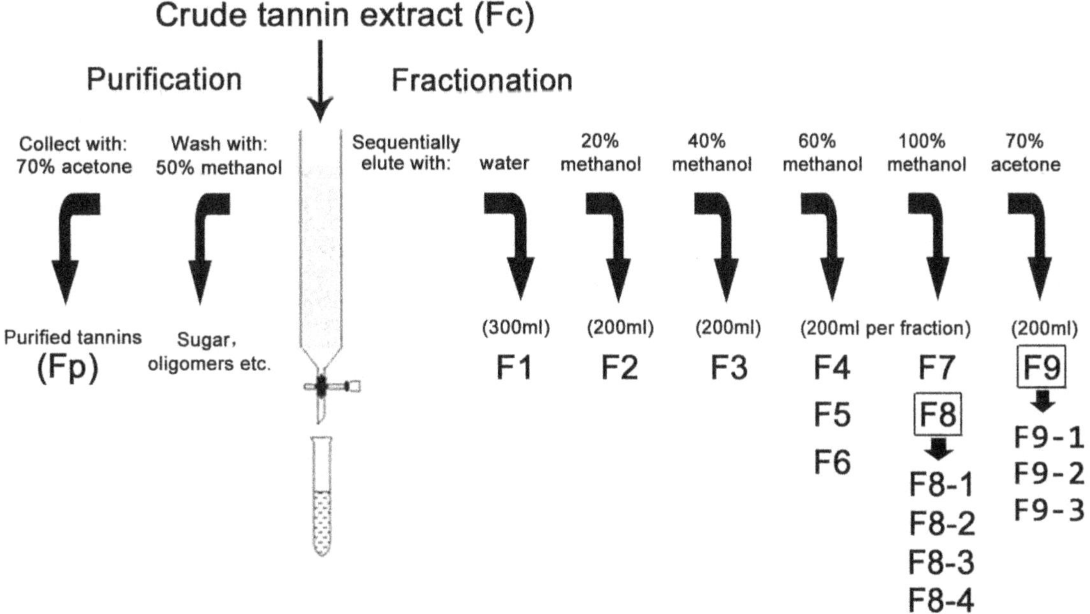

Figure 1. Sephadex LH-20 column chromatography for purification and fractionation of tannins from *C. tagal* leaves. Subfraction 8 (F8) and Subfraction 9 (F9) were further fractionated into F8-1 to F8-4 and F9-1 to F9-3, respectively.

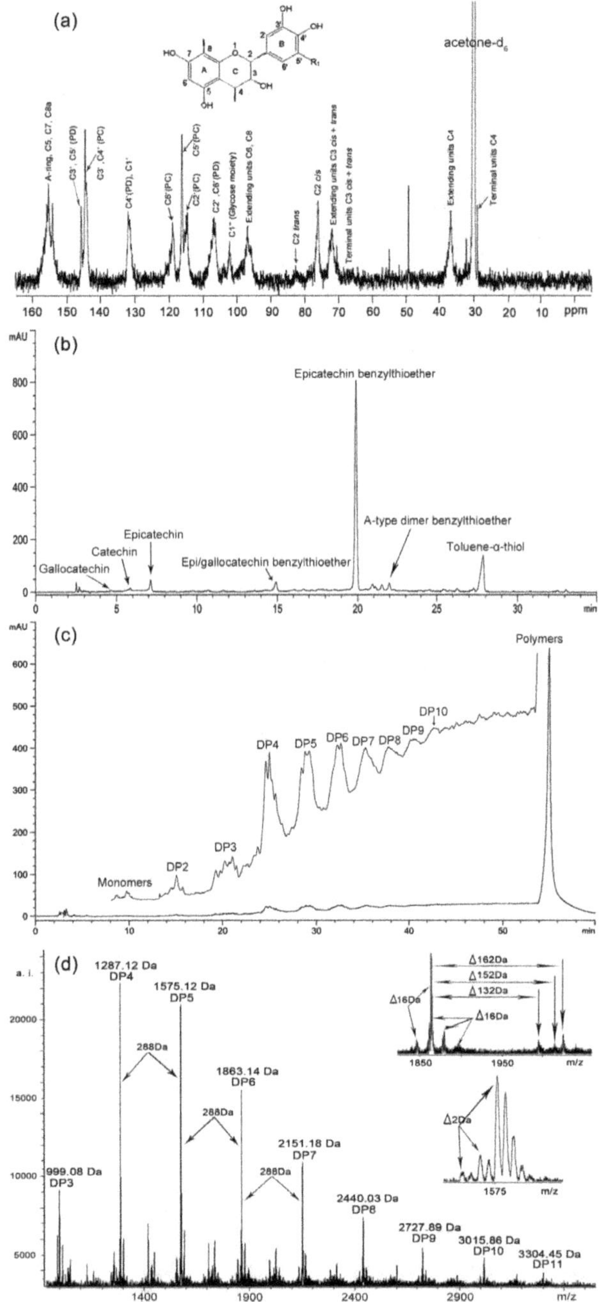

Figure 2. Structural characterization of purified tannins (Fp) from *C. tagal* leaves. Analyzed by (a) ^{13}C-NMR, (b) reversed-phase HPLC-ESI-MS, (c) normal-phase HPLC-ESI-MS and (d) MALDI-TOF MS.

1200 liquid chromatograph system as described above with a 250 mm×4.6 mm i.d. 5.0-μm Silica Luna column (Phenomenex, Darmstadt, Germany).

2.7. MALDI-TOF MS Analysis

The MALDI-TOF MS analyses were performed on a Bruker Reflex III instrument (Germany). The measurements were carried out using the conditions reported in the previous work [29]. Reflectron modes coupled with linear modes were further applied to show the different DP distribution of condensed tannins.

2.8. Determination of Antioxidant Activities

Antioxidant activities of each subfraction were evaluated by DPPH radical scavenging activity (DPPH method) [30] and ferric ion reducing antioxidant power (FRAP method) [31]. Their results were expressed in EC_{50} value (μg mL^{-1}) and mmol ascorbic acid equivalents per gram of each subfraction (mmol AAE g^{-1}), respectively. Procedures were modified according the method of Zhou et al. [32].

2.9. Statistical Analysis

Results of antioxidant activity were expressed as mean ± standard deviation of three independent determinations. A parametric one-way analysis of variance (ANOVA) was used to test any significant difference among subfractions at $P<0.05$, and all the data without transformation fulfilled the two assumptions, normal distribution and homogeneity in variance of the parametric test. All statistical analyses were performed with SPSS 17.0 for Windows.

2.10. Ethics Statement

Field permit issued by The Mangroves of Dongzhai Gang National Reserve in Dongzhai harbor, Hainan, China, allowed us to collect the mature leaves of *C. tagal* in this site. They also confirmed that our study did not involve any endangered or protected species.

Results and Discussion

3.1. Contents of Total Phenolics and Extractable Condensed Tannins

The total phenolics and extractable condensed tannins content of *C. tagal* leaves was 27.4±2.8% and 20.1±2.0%, respectively. Phenolic compounds (including proanthocyanidins) were considered to be the major contributor to the antioxidant activity of vegetables, fruits or medicinal plants [25,33,34]. The high level of total phenolics and extractable condensed tannins in *C. tagal* leaves strongly suggested that some of their pharmacological effects could be attributed to the presence of these valuable constituents.

3.2. Structural Characterization of Purified Tannins (Fp)

Fp from *C. tagal* leaves were initially analyzed by ^{13}C-NMR spectroscopy, and the signal assignment was performed according to the method of Czochanska et al. [35] The ^{13}C-NMR spectrum (Figure 2a) showed that the characteristic ^{13}C peaks are consistent with that of condensed tannins with dominant procyanidin units and a minor amount of prodelphinidins. The signals near 145 ppm, which arise from quaternary C3′ and C5′ in prodelphinidins (PD), and C3′ and C4′ in procyanidin (PC) units were used to estimate the PD:PC ratio [35]. In this experiment, direct integration of the signals at 146.2 ppm (PD units) and 145.2 ppm (PC units) gave a PD:PC ratio of about 8:92. The signals at 115.1 ppm (C2′), 116.1 ppm (C5′) and 119.0 ppm (C6′) further supported the presence of PC, so does signals at 107–108 ppm (C2′ and C6′) and 133.0 ppm (C4′) for PD. The structural diversity of the linkage (A type and B type) and the stereoisomer of catechin and epicatechin units were apparent from the spectrum. Specially, C5, C7, and C8a carbons of procyanidins appeared at 150 to 160 ppm [36]. The region between 70 and 90 ppm was sensitive to the stereochemistry of the C-ring. The ratio of the 2,3-*cis* to 2,3-*trans* isomers could be determined through the distinct differences in their respective C2 chemical shifts. C2 gave a resonance at 76.7 ppm for the *cis* form and

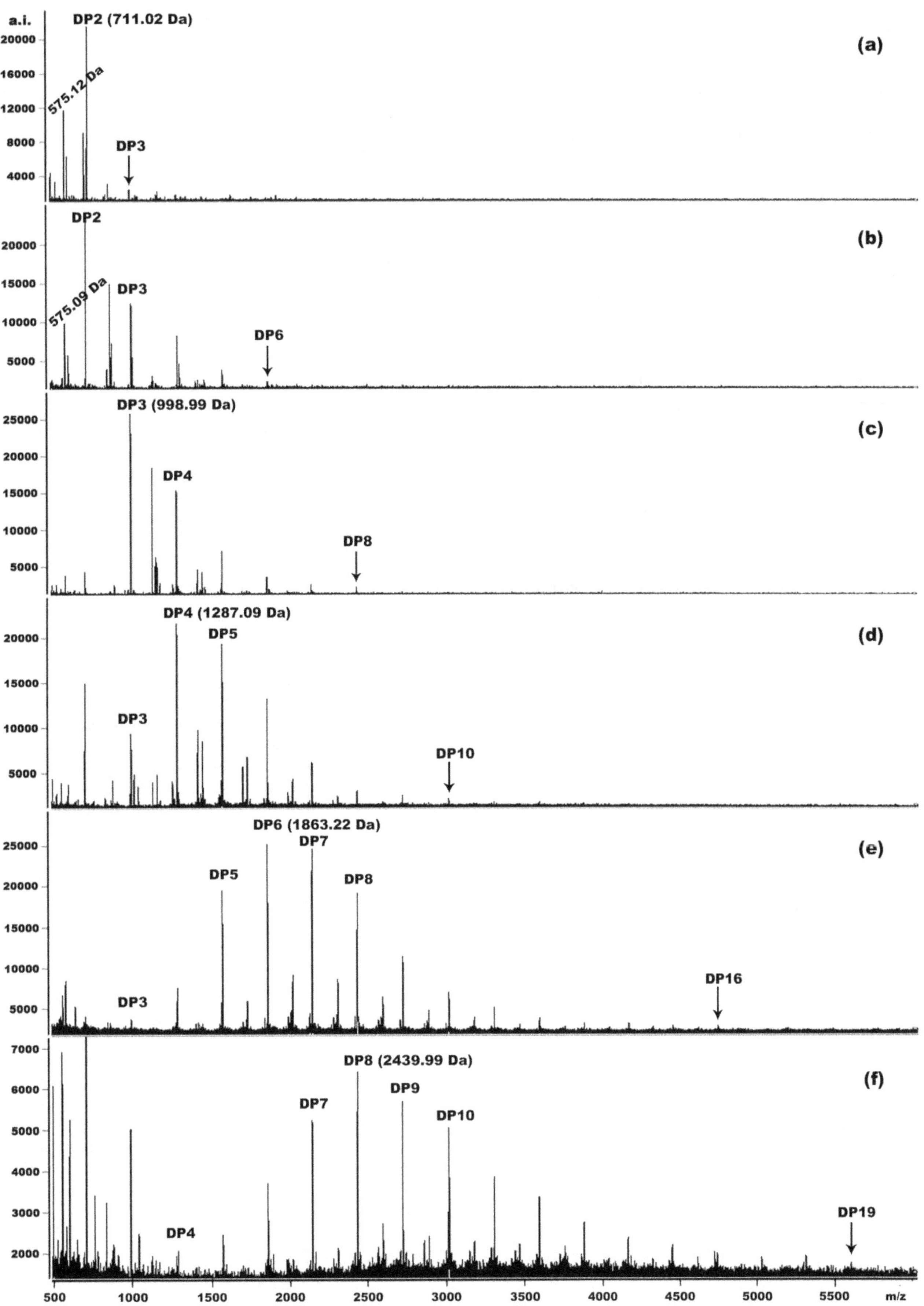
a.i.
DP2 (711.02 Da)
575.12 Da
DP3
(a)
DP2
575.09 Da
DP3
DP6
(b)
DP3 (998.99 Da)
DP4
DP8
(c)
DP4 (1287.09 Da)
DP5
DP3
DP10
(d)
DP6 (1863.22 Da)
DP7
DP5
DP8
DP3
DP16
(e)
DP8 (2439.99 Da)
DP9
DP7
DP10
DP4
DP19
(f)
m/z

Figure 3. Reflectron mode MALDI-TOF mass spectra for tannin subfraction F4 (a) to F9 (f). Degree of polymerization (DP) of the predominant polymers and largest polymer in each mass spectrum of subfraction are labeled.

83.6 ppm for the *trans* form. From the peak areas, it was estimated that the *cis* isomer is dominant. The C3 generally have their chemical shift 66–68 ppm in terminal units, and 72–74 in extending units [37]. Theoretically, the intensity of the C3 signal in terminal units relative to that of the signal in extension units could be used for elucidating the polymer chain length [35]. However, the application of this technique for the quantification of molecular weight suffered from inaccuracy due to the low signal-to-noise ratio [24]. Similarly, the spectra obtained in the present study also showed low signal-to-noise ratio (Figure 2a). The C4 atoms of the extension units showed a broad peak at around 37 ppm, while the terminal C4 exhibits 29 ppm [37], next to the solvent peak acetone-d_6. The signal at 103.1 ppm would attribute to C1″ of the glycose moiety connected to the C3 position, while the signals of C2″ (74.0 ppm), C3″ (76.3 ppm), C4″ (71.4 ppm) and C5″ (75.7 ppm) were overlapped with the chemical shifts 70–80 ppm referred above [38]. These results thus showed that the mangrove condensed tannins of *C. tagal* leaves were predominantly constituted of epicatechin, the main constitutive monomer with some glycosides.

Although the ^{13}C-NMR spectrum revealed complex structural characteristics of the condensed tannins of *C. tagal* leaves, this analytical technique hardly allows the determination of the polymer chain length and the chemical constitution of individual chains. Moreover, the sequential succession of monomer units in individual chains could not be elucidated [17,36,39]. To overcome these problems, further characterization was continued by means of HPLC-ESI-MS and MALDI-TOF MS.

The nature of flavan-3-ol units within proanthocyanidins was analyzed by acid catalysis in the presence of benzylmercaptan as nucleophile, which had been widely used [18,21,36]. In the thiolysis reaction, terminal units of condensed tannins are released as free flavanoids, while external units are distinguished as benzylthioether adducts. The extension and terminal units of proanthocyanidins could be distinguished by reversed-phase HPLC analysis (Figure 2b). The dominant peak observed was epicatechin benzylthioether along with five other smaller peaks for epicatechin, epi/gallocatechin benzylthioether, A-type dimer benzylthioether, catechin and gallocatechin. These results suggested that epicatechin was the dominant flavan-3-ol unit, which occurred as both terminal and extension units in proanthocyanidins of *C. tagal* leaves.

Normal-phase HPLC analysis was able to further reflects the heterogeneity of the proanthocyanidin mixture [28]. Although the peak band became broader in higher DP and was hard to be resolved, this normal-phase HPLC chromatogram of Fp (Figure 2c) indicated the presence of proanthocyanidins with DP from monomers to decamers (DP10). The results suggested the heterogeneity of proanthocyanidins in their constituent units and the linkages between them.

Figure 2d showed the MALDI-TOF mass spectra of Fp, recorded as Cs^+ adducts in the positive ion reflectron mode, and the enlarged spectrums demonstrated the good resolution. These results showed large structural heterogeneity of condensed tannins in the mangrove plant *C. tagal*. The mass increased from DP3 (999.08 Da) to DP11 (3304.45 Da) by 288 Da, which corresponded to one epi/catechin monomer unit. For each multiple, substructures with 16 Da, both higher and lower, could be found (Figure 2d). These masses had been identified as heteropolymers of repeating flavan-3-ol units, which demonstrated the coexistence of epi/gallocatechin and epi/afzelechin [39]. Additionally, there were three series peaks with distances of 162 Da, 152 Da, and 132 Da. The first series (162 Da) represented glucosylated heteropolyflavans containing the flavanone [39]. The second series (152 Da) was corresponding to the addition of one galloyl group at the heterocyclic C-ring [12,24]. And the 132 Da distance may have two different interpretations according to previous works. Reed et al. [39] interpreted it as the substitutions by pentoside (132 Da), while Xiang et al. [40] explained as quasimolecular ions $[M+2Cs^+-H]^+$ that generated by simultaneous attachment of two Cs^+ and loss of a proton. The distinct decrease of 2 Da illustrated the nature of interflavan bonds including A-type and B-type linkages. The data from MALDI-TOF MS well agreed with the results obtained by ^{13}C-NMR, reversed and normal-phase HPLC-ESI-MS analyses.

3.3. MALDI-TOF MS Analysis of Proanthocyanidin Subfractions

For the first step, ten subfractions (F1-F9) were obtained by fractionation on Sephadex LH-20 as described in methods section. After analysis by MALDI-TOF MS, fractions F1-F3 showed a cluster of unresolved peak lower than 500 Da (data not shown), which could be the sugars and other impurities. The spectra obtained by reflectron modes of the seven collected subfractions from fractions F4 (Figure 3a) to F9 (Figure 3f) clearly displayed the distinct predominant polymers and polymer range. The signals of lower oligomers were firstly detected in fraction F4 dominated by dimers (Figure 3a). Fractionation of the lower oligomers that could lead to the saturation of detector can significantly enhance the sensitivity of the detection of large polymers of proanthocyanidins under MALDI-TOF analysis [17]. Therefore, the last subfraction (F9) detected in the present study reached as high as DP19 polymer (Figure 3f), which demonstrated that the fractionation method could improve the detection of high DP polymers compared with the spectra of Fp (Figure 2d), even though it did not completely overcome the discrimination of high molecular weight polymers [17].

Since F8 and F9 still possessed of large DP distribution (with DP3-DP16 and DP4-DP19, respectively), a refined fractionation method was conducted to re-fractionate F8 (Figure 4) and F9 (Figure 5), which yielded other four and three subfractions, respectively, as well as with increasing trends for their predominant polymers and polymer range.

After refined fractionation for F8 and F9, as high as DP27 and DP33 polymer were detected in linear spectra of F8-4 (Figure 4) and F9-3 (Figure 5), respectively. It suggested that proanthocyanidins in *C. tagal* leaves can reach DP33, or even more. The present study clearly demonstrates a higher DP and clearer DP distribution could be obtained than many previous works reporting a high DP in various plant materials using MALDI-TOF MS [17]. In the present study, however, we showed a higher DP and clearer DP distribution than those described. Both the reflectron and linear modes of MALDI-TOF MS have been successfully applied to the analysis of proanthocyanidins. The high mass resolution power of reflectron modes allows distinguishing the mass of the isotopic peaks with enough accuracy to compare with the theoretical calculated mass, while the linear mode provides better information about the mass distribution, especially for the high DP of proanthocyanidins [17,29].

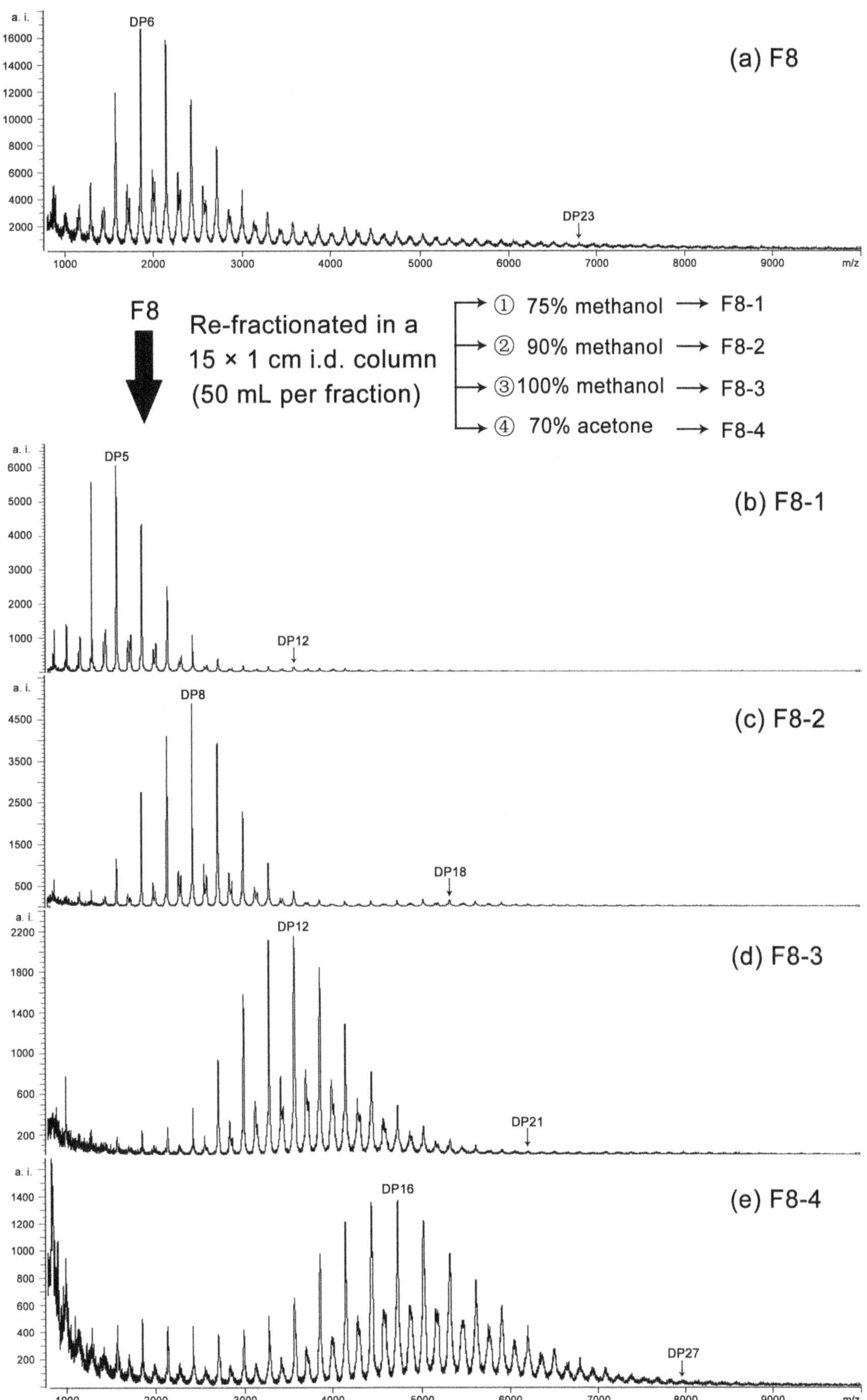

Figure 4. Linear mode MALDI-TOF mass spectra for tannin subfraction F8 (a). F8-1 (b) to F8-4 (e) were obtained by the further re-fractionation of F8. Degree of polymerization (DP) of the predominant polymers and largest polymer in each mass spectrum of subfraction are labeled.

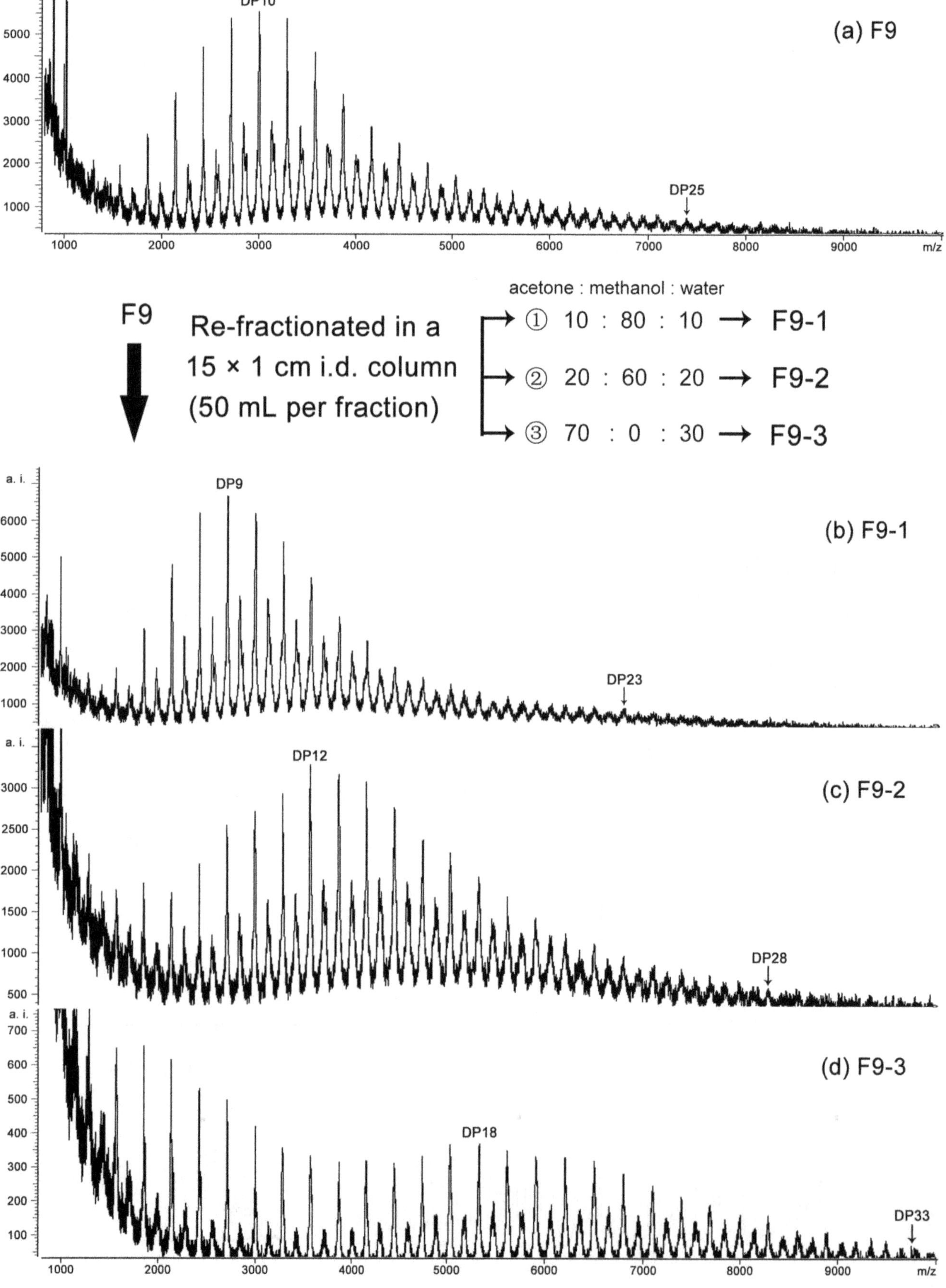
a. i.
DP10
(a) F9
DP25
m/z
acetone : methanol : water
F9
Re-fractionated in a
15 × 1 cm i.d. column
(50 mL per fraction)
① 10 : 80 : 10 → F9-1
② 20 : 60 : 20 → F9-2
③ 70 : 0 : 30 → F9-3
DP9
(b) F9-1
DP23
DP12
(c) F9-2
DP28
DP18
(d) F9-3
DP33

Figure 5. Linear mode MALDI-TOF mass spectra for tannin subfraction F9 (a). F9-1 (b) to F9-3 (d) were obtained by the further re-fractionation of F9. Degree of polymerization (DP) of the predominant polymers and largest polymer in each mass spectrum of subfraction are labeled.

3.4. Mean Degree of Polymerization (mDP) Analysis of Proanthocyanidin Subfractions

Although the above MALDI-TOF mass spectra graphically showed the differences among each proanthocyanidin subfraction, it only qualitatively illustrated the changes of DP. Therefore, mDP was calculated using thiolysis degradation coupled with reversed-phase HPLC-ESI MS. Table 1 listed the mDP of 13 proanthocyanidin subfractions from *C. tagal* leaves. The mDP increased with the eluted order on Sephadex LH-20: ranging from 1.43±0.04 (F4) to 16.14±0.58 (F9), from 7.26±0.31 (F8-1) to 16.14±0.58 (F8-4) and from 14.19±0.50 (F9-1) to 31.77±1.15 (F9-3), which agreed with the results of MALDI-TOF analysis.

These results confirmed that the fractionation method used in this study was effective to gradually separate the plant proanthocyanidins according to DP. The occurrence of proanthocyanidins in complex mixtures makes it difficult to firmly establish the effect of molecular size (or DP) on the biologically important properties of proanthocyanidins [41]. Therefore, the distinct subfractions obtained from this fractionation method were subjected to explore the relationship between DP and antioxidant activities for proanthocyanidins.

3.5. Antioxidant Activities and Their Relationships with mDP

DPPH method is a reliable and reproducible method to determine the radical scavenging activity of proanthocyanidins extracted from plant or fruits [24,42]. FRAP method is based on the redox reaction of ferric ion in the presence of a reducer. The reduction capacity of a compound may serve as a significant indicator of its potential antioxidant activity [43]. Table 1 summarized the antioxidant activities (DPPH and FRAP method) of proanthocyanidin subfractions from *C. tagal* leaves. The highest antioxidant activities were found in F8 with $EC_{50/DPPH}$ value 78.13±1.14 μg ml^{-1} and FRAP value 5.87±0.08 mmol AAE g^{-1}, while the lowest in F4 with $EC_{50/DPPH}$ value 133.45±2.24 μg ml^{-1} and FRAP value 2.58±0.04 mmol AAE g^{-1}. FRAP showed a similar trend to the results of DPPH assay. Although these two antioxidant assays with different mechanisms were used in this study, other antioxidant assays, such as peroxyl radical scavenging capacity or even assays for peroxidative/prooxidative enzymes should be included in further research, such as peroxyl radical scavenging capacity or even assays for peroxidative/prooxidative enzymes. Compared with the results from other plant materials, such as *Delonix regia* [18], *Litchi chinensis* [25] and *Garcinia mangostana* [32], proanthocyanidins from *C. tagal* leaves also exhibited substantial radical scavenging activity and reduction capacity.

Regression analysis was performed to establish the relationship between antioxidant activities and mDP (Figure 6). It demonstrated that antioxidant activities was positively correlative with mDP when mDP <10, while dropped and then remained at a level with around 95 μg ml^{-1} for $EC_{50/DPPH}$ value and 4 mmol AAE g^{-1} for FRAP value (similar to the antioxidant level of mDP = 5). Although some researchers attempted to study this relationship of proanthocyanidins, the work was limited to low oligomers and low DP. For instance, Gaulejac et al. [44] found an increase in the activity of procyanidins, but their work only focused on the oligomers from 1 to 4 units. Es-Safi et al. [24] simply compared the commercial standard substances (catechin and B_3 procyanidin dimer) with pear juice polymeric proanthocyanidins. Jerez et al. [42] showed an increase of the antiradical activity of *Pinus pinaster* and *Pinus radiata* procyanidins up to 6–7 mDP (the mDP obtained by thiolysis with cysteamine). Recently, Zhou et al. [32]

Table 1. Mean degree of polymerization (mDP) and antioxidant activities of each subfraction obtained by fractionation of proanthocyanidins from *C. tagal* leaves.

Fractions	mDP	$EC_{50/DPPH}$ value[I] (μg ml^{-1})	FRAP value[II] (mmol AAE g^{-1})
F4	1.43±0.04 k	133.45±2.24 a	2.58±0.04 h
F5	2.49±0.14 j	114.74±2.24 b	3.60±0.05 g
F6	3.52±0.34 i	103.56±0.79 c	4.05±0.13 ef
F7	7.78±0.29 h	83.60±0.62 g	5.46±0.05 b
F8	10.52±0.62 g	78.13±1.14 h	5.87±0.08 a
F8-1	7.26±0.31 h	84.11±1.41 g	5.19±0.05 c
F8-2	11.72±0.56 f	97.04±1.04 d	3.83±0.07 fg
F8-3	16.84±0.82 d	86.40±0.24 fg	4.28±0.11 e
F8-4	24.62±1.15 b	98.49±1.48 d	3.80±0.01 fg
F9	16.14±0.58 d	87.14±1.68 f	4.52±0.23 d
F9-1	14.19±0.50 e	96.49±1.55 d	3.80±0.16 fg
F9-2	22.01±0.81 c	93.17±2.01 e	4.05±0.08 ef
F9-3	31.77±1.15 a	105.37±1.22 c	3.66±0.09 g

[I]The $EC_{50/DPPH}$ value (μg ml^{-1}) is the concentration of each subfraction at which the scavenging activity is 50%.Values with different letters in the same column are significantly different at $P<0.05$ level.

[II]The FRAP value (mmol AAE g^{-1}) is expressed in mmol ascorbic acid equivalents per gram of each subfraction. Values with different letters in the same column are significantly different at $P<0.05$ level.

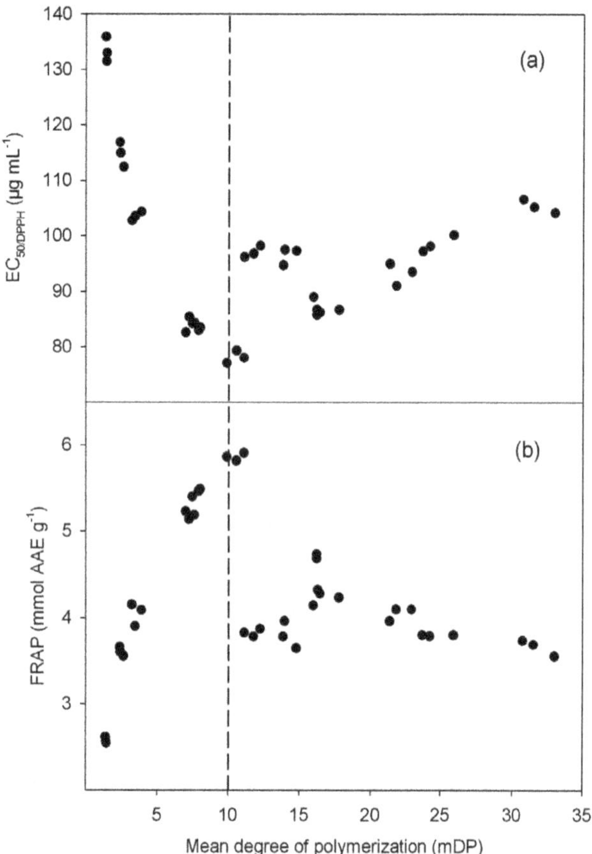

Figure 6. Regression analysis for mean degree of polymerization (mDP) and antioxidant activity. (a) $EC_{50/DPPH}$ value and (b) FRAP value correlated with mDP for each proanthocyanidin subfraction from *C. tagal* leaves.

reported an increasing antioxidant activity of mangosteen tannins with mDP between 2.71 and 9.27, but decreasing in mDP = 16.80, and proposed that 9–10 mDP could be considered as a dividing and critical point for predicting the structure-activity of mangosteen condensed tannins, which was confirmed with the present results. However, all those previous works could not provide more information for the higher DP. In this study, a decrease and then remain at a lower antioxidant level was showed when mDP >10. However, the antioxidant activity based on the DPPH test in the present study could be influenced by the amount of total polyphenols. Zhou et al. [32] also showed linear relationships between DPPH and total polyphenol content. Similarly, antioxidant tests were positively correlated with total polyphenols [32,45,46,47]. So only the mDP data may not be sufficient to explain the differences in EC_{50} and it is better to explain the differences in EC_{50} with reference to total phenolics and ECT in each fraction. Further investigations on the relationships between antioxidant activity and polyphenols in mangrove leaves and their mechanisms are required. Further investigations on the relationships between antioxidant activity and polyphenols in mangrove leaves and their mechanisms are required.

Mangrove tannins from *C. tagal* leaves were successfully characterized by ^{13}C-NMR, MALDI-TOF MS, HPLC-ESI MS and column chromatographic fractionation. They had substantial DPPH free radical scavenging ability and ferric reducing antioxidant power, which could be used as a new source of antioxidants. The major challenge in the research on condensed tannins is probably the difficulty in obtaining them in an individual molecular form. The complete purification of a procyanidin with a DP above five is almost impossible. Therefore, for studying their structures and properties, more or less mixtures polymerized are often employed [48]. More, the synergistic effects of active mixtures make plant extracts and fractions more interesting than the pure compounds for functional food applications [42]. In the present study, an effective fractionation method was established to elucidate more about the relationships between DP and antioxidant activities. The established relationships could be used as a theoretical method for predicting the structure-activity of proanthocyanidins.

Author Contributions

Conceived and designed the experiments: HCZ NFYT. Performed the experiments: HCZ WMC SDW. Analyzed the data: HCZ YML. Contributed reagents/materials/analysis tools: HCZ YML ZHD. Wrote the paper: HCZ NFYT.

References

1. Lin P (2006) Marine higher plant ecology. Science Press, Beijing: 76–80.
2. Rastogi RP, Mehrotra BN (1991) Compendium of Indian medicinal plants. Publications & Information Directorate, New Delhi 1.
3. Chen JD, Qiu Y, Yang ZW, Lin P, Lin YM (2008) Dimeric diterpenes from the roots of the mangrove plant *Ceriops tagal*. Helvetica Chimica Acta 91: 2292–2298.
4. Chacha M, Mapitse R, Afolayan AJ, Majinda RRT (2008) Antibacterial diterpenes from the roots of *Ceriops tagal*. Natural Product Communications 3: 17–20.
5. Feller IC, Whigham DF, O'Neill JP, McKee KL (1999) Effects of nutrient enrichment on within-stand cycling in a mangrove forest. Ecology 80: 2193–2205.
6. Mainoya J, Mesaki S, Banyikwa F (1986) The distribution and socio-economic aspects of mangrove forests in Tanzania. Man in the Mangroves: the Socio-Economic Situation of Human Settlements in Mangrove Forests: 87–95.
7. Kraus TEC, Dahlgren RA, Zasoski RJ (2003) Tannins in nutrient dynamics of forest ecosystems - a review. Plant and Soil 256: 41–66.
8. Lin YM, Liu JW, Xiang P, Lin P, Ye GF, et al. (2006) Tannin dynamics of propagules and leaves of *Kandelia candel* and *Bruguiera gymnorrhiza* in the Jiulong River Estuary, Fujian, China. Biogeochemistry 78: 343–359.
9. Zhou HC, Tam NYF, Lin YM, Wei SD, Li YY (2012) Changes of condensed tannins during decomposition of leaves of *Kandelia obovata* in a subtropical mangrove swamp in China. Soil Biology and Biochemistry 44: 113–121.
10. Zhang LL, Lin YM, Zhou HC, Wei SD, Chen JH (2010) Condensed tannins from mangrove species *Kandelia candel* and *Rhizophora mangle* and their antioxidant activity. Molecules 15: 420–431.
11. Maie N, Pisani O, Jaffe R (2008) Mangrove tannins in aquatic ecosystems: Their fate and possible influence on dissolved organic carbon and nitrogen cycling. Limnology and Oceanography 53: 160–171.
12. Oo CW, Pizzi A, Pasch H, Kassim MJ (2008) Study on the structure of mangrove polyflavonoid tannins with MALDI-TOF mass spectrometry. Journal of Applied Polymer Science 109: 963–967.
13. Hernes PJ, Benner R, Cowie GL, Goni MA, Bergamaschi BA, et al. (2001) Tannin diagenesis in mangrove leaves from a tropical estuary: A novel molecular approach. Geochimica Et Cosmochimica Acta 65: 3109–3122.
14. Wang Y, Zhu H, Tam NFY (2014) Polyphenols, tannins and antioxidant activities of eight true mangrove plant species in South China. Plant and Soil 374: 549–563.
15. Khanbabaee K, van Ree T (2001) Tannins: Classification and definition. Natural Product Reports 18: 641–649.
16. Alasalvar C, Karamac M, Kosinska A, Rybarczyk A, Shahidi F, et al. (2009) Antioxidant Activity of Hazelnut Skin Phenolics. Journal of Agricultural and Food Chemistry 57: 4645–4650.
17. Monagas M, Quintanilla-Lopez JE, Gomez-Cordoves C, Bartolome B, Lebron-Aguilar R (2010) MALDI-TOF MS analysis of plant proanthocyanidins. Journal of Pharmaceutical and Biomedical Analysis 51: 358–372.
18. Chai WM, Shi Y, Feng HL, Qiu L, Zhou HC, et al. (2012) NMR, HPLC-ESI-MS, and MALDI-TOF MS analysis of condensed tannins from *Delonix regia* (Bojer ex Hook.) Raf. and their bioactivities. Journal of Agricultural and Food Chemistry 60: 5013–5022.

19. Chen XX, Shi Y, Chai WM, Feng HL, Zhuang JX, et al. (2014) Condensed tannins from *Ficus virens* as tyrosinase inhibitors: structure, inhibitory activity and molecular mechanism. Plos One 9: e91809.
20. Gu LW, Kelm M, Hammerstone JF, Beecher G, Cunningham D, et al. (2002) Fractionation of polymeric procyanidins from lowbush blueberry and quantification of procyanidins in selected foods with an optimized normal-phase HPLC-MS fluorescent detection method. Journal of Agricultural and Food Chemistry 50: 4852–4860.
21. Li CM, Leverence R, Trombley JD, Xu SF, Yang J, et al. (2010) High molecular weight persimmon (*Diospyros kaki* L.) proanthocyanidin: a highly galloylated, A-linked tannin with an unusual flavonol terminal unit, myricetin. Journal of Agricultural and Food Chemistry 58: 9033–9042.
22. Ariga T, HAMANO M (1990) Radical scavenging action and its mode in procyanidins B-1 and B-3 from azuki beans to peroxyl radicals. Agricultural and Biological Chemistry 54: 2499–2504.
23. Hagerman AE, Riedl KM, Jones GA, Sovik KN, Ritchard NT, et al. (1998) High molecular weight plant polyphenolics (tannins) as biological antioxidants. Journal of Agricultural and Food Chemistry 46: 1887–1892.
24. Es-Safi NE, Guyot S, Ducrot PH (2006) NMR, ESI/MS, and MALDI-TOF/MS analysis of pear juice polymeric proanthocyanidins with potent free radical scavenging activity. Journal of Agricultural and Food Chemistry 54: 6969–6977.
25. Zhou HC, Lin YM, Li YY, Li M, Wei SD, et al. (2011) Antioxidant properties of polymeric proanthocyanidins from fruit stones and pericarps of Litchi chinensis Sonn. Food Research International 44: 613–620.
26. Makkar HPS, Blümmel M, Borowy NK, Becker K (1993) Gravimetric determination of tannins and their correlations with chemical and protein precipitation methods. Journal of Agricultural and Food Chemistry 61: 161–165.
27. Terrill TH, Rowan AM, Douglas GD, Barry TN (1992) Determination of extractable and bound condensed tannin concentrations in forage plants, protein concentrate meals and cereal grains. Journal of the Science of Food and Agriculture 58: 321–329.
28. Hellstrom J, Sinkkonen J, Karonen M, Mattila P (2007) Isolation and structure elucidation of procyanidin oligomers from saskatoon berries (*Amelanchier alnifolia*). Journal of Agricultural and Food Chemistry 55: 157–164.
29. Zhou HC, Lin YM, Chai WM, Wei SD, Liao MM (2011) Characterization of condensed tannins from litchi seed by reflectron modes and linear modes of MALDI-TOF MS. Acta Chimica Sinica 69: 2981–2986.
30. Brand-Williams W, Cuvelier ME, Berset C (1995) Use of a free radical method to evaluate antioxidant activity. Lebensmittel Wissenschaft and Technologie 28: 25–30.
31. Benzie IFF, Strain JJ (1996) The ferric reducing ability of plasma as a measure of "antioxidant power": the FRAP assay. Analytical Biochemistry 239: 70–76.
32. Zhou HC, Lin YM, Wei SD, Tam NFY (2011) Structural diversity and antioxidant activity of condensed tannins fractionated from mangosteen pericarp. Food Chemistry 129: 1710–1720.
33. Cai YZ, Sun M, Xing J, Luo Q, Corke H (2006) Structure-radical scavenging activity relationships of phenolic compounds from traditional Chinese medicinal plants. Life Sciences 78: 2872–2888.
34. Chirinos R, Rogez H, Campos D, Pedreschi R, Larondelle Y (2007) Optimization of extraction conditions of antioxidant phenolic compounds from mashua (*Tropaeolum tuberosum* Ruiz & Pavo) tubers. Separation and Purification Technology 55: 217–225.
35. Czochanska BZ, Foo YL, Newman RH, Porter LJ (1980) Polymeric proanthocyanidin. stereochemistry, structural units, and molecular weight. Journal of the Chemical Society, Perkin Transactions 1: 2278–2286.
36. Fu C, Loo AEK, Chia FPP, Huang D (2007) Oligomeric proanthocyanidins from mangosteen pericarps. Journal of Agricultural and Food Chemistry 55: 7689–7694.
37. Kraus TEC, Yu Z, Preston CM, Dahlgren RA, Zasoski RJ (2003) Linking chemical reactivity and protein precipitation to structural characteristics of foliar tannins. Journal of Chemical Ecology 29: 703–730.
38. Castillo-Munoz N, Gomez-Alonso S, Garcia-Romero E, Gomez MV, Velders AH, et al. (2009) Flavonol 3-O-glycosides series of *Vitis vinifera* cv. *Petit Verdot* red wine grapes. Journal of Agricultural and Food Chemistry 57: 209–219.
39. Reed JD, Krueger CG, Vestling MM (2005) MALDI-TOF mass spectrometry of oligomeric food polyphenols. Philadelphia, PA. Pergamon-Elsevier Science Ltd: 2248–2263.
40. Xiang P, Lin YM, Lin P, Xiang C, Yang ZW, et al. (2007) Effect of cationization reagents on the matrix-assisted laser desorption/ionization time-of-flight mass spectrum of Chinese gallotannins. Journal of Applied Polymer Science 105: 859–864.
41. Taylor AW, Barofsky E, Kennedy JA, Deinzer ML (2003) Hop (*Humulus lupulus* L.) proanthocyanidins characterized by mass spectrometry, acid catalysis, and gel permeation chromatography. Journal of Agricultural and Food Chemistry 51: 4101–4110.
42. Jerez M, Tourino S, Sineiro J, Torres JL, Nunez MJ (2007) Procyanidins from pine bark: Relationships between structure, composition and antiradical activity. Food Chemistry 104: 518–527.
43. Meri S, Kanner J, Akiri B, Hadas SP (1995) Determination and involvement of aqueous reducing compounds in oxidative defense systems of various senescing leaves. Journal of Agricultural and Food Chemistry 43: 1813–1815.
44. de Gaulejac N, Vivas N, de Freitas V, Bourgeois G (1999) The influence of various phenolic compounds on scavenging activity assessed by an enzymatic method. Journal of the Science of Food and Agriculture 79: 1081–1090.
45. Cai YZ, Luo Q, Sun M, Corke H (2004) Antioxidant activity and phenolic compounds of 112 traditional Chinese medicinal plants associated with anticancer. Life Sciences 74: 2157–2184.
46. Silva E, Souza J, Rogez H, Rees J, Larondelle Y (2007) Antioxidant activities and polyphenolic contents of fifteen selected plant species from the Amazonian region. Food Chemistry 101: 1012–1018.
47. Lizcano LJ, Bakkali F, Ruiz-Larrea MB, Ruiz-Sanz JI (2010) Antioxidant activity and polyphenol content of aqueous extracts from Colombian Amazonian plants with medicinal use. Food Chemistry 119: 1566–1570.
48. Guyot S, Marnet N, Drilleau JF (2001) Thiolysis-HPLC characterization of apple procyanidins covering a large range of polymerization states. Journal of Agricultural and Food Chemistry 49: 14–20.

6

Barbs Facilitate the Helical Penetration of Honeybee (*Apis mellifera ligustica*) Stingers

Jianing Wu, Shaoze Yan*, Jieliang Zhao, Yuying Ye

Division of Intelligent and Biomechanical Systems, State Key Laboratory of Tribology, Department of Mechanical Engineering, Tsinghua University, Beijing, P. R. China

Abstract

The stinger is a very small and efficient device that allows honeybees to perform two main physiological activities: repelling enemies and laying eggs for reproduction. In this study, we explored the specific characteristics of stinger penetration, where we focused on its movements and the effects of it microstructure. The stingers of Italian honeybees (*Apis mellifera ligustica*) were grouped and fixed onto four types of cubic substrates, before pressing into different substrates. The morphological characteristics of the stinger cross-sections were analyzed before and after penetration by microscopy. Our findings suggest that the honeybee stinger undergoes helical and clockwise rotation during penetration. We also found that the helical penetration of the stinger is associated directly with the spiral distribution of the barbs, thereby confirming that stinger penetration involves an advanced microstructure rather than a simple needle-like apparatus. These results provide new insights into the mechanism of honeybee stinger penetration.

Editor: Friedrich Frischknecht, University of Heidelberg Medical School, Germany

Funding: This work was funded by the Research Project of the State Key Laboratory of Tribology under Contract no. SKLT11B03. The funders had no role in study design, data collection and analysis, decision to publish, or preparation of the manuscript.

Competing Interests: The authors have declared that no competing interests exist.

* Email: yansz@mail.tsinghua.edu.cn

Introduction

In variable and complex environments, animals are equipped with different organs for accomplishing diverse physical activities. Stingers and needles are found in some insects in the orders Diptera and Hymenoptera, where they play important roles in predation, mating, and defense [1–5]. Various theories have been developed to describe the penetration mechanism of insect stingers and needles [6–9]. A comprehensive understanding of stinger penetration has been obtained gradually, which has attracted the interest of the developers of bio-inspired instruments, e.g., painless insertion for medical care [10] and bionic-based drilling technologies for planetary subsurface exploration [11].

The abdomen of the honeybee (*Apis mellifera*) comprises 10 segments, seven of which are obvious [12–14]. The cavity within the last abdominal segment of the honeybee is called the sting chamber and the entire sting apparatus is enclosed within the chamber when it is not in use, as well as nerve ganglions, various muscles, a venom sac, and the end of the insect's digestive tract [15–16]. The stinger is a small and delicate device, which allows honeybee workers to defend their nest against predators [12]. As shown in Fig. 1, when dangerous enemies are encountered, the sting apparatus receives a signal from the nerve ganglions and the bee bends its abdomen downward due to muscle contractions as it prepares for vertical stinger penetration. During the use of the stinger, two pairs of protractor and retractor muscles move the stinger up and down, which causes a flexible extension of the stinger shaft. Movements of the bee's legs, the muscles of the abdomen, and the effect of the backward pointing barbs combine to produce a thrust that drives the stinger efficiently, and the venom is delivered instantly into tough skin through a channel in the stinger. The first analysis of the stinger penetration mechanism was performed by Dade in 1890s, particularly the coordination between various organs [14].

The stinger comprises two lancets with groups of curved barbs on the outer aspects of their distal ends, which are held in grooves on the stylet [14]. It is well known that the main role of the barb is to provide one-way traction, which allows the stinger to work itself deeper into the flesh [13,14]. The raked structure of the barbs makes it difficult to remove the sting, which might help the bee to continue pumping venom into the flesh via the detached stinger for a relatively long time [14]. In this case, the underlying mechanism of penetration appears relatively simple, i.e., the needle-like stinger is assumed to move axially while piercing the skin, but the possible role of rotation along the stinger shaft has been neglected. This is because the bee stingers measure a few millimeters and the action of stinging occurs within one second, thus the actual penetration behavior cannot be observed easily. Previous research has only considered the morphology at the level of a single barb. However, the potential effects of the distribution of the barbs on the efficiency of penetration have not been identified clearly.

In this study, we explored the penetration mechanism of the honeybee stinger. We investigated the morphology of the barbs on the bee stinger and elucidated the specific factors that determine the rotation of the stinger. Our results showed that the stinger undergoes helical and clockwise rotation during penetration, where the spiral distribution of the barbs is responsible for this phenomenon.

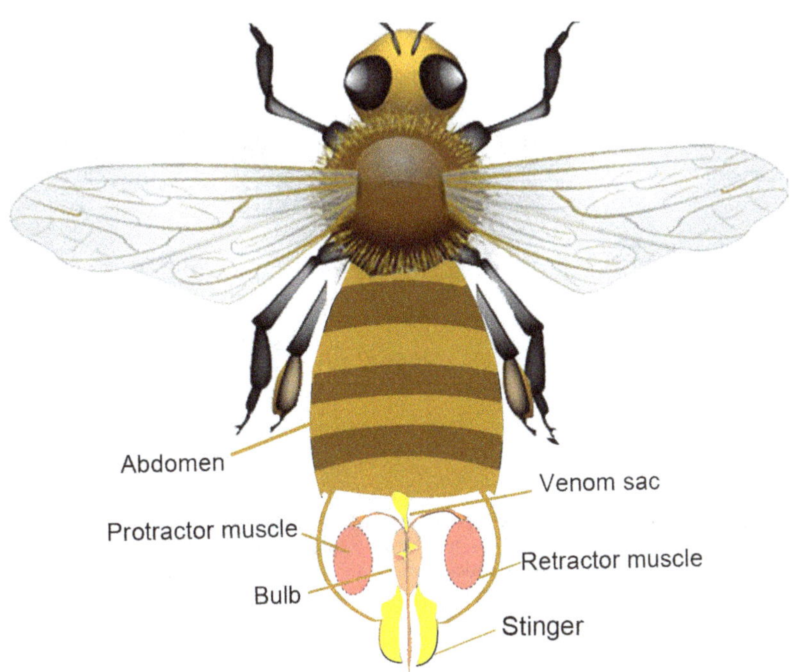

Figure 1. Anatomy of the honeybee's stinger apparatus. The stinger resides in the sting chamber inside the last abdominal segment (not to scale). The sting apparatus mainly comprises the protractor/retractor muscles, the bulb, the stinger, and the venom sac. The protractor muscles drive the stinger to penetrate the wound and the retractor muscles are used in the reverse manner to pull the stinger back into the sting chamber. During penetration, the venom is pumped into the stinger from the bulb, which is also known as the venom reservoir.

Materials and Methods

Experimental method

We studied the penetration characteristics of the stingers of honeybee (*Apis mellifera ligustica*) workers. The samples were collected at Tsinghua University of Beijing, China (40.000153°N, 116.326414°E). No specific permissions were required for these locations/activities. We confirm that the field studies did not involve endangered or protected species. To ensure the reliability and repeatability of the experiments, all of the honeybee samples were captured around wild bee nests and the experiments were conducted within 1 h of collection. In total, 30 fresh stingers from

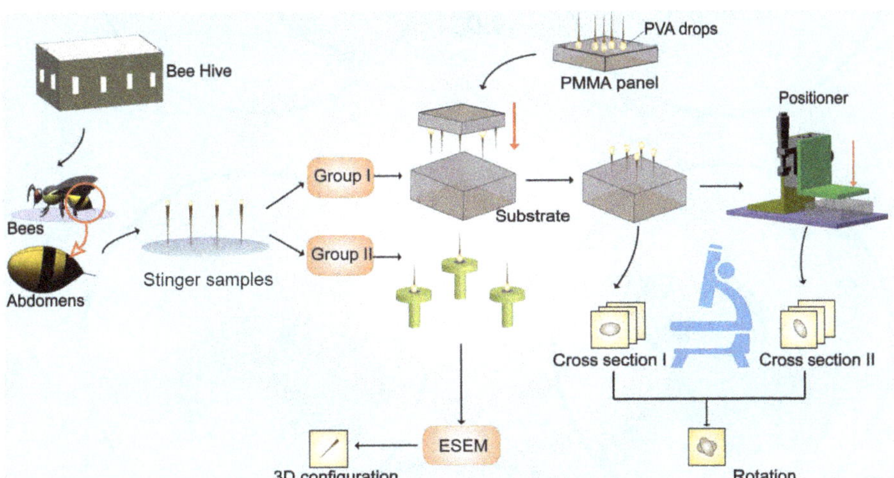

Figure 2. Preparation of the stinger samples and the experiments. The stingers of worker bees were collected and separated into two groups. (1) The first group of stingers were placed onto the polymethyl methacrylate panel using drops of 15% polyvinyl alcohol (0.1 μL), and they were then placed vertically on the substrates (agar, silica gel, soft rubber and paraffin wax), before pushing the stingers into the substrates at a velocity of 6 mm/s using the positioner. The positioner, also called the precision position platform, is a machine that is able to push tiny appendages accurately into the substrates following the planned kinematics, for instance the preset average velocity, the total displacement even the acceleration. (2) The microstructures of the second group of stingers were observed using an environmental scanning electron microscope.

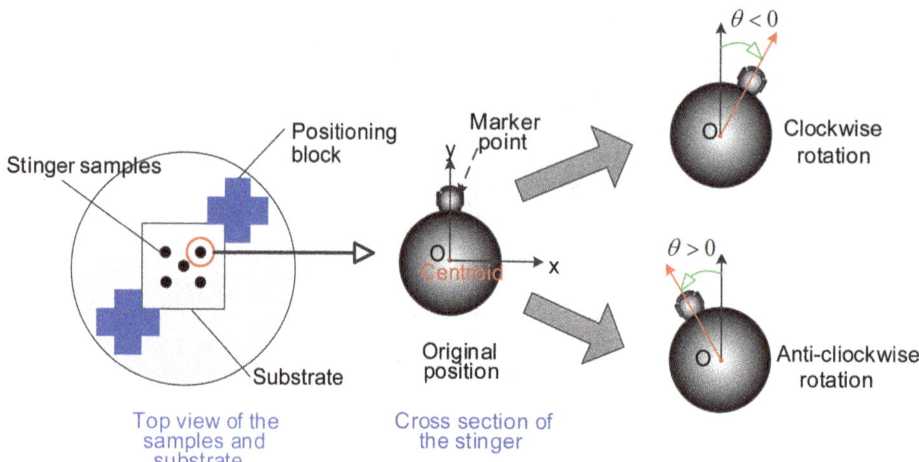

Figure 3. Observation of the stinger cross-sections and calculation of the rotation angles. (A) The four types of cubic substrates were attached to the positioning block assembly under the microscope. (B) Marker points in the stinger cross-section. If the stinger rotated clockwise, the rotation angle was recorded as a negative angle, whereas an anticlockwise rotation was recorded as a positive angle.

worker bees were selected, cleaned, and dehydrated, where the average length was 8.5 mm (Fig. 2).

We performed two types of experiments to elucidate the stinging mechanism. As shown in Fig. 1, we first separated the stingers into two groups, where group I and group II contained 20 and 10 stinger samples, respectively. The 20 stingers in group I were then grouped into subgroups A_1, A_2, A_3, and A_4, each of which comprised five samples. Four types of $10\times10\times10$ mm cubic substrates were prepared, which were made of agar, silica gel, soft rubber, and paraffin wax. The samples in Groups $A_1 \sim A_4$ were fixed to the 20% *Poly Vinyl Alcohol* (PVA) colloid droplets which were firstly dispensed on the PMMA panel. Thereby tips of the

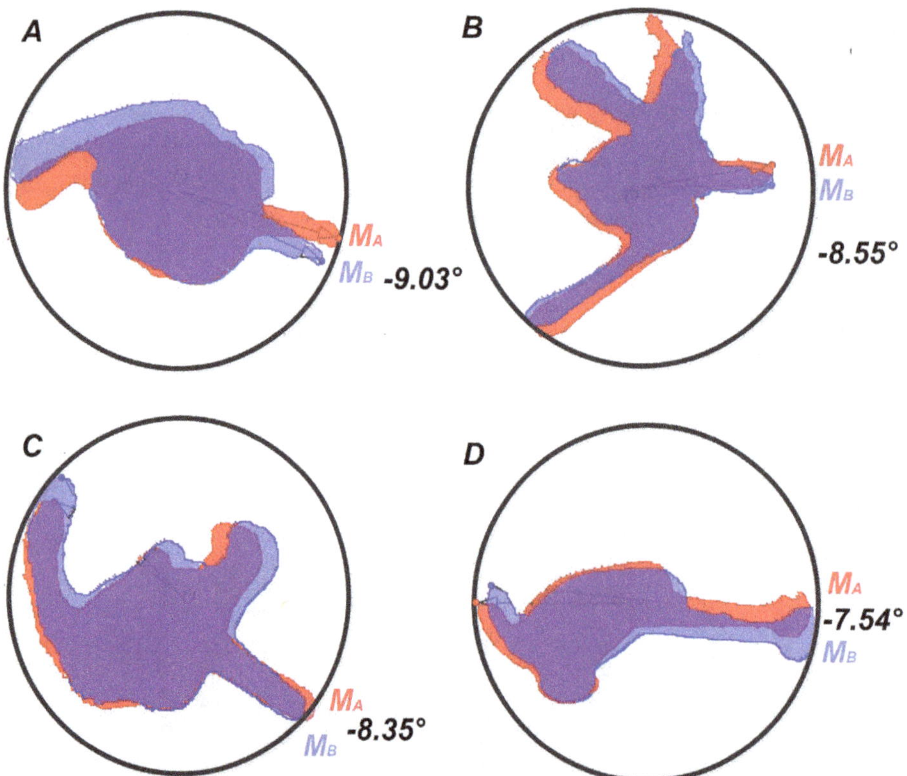

Figure 4. Cross-sections of the stinger shafts. We determined the outlines of the stinger cross-sections by applying the Canny operator. (A)–(D) show the four stinger cross-sections before and after penetration of the substrates, i.e., agar, silica gel, soft rubber, and paraffin wax, respectively. Helical penetration was identified clearly by analyzing the rotation angles of the marker points. We also found that the rotation angle decreased when we used relatively harder substrates.

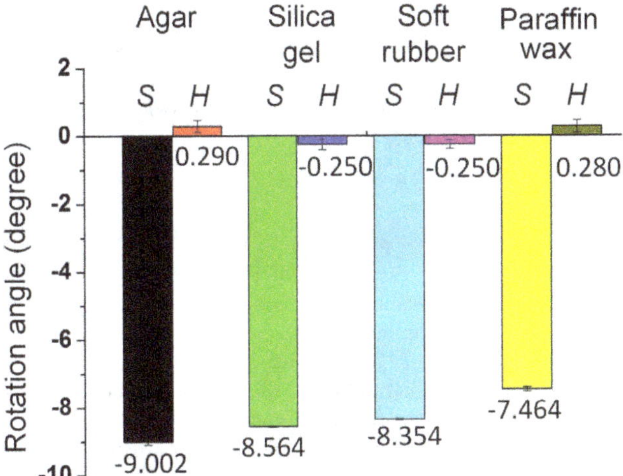

Figure 5. Rotation angles of the stinger shafts. The stinger samples marked with *S* and the hair samples marked with *H* were used for comparison. The two adjacent bars show the rotation angles of the honeybee stingers (left side) and the hair samples (right side). In each type of substrate, the rotation angle of the hair was very small, which demonstrated that the instrument had no significant effect on the rotation angle during pushing. The experimental observations confirmed the existence of rotation during the stinging process. Furthermore, we found that the rotation angle was associated with the stiffness of the substrate.

stings were placed onto the substrate of the cubic block of different materials (See Figure S3 in File S1). All of the stings were pressed for 6.0 mm into the substrate with a precision position platform with the average velocity approximately to 6 mm/s (Fig. 2). In addition, to eliminate any errors caused by the setup, we tested rotation angles of the human hair samples that measured ca 8.5 mm in length for comparison. Morphological images of cross-sections of the stingers and hairs were obtained before and after penetration by microscopy. Notably we observed the natural heads of the stingers directly to determine whether the helical penetration exists or not [17]. With the help of the environmental scanning electron microscope (ESEM), we observed the microstructure of 10 stingers in group I.

Rotation measurement

Fig. 3 shows the method used to observe the stinger cross-sections and to calculate the rotation angles. The cross-sections of the stinger were observed and photographed before and after penetration. We enhanced the microscope so it could locate the cubic substrates by using a positioning block, thereby ensuring that the cubic substrate remained fixed. The rotation angles were measured by comparing the positions of markers in the stinger cross-sections.

The image processing system used to capture the contours of the cross-sections of the stingers was implemented with the Canny operator. We define the equation of the contours as $F(x,y)=0$,

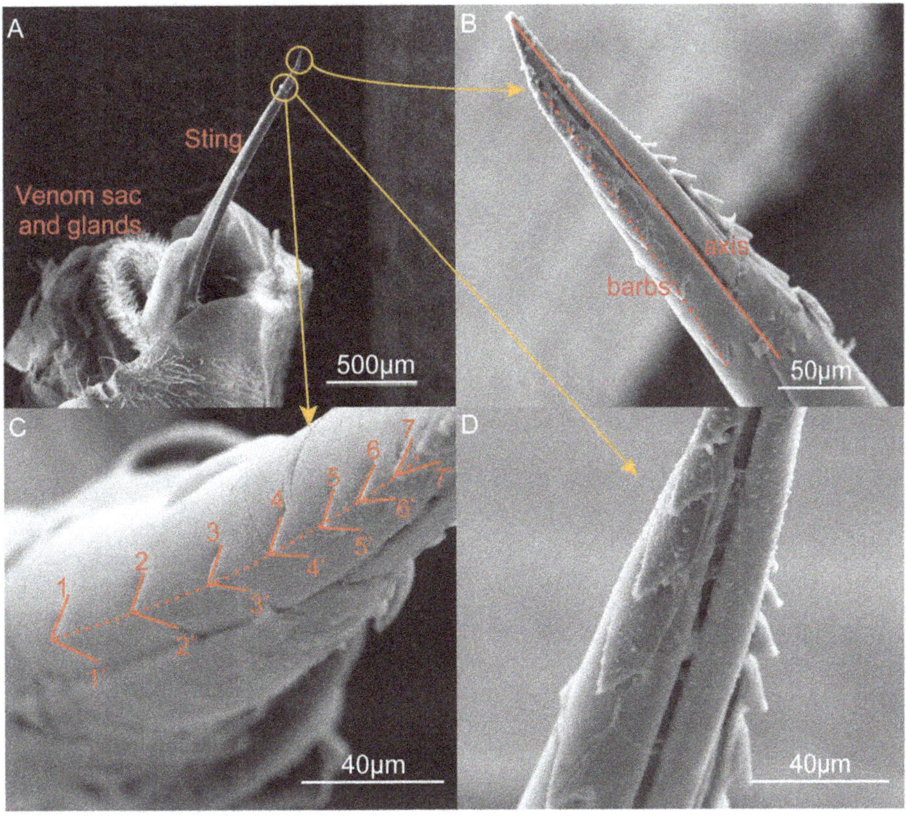

Figure 6. Environmental scanning electron microscope images of the stinger. (A) The needle-like sting, venom sac, and related glands. The stinger is activated by the muscles to penetrate the skin of the victim. (B) Barbs along the axial direction of the sting. The solid line in Figure 3(b) is the axis of the sting which is obtained by connecting the tip of the stinger and the midpoint of the stinger root. The stinger of *Apis mellifera ligustica* has two rows of barbs, each of which comprises about 10 barbs. The angle between the rows of barbs and the axis of the stinger shaft was around 8–9°, according to observations based on 10 samples. The row of barbs was found to form a right-handed helix. (C) Magnified view of the barbs. Seven barbs are marked with the notations 1-1′, 2-2′, etc. Note that the angles of the tips were 90.33°, 89.62°, 80.31°, 72.13°, 72.36°, 59.63°, and 46.19°, thereby demonstrating that the barbs were relatively sharper near the tip of the stinger. (D) Magnified view of two rows of barbs. Viewed in the axial direction, the angles between the two rows of barbs were about 95°.

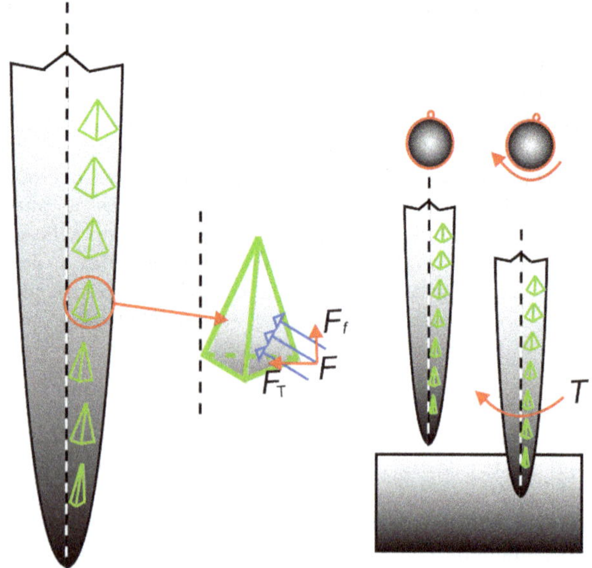

Figure 7. Mechanistic model of stinging. The angle between a row of barbs and the shaft axis was about 9°. The equivalent tetrahedron model mimics the force condition of a single barb (not to scale). During penetration, the contact force on the barb (F) can be decomposed into the friction force (F_f) and the tangential force (F_T). The torque around the stinger shaft is generated by the tangential force, which drives the stinger to rotate clockwise in the line of sight along the stinger shaft toward the tip.

thus the centroid of the cross-section, which is denoted as O (x_O, y_O), can be calculated by

$$x_O = \frac{\iint_\sigma x\,dx\,dy}{\iint_\sigma dx\,dy}, y_O = \frac{\iint_\sigma y\,dx\,dy}{\iint_\sigma dx\,dy} \quad (1)$$

where σ represents the region surrounded by the contour of the stinger shaft. The key point used to measure the change in configuration is defined as $M(x_k, y_k)$, which satisfies

$$\left.\frac{dF(x,y)}{dx}\right|_{\substack{x=x_k\\y=y_k}} = \max\left[\frac{dF(x,y)}{dx}\right]\Bigg|_{\substack{x\in X\\y\in Y}} \quad (2)$$

We define the coordinates of the key points before and after penetration as $M_1(x_{k1}, y_{k1})$ and $M_2(x_{k2}, y_{k2})$, respectively. By comparing the contours of the cross-sections, we determined the amount of rotation before and after penetration. In this case, the rotation angle is θ

$$\theta = \sum_{i=1}^{2} (-1)^{i+1} \cdot arctan[(y_{ki} - y_O)/(x_{ki} - x_O)] \quad (3)$$

where $\theta = 0$indicates that the stinger shaft does not rotate, whereas $\theta > 0$ and $\theta < 0$ indicates that the stinger shaft rotates in clockwise or anticlockwise directions, respectively.

As stated above, the remaining 10 stingers were used to obtain microstructural observations with an ESEM. Particular emphasis was placed on the physical distribution and geometrical properties of the barbs on the stinger.

Results

Helical penetration

Fig. 4 shows four cross-sections of stingers that penetrated the agar, silica gel, soft rubber, and paraffin wax substrates. The features of helical penetration are shown in the comparison chart. In particular, M_A and M_B are the marker points that indicate the rotation angles according to Eqs (1)–(3). We averaged the rotation angles for the subgroups A_1, A_2, A_3, and A_4 which were collected from the experiments, and calculated the standard deviations of the data in different groups to test the data stability. Statistical analysis demonstrated that the data were stable with the standard deviation around 0.046° (See Table S1 in File S1). By comparing the stinger sections before and after penetration, we found that the mean rotation angles of the 20 samples were all negative, which showed that the stingers rotated in a clockwise direction while penetrating the substrates (Fig. 5). By contrast, the rotation angle of the hair section was <0.3°, which demonstrated that the experimental setup did not introduce large errors, thereby validating the results. As shown in Fig. 5, the average rotation angle of the stinger shaft based on all the data was –8.364°. The rotation angle decreased gradually as the hardness of the substrate surface increased, i.e., from the agar and silica gel, to the soft rubber and paraffin wax. The hardness of the silica gel is closest to the human skin, indicating that the stinger may approximately rotate –8.0° when it penetrates the human skin.

Microstructure of the honeybee stinger

This rotation may be attributable to the shape of the stinger and the distribution of the barbs. Understanding the underlying mechanism will help to elucidate the stinging process and related behaviors. Fig. 6 shows the microstructure of the stinger, which was obtained by ESEM. The venom sac and related glands support the stinger. The bee stinger is generally similar to a long needle and its proximal apex is covered by two lines of barbs, which subtend two angles with the stinger axis. The tetrahedron-shaped barbs decrease in size as they approach the tip of the stinger.

Discussion

Our results showed that the stinger shaft rotates when it is pushed into a solid substrate and its rotation angle might be affected by the hardness of the substrate. Thus, it is possible that the helical penetration might improve the puncture process, in a similar manner to the spiral flight of bullets. In addition, this specific penetration behavior increases the difficulty of sting removal, which might increase the amount of venom injected. To better understand the mechanism of the rotation of the stinger shaft, it is important to explore the dynamic performance during penetration. In particular, it is possible that the microstructure of the stinger facilitates the rotation of the stinger shaft, especially the protuberances.

Fig. 7 shows a mechanistic model of stinging that considers the microstructure of the barbs. The tetrahedron model mimics the force condition of a single barb. During penetration, the contact force is not parallel to the stinger axis, therefore the torque generated by the tangential force (marked as F_T in Fig. 7) can be calculated by $T = rF_T$, where r is the corresponding radius of the stinger. By focusing on the right-handed helical distribution of barbs, the angle between the line of barbs and the stinger shaft is about 8.523°, which agrees with the average rotation angle (–8.364°) shown in Fig. 5. The blunter barbs located further from the tip of stinger will expand the wound size during penetration. Therefore, the force analysis demonstrates that the helical distribution of barbs leads to the rotation of the stinger during penetration.

The specific microstructure of the stinger provides new insights, which may facilitate a deeper investigation of the biological significance of stinger penetration. The main appendages of the bee stinger of *Apis mellifera ligustica* are derived from the segmental appendages of the 9th and 10th abdominal segments. The appendages of all segments are organized into segments and branches during the early stages of evolution beyond the primitive worm stage, according to the same plan, after which they have been modified in different ways for specific purposes [14,18]. According to many researchers [16], the worker bees represent the body organs required for nest maintenance, digestion, and defense [15,19–20]. Helical penetration might improve the effectiveness of attack with fewer sacrifices. The development of novel devices may also be inspired by this specific penetration mechanism [21,22].

Supporting Information

File S1 Contains the files: Figure S1. Principle of experiment. **Figure S2.** Samples of the worker bees' stings. **Figure S3.** The precision positioner and the substrate. **Figure S4**. Location of the substrate. **Figure S5.** Morphology of the cross section. **Figure S6.** Rotation angles of the sting shaft. **Table S1.** The experimental results from the observation of the rotational angles.

Acknowledgments

We thank the Center of Biomedical Analysis of Tsinghua University contributed to the data collecting for this project.

Author Contributions

Conceived and designed the experiments: JW SY JZ YY. Performed the experiments: JW SY JZ YY. Analyzed the data: JW JZ YY. Contributed reagents/materials/analysis tools: JW SY JZ YY. Contributed to the writing of the manuscript: JW SY JZ YY.

References

1. Klowden MJ (2013) Physiological systems in insects. London: Academic Press. 90 p.
2. Frasson L, Ko SY, Turner A, Parittotokkaporn T, Vincent JF, et al. (2010) sting: a soft-tissue intervention and neurosurgical guide to access deep brain lesions through curved trajectories. P I Mech Eng H 224: 775–788.
3. Izumi H, Suzuki M, Aoyagi S, Kanzaki T (2011) Realistic imitation of mosquito's proboscis: Electrochemically etched sharp and jagged needles and their cooperative inserting motion. Sensor Actuat A-Phys 165: 115–123.
4. Rahman MH, Fitton MG, Quicke DL (1998) Ovipositor internal microsculpture and other features in doryctine wasps (Insecta, Hymenoptera, Braconidae). Zool Scr 27: 333–343.
5. Kong XQ, Wu CW (2010) Mosquito proboscis: An elegant biomicroelectromechanical system. Phys Rev E 82(1), 011910.
6. Vilhelmsen L, Isidoro N, Romani R, Basibuyuk HH, Quicke DLJ (2001) Host location and oviposition in a basal group of parasitic wasps: the subgenual organ, ovipositor apparatus and associated structures in the Orussidae (Hymenoptera, Insecta). Zoomorphology 121: 63–84.
7. Duan JJ, Messing RH (2000) Effects of host substrate and vibration cues on ovipositor-probing behavior in two larval parasitoids of tephritid fruit flies. J Insect Behav, 13: 175–186.
8. Brown PE, Anderson M (1998) Morphology and ultrastructure of sense organs on the ovipositor of *Trybliographa rapae*, a parasitoid of the cabbage root fly. J Insect Physiol, 44: 1017 1025.
9. Wu E, Sheen HJ, Chen YC, Chang LC (1994) Penetration force measurement of thin plates by laser Doppler anemometry. Exp Mech, 34: 93–99.
10. Aoyagi S, Izumi H, Fukuda M (2008) Biodegradable polymer needle with various tip angles and consideration on insertion mechanism of mosquito's proboscis. Sensor Actuat A-Phys: Phys. 143: 20–28.
11. Gao Y, Ellery A, Sweeting M, Vincent J (2007) Bioinspired drill for planetary sampling: literature survey, conceptual design, and feasibility study. J Spacecraft Rockets. 44: 703–709.
12. Visscher PK, Vetter RS, Camazine S (1996) Removing bee stings. Lancet, 348: 301–302.
13. Snodgrass RE (1956) Anatomy of the honey bee. Ithaca: Cornell University Press. 37 p.
14. Dade HA (1994) Anatomy and dissection of the honey bee. International Bee Research Association, 17 p.
15. Tautz J (2008) The buzz about bees: biology of a super organism. Berlin: Springer. 71 p.
16. Winston ML (1991) The biology of the honey bee. Boston: Harvard University Press. 13 p.
17. Robbins C R, Robbins C R (2002) Chemical and physical behavior of human hair. New York: Springer. 225 p.
18. Schilthuizen M, Nordlander G, Stouthamer R, Van Alphen J (1998) Morphological and molecular phylogenetics in the genus Leptopilina (Hymenoptera: Cynipoidea: Eucoilidae). Syst Entomol 23: 253–264.
19. Van Dooremalen C, Gerritsen L, Cornelissen B, van der Steen JJ, van Langevelde F (2012) Winter survival of individual honey bees and honey bee colonies depends on level of Varroa destructor infestation. PloS one 7: e36285.
20. Cardinal S, Danforth BN (2011) The antiquity and evolutionary history of social behavior in bees. PLoS One 6: e21086.4.
21. Choo YM, Lee KS, Yoon HJ, Kim BY, Sohn M, et al (2010) Dual function of a bee venom serine protease: prophenoloxidase-activating factor in arthropods and fibrin (ogen) olytic enzyme in mammals. PLoS One 5: e10393.
22. Srinivasan MV, Zhang S, Altwein M, Tautz J (2000) Honey bee navigation: nature and calibration of the "odometer". Science 287: 851–853.

7

A Fractal Nature for Polymerized Laminin

Camila Hochman-Mendez[1,2], Marco Cantini[3], David Moratal[4], Manuel Salmeron-Sanchez[3], Tatiana Coelho-Sampaio[2]*

1 Institute of Biomedical Sciences, Federal University of Rio de Janeiro, Rio de Janeiro, Brazil, 2 Institute of Biophysics Carlos Chagas Filho, Federal University of Rio de Janeiro, Rio de Janeiro, Brazil, 3 Division of Biomedical Engineering, School of Engineering, University of Glasgow, Glasgow, United Kingdom, 4 Center for Biomaterials and Tissue Engineering, Universitat Politècnica de València, València, Spain

Abstract

Polylaminin (polyLM) is a non-covalent acid-induced nano- and micro-structured polymer of the protein laminin displaying distinguished biological properties. Polylaminin stimulates neuritogenesis beyond the levels achieved by ordinary laminin and has been shown to promote axonal regeneration in animal models of spinal cord injury. Here we used confocal fluorescence microscopy (CFM), scanning electron microscopy (SEM) and atomic force microscopy (AFM) to characterize its three-dimensional structure. Renderization of confocal optical slices of immunostained polyLM revealed the aspect of a loose flocculated meshwork, which was homogeneously stained by the antibody. On the other hand, an ordinary matrix obtained upon adsorption of laminin in neutral pH (LM) was constituted of bulky protein aggregates whose interior was not accessible to the same anti-laminin antibody. SEM and AFM analyses revealed that the seed unit of polyLM was a flat polygon formed in solution whereas the seed structure of LM was highly heterogeneous, intercalating rod-like, spherical and thin spread lamellar deposits. As polyLM was visualized at progressively increasing magnifications, we observed that the morphology of the polymer was alike independently of the magnification used for the observation. A search for the Hausdorff dimension in images of the two matrices showed that polyLM, but not LM, presented fractal dimensions of 1.55, 1.62 and 1.70 after 1, 8 and 12 hours of adsorption, respectively. Data in the present work suggest that the intrinsic fractal nature of polymerized laminin can be the structural basis for the fractal-like organization of basement membranes in the neurogenic niches of the central nervous system.

Editor: Heidar-Ali Tajmir-Riahi, University of Quebec at Trois-Rivieres, Canada

Funding: This work was supported by a grant from the Brazilian National Research Council (CNPq; 476772/2008-7) to TCS. MSS acknowledges support from the European Research Council through ERC - 306990. The funders had no role in study design, data collection and analysis, decision to publish, or preparation of the manuscript.

Competing Interests: The authors have declared that no competing interests exist.

* Email: tcsampaio@histo.ufrj.br

Introduction

Laminin is the major molecular component of the basement membrane, a specialized type of extracellular matrix characterized by a flat sheet-like geometry. Laminin regulates a variety of biological phenomena, including the provision of boundaries between neighboring tissues, the establishment of molecular filters and the modulation of cell behavior [1-3]. The quaternary structure of laminin is given by three different polypeptide chains, which associate to form a cross-shaped heterotrimer, exhibiting one long and three shorter arms. In the three-dimensional space laminin has the shape of a three-leafed clover, in which the leaves correspond to the three short arms while the stem corresponds to the long arm. This 3-D structure is particularly adequate to favor the assembly of a sheet-like polymer, where the three short arms simultaneously interact with each other within a single spatial plane, while the long arm is left available to interact with the surface of contiguous cells [4].

As a consequence of its structural properties, laminin can spontaneously self-polymerize in a test tube, requiring either a minimal protein concentration [5] or a decrease in the solution pH [6,7]. Polymers formed upon pH acidification, designated as polylaminin (polyLM[1]), present specific signaling properties and have been shown to stimulate the outgrowth of neurites with at least twice the efficiency of ordinary laminin (LM), namely a matrix obtained by adsorbing the protein diluted in neutral pH onto a glass coverslip [8]. PolyLM was also shown to reverse the loss of migratory and neuritogenic potentials of cortical neurons and to promote the survival and the proliferation of axotomized retinal ganglion cells, both isolated from newborn rodents [9]. Finally, it was demonstrated that polyLM, but not the laminin protein diluted in neutral buffer, promoted axonal regeneration and functional recovery after spinal cord injury in rats [10].

The morphology of polyLM has previously been studied both at the micro and at the nanometer scales. Using negative staining followed by transmission electron microscopy it was possible to characterize it as a regular polygonal network displaying the same features of the natural laminin networks assembled by living cells [11,12]. In such polymers the unit polygon was a hexagon of approximately 30 nm of side, which well corresponded to the size of each short arm in the laminin molecule. Curiously, a polygonal array of comparable features was observed at a three orders of magnitude larger scale when immunolabeled polyLM was analyzed by fluorescence microscopy [7,8,13].

The fact that polyLM observed at different magnifications showed similar structures suggested that the polymer could be a fractal structure. Nevertheless, the lack of images obtained at intermediary magnifications, as well as the insufficient resolution of previous epifluorescence photomicrographs, prevented the search for a fractal dimension of polyLM. In the present work we used confocal fluorescence microscopy (CFM), scanning electron microscopy (SEM) and atomic force microscopy (AFM) to obtain a detailed structural characterization of polyLM that would permit assessment of its putative fractal nature. We showed that polyLM presented a fractal dimension, which increased its complexity upon accumulation of the polymer on a flat substrate. These findings may have important implications as they can provide an intrinsic molecular basis for the fractal-like organization of the basement membranes present at the neurogenic niches in the adult central nervous system.

Experimental Procedures

Preparation of laminin matrices

PolyLM was produced by diluting EHS laminin (laminin 111; Invitrogen) to 50 μg/mL with 20 mM sodium acetate (pH 4), containing 1 mM $CaCl_2$. The laminin protein has previously been shown to polymerize in solution within a few minutes after dilution in acidic buffer, independently of its concentration [6]. The polymers produced in solution (polyLM) were adsorbed to glass coverslips to produce the matrices used here for microscopic analyses. LM was produced by diluting EHS laminin to 50 μg/mL with 20 mM Tris-HCl (pH 7) containing 1 mM $CaCl_2$. This concentration was below the critical protein concentration necessary to trigger laminin polymerization in solution at neutral pH [5], so that the LM matrix was formed as the protein decanted and raised its concentration at the glass surface. In order to avoid unwanted polymerization in the highly concentrated stock solution of EHS laminin (0.5–1.5 mg/mL), working aliquots (2–10 μL) were stored frozen and individually thawed in ice immediately before dilution. Unless otherwise indicated, incubations with glass coverslips were carried out at 37°C for 12 hours, which is known to be sufficient to warrant that at least 60% of the protein would decant and adsorb regardless of the solution pH [8].

Immunostaining and confocal microscopy

Laminin matrices adsorbed on glass coverslips were fixed with paraformadehyde 4% for 20 min and prepared for indirect immunofluorescence analysis. Coverslips were washed 3 times for 5 min in PBS and incubated with bovine serum albumin 5% in PBS (PBS-BSA 5%) for 30 min. The primary antibody was a polyclonal rabbit anti-laminin antibody (1:30, Sigma-Aldrich, no. L9393). After overnight incubation at 4°C, coverslips were washed 3 times for 5 min in PBS and incubated with an Alexa Fluor 488 anti-rabbit secondary antibody (1:300; Life Technology, no. A-11001) for 1 hour at room temperature. They received three 5-min washes with PBS and one with distilled water before being mounted in n-propyl gallate in 80% glycerol (Sigma-Aldrich). Confocal images were obtained in a Leica TCS-SP5 confocal laser scanning microscope using a HCX PL APO lambda blue 63X objective for oil immersion (1.4 of numerical aperture). Images correspond to renderized stacks of 74 optical slices obtained with a zoom of 3.2 at each 21.4 μm (total width of 1530.8 μm).

Scanning electron microscopy

Laminin matrices attached to coverslips were fixed in Karnowsky reagent (4% PA and 0.5% glutaraldehyde in 0.1 M cacodylate buffer, pH 7.2) for 2 hours, washed three times with sodium cacodylate buffer 0.1 M, pH 7.2, dehydrated through increasing concentrations of ethanol and dried in E300 (Polaron, Quorum Technologies Ltd, Laughton, United Kingdom) critical point. The samples were then coated with a thin layer of gold sputter (Leica EM MED020) and viewed under a scanning electron microscope Jeol JSM6300.

Transmission electron microscopy

Transmission electron microscopy after negative staining was carried out as previously described [11]. Briefly, 5 μl of laminin in acidic buffer (polyLM) was deposited on a Formvar-coated copper grid and a 5 μl drop of 2% uranyl acetate was added over it. Samples were visualized in a Zeiss EM 900 transmission electron microscope operated at 80 kV.

Atomic force microscopy

Laminin matrices were fixed in 4% paraformaldehyde for 20 minutes and dried in critical point dryer E300 (Polaron, Quorum Technologies Ltd, Laughton, United Kingdom). AFM analyses were performed using a Multimode AFM equipped with a NanoScope IIIa controller (Bruker) operating in tapping mode in air; the Nanoscope 5.30r2 software version was used for image processing and analysis. Si-cantilevers MPP-21120 from Bruker were used, with force constant of 3 $N.m^{-1}$ and resonance frequency of 75 kHz. The phase signal was set to zero at a frequency 5–10% lower than the resonance one. Drive amplitude was 200 mV and the amplitude set-point (A_{sp}) was 1.4 V. The ratio between the amplitude set-point and the free amplitude (A_{sp}/A_0) was kept equal to 0.7.

Quantification of laminin adsorption

The amount of adsorbed laminin was measured using a Micro BCA Protein Assay Kit (23235#, Thermo Scientific Pierce). Following the standard protocol, the working reagent (WR) was prepared from 25 parts of MA (sodium carbonate, sodium bicarbonate and sodium tartrate in 0.2 N NaOH), 24 parts of MB (4% BCA in water) and 1 part of MC (4% cupric sulfate, pentahydrate in water). As standards, nine bovine serum albumin (BSA) solutions with concentrations ranging from 0.0 to 200 μg/mL were prepared by dissolving BSA in the buffers used for laminin adsorption (acidic buffer and neutral buffer). The amount of adsorbed protein was calculated by measuring the amount of protein remaining in the supernatant at each time point. The samples and the standards were incubated with WR 1:1 at 37°C for two hours before cooling to room temperature. Then, the absorbance at 562 nm was measured with the plate reader Victor3 (PerkinElmer, Waltham, Massachusetts). All the absorbance values were corrected by the average 562 nm absorbance reading of the blank standard replicates. Each measurement was performed in duplicate.

Calculus of the fractal or the Hausdorff dimension

All image processing and analysis was done using an in-house software developed under MATLAB R2006a (The MathWorks, Inc., Natick, MA). The fractal dimension was determined using a box-counting dimension estimate of the Hausdorff dimension, which is a descriptor of the complexity of geometry of a given set [14]. As the definition of the Hausdorff dimension does not offer any guideline to an estimate calculation, the box-counting dimension has been used. The box-counting dimension is an estimate of the Hausdorff dimension based on covering the investigated set with a fixed grid of size r [15,16]. This box-counting dimension can be calculated using the equation (1):

$$D_H = \frac{-\Delta[\log k(r)]}{\Delta \log r} \quad (1)$$

where $k(r)$ is a number of grid boxes that contain any part of the investigated set, and where the process is repeated with different values of the grid size. D_H is known as the Hausdorff dimension, the Minkowski-Bouligand dimension, the Kolmogorov capacity or dimension, or simply the box-counting dimension.

Before applying the box-counting algorithm, the grayscale image histogram was equalized and the resulting image was binarized using the Otsu's method [17]. Finally, the box-counting estimate was calculated on this image. All this process was performed on ten different image crops of the same size (100 pixels×100 pixels) of the original images, and all these image crops were rescaled ten times to obtain an image big enough to apply the box-counting algorithm.

Results

Three-dimensional structures of polyLM and LM assessed by confocal fluorescence and scanning electron microscopy

Matrices of adsorbed laminin obtained at acidic (polyLM) or neutral pH (LM) were initially analyzed by reconstructing a series of 74 confocal optical slices renderized to reveal their 3-D structure. PolyLM corresponded to a sponge-like network of an apparently homogeneous density (Fig. 1A). Labeling was equally distributed, indicating that the antibody could evenly access protein epitopes within the polymer. On the other hand, LM presented a branched morphology, resembling that of a marine coral, which protruded from the glass surface (Fig. 1B). The antibody did not penetrate the spherical protein clumps and only their contours were brightly stained. Noteworthy was the observation that in LM a significant amount of protein adsorbed directly to the glass coverslip. This likely corresponds to the protein not incorporated to the aggregates. As a comparison, the background of the image depicting polyLM was dark, indicating that virtually all laminin protein engaged into that polymer. The 3-D structures of the two laminin matrices can be better appreciated in the animation movies presented as Movies S1 and S2.

The 3-D structures of polyLM and LM were additionally investigated by using SEM. While polyLM was again seen as a homogeneous mesh, LM showed at least two structural components (Fig. 1C and D). Besides the spherical aggregates already identified by CFM, it was possible to devise the presence of rod-like structures. In addition, the tips of these rods possessed lamellar terminations, suggesting the occurrence a third structural component of the LM matrix (arrows in Fig. 1D).

Kinetics of formation of laminin matrices

The amount of adsorbed laminin was calculated by measuring the concentration of protein remaining in solution at 20, 40 and 60 min and after 4, 8 and 12 hours of adsorption (Fig. 2A). Laminin decanted more quickly at neutral pH to form the LM matrix. More than 60% of the protein was absorbed after 20 min of incubation and such amount slightly increased up to 80% at 12 hours. On the other hand, at acidic pH the kinetic of adsorption was more linear, whereas 45% of the protein was adsorbed at 20 min and such proportion increased to 90% at 12 hours. The morphologies of the two matrices were analyzed at 1, 8 and 12 hours by SEM (Fig. 2B–G). Already at 1 hour polyLM decanted exhibiting a structured 2-D morphology consistent with the formation of the meshwork observed at 12 hours (Fig. 2B, D).

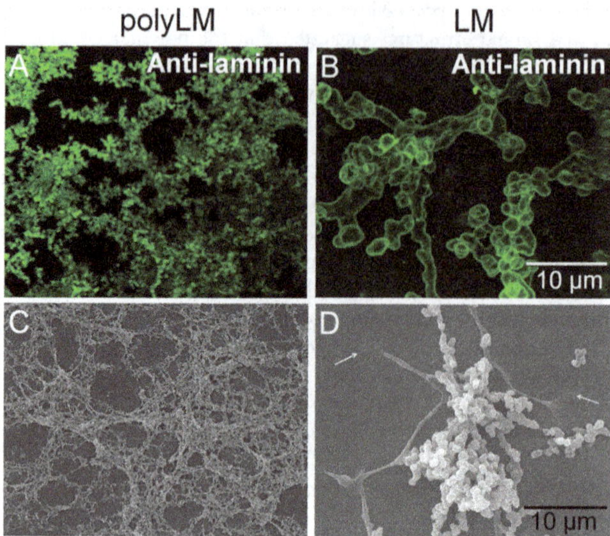

Figure 1. Three-dimensional structure of laminin polymers under confocal fluorescence microscopy and scanning electron microscopy. Laminin was incubated on glass coverslip for 12 hours in acidic (polyLM) or in neutral buffer (LM). (A, B) Indirect immunofluorescence was performed using a polyclonal antibody against laminin. The images depict z-stacks obtained by the superposition of 74 confocal slices renderized using the software 7.2.3 (Bit-plane; free trial). (C, D). Scanning electron micrographies (SEM) of the polymers shown at a similar magnification. Arrows in D point to lamellar deposits of laminin. The scale bars apply to panels A–D and represent 10 μm.

Characterization of the seed units of polyLM and LM

When the seed unit of polyLM was observed at higher magnification one could see that it was itself composed of polygons in a planar organization. Such seed unit was consistently present in matrices obtained within one hour of incubation (Fig. 3A–C). The extent of their longest axes ranged between 14 and 28 μm (Fig. 3D). After 8 and 12 hours of incubation we could already observe the presence of a mesh-like network whose morphology was compatible with the overlay of the seed units observed at one hour; these units however could no longer be distinguished within the meshwork (Fig. 3E, F). In LM we could observe lamellar-like deposits adsorbed directly to the coverslips (pseudo colored green in Fig. 3). These lamellar deposits were either located at the end of rod-like structures (yellow) or they appeared as individual patches (arrowheads in Fig. 3I). The seed unit was pseudo colored as to reveal the three types of deposits seen in LM, namely spheres (pink), rods (yellow) and lamellar deposits (green) (Fig. 3G–I).

Three-dimensional structures of polyLM and LM assessed by atomic force microscopy

We next examined the 3-D features of polyLM and LM using AFM. When areas of 50×50 μm were scanned, the overall appearances of polyLM and LM were comparable to those visualized by CFM and SEM (Fig. 1). PolyLM displayed the morphology of a multilayered meshwork containing homogeneous struts (Fig. 4A), while LM exhibited rods and lamellar deposits (Fig. 4B). The spherical aggregates previously seen under confocal fluorescence (Fig 1B) and SEM (Fig 1D and 3G–I) were not

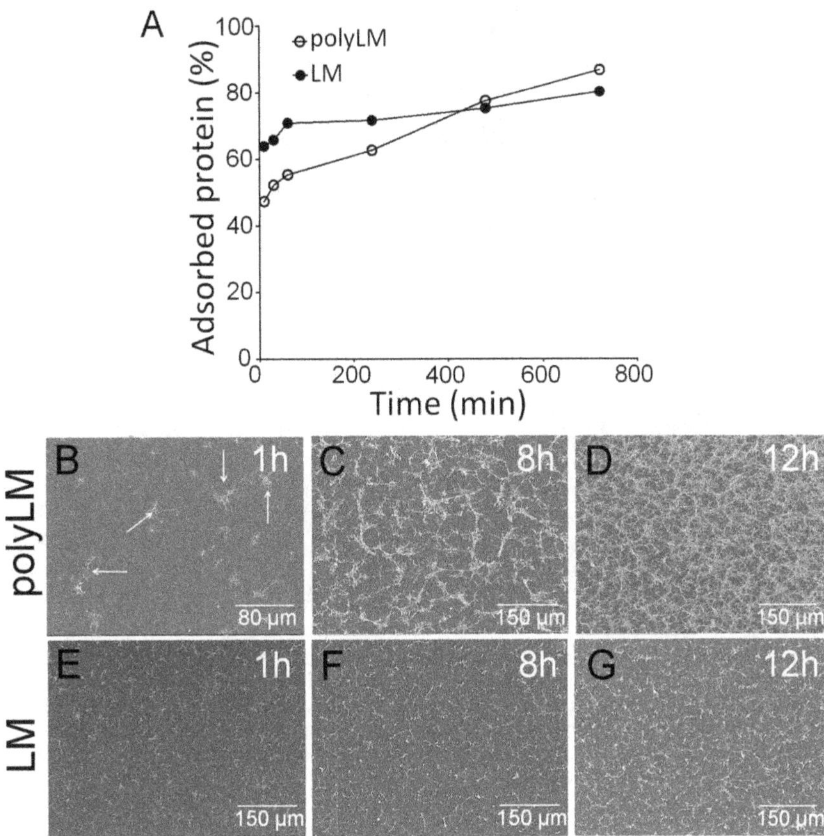

Figure 2. Kinetics of adsorption of polyLM and LM. (A) Laminin was incubated in acidic (polyLM) or neutral buffer (LM) and a kinetic of adsorption was carried out by collecting aliquots of the supernatant at 10 minutes, 30 minutes, 1 hour, 4 hours, 8 hours and 12 hours for quantification of the protein content remaining in solution. Open symbols represent polyLM and closed symbols represent the LM. (B–G) SEM images show the polymers obtained in acidic (B–D) or neutral (E–G) buffers at the indicated times. The arrows (B) point to structured polymers observed at 1 hour of incubation.

observed under AFM due to their large size, which was beyond the Z scan range of the AFM.

To further characterize the homogeneity of the struts in polyLM, the matrix was scanned at higher magnifications (Fig. 5A, B). In fields of 0.5×0.5 µm, we measured the heights of 10 individual struts of the mesh, chosen as those in direct contact with the glass support, i.e., those in the bottom layer of the mesh. We found values ranging between 50 and 73 nm, with a mean height of 60.25±1.764 nm (Fig. 5C, D). In the heterogeneous LM matrix, we measured the heights of rods and lamellar deposits, which displayed average heights of 1213±134.6 nm (Fig. 5E–H) and 125.1±13.51 nm (Fig.5I–L), respectively.

Evidence of a fractal nature for polyLM

When polyLM was observed at a higher magnification with AFM it was possible to observe the occurrence of figures that matched the hexagon-like shape of polymerized laminin at the molecular level (Fig. 6A–C). Such figures were still one order of magnitude larger than the basic hexagons formed by the association of the short arms of individual laminin molecules (Fig. 6G, H) [4,12]. In that structure, each side of the hexagon possesses ~30 nm, resulting from the interaction of the laminin short arms (35–50 nm long). The sides of the putative hexagons observed here were larger and their length was in the range of a few hundreds of nanometers. Nevertheless, these polygons were made out of small globules with a diameter and a spacing around 30–40 nm, compatible with the characteristic length of laminin polymerized through interaction between short arms (Fig. 6D–F). These structures could not be further resolved via AFM.

The results obtained up this point suggested that the hexagonal network formed by the association of individual laminin molecules could reproduce itself at higher levels of organization. In order to investigate this hypothesis we compared images of polyLM obtained with SEM and with transmission electron microscopy after negative staining (Fig. 7). Surprisingly, the morphologies of polyLM were very similar under SEM (Fig. 7A) and transmission electron microscopy (Fig. 7B) regardless of a difference in magnification of 1,000 fold. This observation suggests that polyLM presents a fractal nature.

Determination of the fractal dimension of polyLM

Based on evidence that polyLM possessed a fractal structure, we analyzed images of polyLM obtained at increasing incubation times in search for its fractal dimension. Fractal or Hausdorff dimensions of increasing complexities were found for polyLM as the adsorption time increased from 1 to 12 h (Fig. 8). The calculated values were 1.55, 1.62 and 1.70 after 1, 8 and 12 hours of adsorption, respectively. By contrast, the LM matrix did not present a fractal structure from which a fractal dimension could be obtained.

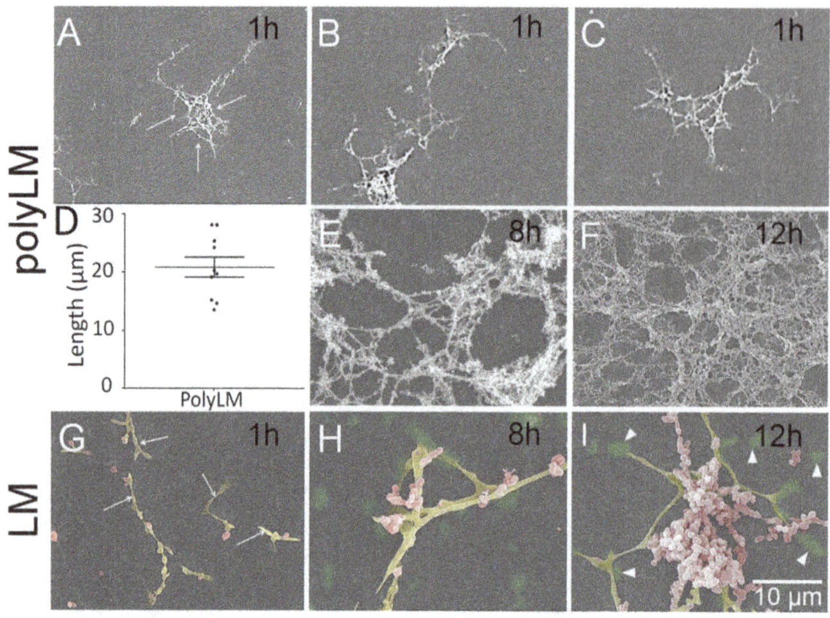

Figure 3. Characterization of polymer units in polyLM and LM. Laminin polymers were analyzed at high magnification in order to characterize the morphologies of the seed units of each polymer. (A–C) At 1 hour polyLM forms star-like 2D structures as exemplified in the three panels. (D) The sizes of the longer axes in these structures were quantified and shown to average at 20.84±5.449 µm. (E, F) High magnification images of polyLM at 8 and 12 hours show a meshwork pattern compatible with the deposition of the star-like structures. (G–I) LM observed at high magnification reveals three types of seed structures: rods (pseudocolored yellow), spheres (pseudocolored pink) and lamellas (pesudocolored green). The scale bar in I applies to all panels and represents 10 µm.

Discussion

In the present work we described that two matrices of laminin, polyLM and LM, obtained in different conditions presented highly different structures when observed at a wide range of magnifications. In a previous study [11], we had already shown that one of these polymers, polyLM, displayed the same nanostructure reported for laminin arrays secreted by cells [12]. Moreover, we had described that polyLM and LM presented different morphologies in the range of tens of micrometers [8]. Nevertheless intermediary magnifications between these two ranges of sizes had never been assessed before. Using confocal, electronic and atomic force microscopy we filled in this gap and found that surprisingly polyLM presented similar structures independently of the magnification used to observe it. Since this property is a feature of fractals, we searched for a possible fractal dimension and confirmed that polyLM indeed corresponded to a fractal structure. Conversely, the second polymer, LM, was more heterogeneous and did not present a fractal nature.

Before addressing the biological significance of the present findings it is important to recapitulate the features of the polymers studied here. The term ordinary laminin (LM) is used to refer to

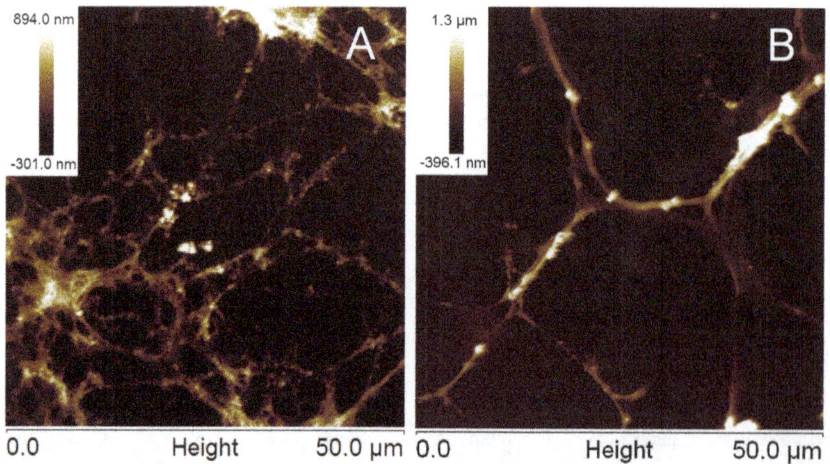

Figure 4. Overall morphology of polyLM and LM under AFM. Atomic force microscopy images of polyLM (A) and LM (B) are shown in height mode after critical-point drying of the samples. Both matrices were obtained by incubating laminin with glass coverslips in the appropriate buffers for 12 hours. The scanned area was 2500 µm2.

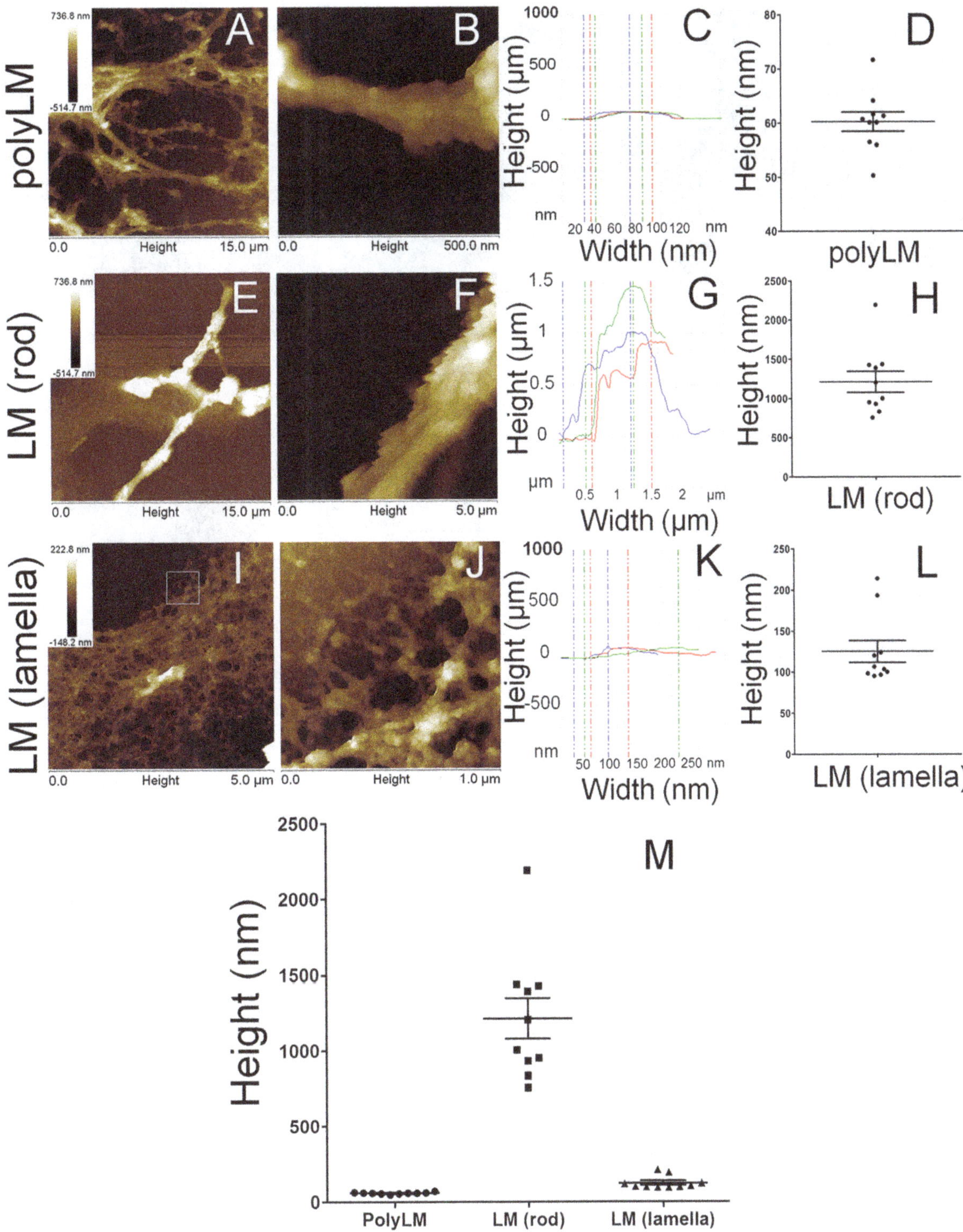

Figure 5. AFM analysis of polyLM and LM at increasing magnifications. PolyLM (A, B) and LM (E, F, I, J) obtained as described in Figure 4 were scanned in areas of 225 µm2 (A, E, I) or 0.25 µm2 (B, F, J) and shown in height mode. In order to determine the thickness of the structural units forming each polymer, the heights of 10 struts were calculated in the fields depicted in B (struts of the polyLM mesh), F (rods in LM) and J (lamellas in LM). Considering that both matrices were multilayered, each structure selected for measurement followed the criteria of being the closest possible to the support (glass coverslip). Panels C, G and K depict examples of three measurements and panels D, H and L show the distribution of the values obtained for each 10 structures. The white square in I represents an area at the edge of the lamellar structure used for the height measurement. Panel M shows the distribution of heights obtained at each condition all together for comparison.

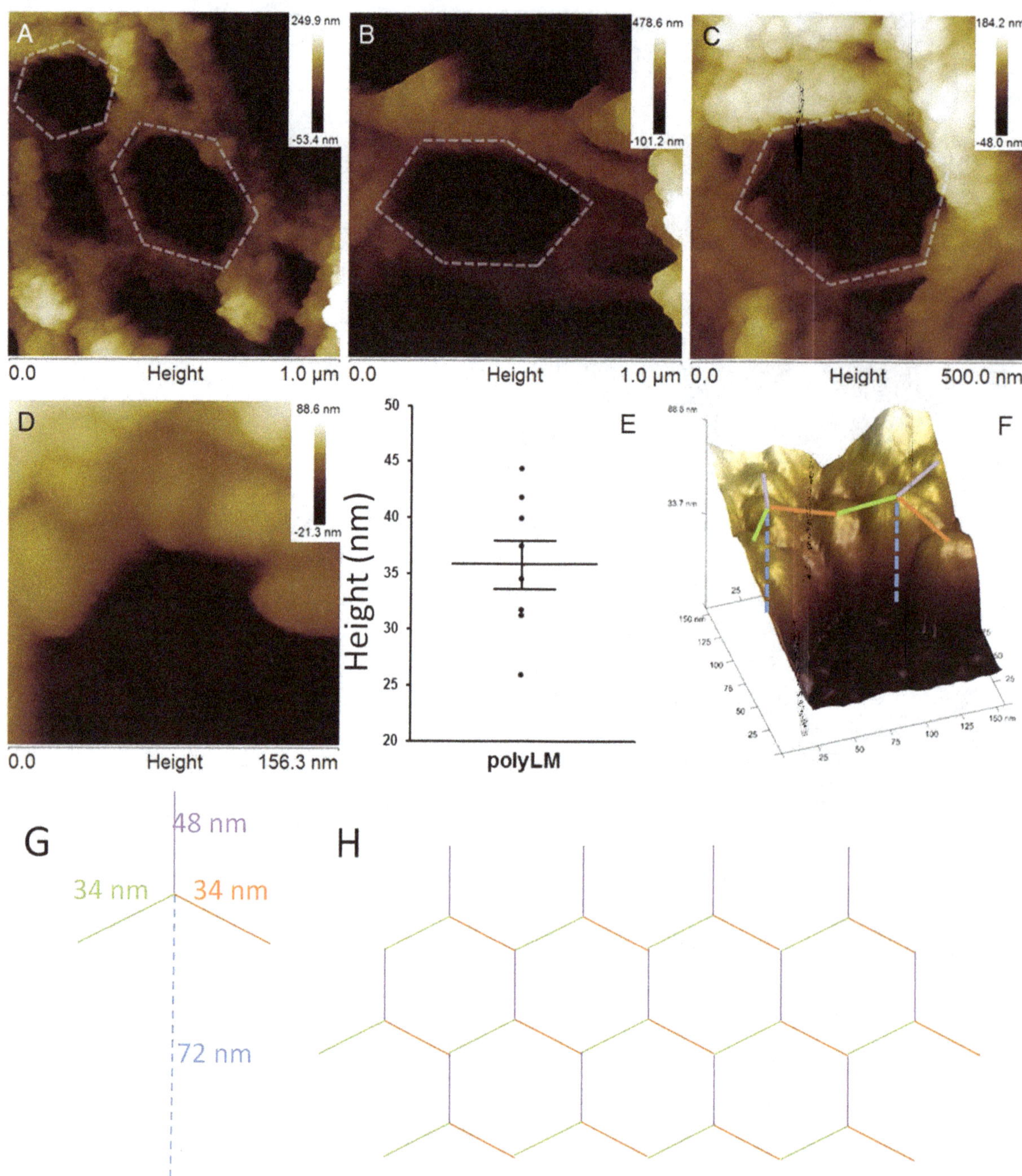

Figure 6. Atomic force microscopy reveals the occurrence of hexagonal-like figures in polyLM. AFM was performed on polyLM matrices obtained as described in Figure 4 and areas of 1 (A, B) or 0.25 μm^2 (C) were scanned in height mode. Hexagons-like figures similar to those occurring in natural laminin polymers [12] were identified. These hexagons were visible at different magnifications (A–C) and presented variable side lengths (sketched with white dashed lines), but they were never as short as 30 nm as they should be to correspond to the short arm of a laminin molecule. The smallest distinguishable structures contained within the sides of the hexagons were little globules (D) whose size and spacing was measured in images of 0.02 μm^2 (D). Panel E shows the distribution of spacing values, which are compatible with the characteristic length (~30 nm) of laminin polymerized via the short arms. Panel F depicts a three-dimensional reconstruction of the same area shown in panel D, with superposition of compatible locations of laminin molecules. Schemes of one individual laminin molecule (long arm dashed and short arms colored blue, green and orange), with indication of its characteristic dimensions (G) and of the hexagonal polymer generated by the interaction between individual laminin molecules (H) are also shown.

clusters of laminin adsorbed onto a glass surface at a concentration below the critical concentration of 60–100 μg/ml, previously shown to induce solution polymerization at pH 7 [5]. In this condition the protein does not self-assemble in solution but it tends to form clusters as its concentration increases at the glass surface upon decantation/adsorption. Since laminin is used as a coating substrate for cell attachment at concentrations below the critical concentration (typically between 1 and 20 μg/ml), ordinary

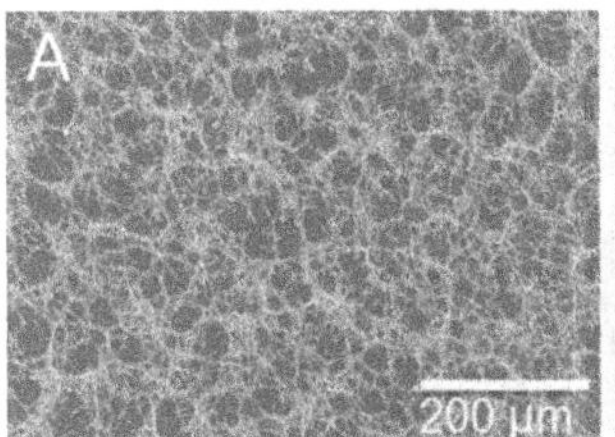

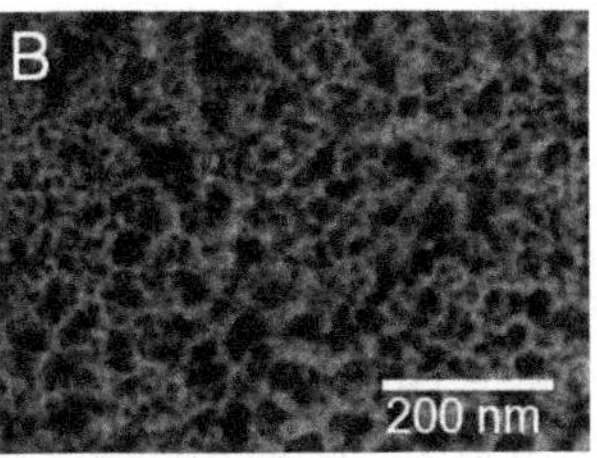

Figure 7. PolyLM displays similar morphologies at both 200 and 200,000 fold magnifications. Images of polyLM were obtained using SEM (A) or transmission electron microscopy (B) after negative staining. Under SEM the magnification was 200 fold while under TEM it was 200,000 fold. Note that the observed patterns were alike despite the 1000 fold increase in magnification.

laminin can be considered as the standard form of the protein referred to in the literature. On the other hand, polylaminin (polyLM) is an artificial polymer generated upon pH acidification. It is formed independently of the protein concentration and it is not disrupted after increasing the pH to 7 [8]. It was initially described as a high molecular weight entity observed in solution by monitoring a spectroscopic parameter, namely static light scattering [6]. At a given medium and at a fixed wavelength, the intensity of light scattering is related to the size of the particles in solution, which allows for the use of this technique to follow the aggregation state of proteins in the presence or not of ligands or other interacting particles of biological interest [18]. Therefore, the 40-fold increase in light-scattering intensity observed upon transferring laminin from acidic to neutral buffer reflected an increase in the volume of protein particles in solution [6]. These particles were subsequently called "polymers", instead of "aggregates", which would suggest that they corresponded to clusters of denaturated protein. The term "polymer" was employed due to evidence that 1) the tertiary structure of laminin was preserved within the clustered particles [6]; 2) decanted/adsorbed particles formed biomimetic matrices both at the nano [11] and at the micrometer scales of size [8], and 3) key signaling properties of laminin were preserved and even augmented after the acid-induced assemblage, which was demonstrated mainly for neurons [8], but also for other cells types as glial [9] and thyroid cells [19].

The concept that the polymeric structure of proteins can influence their biological properties has gained increasing confirmation in recent years. In particular, laminin, a protein that occurs in the polymeric form in natural basement membranes, is known to have their signaling properties dependent on the establishment of the polymeric array [3,4]. The interaction of laminin with other components of basement membranes, such as nidogens and/or the proteoglycan perlecan, is also influenced by polymerization, which is postulated to create new interacting sites at the nanoscale that did not exist in the individual molecule [20]. These studies however considered polymerization as having only two states (non-polymerized and polymerized). In other words, the polymeric state corresponded to a single entity, a supramolecular array in which the laminin trimers interacted with each other to form a sheet-like polymer anchored to the plasma membrane through cellular receptors. Such sheet-like polymer corresponds to the internal layer of basement membranes. However, there is evidence that certain tissues can produce other types of laminin structures. For instance, skeletal muscle fibers *in vitro* display membrane-bound deposits of laminin of two distinct morphologies [21], a reticular and a fibrillar one. These two morphologies were assigned to result from interactions with different cellular receptors, which, by presenting distinct regional distributions on the membrane, would lead to the formation of each laminin deposit. One complementary explanation, however, is that the morphologies of the two deposits at the micrometer scale would reflect specific molecular interactions at the nanometer range and therefore could involve interaction with the same receptor. In the nervous system, deposits of laminin appear with four different shapes throughout the development of the brain [22,23]. In consonance with these *in vivo* findings, neurons isolated from lateral and medial regions of the midbrain have been shown to secrete either punctate or fibrillar laminin matrices, respectively [24]. Neurons from embryonic and neonatal brain cortex have also been shown to produce distinct laminin matrices, whereas one remained associated to the cell membrane, the other extended away from the cell surface, exhibiting the appearance of an array of tangled threads [13]. Finally, laminin deposits found in the neural stem cell niche, present two different morphologies described as puncta alternated with thin linear membranous structures anchored to blood vessels [25,26]. We propose that the two laminin matrices observed in the present work reflect two alternative manners of protein assembly.

One important question to address is how this fractal would be generated and what would determine the self-assembly of laminin units into polyLM, LM or any other polymer. In a biological setting, the interaction with integrins and other laminin receptors would guide the process. As most of them bind to the long arm of laminin, it can be predicted that the simultaneous interaction of laminin trimers with membrane receptors will influence the formation of the type of polymerization. The ideal distance between two neighboring receptors to selectively favor interactions only among the short arms of laminin should correspond to the distance between the centers of the laminin molecules in the hexagonal network, *i.e.*, approximately 52 nm (Fig. 6). If the distance is such or shorter a flat polymer should be favored. On the other hand, as the spacing between receptors increased, other arrangements would be allowed. Interestingly, it has been shown that in fibroblasts and mesenchymal cells the ideal distance for signaling through integrins was in the range of 50–70 nm [27,28].

In the case of a polymer generated in a cell-free system as polyLM the lack of cellular receptors demands an alternative explanation. By analyzing the distribution of surface charges, we have previously shown that the pH acidification necessary to trigger the formation of polyLM rends the distal portion of the long arm (fragments LG4 and LG5) completely positive [11]. We proposed that this would be the determinant for the prevention of the interaction between two long arms, which is the predominant laminin-laminin interaction in the absence of other molecules [5]. Therefore, polyLM could give rise to a hexagonal polymer mimicking the sheet-like polymer assembled on the cell surface even before adsorption to the coverslip. The previous report that polyLM increased the light scattering of the solution containing it, while LM in similar conditions did not, supports this notion [6]. The seed polymer already decants as a fractal structure within one hour of incubation and the complexity of the fractal increases as deposition proceeds (Fig. 8D). The height of the lowest structures found in polyLM and LM were around 60 nm and 120 nm, respectively (Fig. 5D, L). Interestingly, the former is the approximate size of the extended long arm of laminin (Fig. 6G). Such an extended conformation of individual laminin molecules has been previously detected by using electron microscopy after rotatory shadowing [5] and AFM [29]. Thus, it is likely that polyLM sediments as flat 2-D polymers in which only interactions among the short arms occur, as also suggested by the AFM measurements (Fig. 6D–F). On the other hand, in LM at least two layers of flat polymers mediated by interactions between long arms would be

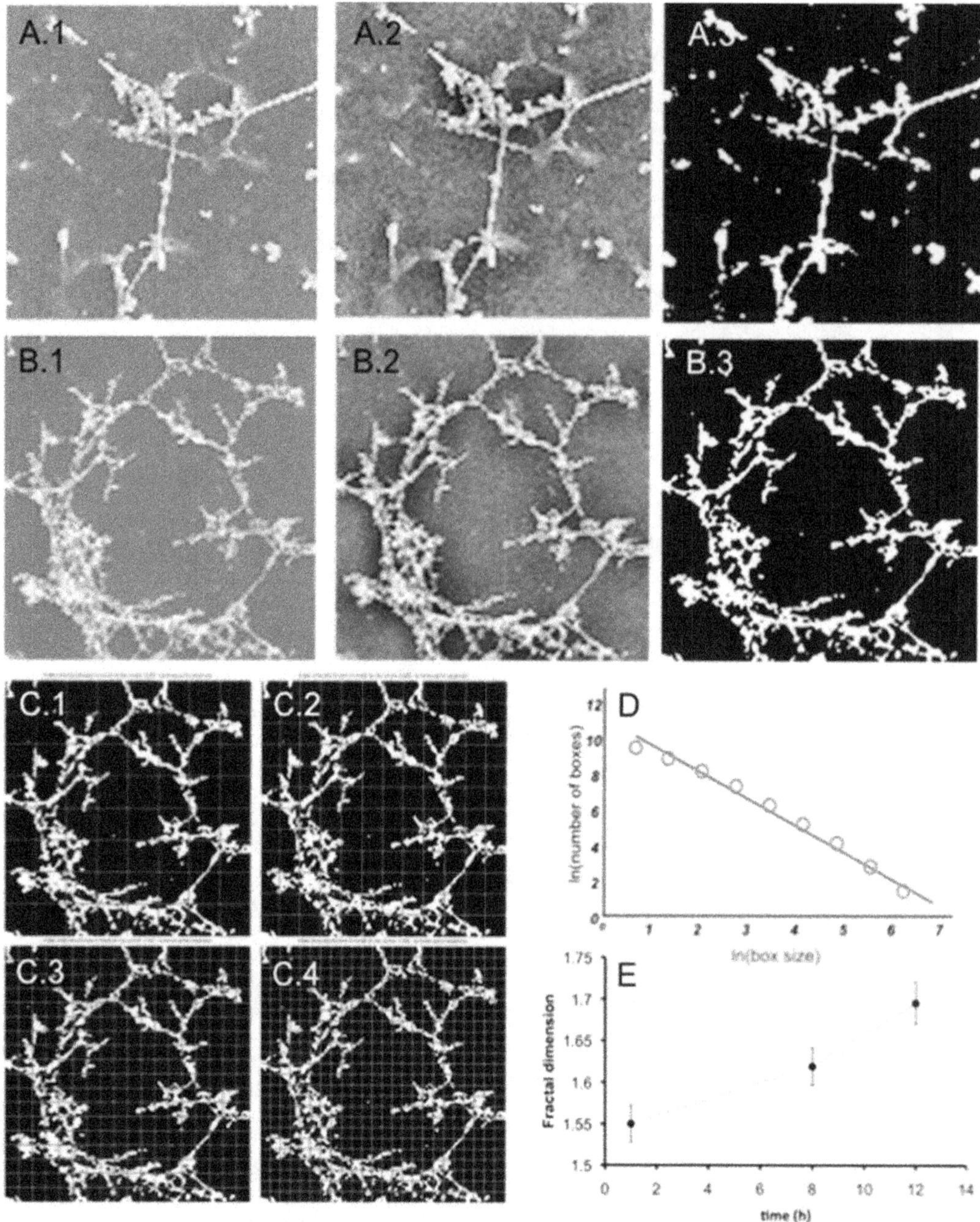

Figure 8. Calculation of the fractal dimension (Hausdorff estimate). (A, B) Image processing in order to prepare the image for the box-counting algorithm for LM (A) and polyLM (B). 1, original image; 2, original image in which the histogram has been equalized; 3, binarized image using Otsu's method). (C) From images A.3 and B.3 the Hausdorff dimension estimates can be calculated superimposing a grid of variable size (C.1-C.4, examples of the same image on which a grid of variable size has been superimposed). (D) Repeating the previous process for different values of grid size and computing the number of grid boxes that contain any part of the investigated set, the Hausdorff dimension or simply the box-counting dimension can be calculated. (E) Fractal dimension calculated for polyLM structures as a function of time.

necessary to account for the 120 nm height observed in the lowest deposits.

In the present work we demonstrate that the protein laminin can give rise to a fractal structure. The observation that the supramolecular organization of a pure protein (polyLM) is fractal implies that the information contained ultimately within its primary sequence is sufficient to determine the morphology of larger structures that will spatially organize tissue compartments. This is particularly interesting because it correlates with the "fractal-like" organization of the niche for stem cells in the subventricular zone of the adult brain. In this case, a laminin-rich basement membrane, named "fractone", has been proposed to orient the binding of proteoglycans, which, in turn, organize the distribution of the growth factors controlling the maintenance of the stem cell niche [26,30], as shown to be the case for bFGF [31].

Although fractones have been described only in the central nervous system, it is well possible that a similar fractal-like extracellular matrix is present in other stem cell niches. Adult stem cell niches manifest under certain restricted microenvironments that host tissue-specific stem cells and regulate their physiology. They exhibit complex cytoarchitectures composed of stem cells, progenitor cells, supportive cells and a laminin-rich basement membrane. The basement membrane regulates cell division and differentiation within the niche due to several of its properties [32]. First, its molecular components interact with integrins to regulate the cytoskeletal assembly. It also harbors and controls the availability of growth factors and cytokines. A basement membrane provides an orienting surface for asymmetric cell division. Finally, it guides the traffic of progenitor cells within the niche. It is known that the assembly of basement membranes is initiated and dictated by laminin secretion and polymerization at the cell surface [33]. In addition, laminin is able to interact with integrins, to bind heparan sulfate proteoglycans, which, in turn, present soluble factors. It is thus very likely that laminin is the key component in basement membranes to play a regulatory role controlling the physiology of the stem cell niche. The importance of laminin for the maintenance of the stem cell niches has been demonstrated in several cases such as in pancreatic islets [34,35], the germline niche [36], skeletal muscle [37] and the embryonic neocortical stem cell niche [38].

It is tempting to speculate that the fractal-like organization of stem cell niches can not only control the distribution of growth factors but also provide a physical constrain for progenitor cells during differentiation. The representation of a differentiation pathway as a tree, in which the stem corresponds to the stem cell and the branches, to progenitors is *per se* a fractal. In this scenario, the propagation of laminin polymerization throughout lager size scales could be a key step for the organization of multicellular organisms, which is in line with the observations that laminin-containing basement membranes are the first assembled extracellular matrix appearing during mammalian development and that laminin is present in virtually all metazoans [39].

Supporting Information

Movie S1 Animated view of the three-dimensional structure of polyLM. The animation was generated from a series of confocal optical slices (the same shown in Figure 1A), using the software Imaris, version 7.2.3.

Movie S2 Animated view of the three-dimensional structure of LM. The animation was generated from a series of confocal optical slices (the same shown in Figure 1B), using the software Imaris, version 7.2.3.

Acknowledgments

We thank Laina Cunha for the excellent technical assistance.

Author Contributions

Conceived and designed the experiments: CHM MSS TCS. Performed the experiments: CHM MC DM. Analyzed the data: CHM MC TCS MSS DM. Contributed reagents/materials/analysis tools: TCS MSS DM. Wrote the paper: TCS MSS DM.

References

1. Durbeej M (2010) Laminins. Cell Tissue Res 339: 259–268.
2. Miner JH, Yurchenco PD (2004) Laminin functions in tissue morphogenesis. Annu Rev Cell Dev Biol 20: 255–284.
3. Yurchenco PD (2011) Basement membranes: cell scaffoldings and signaling platforms. Cold Spring Harb Perspect Biol 3: a004911
4. Hohenester E, Yurchenco PD (2013) Laminins in basement membrane assembly. Cell Adhes Migr 7: 56–63.
5. Yurchenco PD, Tsilibary EC, Charonis AS, Furthmayr H (1985) Laminin polymerization in vitro. Evidence for a two-step assembly with domain specificity. J Biol Chem 260: 7636–7644.
6. Freire E, Coelho-Sampaio T (2000) Self-assembly of laminin induced by acidic pH. J Biol Chem 275: 817–822.
7. Freire E, Barroso MM, Klier RN, Coelho-Sampaio T (2012) Biocompatibility and structural stability of a laminin biopolymer. Macromol Biosci 12: 67–74.
8. Freire E, Gomes FC, Linden R, Neto VM, Coelho-Sampaio T (2002) Structure of laminin substrate modulates cellular signaling for neuritogenesis. J Cell Sci 115: 4867–4876.
9. Hochman-Mendez C, de Menezes JR, Sholl-Franco A, Coelho-Sampaio T (2014) Polylaminin recognition by retinal cells. J Neurosci Res 92: 24–34.
10. Menezes K, de Menezes JR, Nascimento MA, Santos RS, Coelho-Sampaio T (2010) Polylaminin, a polymeric form of laminin, promotes regeneration after spinal cord injury. FASEB J 24: 4513–4522.
11. Barroso MM, Freire E, Limaverde GS, Rocha GM, Batista EJ, et al. (2008) Artificial laminin polymers assembled in acidic pH mimic basement membrane organization. J Biol Chem 283: 11714–11720.
12. Yurchenco PD, Cheng YS, Colognato H (1992) Laminin forms an independent network in basement membranes. J Cell Biol 117: 1119–1133.
13. Freire E, Gomes FC, Jotha-Mattos T, Neto VM, Silva Filho FC, et al. (2004) Sialic acid residues on astrocytes regulate neuritogenesis by controlling the assembly of laminin matrices. J Cell Sci 117: 4067–4076.
14. Hausdorff F (1919) Dimension und äuβeres Maβ. Mathematische Annalen 79: 157–159.
15. Soille P, Rivest JF (1996) On the validity of fractal dimension measurements in image analysis. J Vis Commun Image Repres 7: 217–229.
16. Theiler J (1990) Estimating fractal dimension. J Opt Soc Am A 7: 1055–1073.
17. Otsu N (1979) A threshold selection method from gray-level histograms. IEEE Trans Syst Man Cybern 9: 62–66.
18. Hediyeh I, Rajabi O, Salari R, Chamani J (2012) Probing the interaction of human serum albimuni with ciprofloxacin in the presence of silver nanoparticles of three sizes: multispectroscopic and ζ potential investigation. J Phis Chem B 116: 1951–1964.
19. Palmero CY, Miranda-Alves L, Sant'Ana Barroso MM, Souza EC, Machado DE, et al. (2013) The follicular thyroid cell line PCCL3 responds differently to laminin and to polylaminin, a polymer of laminin assembled in acidic pH. Mol Cell Endocrinol 376: 12–22.
20. Behrens DT, Villone D, Koch M, Brunner G, Sorokin L, et al. (2012) The epidermal basement membrane is a composite of separate laminin- or collagen IV-containing networks connected by aggregated perlecan, but not by nidogens. J Biol Chem 287: 18700–18709.
21. Colognato H, Winkelmann DA, Yurchenco PD (1999) Laminin polymerization induces a receptor-cytoskeleton network. J Cell Biol 145: 619–631.
22. Liesi P, Silver J (1988) Is astrocyte laminin involved in axon guidance in the mammalian CNS? Dev Biol 130: 774–785.
23. Zhou FC (1990) Four patterns of laminin-immunoreactive structure in developing rat brain. Brain Res Dev Brain Res 55: 191–201.
24. Garcia-Abreu J, Cavalcante LA, Moura Neto V (1995) Differential patterns of laminin expression in lateral and medial midbrain glia. Neuroreport 6: 761–764.
25. Kazanis I, Ffrench-Constant C (2011) Extracellular matrix and the neural stem cell niche. Dev Neurobiol 71: 1006–1017.
26. Mercier F, Schnack J, Chaumet MSG (2011) Chapter 4 Fractones: home and conductors of the neural stem cell niche. In: Seki, T., Sawamoto, K., Parent, J. M., Alvarez-Buylla, A., (Eds.) Neurogenesis in the adult brain I: neurobiology. Springer. pp 109–133.
27. Cavalcanti-Adam EA, Micoulet A, Blümmel J, Auernheimer J, Kessler H, et al. (2006) Lateral spacing of integrin ligands influences cell spreading and focal adhesion assembly. Eur J Cell Biol 85: 219–224.
28. Frith JE, Mills RJ, Cooper-White JJ (2012) Lateral spacing of adhesion peptides influences human mesenchymal stem cell behavior. J Cell Sci 125: 317–27.
29. Rodríguez Hernández JC, Salmerón-Sánchez M, Soria JM, Gómez Ribelles JL, Monleón Pradas M (2007) Substrate chemistry-dependent conformations of single laminin molecules on polymer surfaces are revealed by the phase signal of atomic force microscopy. Biophys J 93: 202–207.
30. Chyba M, Mercier F, Rader J, Douet V, Arikawa-Hirasawa E, et al. (2011) Dynamic mathematical modeling of cell-fractone interactions. Journal of Math for Industry 3: 79–88.
31. Douet V, Kerever A, Arikawa-Hirasawa E, Mercier F (2013) Fractone-heparan sulphates mediate FGF-2 stimulation of cell proliferation in the adult subventricular zone. Cell Prolif 46: 137–145.
32. Nikolova G, Strilic B, Lammert E (2007) The vascular niche and its basement membrane. Trends Cell Biol 17: 19–25.

33. Yurchenco PD, Amenta PS, Patton BL (2004) Basement membrane assembly, stability and activities observed through a developmental lens. Matrix Biol 22: 521–538.
34. Nikolova G, Jabs N, Konstantinova I, Domogatskaya A, Tryggvason K, et al. (2006) The vascular basement membrane: a niche for insulin gene expression and beta cell proliferation. Dev Cell 10: 397–405.
35. Qu H, Liu X, Ni Y, Jiang Y, Feng X, et al. (2014) Laminin 411 acts as a potent inducer of umbilical cord mesenchymal stem cell differentiation into insulin-producing cells. J Transl Med 12: 135.
36. Kanatsu-Shinohara M, Shinohara T (2013) Spermatogonial stem cell self-renewal and development. Annu Rev Cell Dev Biol 29: 163–187.
37. Lander AD, Kimble J, Clevers H, Fuchs E, Montarras D, et al. (2012) What does the concept of the stem cell niche really mean today? BMC Biol 10: 19.
38. Loulier K, Lathia JD, Marthiens V, Relucio J, Mughal MR, et al. (2009) Beta1 integrin maintains integrity of the embryonic neocortical stem cell niche. PLoS Biol 7: e1000176.
39. Miner J, Yurchenco PD (2004) Laminin functions in tissue morphogenesis. Annu Rev Cell Dev Biol 20: 255–284.

8

Mycobacterium tuberculosis Rho Is an NTPase with Distinct Kinetic Properties and a Novel RNA-Binding Subdomain

Anirban Mitra[1], Rachel Misquitta[1], Valakunja Nagaraja[1,2]*

1 Department of Microbiology and Cell Biology, Indian Institute of Science, Bangalore, India, 2 Jawaharlal Nehru Centre for Advanced Scientific Research, Bangalore, India

Abstract

Two mechanisms – factor independent and dependent termination – ensure the completion of RNA synthesis in eubacteria. Factor-dependent mechanism relies on the Rho protein to terminate transcription by interacting with RNA polymerase. Although well studied in *Escherichia coli*, the properties of the Rho homologs from most bacteria are not known. The *rho* gene is unusually large in genus *Mycobacterium* and other members of actinobacteria, having ~150 additional residues towards the amino terminal end. We describe the distinct properties of Rho from *Mycobacterium tuberculosis*. It is an NTPase with a preference for purine nucleoside triphosphates with kinetic properties different from *E. coli* homolog and an ability to use various RNA substrates. The N-terminal subdomain of MtbRho can bind to RNA by itself, and appears to contribute to the interaction of the termination factor with RNAs. Furthermore, the interaction with RNA induces changes in conformation and oligomerization of MtbRho.

Editor: Dipankar Chatterji, Indian Institute of Science, India

Funding: Support from J.C. Bose Fellowship from Department of Science and Technology, Government of India to VN. The funders had no role in study design, data collection and analysis, decision to publish, or preparation of the manuscript.

Competing Interests: The authors have declared that no competing interests exist.

* Email: vraj@mcbl.iisc.ernet.in

Introduction

Termination of transcription of bacterial RNA polymerase (RNAP) is achieved either by intrinsic terminators or protein factor Rho. For factor-mediated termination, Rho binds to the nascent transcript emerging from the ternary elongation complex, translocates along the RNA by ATP-powered steps and finally enforces dissociation of the complex [1,2,3,4]. The RNA-binding and ATPase properties of the prototype Rho factor from *Escherichia coli* have been studied extensively [5,6,7]. Briefly, *E. coli* Rho (EcRho) is functionally a homohexameric molecule [8] [9] that preferentially binds to an unstructured, C-rich RNA. This interaction induces transition from an 'open' ring to 'close' ring state [10]. The closed ring is proficient in ATP hydrolysis and translocation along RNA. Once it catches up with the transcribing or paused RNAP, the interaction triggers termination, dissociation of RNAP from the template and release of the transcript [11,12,13,14]. Several studies on EcRho have unraveled the biochemical and structural basis for its preference for C-rich RNA [15,16,17,18,19,20]. However, in spite of its key cellular role and its presence in a large number of diverse bacterial families [21], very few Rho homologs have been studied. Characterization of the properties and understanding Rho-mediated termination is of paramount importance in organisms such as *Mycobacterium tuberculosis* (*Mtb*) [22] which is the causative agent of the number-one killer disease worldwide.

In addition to *Mtb*, the genus *Mycobacterium* includes some of the well-known Actinobacteria, such as the well-studied model organism *Mycobacterium smegmatis*, *Mycobacterium abscessus*, *Mycobacterium leprae*, and a large number of *Streptomyces* species. In recent years, studies on transcription initiation, elongation and termination have been carried out in *Mtb* and other members. These studies, though not as exhaustive as in *E. coli*, have revealed considerable differences from the *E. coli* paradigm. For example, the *Mtb* and other mycobacterial genomes code for a larger number of sigma factors [23] as compared to *E.coli* and also for the several transcription factors unique to mycobacteria [24,25]. Several promoters and the mechanism of gene expression regulation have been studied [26,27,28,29,30]. Absence of AT-rich UP elements and GC-rich sequences in discriminator sequences in the promoters contribute to the differences in promoter-polymerase interaction and its regulation [31,32,33,34,35]. Additionally, attempts have been made to elucidate features of RNAP from *Mycobacterium* species [36,37,38] and the transcription elongation rates also appear to vary between different RNAPs [39]. Furthermore, the scarcity of canonical intrinsic terminators and an abundance of non-canonical intrinsic terminators across mycobacteria also suggest differences in the transcription termination machinery [40,41].

Given the dissimilarities in various steps of transcription between mycobacteria and *E. coli*, it is likely that mycobacterial Rho homologs have also evolved to function differently and optimally for their specific cellular context. Notably, sequence analysis showed that the Rho homologs in *Mycobacterium* species and other actinobacteria are larger than EcRho mainly due to an 'extra-stretch' of ~150–200 residues in their RNA-binding domains [5,7,21,22,42,43]. In this manuscript, we present results demonstrating that purified *M. tuberculosis* Rho (MtbRho) can hydrolyse purine nucleoside triphosphates(NTP) – ATP and GTP- in presence of mycobacterial RNA. The extended N-terminal region of MtbRho, having a distinct RNA-binding 'subdomain', can itself interact with RNA and may contribute to the overall interaction. The MtbRho-RNA interactions are stable and the interactions induce changes in the conformation and oligomerization status of the protein.

Results

MtbRho can hydrolyse NTP in presence of mycobacterial RNA

The presence of Rho homologs in diverse bacterial lineages (Figure S1A) indicated the functional importance of factor-dependent termination in bacterial gene expression. Analysis of Rho sequences showed that while the C-terminal half is well-conserved, the N-terminal half is more variable. Notably, the Rho homologs from actinobacteria form a distinct branch and the N-terminal halves of actinobacterial rho proteins contain an 'extra-stretch/subdomain' of 150–200 residues. The sequence composition of this fragment includes a large number of Arg, Asp, Asn and Gly residues but very few hydrophobic and aromatic amino acids [5,7,22]. Although no function can be conclusively predicted from the sequence analysis, its role in interacting with RNA is postulated from the presence of a large number of basic amino acid residues [7,42] (Figure S1B). To understand the properties of the Rho protein from *M. tuberculosis*, we expressed recombinant MtbRho in BL21(DE3) cells and purified it to homogeneity. MtbRho has a monomeric size of 65 kDa (Figure S2), as estimated from sequence analysis and also shown by mass spectrometry. However, the protein showed anomalous migration at ~80 kDa on SDS-PAGE [22], probably due to the presence of clusters of polar residues in the subdomain.

In the several steps involved in Rho-mediated transcription termination, the first step of Rho's action is its binding to the rut (rho utilization) site [7]. EcRho is known to have a preference for C-rich, unstructured RNA for initial Rho binding [16] and polycytidylic acid (polyC) has been used to study Rho activity *in vitro* [15,19,44,45,46,47,48]. The residues of EcRho that have been implicated in binding to a C-rich sequence (5′-CC YC-3′) [18] the motifs involved in RNA-dependent ATPase activity are similar in MtbRho. The ATPase activity of MtbRho in presence of synthetic homopolymeric polyC, polyA and polyU is shown in Figure 1A. While polyC is, not unexpectedly, the best substrate, polyA and polyU also stimulate the hydrolysis of ATP(Figure 1A). Homopolymeric polyC, polyA and polyU are, however, not natural substrates of MtbRho. *In vivo*, MtbRho would function in presence of various mycobacterial RNAs and it is likely that MtbRho has evolved a greater ability to interact with its natural substrates. To assess if MtbRho could use mycobacterial RNA as substrate for ATPase, cellular RNA from *M. smegmatis* mc^{2}155 was used. The results presented in Figure 1B show that MtbRho can hydrolyse ATP in presence of mycobacterial RNAs. The ATPase activity was specific to MtbRho as it was inhibited by Bicyclomcycin. To study if a specific mycobacterial RNA molecule could be used as a substrate for ATPase activity, we used a RNA corresponding to the region downstream of the *sdaA* gene of *M. tuberculosis* genome. This RNA was chosen as *in silico* analysis revealed the absence of intrinsic terminator downstream of the

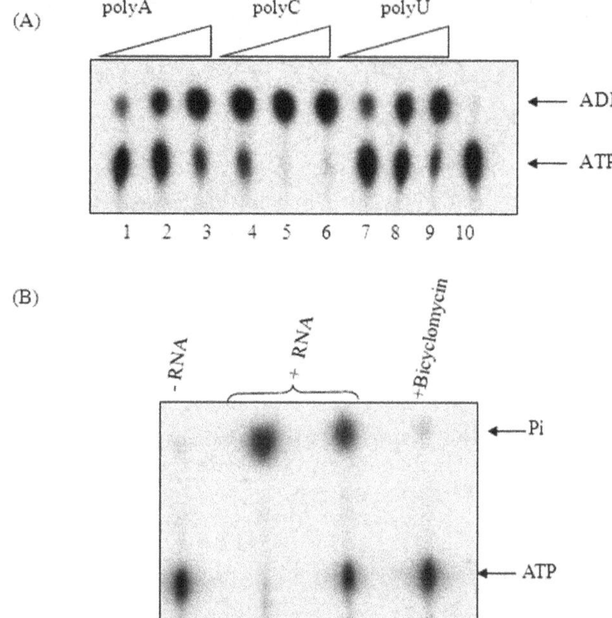

Figure 1. ATPase activity of MtbRho. (A) MtbRho hydrolyses ATP in presence of increasing concentrations of homopolymeric RNA – polyA (lanes1–3), polyC (lanes4–6) and polyU (lanes 7–9). No hydrolysis was observed in absence of RNA (lane 10). (B) MtbRho hydrolyzes ATP in presence of mycobacterial RNA. No hydrolysis was observed in absence of RNA (lane 1); 2 and 1 µg of *M. smegmatis* RNA stimulated ATP hydrolysis (lanes 2,3); the reaction is inhibited by Bicyclomycin (lane 4). ATPase assay was carried out as described in Methods. [1 mM unlabeled ATP was used as substrate, along with 100 nCi of α-^{32}P-ATP (Panel A) or 100 nCi of γ-^{32}P-ATP(Panel B), as tracer. Hydrolysis resulted in formation of α-^{32}P-ADP (Panel A) or ^{32}Pi (Panel B) which were visualized using phosphorimager].

sdaA gene [49] and hence it is likely target for Rho-dependent termination. The results presented in Figure 2A and B show that MtbRho can hydrolyse ATP in presence of *sdaA* RNA.

But, MtbRho is inherently a weaker ATPase [22]. The rate of ATP hydrolysis by MtbRho in presence of polyC was >10-fold less when compared to that of EcRho (Figure 2A). Also, as previous studies have shown, MtbRho exhibited a higher K_m (75.2 µM) for ATP than EcRho (10 µM) [22]. The slow rate of ATP hydrolysis indicates the intrinsically low activity of the enzyme. However, when mycobacterial cellular RNA was used, the ATPase rate of MtbRho became similar to previous observations where *E. coli* terminator RNAs had been used [22]. Remarkably, the rates of MtbRho and EcRho became comparable in presence of mycobacterial cellular RNA (Figure 2A). This indicated that MtbRho could be more proficient in using mycobacterial RNA as substrate than its *E. coli* homolog. The superior ability of MtbRho to utilize mycobacterial RNA was further evident when, in presence of *sdaA* RNA, a specific *Mtb* RNA, MtbRho hydrolysed ATP at a rate that is 2-fold higher than that reported in presence of the *E. coli* terminators [22], while EcRho was unable to use the *sdaA* RNA for ATPase activity (Figure 2A). Thus, although a weaker ATPase in presence of polyC, MtbRho seems to be more efficient in catalysis in presence of a specific RNA from *M. tuberculosis* that is likely to be its natural substrate (Figure 2A). Besides ATP, MtbRho hydrolysed GTP while RNA-dependent CTP and UTP hydrolysis was undetectable (Figure 2B) [22]. Thus, MtbRho can be considered

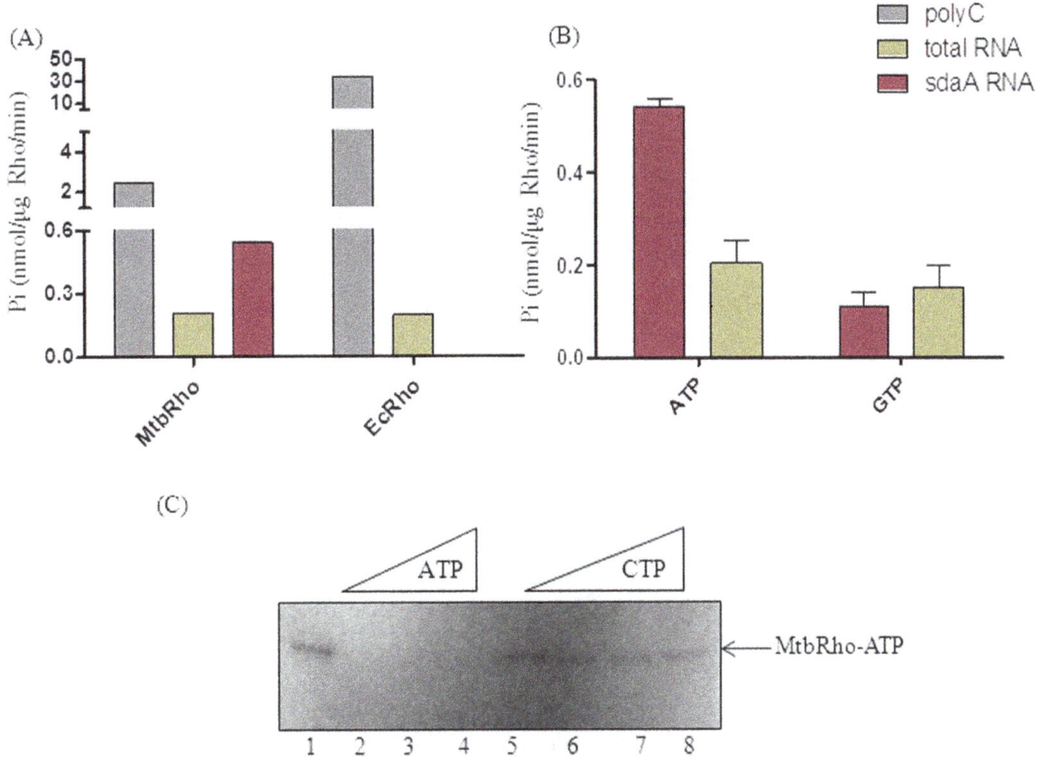

Figure 2. RNA preference by MtbRho. (A) Rates of ATPase activity of MtbRho and EcRho in presence of different RNAs. MtbRho is a weaker ATPase when polyC)is used(grey bars). But, in presence of mycobacterial RNA (brown bars), they have comparable rates of hydrolysis, while only MtbRho can hydrolyse ATP in presence of *sdaA* RNA(red bars). (B) Differential NTP hydrolysis by MtbRho. MtbRho can hydrolyze ATP and GTP in presence of the mycobacterial total RNA and *sdaA* RNA. Colorimetric assay was carried out as described in Methods (C) Differential ability of MtbRho to bind ATP and CTP. γ-^{32}P-ATP alone can be UV-crosslinked to MtbRho (lane 1), and visualized on 8% SDS-PAGE. Addition of unlabeled ATP to the reaction competes out the γ-^{32}P-ATP (lanes 2–4), but unlabeled CTP fails to do so (lanes 5–8). The reaction conditions were similar to ATPase assay, but without the addition of RNA.

an NTPase, with a substrate preference for ATP and GTP, when provided with its cognate RNAs. The ability to hydrolyze the various NTPs at different levels was mirrored by MtbRho's ability to bind the various NTPs with varying efficiency. When UV-radiation was used to crosslink γ-32P-ATP to MtbRho, in presence of unlabeled NTPs, ATP could compete out the crosslinking of γ-32P-ATP, but CTP could not (Figure 2C).

N-terminal subdomain binds to RNA

All actinobacterial Rho homologs carry a 150–200 amino acid long subdomain in the N-terminal region, located immediately upstream of the canonical RNA-binding motifs, which has been implicated in assisting to bind RNA [7]. By sequence comparison, the MtbRho's subdomain spans the residues 76–230. To understand its role in the functionality of MtbRho, the N-terminal 229 residues (N-229), which includes the additional sequences was expressed and purified. Notably, although the N-229 lacks the canonical RNA-binding motifs of Rho, *in silico* analysis predicted that several patches of amino acids in this fragment may interact with a RNA substrate (Figure 3A) [50]. To confirm the interaction of N-229 with mycobacterial RNA, binding was monitored with a defined mycobacterial RNA molecule. The substrate RNA was *in vitro* transcribed ^{32}P-labeled *sdaA* RNA. Predictably, full-length MtbRho bound to *sdaA* RNA (Figure 3C). Although N-229 lacks both the primary and secondary RNA-binding residues present in all Rho homologs, it could stably interact with the 351-nucleotide long *Mtb sdaA* RNA(Figure 3D). However, the binding pattern with 80-mer polydC-80 was significantly different. Although MtbRho would bind to poly-dC80 (Figure 3C), in contrast, N-229 showed very weak interaction with poly-dC80 (Figure 3D). The binding of the N-terminal fragment of MtbRho could be visualized only if the complex is stabilized by the addition of 20% glycerol to the gel matrix (Figure S3). Accordingly, in the MtbRho-dependent ATPase assays, the N-terminal fragment inhibited ATPase activity to a low extent (data not shown).

Conformational changes in MtbRho associated with RNA binding

The presence of the mycobacterial RNAs conferred protection to MtbRho when probed with V8 protease (Endoproteinase Glu-C) (Figure S4A). Both cellular RNA, which contains several large RNA species and smaller tRNA (average size ~100 nucleotides) protected MtbRho from cleavage by V8 protease. The RNA-induced protection is indicative of conformational changes in MtbRho upon RNA-binding rendering certain scissile sites inaccessible to the protease. The RNA-based protection also shows close interaction between the substrate RNA and MtbRho, which sterically hinders cleavage by V8 protease action. The addition of ATP along with RNA did not induce further differences in protease protection pattern and ATP, by itself, did not confer any protection. In order to investigate conformational changes induced in MtbRho by RNA, circular dichroism studies (CD) were carried out. There was a distinct alteration of secondary structures in presence of RNA (Figure S4B). In contrast, CD

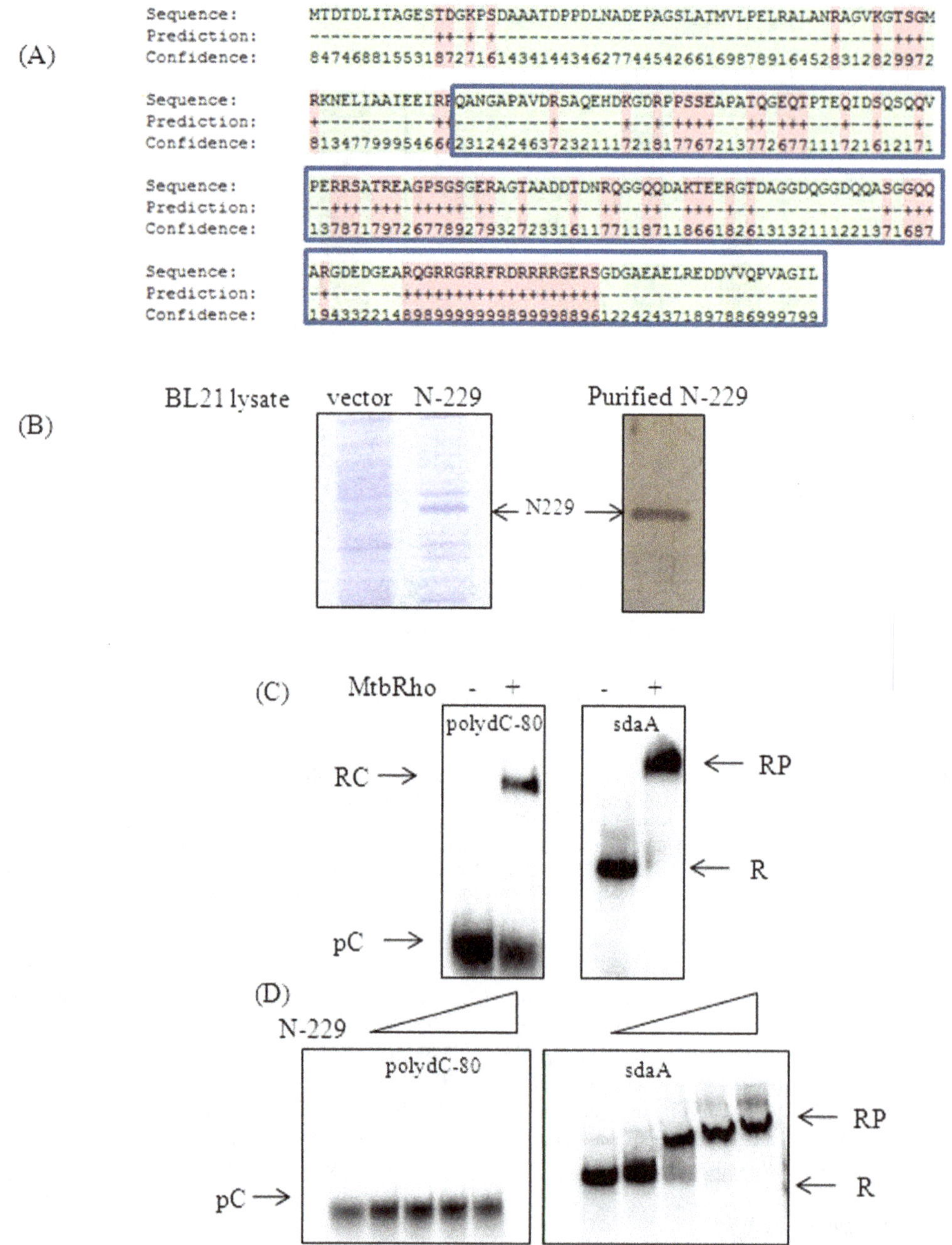

Figure 3. The subdomain of MtbRho contributes to RNA-binding. (A) Sequence of the N-229. The extra-stretch is shown within blue-boxes. The residues predicted to bind to RNA are labeled with '+' and 'red'. Values 0–9 indicate a gradient of confidence in prediction. The *in silico* analysis was carried out using the BindN web-server(50). (B) N-229 fragment was overexpressed from the pET20b clone in BL21 cells. Lysate of pET20b-N229 shows expression (N-229), as compared to control lysate (vector). N-229 was purified to homogenity, shown by silver staining. (C) MtbRho binding to RNA. MtbRho binds to both poly-dC80 and mycobacterial *sdaA* RNA. (D) N-229 binding to RNA. N-229 binds only to *sdaA* RNA. The RNA-protein complexes(RP, RC) were resolved from free RNA (R, pC) by 4% native PAGE.

spectra of MtbRho in presence and absence of its substrate ATP was unchanged, indicating that ATP alone did not induce any significant conformational changes. Thus, both protease protection and CD studies, show RNA-induced conformational changes in the protein.

Treating MtbRho with glutaraldehyde, a bifunctional crosslinking agent, showed that crosslinked MtbRho migrates with a mobility corresponding to a mass of ~400 kDa. Since the MtbRho monomer is 65 kDa, the crosslinked product observed is likely to be hexameric (Figure 4). The functional state of all Rho homologues studied so far is hexameric in nature [8,9,10]. Dynamic light scattering (DLS) and analytical gel filtration studies further highlighted the flexible nature of the protein. The DLS results indicated that, in absence of substrate RNA, >90% of MtbRho existed as a monomer in solution (RH 3.2 nm) (Figure 5A). In contrast, addition of RNA shifted 87% of the molecules towards a RH value of 7.7 nm, which corresponds with a hexameric form. Thus, RNA induced MtbRho monomers to stably adopt the hexameric conformation (Figure 5A). Similarly, analytical gel filtration showed that MtbRho existed both as monomer as well as oligomers in solution. However, a distinct shift towards formation of oligomeric forms was seen in the presence of RNA (Figure 5B), indicating that MtbRho is a dynamic molecule and interaction with RNA promotes the formation of the functionally active hexamer.

MtbRho can terminate transcription by EcRNAP, but Mtb rho cannot complement E. coli rho

Since all the RNA-binding and ATPase motifs in EcRho and MtbRho are similar, the pTrc99C-*Mtbrho* construct was used to complement the *E. coli* AMO14. The AMO14 genome has an inactivated *rho* gene [48] and a functional copy of *E. coli rho* is supplied on a temperature-sensitive plasmid. The strain is viable at non-permissive temperature (42°C) only when a functional *rho* allele is supplied *in trans*. However, the *E. coli* AMO14 competent cells, transformed with pTrc99C-*mtbrho*, showed no growth at non-permissive temperature (Figure 6A) [22]. Since the presence of the N-terminal subdomain makes MtbRho a larger protein (602residues) than EcRho (419residues) (Figure S1B) [21,43], it seemed plausible that this additional region of MtbRho could be a hindrance to complementation. However, a deletant of the *MtbRho* gene which lacked the 158-amino acid region (residues 87–241) also failed to complement(data not shown). The failure to complement was not due to lack of expression (Figure 6B). Moreover, as has been reported [22], MtbRho terminated transcription by *E. coli* RNAP *in vitro*. Run-off transcription on a template that contained a pTrc promoter was significantly

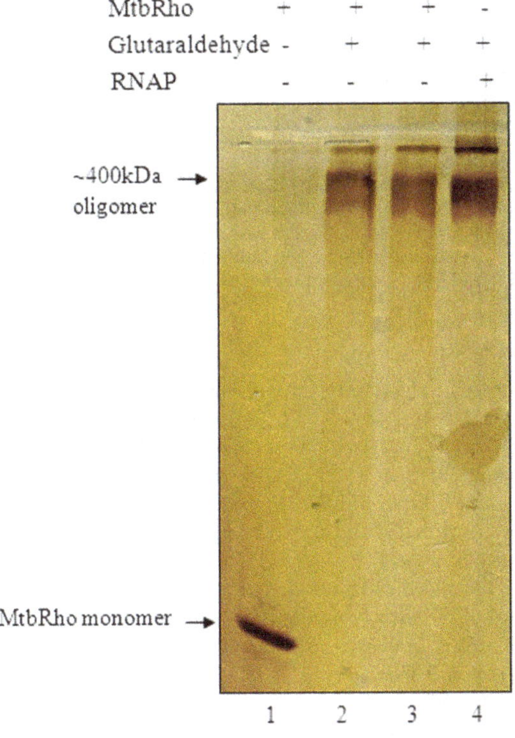

Figure 4. Oligomerization status of MtbRho. 2 µg MtbRho crosslinked by glutaraldehyde (0.05%) for 5 min and 10 min (lanes 2,3). Uncrosslinked MtbRho migrated as monomer (lane 1). RNAP was crosslinked and used as a marker of MW ~400 kDa (lane 4). Reaction buffer was similar to that in ATPase assay, except that KCl replaced K-Glutamate. Products were resolved in gradient SDS-PAGE.

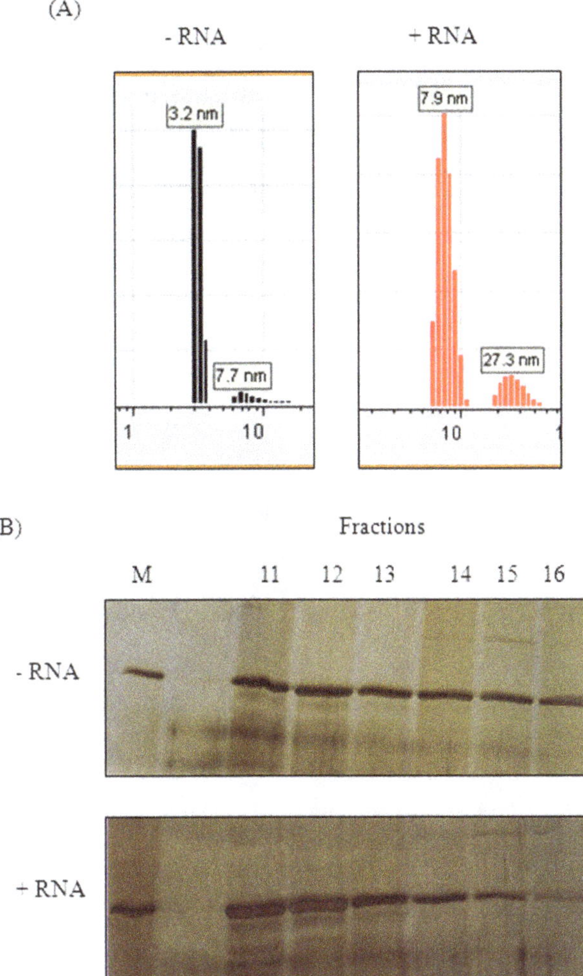

Figure 5. RNA induces hexameric form of MtbRho. (A) Dynamic Light Scattering profile of MtbRho. MtbRho was subjected to DLS alone (-RNA) or in presence of RNA (+RNA). A 10:1 molar ratio of RNA: Rho was maintained. The histogram represents the intensity distribution of the MtbRho. The RH values of 3.2 and 7.7–7.9 nm corresponds to MtbRho monomer and hexamer respectively; 27.3 nm corresponds to protein aggregate. (B) Analysis of Gel filtration fractions. SDS-PAGE profile of fractions eluted from by Superdex S200 column, with MtbRho alone (-RNA), or after incubating with RNA and ATP (+RNA). MtbRho elutes between fractions 11–16 in both cases, but is shifted towards higher oligomers in presence of RNA.

reduced in presence of MtbRho (Figure 6C). Not only did MtbRho show greater termination efficiency compared to equimolar amounts of EcRho, but also terminated at a region upstream to where EcRho could function. The smaller transcripts were products of *bona fide* Rho-dependent termination as pre-incubating MtbRho with bicyclomycin abolished them and restored run-off transcript levels. The results suggest that inability to interact with *E. coli* transcription machinery is not the reason for failure to complement.

Discussion

MtbRho is an NTPase that can hydrolyse ATP and GTP efficiently in presence of GC-rich RNA. Although a weaker ATPase compared to the well-characterized EcRho, MtbRho is more efficient in utilizing ATP in presence of its natural substrate

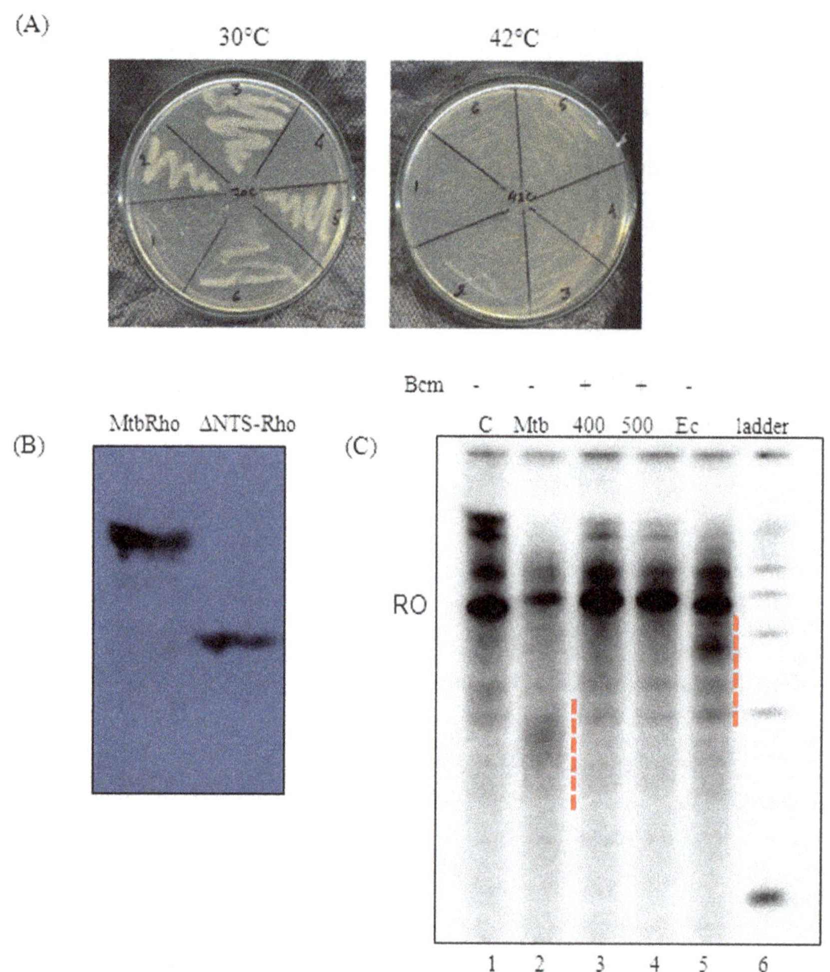

Figure 6. MtbRho is specific to *M. tuberculosis*. (A) Complementation of *E.coli* AMO14 strain attempted with Mtb *rho*. Transformants harbouring pTrc99C or pTrc99C-*Mtbrho* were grown to OD_{600} 0.6 at 30°C, streaked on IPTG- containing plates and incubated at permissive (30°C) or non-permissive (42°C) temperature. Untransformed cells(1,4); transformants with pTrc99C (2,5); transformants with pTrc99C-*Mtbrho*(3,6).(B) Expression of MtbRho and ΔNTS-MtbRho in *E.coli* AMO14 transformants when induced with 0.3 mM IPTG at permissible temperature. (C) MtbRho terminates transcription by *E. coli* RNAP *in vitro*. Run-off transcript (RO) was reduced and several smaller transcripts appeared when transcription was done in presence of MtbRho (Mtb; lane 2) compared to that in absence of MtbRho (C; lane 1). The effect was reversed when MtbRho has been pre-incubated with 400 or 500 μM bicyclomycin (400,500; lanes 3, 4). EcRho also caused termination (Ec; lane 5), at a different region of transcript. The terminated transcripts are indicated with dashed lines.

i.e. mycobacterial RNAs, compared to EcRho. The intrinsically weak ATPase activity of MtbRho suggests that the enzyme could be a poor translocase as ATP hydrolysis is necessary for powering translocation along the nascent transcript towards RNAP [6,14,22,51,52]. Notably, it is not the first mycobacterial enzyme shown to have slow catalytic rates. The rates of chain elongation of both *M. tuberculosis* RNAP and DNA Polymerase is significantly lower than the corresponding *E. coli* enzymes [39,53]and the slow rates are considered to be optimized to the slow-growing lifestyle of the bacteria. Thus, the ATPase rate of MtbRho could be an optimized rate evolved so that termination is functionally *in sync* with the slowly transcribing MtbRNAP. A stronger MtbRho ATPase, in contrast, could possibly result in swift and premature termination, which would be catastrophic for gene expression. The ability of MtbRho to hydrolyse ATP in presence of polyA and polyU, RNA substrates that are inaccessible to EcRho, shows that such broad-RNA specificity is a conserved feature of actinobacterial Rho [54,55]. Such broad specificity for RNAs could be useful in interacting with a larger number of RNA molecules, both for normal, intergenic rho-dependent termination as well as silencing of xenogenic DNA [3,56]. It may be indicative of the greater importance of Rho-assisted termination in *Mtb* and other actinobacteria, which have fewer intrinsic terminators [41]. The inefficient use of CTP by MtbRho also parallels earlier studies with Rho homologs from *M. luteus* and *Streptomyces lividans*. The latter two enzymes also hydrolyzed CTP very inefficiently [54,55]. This strategy could be possibly an adaptation to spare CTP for transcription of GC-rich transcriptomes of these actinobacteria.

The conserved amino acid composition of the N-terminal additional region is indicative of its functional importance and it has been hypothesized to have a role in binding to GC-rich RNA [7]; *M. luteus* Rho has been shown to terminate transcription by *E. coli* RNAP at sites where EcRho cannot terminate, and this is considered indicative that the 'larger' Rho are more efficient for terminating on GC-rich RNAs [42,54]. Our biochemical assays directly validate the long-held assumption that the N-terminal subdomain of actinobacterial Rho can indeed interact with its natural RNA substrate [6,7]. The functional importance of the additional N-terminal fragment in MtbRho is further evident from the inability of the truncated protein, which lacks the subdomain, to complement EcRho. It is notable that while N-229 can stably interact with mycobacterial RNA, it failed to bind efficiently to poly-dC80. In contrast, full-length MtbRho can interact with both the substrates. The primary RNA-binding motifs of Rho are known to that have a preference for C-rich, unstructured RNA [18,57] and their absence in N-229 could have compromised its ability to interact with poly-dC80. It is also possible that the extended N-terminal subdomain has evolved to bind efficiently to longer RNA molecules, such as the *sdaA* RNA in this study. Thus, it seems that MtbRho, while retaining ability to bind to the canonical, C-rich substrates of Rho, has a gain-of-function whereby the extended N-terminal region can facilitate binding to mycobacterial RNA. Since MtbRho is a weaker ATPase, interacting with long RNA could result in efficient spooling of RNA around itself and facilitate catching up with RNAP to bring about termination in a largely ATP-independent manner [22].

The inability of MtbRho to complement the strain *E.coli* AMO14 could be attributed to several reasons. The N-terminal subdomain of MtbRho can facilitate binding to RNA that is not a natural substrate for EcRho. This, in turn, could lead to spurious unregulated termination by MtbRho in the *E. coli* strain. Alternatively, the lower ATPase activity of MtbRho could result in inefficient termination, especially since rate of transcription elongation by *E.coli* RNAP is considerably faster than that of mycobacterial RNAP [39]. Approximately 50% of *E. coli* genes rely on Rho for termination [2]. Hence, delayed or imprecise termination could lead to run-off transcripts, ectopic expression from genomic islands and cryptic prophages [3,56,58].

The formation of a distinct hexameric (~400 kDa) species by glutaraldehyde crosslinking indicates that MtbRho can exist as a hexamer even in absence of its substrate RNA and ATP. DLS and analytical gel filtration studies confirm that at least a fraction of MtbRho exists as hexamer in solution. This hexameric population could provide a 'readily available' termination-proficient MtbRho within *Mtb* cells [22]. This is in contrast to EcRho which forms hexamer only under catalytic conditions [9]. Our results show that the oligomerization status of MtbRho is significantly influenced by its interaction with RNA. Although a fraction of MtbRho always exists as a ~400 kDa hexamer, it becomes the predominant form in presence of RNA. The availability of intracellular RNA would thus, act as a cue and 'activate' MtbRho to its functional hexameric form. This, in turn, could achieve a functional coordination between MtbRNAP and MtbRho to ensure an orchestrated transcription elongation and termination. In this context, it is noteworthy that the hexamers formed by two *Mtb* elongation/termination factors seem to have different functional consequences. A monomer-hexamer equilibrium, similar to that observed for MtbRho, has also been seen for *Mtb* Mfd protein although the trigger that causes a shift towards monomer or hexamer was not yet identified. The monomeric form of *Mtb* Mfd protein is considered to be the functional one, while the hexamer is likely to be a 'storage state' [59]. In case of MtbRho, it appears that nascent RNA could be driving the oligomerization of the termination factor to its functional form.

Methods

Bacterial strains, plasmids, construction of clones and chemicals

E. coli strains DH10B and BL21(DE3) were used for cloning and overexpression of proteins respectively. The temperature-sensitive *E. coli* AMO14 strain was used for complementation studies [49](Table S1). The *rho* gene was PCR-amplified using *M. tuberculosis* H37Ra genomic DNA as template, primers MtbRhoF and MtbRhoR (Table S1) and Pfu DNA polymerase. The PCR product was first cloned into the EcoRV site of pACYC184 and subsequently mobilized into the NcoI-BamHI sites of pET11d by digesting with RcaI and BglII. For complementation studies, *rho* was similarly cloned into the NcoI-BamHI sites of pTrc99C vector(Table S1). A deletant of MtbRho that lacks the subdomain (ΔNTSrho) was cloned by megaprimer mutagenesis. Primers (ΔNTSXhoF and ΔNTSXhoR) were used to introduce XhoI sites at positions 261 and 693 respectively of the *rho* coding sequence in pTrc99C clone. Digestion with XhoI and ligation resulted in an in-frame deletant that lacked residues 87–231. N-229, the N-terminal subdomain of MtbRho, was amplified using primers NTSfwd and NTSrev and cloned into the NdeI and XhoI sites of pET20b. Competent cells of *E. coli* AMO14 strain were transformed with pTrc99C (vector) or pTrc99C-rho DNA. Cultures were grown at permissive (30°C) and non-permissive (42°C and 37°C) temperatures. All restriction enzymes and DNA modifying agents were from New England Biolabs. DNA ligase was obtained from Roche Applied Science.

Expression and purification of proteins, RNA and polydC-80

The recombinant MtbRho was overexpressed from the pET11d-rho clone in BL21(DE3) cells. Cultures were grown till OD_{600} 0.6, induced with 0.3 mM IPTG for 3 hrs at 37°C and harvested. Cells were resuspended in buffer (20 mM Tris-HCl pH 8, 200 mM KCl, 20 mM EDTA, 5 mM 2-mercaptoethanol, 5% glycerol), lysed by sonication and centrifuged at 100000 g for 2 hrs at 4°C. This was followed by 0.5% polyethyleneimine (PEI)-based precipitation and then 0–40% ammonium sulphate fractionation. The pellet was resuspended in lX TGED buffer (10 mM Tris-HCl pH 7.6, 5% glycerol, 0.1 mM EDTA and 0.1 mM DTT) containing 150 mM KCl, dialysed against the same buffer to remove the excess salt and then loaded onto a pre-equilibrated Hitrap Heparin-sepharose column. Fractions were eluted in a salt range from 150 to 1000 mM KCl. The fractions containing MtbRho were pooled, dialyzed with 1× TGED+ 150 mM KCl and applied onto a Hitrap SP-sepharose column. The purity of the protein was assessed by electrophoresing on 8% SDS-PAGE, silver staining and mass spectrometry. Fractions which contained purified MtbRho were pooled and dialyzed overnight in storage buffer (1× TGED+100 mM K-glutamate+ 50%glycerol). EcRho and N-229, both with a C-terminal hexahistidine tag, were expressed in BL21(DE3) cells. The respective lysate was passed through Ni-NTA column (Qiagen), eluted with 500 mM imidazole and further purified to homogeneity using Hitrap-Heparin column. Concentration of the proteins was estimated by Bradford method. The Ni-NTA beads were from Sigma, all columns and materials were form Amersham-GE Healthcare. Mycobacterial RNA was purified as described in [60] and quality assessed by gel electrophoresis and absorbance ratio (260/280 nm). For generation of *sdaA* RNA, a 330 bp region(genomic coordinates: c76257–c75907) downstream of *Mtb sdaA* gene(Rv0069c) was PCR-amplified using forward primer, that contains T7 RNAP promoter sequence (*sdaA*RNAfwd) and *sdaA*RNArev. 1.5 µg of this template was used to transcribe *sdaA* RNA *in vitro* in presence of 0.5 mM NTP(final) and 20 µCi α-^{32}P-UTP (Jonaki, BRIT) and T7 RNAP (Fermentas). Unlabeled *sdaA* RNA was synthesized similarly with only 4 mM 'cold' NTPs as precursors. Following DNaseI treatment, the product was purified by phenol-chloroform extraction, alcohol precipitation and passing through Sephadex G-50 minicolumn, and its homogeneity was assessed on denaturing-PAGE. Poly-dC80 oligonucleotide was end-labeled with 15 μCi γ-^{32}P-ATP (Jonaki, BRIT) by Polynucleotide Kinase(NEB) as per standard protocol and eluted through Sephadex G-25 minicolumn.

NTPase activity

The hydrolysis of ATP and other NTPs in presence of different RNA substrates was assessed either by thin-layer chromatography (TLC) or by a Malachite Green-based colorimetric assay. Assays were carried out in T-Buffer (Tris HCl pH8 50 mM, MgOAc 3 mM, K-Glutamate 100 mM, DTT 0.1 mM, EDTA 0.1 mM, BSA 0.1 mg/ml, Glycerol 5%) with 100 nM MtbRho and the appropriate RNA in 10 μl reaction volume. Qualitative assays were carried out for 30 min in presence of 5,10 and 20 ng of polyC, polyA, polyU, 1 and 2 μg *M. smegmatis* total RNA. To test the effect of bicyclomycin, the MtbRho was pre-incubated with 500 μM bicyclomycin at 37°C for 10 min. For quantitative assays, 40 ng polyC, 1 μg *M. smegmatis* total RNA and 300 nM *sdaA* RNA were found to be optimal. Protein concentrations used were with respect to hexameric form of proteins. Reaction was initiated by addition of 1 mM of the ATP/NTP. 100 nCi of γ^{32}P-ATP (NEN) was used as tracer in radiometric assay to detect formation of inorganic phosphate (Pi), while α^{32}P-ATP(NEN) was used to observe if any AMP- or PPi-generating activity was present. The reactions were stopped by adding chloroform. Following centrifugation, the aqueous phase was used to estimate the release of inorganic phosphate(Pi) by resolving on Polyethyleneimine TLC sheets using LiCl (1.2 M) and EDTA(0.1 mM) as mobile phase and visualized with the Typhoon 9200 phosphorimager. For estimating the hydrolysis of the various NTPs in non-radioactive quantitative assays, a plate-based colorimetric assay [61] was used. Quantitation was carried out by scanning in a microplate ELISA reader at 655 nm. The amount of Pi released was calculated from a calibration curve using known concentrations of KH_2PO_4. Reactions were carried till 30–40% of ATP got hydrolysed. Kinetic parameters were obtained by plotting values on a Lineweaver Burk plot. For UV crosslinking, MtbRho was incubated with γ-^{32}P-ATP alone, or in presence of unlabeled ATP or CTP at 37°C for 10 min, followed by crosslinking on ice for 30 min using a handheld UV-torch (254 nm). No RNA was added.

In vitro transcription

For transcription by *E. coli* RNAP, template containing Trc promoter was amplified using primers FwdTrc and RhoR-evpTrc310, using pTrc99C-rho plasmid as template. Reaction was carried out by incubating 50 nM template with 100 nM *E. coli* RNAP in T-buffer, first on ice and then at 37°C for 10 min each, followed by incubation with 175 nM MtbRho or EcRho for 5 min. Transcription was initiated with 100 μM rNTP and 1 μCi α-^{32}P-UTP and continued for 30 min before terminating by addition of phenol-chloroform. Following centrifugation, equal volume of gel-loading dye (95% deionized Formamide, 0.05% Bromophenol Blue and 0.05% Xylene cyanol) was added to the aqueous phase, heated at 90°C for 2 min, chilled and resolved on 6% denaturing PAGE. To test the effect of bicyclomycin, the MtbRho was pre-incubated with 400 or 500 μM bicyclomycin at 37°C for 10 min before addition to RNAP-DNA complex.

Electrophoretic Mobility Shift assay

For qualititative RNA-binding assays, 100 nM of MtbRho and 100, 200, 400 and 800 nM of N-229 fragment were incubated with 15000 cpm of the ^{32}P-labeled *sdaA* RNA or polydC-80 on ice for 10 min. The 10 μl reactions were carried out in T-buffer. The RNA-protein complexes resolved from free RNA by 4 or 6% native PAGE (30:0.6::acrylamide:bisacrylamide). Protein concentrations used were with respect to hexameric form.

Protease protection assays

1 μg MtbRho was incubated in T-buffer with of *E. coli* tRNA (USB) or *M. smegmatis* SN2 total RNA on ice for 10 min and then at 37°C for 5 min, followed by addition of V8 protease and incubation for different time-points. The reaction was stopped by the addition of inhibitor Tosyl Lysine Chlromethyl Ketone (TLCK, Sigma) and 6× SDS gel loading dye, resolved on 8% SDS-PAGE and visualized with silver staining method.

Chemical crosslinking

2 μg MtbRho was crosslinked by the addition of glutaraldehyde, to a final concentration of 0.05%. The reactions were carried on for 5 and 10 min at 37°C in T-buffer (using KCl, instead of K-Glutamate) and terminated by addition of 2 M Tris, followed by resolving products on a 3.5-8% gradient SDS-PAGE.

CD spectroscopy

The CD spectra were recorded from 250–200 nm using a JASCO J-810 spectropolarimeter (Japan Spectroscopic Co Inc. Japan) in a 0.2 mm path-length quartz cuvette. T-buffer(identical to that used in NTPase experiments, but without Glycerol) was used and readings were recorded at room temperature.

Dynamic Light Scattering(DLS) experiments

DLS analysis were carried out on a Viscotex 802 DLS instrument in 12 μl volume quartz cuvette. T-buffer used for the assay was identical to that used for *in vitro* experiments described above, except that it contained no glycerol. Protein concentration of 300 ng/ul was found to be sufficient for the assay. A 10:1 molar ratio of tRNA: Rho was maintained. Data was analysed using the software Omnisize 3.0, associated with the DLS machine.

Analytical gel filtration chromatography

Typically, 100 μg MtbRho was loaded onto Superdex S200 column (bed volume 24 ml, fractionation range 10–600 kDa) alone, or after pre-incubating with tRNA (molar ratio 1:2) and 1 mM ATP, and 1 ml fractions were collected. After electrophoresing fractions on 8% SDS-PAGE, the eluted protein was assessed by silver staining.

Funding

V.N. is recipient of J.C. Bose Fellowship from Department of Science and Technology, Government of India. The funders had no role in study design, data collection and analysis, decision to publish, or preparation of the manuscript.

Supporting Information

Figure S1 Occurrence and divergence of Rho in bacteria. (A) Phylogenetic distribution of Rho from several bacterial phyla. (B) Alignment of representative Rho homologues from different bacteria. The 150–200 residue-long subdomain present in the N-terminal half of the actinobacterial homologs is enclosed in red. The alignment was done using the MACAW software. Conserved motifs and residues are shown as boxes and shaded regions.

Figure S2 Expression and purification of MtbRho. MtbRho was expressed from pET11d-*rho* clone in BL21(DE3) cells by induction with 0.3 mM IPTG (lane 2). MtbRho purification was carried as described in Methods. The purity of the protein was assessed by silver staining and mass spectrometry.

Figure S3 Inefficient binding of N-229 to poly-dC80. N-229 was incubated with the ^{32}P-labelled polynucleotide, and the complexes (RC) resolved from free polynucleotide (pC) by 6% native PAGE with 20% glycerol included as stabilizer both in gel and running buffer. No protein (lane 1); 0.6, 1.2 and 1.8 µM of N-229 was used (lanes 2,3,4); presence of excess unlabeled poly-dC80 competed with binding (lane 5).

Figure S4 Conformational changes of MtbRho upon binding to RNA and ATP. (A) MtbRho was incubated with *M. smegmatis* RNA and *E. coli* tRNA were used alone, or with ATP and probed with V8 protease. The presence of RNA (lanes 4 to 9; thin arrows) conferred substantial protection against proteolysis, when compared to only MtbRho (lane 3; bold arrows). The addition of ATP along with RNA (lanes 10,11) did not result in any additional or altered protection and ATP alone (lane 12) did not protect MtbRho from protease. (B) Changes in secondary structure were monitored at wavelengths between 200–250 nm in presence of RNA (I) or ATP (II). The molar ellepticity of CD spectra are shown for MtbRho alone (black), in presence of RNA(blue) and in presence of ATP (red).

Table S1 Strains, plasmids and oligonucleotides used in this study.

Acknowledgments

The pET21a-EcRho clone was kindly provided by Dipak Dutta (IMTECH, INDIA). We thank Prof.J.P.Richardson(Indiana University, USA) for the temperature-sensitive *E. coli* AMO14 strain. Priyanka Tare, Swapna Ganduri and Krishan Gopal Thakur are acknowledged for providing *E. coli* RNAP and assistance in DLS experiments respectively.

Author Contributions

Conceived and designed the experiments: AM RM VN. Performed the experiments: AM RM. Analyzed the data: AM VN. Contributed reagents/materials/analysis tools: VN. Contributed to the writing of the manuscript: AM VN.

References

1. Adhya S, Gottesman M (1978) Control of transcription termination. Annu Rev Biochem 47: 967–996.
2. Peters JM, Vangeloff AD, Landick R (2011) Bacterial transcription terminators: the RNA 3'-end chronicles. J Mol Biol 412: 793–813.
3. Cardinale CJ, Washburn RS, Tadigotla VR, Brown LM, Gottesman ME, et al. (2008) Termination factor Rho and its cofactors NusA and NusG silence foreign DNA in *E. coli*. Science 320: 935–938.
4. Roberts JW (1988) Phage lambda and the regulation of transcription termination. Cell 52: 5–6.
5. Banerjee S, Chalissery J, Bandey I, Sen R (2006) Rho-dependent transcription termination: more questions than answers. J Microbiol 44: 11–22.
6. Boudvillain M, Nollmann M, Margeat E (2011) Keeping up to speed with the transcription termination factor Rho motor. Transcription 1: 70–75.
7. Richardson JP (1996) Structural organization of transcription termination factor Rho. J Biol Chem 271: 1251–1254.
8. Geiselmann J, Yager TD, Gill SC, Calmettes P, von Hippel PH (1992) Physical properties of the *Escherichia coli* transcription termination factor rho. 1. Association states and geometry of the rho hexamer. Biochemistry 31: 111–121.
9. Finger LR, Richardson JP (1982) Stabilization of the hexameric form of *Escherichia coli* protein rho under ATP hydrolysis conditions. J Mol Biol 156: 203–219.
10. Skordalakes E, Berger JM (2006) Structural insights into RNA-dependent ring closure and ATPase activation by the Rho termination factor. Cell 127: 553–564.
11. Steinmetz EJ, Platt T (1994) Evidence supporting a tethered tracking model for helicase activity of *Escherichia coli* Rho factor. Proc Natl Acad Sci U S A 91: 1401–1405.
12. Epshtein V, Dutta D, Wade J, Nudler E (2010) An allosteric mechanism of Rho-dependent transcription termination. Nature 463: 245–249.
13. Santangelo TJ, Artsimovitch I (2011) Termination and antitermination: RNAP runs a stop sign. Nature Reviews Microbiology 9: 319–329.
14. Kalyani BS, Muteeb G, Qayyum MZ, Sen R (2011) Interaction with the nascent RNA is a prerequisite for the recruitment of Rho to the Transcription Elongation Complex in vitro. Journal of Molecular Biology 413: 548–560.
15. Ciampi MS (2006) Rho-dependent terminators and transcription termination. Microbiology 152: 2515–2528.
16. Hart CM, Roberts JW (1991) Rho-dependent transcription termination. Characterization of the requirement for cytidine in the nascent transcript. J Biol Chem 266: 24140–24148.
17. Hart CM, Roberts JW (1994) Deletion analysis of the lambda tR1 termination region. Effect of sequences near the transcript release sites, and the minimum length of rho-dependent transcripts. J Mol Biol 237: 255–265.
18. Bogden CE, Fass D, Bergman N, Nichols MD, Berger JM (1999) The structural basis for terminator recognition by the Rho transcription termination factor. Mol Cell 3: 487–493.
19. Chen X, Stitt BL (2004) The binding of C10 oligomers to *Escherichia coli* transcription termination factor Rho. J Biol Chem 279: 16301–16310.
20. Guerin M, Robichon N, Geiselmann J, Rahmouni AR (1998) A simple polypyrimidine repeat acts as an artificial Rho-dependent terminator *in vivo* and *in vitro*. Nucleic Acids Res 26: 4895–4900.
21. D'Heygere F, Rabhi M, Boudvillain M (2013) Phyletic distribution and conservation of the bacterial transcription termination factor Rho. Microbiology 159: 1423–1436.
22. Kalarickal NC, Ranjan A, Kalyani BS, Wal M, Sen R (2009) A bacterial transcription terminator with inefficient molecular motor action but with a robust transcription termination function. J Mol Biol 395: 966–982.
23. Rodrigue S, Provvedi R, Jacques PE, Gaudreau L, Manganelli R (2006) The sigma factors of *Mycobacterium tuberculosis*. FEMS Microbiol Rev 30: 926–941.
24. Cole ST, Brosch R, Parkhill J, Garnier T, Churcher C, et al. (1998) Deciphering the biology of *Mycobacterium tuberculosis* from the complete genome sequence. Nature 393: 537–544.
25. Unniraman S, Chatterji M, Nagaraja V (2002) DNA gyrase genes in *Mycobacterium tuberculosis*: a single operon driven by multiple promoters. J Bacteriol 184: 5449–5456.
26. Bagchi G, Chauhan S, Sharma D, Tyagi JS (2005) Transcription and autoregulation of the Rv3134c-devR-devS operon of *Mycobacterium tuberculosis*. Microbiology 151: 4045–4053.
27. Pashley CA BA, Robertson D, Parish T (2006) Identification of the *Mycobacterium tuberculosis* GlnE promoter and its response to nitrogen availability. Microbiology 152: 2727–2734.
28. Chowdhury RP, Gupta S, Chatterji D (2007) Identification and characterization of the dps promoter of *Mycobacterium smegmatis*: promoter recognition by stress-specific extracytoplasmic function sigma factors sigmaH and sigmaF. J Bacteriol 189: 8973–8981.
29. Schuessler DL, Parish T (2012) The promoter of Rv0560c is induced by salicylate and structurally-related compounds in *Mycobacterium tuberculosis*. PLoS One 7: e34471.
30. Agarwal N, Tyagi AK (2003) Role of 5'TGN3' Motif in the Interaction of Mycobacterial RNA Polymerase with a Promoter of "Extended −10" class. FEMS Microbiol Lett 225: 75–83.
31. Arnvig KB, Gopal B, Papavinasasundaram KG, Cox RA, Colston MJ (2005) The mechanism of upstream activation in the rrnB operon of *Mycobacterium smegmatis* is different from the Escherichia coli paradigm. Microbiology 151: 467–473.
32. China A, Tare P, Nagaraja V (2010) Comparison of promoter-specific events during transcription initiation in mycobacteria. Microbiology 156: 1942–1952.
33. Tare P, China A, Nagaraja V (2012) Distinct and Contrasting Transcription Initiation Patterns at *Mycobacterium tuberculosis* Promoters. PLoS One 7: e43900.
34. Tare P, Mallick B, Nagaraja (2013) V Co-evolution of specific amino acid in sigma 1.2 region and nucleotide base in the discriminator to act as sensors of small molecule effectors of transcription initiation in mycobacteria. Mol Microbiol 90: 569–583.
35. Josaitis CA, Gaal T, Gourse RL (1995) Stringent control and growth-rate-dependent control have nonidentical promoter sequence requirements. Proc Natl Acad Sci U S A 92: 1117–1121.
36. Mathew R, Ramakanth M, Chatterji D (2005) Deletion of the gene rpoZ, encoding the omega subunit of RNA polymerase, in *Mycobacterium smegmatis* results in fragmentation of the beta' subunit in the enzyme assembly. J Bacteriol 187: 6565–6570.

37. China A, Nagaraja V (2010) Purification of RNA polymerase from mycobacteria for optimized promoter-polymerase interactions. Protein Expr Purif 69: 235–242.
38. Agarwal N, Tyagi AK (2006) Mycobacterial Transcriptional Signals: Requirements for Recognition by RNA polymerase and Optimal Transcriptional Activity. Nucleic Acid Research 34: 4245–4257.
39. Harshey RM, Ramakrishnan T (1977) Rate of ribonucleic acid chain growth in *Mycobacterium tuberculosis* H37Rv. J Bacteriol 129: 616–622.
40. Mitra A, Angamuthu K, Nagaraja V (2008) Genome-wide analysis of the intrinsic terminators of transcription across the genus *Mycobacterium*. Tuberculosis (Edinb) 88: 566–575.
41. Czyz A, Mooney RA, Iaconi A, Landick R (2014) Mycobacterial RNA Polymerase Requires a U-Tract at Intrinsic Terminators and Is Aided by NusG at Suboptimal Terminators. MBio 5: e00931–00914.
42. Nowatzke WL, Burns CM, Richardson JP (1997) Function of the novel subdomain in the RNA binding domain of transcription termination factor Rho from *Micrococcus luteus*. J Biol Chem 272: 2207–2211.
43. Opperman T, Richardson JP (1994) Phylogenetic analysis of sequences from diverse bacteria with homology to the Escherichia coli rho gene. J Bacteriol 176: 5033–5043.
44. Chen CY, Galluppi GR, Richardson JP (1986) Transcription termination at lambda tR1 is mediated by interaction of rho with specific single-stranded domains near the 3' end of cro mRNA. Cell 46: 1023–1028.
45. Chen CY, Richardson JP (1987) Sequence elements essential for rho-dependent transcription termination at lambda tR1. J Biol Chem 262: 11292–11299.
46. Lowery C, Richardson JP (1977) Characterization of the nucleoside triphosphate phosphohydrolase (ATPase) activity of RNA synthesis termination factor p. II. Influence of synthetic RNA homopolymers and random copolymers on the reaction. J Biol Chem 252: 1381–1385.
47. Galluppi GR, Richardson JP (1980) ATP-induced changes in the binding of RNA synthesis termination protein Rho to RNA. J Mol Biol 138: 513–539.
48. Martinez A, Opperman T, Richardson JP (1996) Mutational analysis and secondary structure model of the RNP1-like sequence motif of transcription termination factor Rho. J Mol Biol 257: 895–908.
49. Mitra A, Kesarwani AK, Pal D, Nagaraja V (2011) WebGeSTer DB–a transcription terminator database. Nucleic Acids Res 39: D129–135.
50. Wang L, Brown SJ (2006) BindN: a web-based tool for efficient prediction of DNA and RNA binding sites in amino acid sequences. Nucleic Acids Res 34: W243–W248.
51. Thomsen ND, Berger JM (2009) Running in reverse: the structural basis for translocation polarity in hexameric helicases. Cell 139: 523–534.
52. von Hippel PH, Yager TD (1991) Transcript elongation and termination are competitive kinetic processes. Proc Natl Acad Sci U S A 88: 2307–2311.
53. Hiriyanna KT, Ramakrishnan T (1986) DNA replication time in *M. tuberculosis* H37Rv. Archives of Microbiology 144: 105–109.
54. Nowatzke WL, Richardson JP (1996) Characterization of an unusual Rho factor from the high G + C gram-positive bacterium *Micrococcus luteus*. J Biol Chem 271: 742–747.
55. Ingham CJ, Hunter IS, Smith MC (1996) Isolation and Sequencing of the rho gene from Streptomyces lividans ZX7 and Characterization of the RNA-dependent NTPase Activity of the Overexpressed Protein. Journal of Biological Chemistry 271: 21803–21807.
56. Peters JM, Mooney RA, Kuan PF, Rowland JL, Keles S, et al. (2009) Rho directs widespread termination of intragenic and stable RNA transcription. Proc Natl Acad Sci U S A 106: 15406–15411.
57. Skordalakes E, Berger JM (2003) Structure of the rho transcription terminator: mechanism of mRNA recognition and helicase loading. Cell 114: 135–146.
58. Peters JM, Mooney RA, Grass JA, Jessen ED, Tran F, et al. (2012) Rho and NusG suppress pervasive antisense transcription in Escherichia coli. Genes Dev 26: 2621–2633.
59. Prabha S, Rao DN, Nagaraja V (2011) Distinct Properties of Hexameric but Functionally Conserved *Mycobacterium tuberculosis* Transcription Repair Coupling Factor. pLoS One 6: e19131.
60. Unniraman S, Nagaraja V (1999) Regulation of DNA gyrase operon in *Mycobacterium smegmatis*: a distinct mechanism of relaxation stimulated transcription. Genes Cells 4: 697–706.
61. Janscak P, Abadjieva A, Firman K (1996) The Type I Restriction Endonuclease R.EcoR124I: Over-production and Biochemical Properties. 257: 977–991.

9

Biohybrid Polymer-Antimicrobial Peptide Medium against *Enterococcus faecalis*

Lea H. Eckhard[1], Asaf Sol[2], Ester Abtew[3], Yechiel Shai[4], Abraham J. Domb[3], Gilad Bachrach[2¶], Nurit Beyth[1*¶]

1 Department of Prosthodontics, the Hebrew University – Faculty of Dental Medicine, Jerusalem, Israel, **2** Institute of Dental Science, the Hebrew University – Faculty of Dental Medicine, Jerusalem, Israel, **3** Institute for Drug Research, School of Pharmacology, Faculty of Medicine, the Hebrew University, Jerusalem, Israel, **4** Department of Biological Chemistry, the Weizmann Institute of Science, Rehovot, Israel

Abstract

Antimicrobial peptides (AMPs) are conserved evolutionary components of the innate immune system that are being tested as alternatives to antibiotics. Slow release of AMPs using biodegradable polymers can be advantageous in maintaining high peptide levels for topical treatment, especially in the oral environment in which dosage retention is challenged by drug dilution with saliva flow and by drug inactivation by salivary enzymatic activity. *Enterococcus faecalis* is a multidrug resistant nosocomial pathogen and a persistent pathogen in root canal infections. In this study, four ultra-short lipopeptides (C16-KGGK, C16-KLLK, C16-KAAK and C16-KKK) and an amphipathic α-helical antimicrobial peptide (Amp-1D) were tested against *E. faecalis*. The antibacterial effect was determined against planktonic bacteria and bacteria grown in biofilm. Of the five tested AMPs, C16-KGGK was the most effective. Next C16-KGGK was formulated with one of two polymers poly (lactic acid co castor oil) (DLLA) or ricinoleic acid-based poly (ester-anhydride) P(SA-RA). Peptide-synthetic polymer conjugates, also referred to as biohybrid mediums were tested for antibacterial activity against *E. faecalis* grown in suspension and in biofilms. The new formulations exhibited strong and improved anti- *E. faecalis* activity.

Editor: Tilmann Harder, University of New South Wales, Australia

Funding: This research was supported by The Legacy Heritage Clinical Research Initiative of the Israel Science Foundation (grant No.1764/11) (NB). The funders had no role in study design, data collection and analysis, decision to publish, or preparation of the manuscript.

Competing Interests: The authors have declared that no competing interests exist.

* Email: nuritb@ekmd.huji.ac.il

¶ These authors contributed equally to this work.

Introduction

The widespread use of antibiotics leads to the emergence of more resistant and virulent strains of microorganisms. Consequently, the development of new antimicrobial agents becomes paramount for novel treatment options [1]. Moreover, bacterial resistance to antibiotics becomes even more complicated when dealing with bacterial biofilms. Interestingly, potent antimicrobial components against a wide range of pathogens can be found in the innate immune system of various organisms including humans. An example is host-defense cationic antimicrobial peptides (AMPs) which are conserved evolutionary components that possess the capacity to kill invading microbes [2]. It is generally accepted that AMP-mediated killing typically occurs through microbial membrane disruption resulting in irreparable damage.

AMPs exhibit broad-spectrum activity against a wide range of microorganisms including Gram-positive and Gram-negative bacteria, protozoa, yeast, fungi and viruses [4]. Furthermore, whereas conventional antibiotics are becoming less effective, bacteria do not appear to develop resistance to AMPs.

AMPs function and mode of action is a direct derivative of their structure and electric charge. They differ in amino acid length (12–50 aa), sequence and dimensional structure, they are composed of about 50% hydrophobic amino acids and their electric charge is positive. This unique amphipathic structure allows AMPs to bind to the negatively charged outer surfaces of microorganisms and to disrupt and permeate their cell membranes [3]. Consequently, their advantage over conventional antibiotics is a non-specific but antimicrobial selective mode of action. Several resistance strategies to avoid AMPs function were reported. These included, degradation with extracellular proteases [4], altering the net surface charge [5], transporting AMPs into the cytoplasm and degrading them [6] and exporting AMPs by efflux pumps [7].

AMPs are one of the reasons why humans stay healthy [2]. In humans, one of the known potent AMPs groups is the defensins. Specifically, human beta-defensin 3 (hBD3). hBD3 is considered the most potent β-defensin peptide described so far [8]. An additional group of potent antimicrobial peptides, produced in bacteria and fungi are the lipopeptides. They are composed of specific lipophilic moieties attached to anionic peptides. Unfortunately, native lipopeptides are non-cell selective and thus can be toxic to mammalian cells. Interestingly, all of the structural advantages of the native AMPs can be recruited to synthesize improved antimicrobial agents, *e.g.* ultra-short lipopeptides and amphipathic α-helical antimicrobial peptide (Amp-1D). Recently, it was reported that ultra-short lipopeptides composed of only four amino acids conjugated to an aliphatic acids chain (16C, palmitate) can achieve potent antimicrobial activity without compromising biocompatibility.

Enterococcus faecalis is a microorganism residing in the gastro-intestinal tract. None the less, *E. faecalis* can cause life-threatening infections such as: endocarditis, bacteremia, urinary tract infection and meningitis. These complications are mostly associated with the acquisition of resistance to antibiotics. In dentistry, *E. faecalis* is considered a persistent root canal pathogen [9].

Integration of the structural and functional properties of peptides and proteins with the versatility of synthetic polymers has gained significant interest in material design and application [10]. Peptides and proteins have unique structures that convey their ability to function in specific biological activities. Hybrid molecules of peptides conjugated to polymers can be used for various applications with the advantages of being resistant to enzymatic cleavage and less cytotoxic to human cells [11]. Peptide-synthetic polymer conjugates, also referred to as biohybrid medium, consist of biologically relevant peptides and synthetic polymers, aiming to combine the advantages of the two components, namely biological function (biological component) and process-ability (synthetic component). A slow release mechanism can enable high concentration maintenance of therapeutic agents for prolonged periods of time. Examples of biodegradable polymers that were previously described as efficient controlled delivery mediums include the fatty acid-based polymer poly (lactic acid co castor oil) and the ricinoleic acid-based poly (ester-anhydride) [11–15]. Herein, biohybrid medium consisted of two different assembled components, including a polymer matrix, which was responsible for the sustained release function and an antimicrobial agent, *i.e.* AMPs, which was responsible for the potency of the formulation. In the present study novel formulations of biodegradable polymers integrating AMPs were evaluated against *E. faecalis* in planktonic bacteria and biofilm growth.

Materials and Methods

Test materials

Antimicrobial peptides. Human recombinant β-defensin 3 (hBD3) (GIINTLQKYY CRVRGGRCAV LSCLPKEEQI GKCSTRGRKC CRRKK) was obtained from PeproTech (Lot #0108210, Rocky Hill, NJ, USA). Five different synthetic AMP candidates were tested. amphipathic α-helical antimicrobial peptide (Amp-1D) and four ultra short lipopeptides which were synthesized, purified and confirmed as described before [16,17].

Biodegradable polymer synthesis. Poly (lactic acid co castor oil) (DLLA) and ricinoleic acid-based poly (ester-anhydride) P(SA-RA) were synthesized as previously describe [11–13,15,18]. Briefly, a poly (ester-anhydride) copolymer of sebacic acid (SA) and ricinoleic acid (RA) in a 3:7 w/w ratio [P(SA-RA) 3:7] was synthesized by transesterification followed by anhydride melt condensation. SA was used as supplied by Sigma-Aldrich (St. Louis, MO, USA) without any additional purification. RA (>98%) was isolated from castor oil by fractional precipitation based on salt-solubility. Poly (DL lactic acid co castor oil) 4:6 and 3:7 designated P(DLLA:CO) 4:6 and P(DLLA:CO) 3:7 was prepared using racemic mixture (DL) lactic acid. The synthesized polymers were characterized by Infrared (IR) spectroscopy and nuclear magnetic resonance (NMR) spectroscopy. Gel permeation chromatography (GPC) was used to estimate the molecular weight.

Formulation of AMP-based biohybrid media. The peptide powder was mixed with a pasty polymer to form a uniform homogeneous paste. A novel formulation of peptide and biodegradable polymer was prepared at a ratio of 100 µg peptide integrated in 100 mg polymer. The two ingredients were mixed manually with a spatula.

Preparation of bacterial suspension

E. faecalis (ATCC #v583), was cultured overnight in 5 ml brain-heart infusion (BHI) (Difco, Detroit, MI, USA) broth supplemented with 2 mg/ml vancomycin (Sigma-Aldrich), at 37°C under aerobic conditions. The top 4 ml were transferred to a fresh test tube and the optical density (OD) was determined according to the specific experiment.

Antibacterial activity

Minimal inhibitory concentration. The antibacterial activity of the peptides was examined using the microdilution assay [19]. Briefly, the bacterial suspension (at OD 0.3) was diluted at a ratio of 1:1000. Aliquots of 150 µl of bacterial suspension were added to 50 µl of peptide dilutions in phosphate buffered saline (PBS) (Sigma-Aldrich) (in triplicate for each concentration) in a 96-well plate (Nunc 96 microtiter plates, Roskilde, Denmark). The optical density (595 nm) in each well was recorded every 20 min using a microplate reader (VERSAmax tunable microplate reader, molecular devices, Sunnyvale, CA, USA) at 37°C for 18–24 hrs. The minimal inhibitory concentration (MIC) was determined as the concentration which inhibited visible growth after 18–24 hrs.

Antibacterial activity of controlled released peptide. A total 10 mg of formulation was placed on the side walls of each of 6 wells in a 96 microtiter plate and then 270 µl of medium (BHI supplemented with vancomycin) were added Every 24 hrs the medium was collected and transferred to a new set of 6 wells in the same 96-well-plate and fresh medium was added to the 6 original wells containing the tested formulation. After one week, a 10 µl volume of *E. faecalis* suspension was added to each of the 6 wells and bacterial outgrowth was recorded. The plate was incubated at 37°C in a VERSAmax microplate reader and turbidity (OD_{650} nm) changes were recorded, every 20 min for 18–24 hrs.

Antibiofilm activity

Antibiofilm activity was tested on *E. faecalis* biofilms grown for 72 hrs. Biofilm was formed in microtiter plates (24 well plates for the ATP bioluminescence assay and 96 well plates for the crystal violet biomass assay and confocal laser spectroscopy). Saliva was collected from one donor and DL-Dithiolthreitol (DTT) (Thermo Scientific, Abu-Gosh, Israel) was added to 2.5 mM. The suspension was kept at 4°C for 10 min and then centrifuged for 15 min at 6,500×g. The supernatant was transferred to a fresh sterile tube and diluted to 25% with sterile double distilled water (DDW). The diluted saliva was disinfected using a 0.2 µm vacuum-driven filter (0.22 µm, 250 µl, Jet biofil, Belgium). Wells in the microtiter plate were coated with clarified saliva by adding the saliva to the wells for 1 hr at 37°C (150 µl of saliva in the 24 well plate and 50 µl in the 96 well plate). Unbound saliva was removed and the wells were washed gently with PBS. The polymer peptide formulations were placed on the side walls of the wells and 10 µl of bacterial suspension (prepared as described above) were placed in the center of each well not touching the coated sidewall. The saliva coating was used to cover the entire well surface, followed by formulation placement on the sidewall of the wells. The bacterial inoculum was placed in the center of each well, not touching the formulation. After 1 hr incubation at 37°C BHI broth was added (1 ml in the 24 well plate and 100 µl in the 96 well plate). BHI broth was added every 24 hrs during 72 hrs. After 3 days the medium was discarded and the wells were washed gently with PBS. Bacterial metabolism in the attached biofilm was assessed using ATP bioluminescence. Biofilm mass was measured using crystal violet as described below.

ATP bioluminescence. Bacterial killing was evaluated by measuring intracellular ATP levels, an energy parameter commonly used as an indicator of cell injury and viability [19]. The 72 hr biofilm formed on the bottom of the wells was scraped using a pipette tip and collected into a set of 15 ml tubes. The cells were then centrifuged (6,500×g, 5 min), resuspended in 1 ml Lysis Buffer (2 mM DTT, 2 mM trans 1,2 Diaminocyclohexane NNNN Tetraacetic acid, 0.5 mM EDTA, 1% Triton, 25 mM Tris, 25 mM K_2HPO_4, 10% glycerol) and transferred to a 2 ml microcentrifuge tube containing glass beads (Lysing Matrix tubes, 0.1 mm silica spheres; MP Biomedicals, Eschwege, Germany). The cells were disrupted with the aid of a FastPrep cell disrupter (MP Biomedicals, Irvine, CA, USA). The tube was centrifuged for 10 min (4°C, 13,400×g). ATP levels were determined using an ATP bioluminescence assay kit (CLS 2, Roch Diagnostics, Mannheim, Germany). In a 96 microtiter plate designed for luminescence assay (Thermo Scientific, NUNC, 96-well optical Btm Plt white, Rochester, NY, USA) a 100 µl volume of the samples was added to 6 wells for each tested group. Then 100 µl luciferase (from the kit) were added to the same wells. The plate was inserted in a GENios reader (TECAN, Salzburg, Austria) and luminescence was measured using the Magelan program (TECAN, V6.6, 2009). ATP calibration was performed using ATP and luciferase from the kit.

Crystal violet. The total biofilm yield was assessed using crystal violet staining as follows. Biofilm fixation was performed using 200 µl methanol (MERCK, Darmstadt, Germany) that were added to each well for 20 min. The biofilm was then stained using 200 µl 1% crystal violet (Merck) for 20 min. Then the wells were washed gently 3x with PBS, and 200 µl of 30% acetic acid (GADOT, Netanya, Israel) were added to the wells. The acetic acid was transferred to wells of a new 96-well microtiter plate that was placed in a microplate reader and absorbance (OD_{595} nm) was measured.

Confocal microscopy. Confocal laser scanning microscopy (CLSM) was used to explore the vitality of bacteria in the different depth layers of the biofilm. Bacteria were stained using a live/dead kit (Live/Dead BacLight viability kit, Molecular Probes, OR, USA) as described before [20]. Briefly, wells were washed, incubated for 15 min in a solution containing propidium iodide and SYTO 9 and washed again. To read the results directly, the wells were coated with emulsion oil to prevent dehydration. Fluorescence emission was detected using a Zeiss LSM 410 confocal laser scanning microscope (Carl Zeiss Microscopy, Jena, Germany). Red fluorescence was measured at 630 nm and green fluorescence at 520 nm; objective lenses: x60/oil, 1.4 numerical aperture. Horizontal plane (x-y axes) optical sections were made at 700 µm intervals from the surface outwards and images were displayed individually. The biofilm was quantified by measuring the area occupied by the bacteria with the aid of Image Pro 4.5 software (Media Cybernetics, Rockville, MD, USA).

Statistical analysis

The presented data are the mean and standard deviation of triplicates of a representative experiment repeated three times. The growth mean, and multiple comparisons of growth inhibition by AMP (compared with the growth of untreated bacteria) were calculated from each growth curve using Student's t-test. The level of significance was $p<0.01$.

Results

Antibacterial activity

Minimal inhibitory concentration. The MICs results for each of the tested AMPs are summarized in Table 1. Growth of *E. faecalis* was not inhibited by hBD3 at concentrations of up to 20 µg/ml. Amp-1D did not affect bacterial growth at concentrations of up to 25 µg/ml. C16-KGGK, C16-KKK, C16-KAAK and C16-KLLK completely inhibited bacterial growth at concentrations ranging between 5 and 25 µg/ml. The most potent AMP was the lipopeptide C16-KGGK that caused complete growth inhibition at 5 µg/ml (Fig. 1). As a result, all further formulations were tested using the C16-KGGK lipopeptide. Surprisingly, at concentrations below 5 µg/ml the growth of *E. faecalis* was not inhibited by C16-KGGK but was actually accelerated.

Sustained release and anti-*E. faecalis* activity. Release of the C16-KGGK lipopeptide from two biodegradable polymers was monitored over one week in two modes. In the first, the antibacterial action of C16-KGGK released into the medium that came in contact with the formulation every 24 hrs was measured (see Fig. 2A, B). In the second, the bacteria were added to the wells with C16-KGGK that was released from the polymer and accumulated for one week (see Fig. 2C). The anti-*E. faecalis* activity of C16-KGGK released from each formulation was reflected by: 1. The final optical density of the treated bacteria, which was lower than that of the untreated ones. 2. The slope of the curve (generation time, see Fig. 2C), which was more moderate in the treated bacteria.

The anti-*E. faecalis* activity in the medium exposed to the formulation for an entire week (see Fig. 2D, E) generated a longer generation time, especially with P(SA:RA). Bacteria treated with the DLLA formulation exhibited a 27% reduction in growth and those treated with P(SA:RA) a 60% reduction compared with the non treated bacteria.

Anti-biofilm effect

Crystal violet dye. Crystal violet was used to stain and measure biofilm mass so that inhibition of biofilm formation in the presence of C16-KGGK formulated with P(SA-RA) (Fig. 3A) or DLLA (Fig. 3B) or in the presence of the soluble tested AMPs (Fig. 3C) could be determined. A significant anti biofilm effect was obtained with C16-KGGK using both formulations but not with the soluble C16-KGGK (Fig. 3A–B). The vehicle formulation itself does not possess anti-biofilm activity. From the other peptides tested in suspension, only C16-KKK showed anti-biofilm activity (Fig. 3C).

ATP bioluminescence assay. The level of ATP indicates the active metabolism of a cell. ATP levels in *E. faecalis* biofilms treated with soluble C16-KGGK were relatively low compared with that in the untreated control (*E. faecalis* alone) (see Fig. 4). The P(SA:RA) formulation (without C16-KGGK) also reduced bacterial viability. As opposed to P(SA:RA), DLLA had the reverse effect and the luminescence values were much higher than that of the positive control. These findings led to the question whether the luminescence values are derived from bacterial number, the metabolic status or both. In addition, a similar experiment was performed in which the biofilm was first grown for 48 hrs and then the tested materials were added to verify if C16-KGGK can affect an already constructed biofilm. The results were similar to those above (where the materials were added immediately after inoculating the bacteria). This may indicate that the formulation and the peptide (each) have an anti-metabolic effect even after the biofilm is formed.

Table 1. Anti *E. faecalis* MICs of the AMPs investigated.

AMP	Amino acid sequence	MIC [μg/ml]
hBD3	GIINTLQKYY CRVRGGRCAV LSCLPKEEQI GKCSTRGRKC CRRKK	>20
Amp-1D	LKLLKKLLKKLLKLL-NH_2	>25
C16-KGGK	$CH_3(CH_2)_{14}CO$-KGGK-NH_2	4–5
C16-KKK	$CH_3(CH_2)_{14}CO$-KKK-NH_2	6–12.5
C16-KAAK	$CH_3(CH_2)_{14}CO$-KAAK-NH_2	12.5–25
C16-KLLK	$CH_3(CH_2)_{14}CO$-KLLK-NH_2	6–12.5

Underlined amino acids are D-enantiomers.

Bacterial vitality. To test the vitality of the bacteria within the biofilm by a different, independent method, live/dead staining followed by confocal microscopic analysis was performed. The differences between the four tested groups are clearly evident for both P(SA:RA) and DLLA incorporated C16-KGGK (Fig. 5). Soluble C16-KGGK induced death in the biofilm bacteria (Fig. 5A–B, *E. faecalis* + KGGK). However, C16-KGGK in both formulations was more effective than the soluble peptide alone. The P(SA:RA) polymer had a strong inhibitory activity against biofilm formation as seen by the reduction in bacterial load.

Discussion

Antimicrobial peptides are one of nature's solutions to bacterial invasion. Their nonspecific mode of action, which is based on physical membrane disruption, is effective against various bacteria

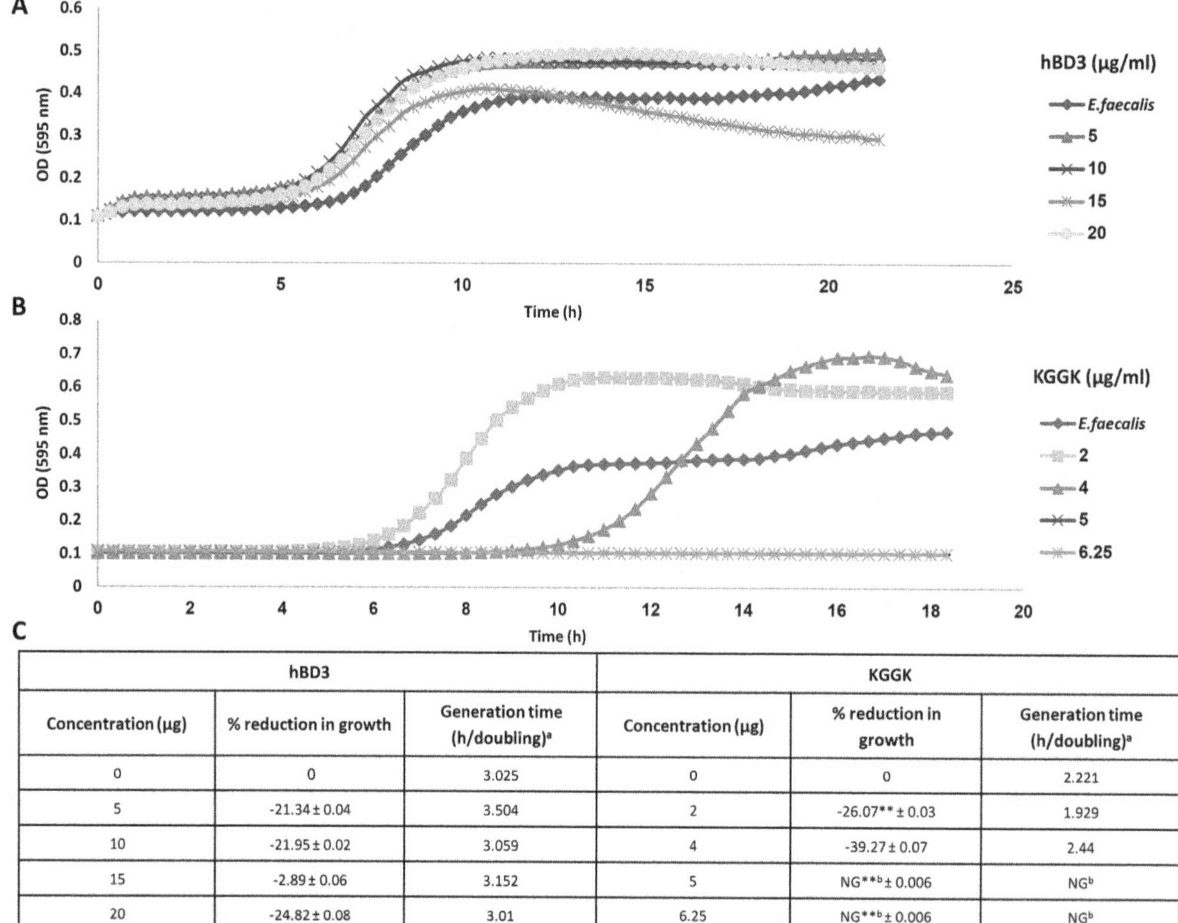

hBD3			KGGK		
Concentration (μg)	% reduction in growth	Generation time (h/doubling)[a]	Concentration (μg)	% reduction in growth	Generation time (h/doubling)[a]
0	0	3.025	0	0	2.221
5	-21.34 ± 0.04	3.504	2	-26.07** ± 0.03	1.929
10	-21.95 ± 0.02	3.059	4	-39.27 ± 0.07	2.44
15	-2.89 ± 0.06	3.152	5	NG**[b] ± 0.006	NG[b]
20	-24.82 ± 0.08	3.01	6.25	NG**[b] ± 0.006	NG[b]

[a] calculated at logarithmic phase at each tested condition ** P<0.01 compared with control without AMP, determined using Student's t-test [b] No growth

Figure 1. *E. faecalis* growth is inhibited by C16-KGGK but not by hBD3. Growth of *E. faecalis* was measured (see Materials and Methods) in the presence of increasing concentrations of hBD3 (A) or of the lipopeptide KGGK (B). Percent growth inhibition was calculated compared with that of untreated bacteria during the logarithmic phase of the non treated bacteria. Generation time was calculated from each curve using the section representing the exponential growth phase (C).

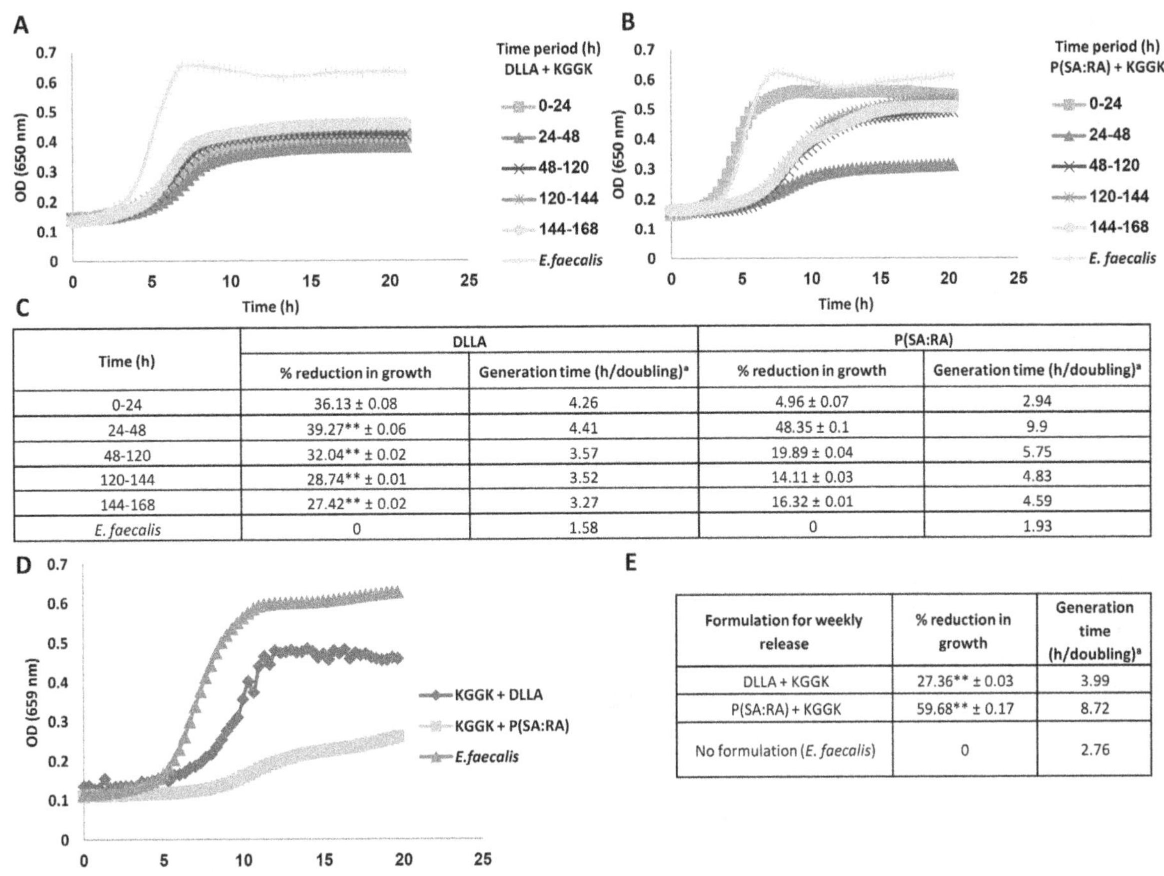

C

Time (h)	DLLA		P(SA:RA)	
	% reduction in growth	Generation time (h/doubling)[a]	% reduction in growth	Generation time (h/doubling)[a]
0-24	36.13 ± 0.08	4.26	4.96 ± 0.07	2.94
24-48	39.27** ± 0.06	4.41	48.35 ± 0.1	9.9
48-120	32.04** ± 0.02	3.57	19.89 ± 0.04	5.75
120-144	28.74** ± 0.01	3.52	14.11 ± 0.03	4.83
144-168	27.42** ± 0.02	3.27	16.32 ± 0.01	4.59
E. faecalis	0	1.58	0	1.93

E

Formulation for weekly release	% reduction in growth	Generation time (h/doubling)[a]
DLLA + KGGK	27.36** ± 0.03	3.99
P(SA:RA) + KGGK	59.68** ± 0.17	8.72
No formulation (*E. faecalis*)	0	2.76

[a] calculated at logarithmic phase at each tested condition ** P<0.01 compared with control without AMP, determined using Student's t-test

Figure 2. Growth inhibition of *E. faecalis* by KGGK released from P(SA:RA) or from DLLA. The side walls of 6 wells from line A of a 96 microwell plate were coated with the tested formulation (100 μg peptide+100 mg polymer, ratio 1:1000). Fresh medium was added to the first line of wells and was transferred every 24 hrs to a new line below for a week. Then the bacteria were added to the tested wells and the plate was incubated at 37°C in a VERSAmax microplate reader and OD_{650} in each well was followed automatically for 20 hrs. (A, C) KGGK+DLLA. (B, C) KGGK+ P(SA:RA) (D, E) weekly release of both formulations. Percent growth inhibition calculated compared with that of the non- treated bacteria during the logarithmic phase of the non treated bacteria. Generation time was calculated from each curve using the section representing the exponential growth phase (C, E).

and is less likely to induce bacterial resistance than antibiotics. Recently, synthetic AMPs mimicking these strategic antibacterial agents have been gaining interest. Combining sustained release and an antimicrobial compound holds many advantages and has proved itself in the past. In this study a potent antimicrobial agent was identified against *E. faecalis*, and then incorporated in two candidate biodegradable polymers. The most efficient of the six investigated antimicrobial peptides was the lipopeptide C16-KGGK.

The antibacterial effect in this study was tested against *E. faecalis*. We chose this bacterium as an example of a pathogen that causes severe nosocomial infections and as an example of a strongly forming biofilm bacterium. *E. faecalis* can grow and survive in a wide range of environments (wide range of temperatures and pH) affording it the ability to surmount many obstacles [9]. Interestingly, root canal treated teeth are about nine times more likely to harbor *E. faecalis* than are primary infections. *E. faecalis* has been found in root canal-treated teeth in 30% to 90% of the cases. This frustrating rate of post treatment disease is mainly attributed to the limitations of the present technology that offers no tool to combat intra-canal infection following the cleaning and shaping stage of the endodontic treatment [21].

The tested antibacterial peptides were first assayed in suspension against planktonic *E. faecalis*. Although hBD3 was previously reported as being a highly potent antibacterial AMP against *E. faecalis* [8,22–24], in the present study it showed an antibacterial effect against *E. faecalis* only when used at high concentrations. This may be due to the differences in *E. faecalis* strains and the hBD3 chemical synthesis. As hBD3 is a costly peptide, high concentrations are predestined to be irrelevant as a conventional therapeutic agent and thus were not tested further. Screening of the AMPs' MICs demonstrated that the C16-KGGK lipopeptide was the most potent against *E. faecalis* and it was further investigated and formulated into biodegradable polymers. Interestingly, in some experiments at low concentrations bacterial growth was not inhibited but rather accelerated. As this phenomenon may compromise the antimicrobial effect, further investigation of the peptides' mode of release is required. The exact mechanism of this opposite outcome is unknown, but the main assumption is that somehow the bacteria overcome lower concentrations of the lipopeptide and show accelerated growth compared with the untreated bacteria. This phenomenon needs to be considered when dealing with the amount of peptides that are released from the polymer. The new biohybrid medium incorpo-

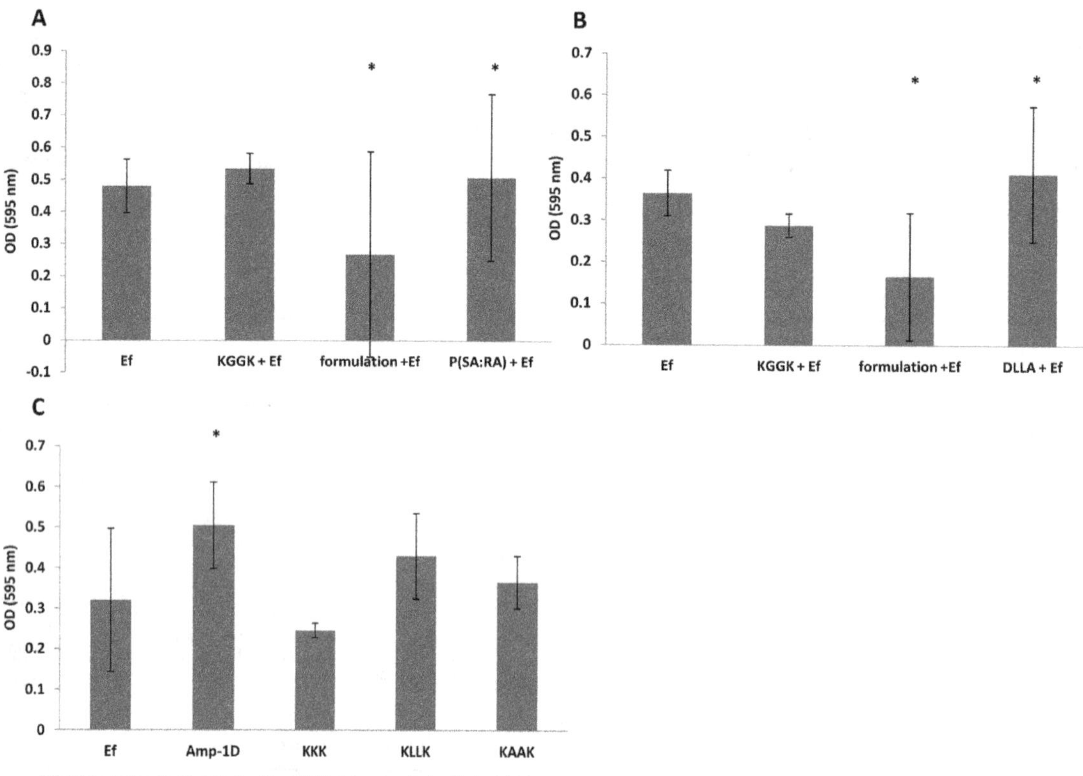

Figure 3. Effect of the antimicrobial peptides on the development of *E. faecalis* biofilms. *E. faecalis* biofilms were grown in 96 microtiter plate wells for 72 hrs in the presence of KGGK formulated with P(SA:RA) (panel A), or formulated with DLLA (B,) or with soluble peptides (C). *Ef* represents the non-treated bacteria, *KGGK+EF* - bacteria treated only with peptide, *formulation+EF* - bacteria treated with polymer and peptide and *Ef + polymer* - bacteria treated only with polymer control). The biofilm was stained with 1% crystal violet measured at OD 595 nm (see Materials and Methods). The optical density of the polymers alone without the bacteria was subtracted from the results of the biofilm that came in contact with the formulation and the polymer.

rating C16-KGGK results in an anti- *E. faecalis* effect when tested against planktonic bacteria. Indeed calculation of bacterial number (using a calibration curve) revealed that the final bacterial load was lower by one order of magnitude in the treated wells. Additionally, the slope of the curve representing the bacterial growth rate (generation time) was more moderate in the treated bacteria, showing that the peptide is released into the medium. The generation time of the bacteria treated with each of the

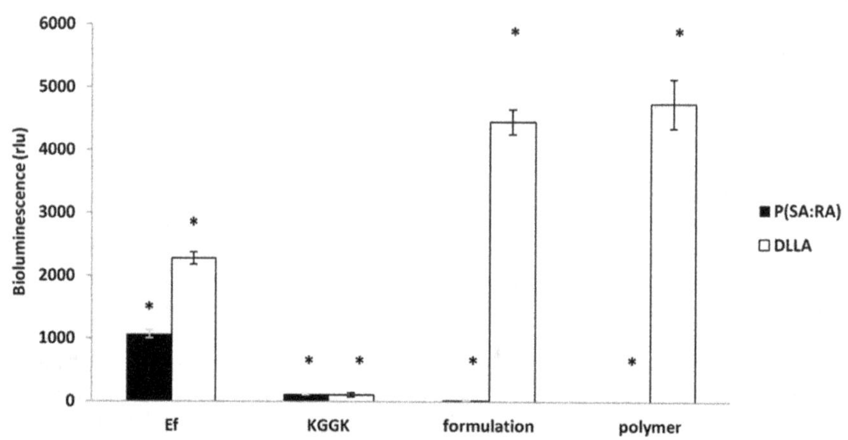

Figure 4. Effect of KGGK incorporated in biodegradable polymer on ATP in *E. faecalis* biofilm. Biofilm was exposed to the formulation for 72 hrs and ATP was measured as described in Materials and Methods. *Ef* represents the untreated bacteria as control, *KGGK* - bacteria treated only with peptide; *formulation* - bacteria treated with sustained release peptide; *polymer* - bacteria treated only with polymer as control.

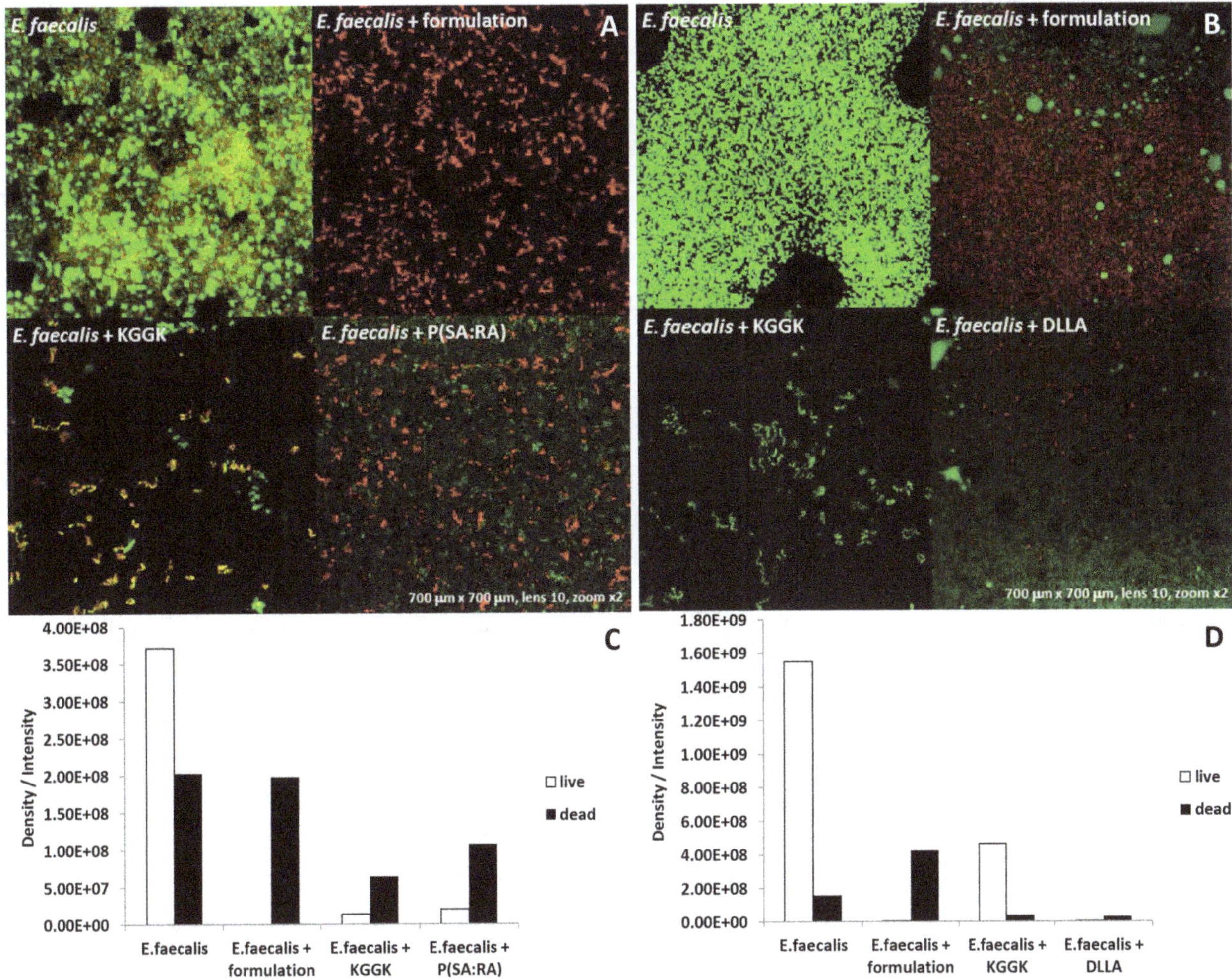

Figure 5. Live/dead assay. *E. faecalis* came in contact with the examined materials for 72 hrs to form biofilm. The medium was discarded and the wells were washed gently with PBS. The live bacteria were stained with green dye, the dead bacteria were stained with a red dye. A 5 ml volume of each dye from the dead/live dying kit was added to 450 µl PBS using an Eppendorf and 30 µl of the solution were added in each well. Images were taken using an Olympus confocal microscope [A, B]. The black column represents the dead bacteria, the white column represents the live bacteria. The biofilm was quantified by measuring the area occupied by the bacteria with the aid of Image Pro 4.5 software (Media Cybernetics) [C, D].

formulations and especially with P(SA:RA) was longer compared with that of the non-treated bacteria.

Bacteria grow naturally as biofilm, especially *E. faecalis* within the root canal. Moreover, *E. faecalis* is known to form biofilms that greatly increase its resistance to phagocytosis, antibodies and antimicrobials [21]. Therefore, in the second part of the study the anti-biofilm effect was tested. Three approaches were used to test the activity of the soluble AMPs and the new controlled release C16-KGGK lipopeptide formulations against *E. faecalis* biofilms. In the first, crystal violet was used to stain and measure biofilm mass. In the second, an ATP bioluminescence assay was performed and used as a viability indicator. In the third, the vitality of bacteria grown in a biofilm was tested using a dead/live stain. All three experiments revealed inhibition of biofilm formation when *E.* faecalis was exposed to the novel formulation. The three aspects examined were the amount of biofilm, its metabolic state and bacterial viability. Interestingly, the formulations were effective against a biofilm in the process of formation and against an established biofilm (mature biofilm). This is an important finding considering the fact that mature biofilm is much harder to treat because of its virulence factors. Moreover, we specifically tested the formulations' potency against ATCC v583 strain due to its high known resistance to several antibiotics (among them vancomycin), compared to other strains such as ATCC 29212 [25]. It can be suggested that a formulation that was shown to be active against ATCC v583 is likely to be potent against other *E. faecalis* strains.

The polymer candidates which contain fatty acids have several advantages over other biodegradable polymers such as: flexibility, low melting point, improved handling and provide better degradation and release profiles [26]. As previously reported, biodegradable polyanhydrides and polyesters are useful materials for controlled drug delivery. They have a hydrophobic backbone with hydrolytically labile anhydride and/or ester that may be hydrolyzed to dicarboxylic acids and hydroxy acid monomers when placed in an aqueous medium. Fatty acids are suitable candidates for the preparation of biodegradable polymers, as they are natural body components and hydrophobic, and thus may retain an encapsulated drug for longer time periods when used as drug carriers [11]. Moreover, it was shown that these polymers are biocompatible [15]. As described before, this polymer-peptide interaction may hold many advantages over the peptide by itself, such as improved solubility, reduced immunogenicity, increased stability against degradation and prolonged biological activity [27]. Two different polymers were tested as delivery media and led to different results in their activity and mode of action. In the sustained release experiments, DLLA showed similar bacterial kinetic growth curves whereas P(SA:RA) did not, indicating that the two have separate modes of release. Furthermore, in the ATP bioluminescence assay, DLLA presented higher levels of luminescence and accordingly higher levels of ATP, suggesting that this polymer elevates the metabolic state of the biofilm, compared with P(SA:RA) which had the opposite effect. In the live/dead assay, the main difference between the two polymers is that in P(SA:RA) a larger amount of dead bacteria appeared, reinforcing our previous findings that P(SA:RA) itself may be an antibacterial agent. Thus, P(SA:RA) is apparently a more suitable delivery medium for this purpose.

Within the sustained release field wider experiments should be performed in order to learn the exact amount of peptide released from the polymer and the kinetics of the release. Furthermore, the period tested in this study was one week so that additional experiments should be performed using various periods of time.

Conclusions

Synthetic AMPs were shown to have an effective antimicrobial activity against *E. faecalis*. A peptide that allows selective killing of *E. faecalis* would be a good candidate for endodontic treatment. Here, we show that a synthetic lipopetide can be highly effective against *E. faecalis*. Moreover, this lipopeptide when formulated in a biohybrid polymer medium has an increased antibiofilm effect. Thus, the novel effective formulation presented here can be advantageous in root canal treatment for the prevention of endodontic failure due to *E. faecalis*.

Author Contributions

Conceived and designed the experiments: LHE NB GB AS. Performed the experiments: LHE. Analyzed the data: LHE NB GB AS. Contributed reagents/materials/analysis tools: AJD EA YS NB GB AS. Contributed to the writing of the manuscript: LHE NB GB AS.

References

1. Blondelle SE, Perez-Paya E, Houghten RA (1996) Synthetic combinatorial libraries: novel discovery strategy for identification of antimicrobial agents. Antimicrob Agents Chemother 40: 1067–1071.
2. Boman HG (2003) Antibacterial peptides: basic facts and emerging concepts. J Intern Med 254: 197–215.
3. Reddy KV, Yedery RD, Aranha C (2004) Antimicrobial peptides: premises and promises. Int J Antimicrob Agents 24: 536–547.
4. Devine DA, Marsh PD, Percival RS, Rangarajan M, Curtis MA (1999) Modulation of antibacterial peptide activity by products of *Porphyromonas gingivalis* and *Prevotella* spp. Microbiology-Uk 145: 965–971.
5. Peschel A, Otto M, Jack RW, Kalbacher H, Jung G, et al. (1999) Inactivation of the dlt operon in *Staphylococcus aureus* confers sensitivity to defensins, protegrins, and other antimicrobial peptides. J Biol Chem 274: 8405–8410.
6. Shelton CL, Raffel FK, Beatty WL, Johnson SM, Mason KM (2011) Sap transporter mediated import and subsequent degradation of antimicrobial peptides in *Haemophilus*. PLoS Pathog 7: e1002360.
7. Nikaido H (1996) Multidrug efflux pumps of gram-negative bacteria. J Bacteriol 178: 5853–5859.
8. Joly S, Maze C, McCray PB Jr, Guthmiller JM (2004) Human beta-defensins 2 and 3 demonstrate strain-selective activity against oral microorganisms. J Clin Microbiol 42: 1024–1029.
9. Murray BE (1990) The life and times of the *Enterococcus*. Clin Microbiol Rev 3: 46–65.
10. Al-Tahami K, Singh J (2007) Smart polymer based delivery systems for peptides and proteins. Recent Pat Drug Deliv Formul 1: 65–71.
11. Shikanov A, Vaisman B, Krasko MY, Nyska A, Domb AJ (2004) Poly(sebacic acid-co-ricinoleic acid) biodegradable carrier for paclitaxel: In vitro release and in vivo toxicity. Journal of Biomedical Materials Research Part A 69A: 47–54.
12. Shikanov A, Domb AJ (2006) Poly(sebacic acid-co-ricinoleic acid) biodegradable injectable in situ gelling polymer. Biomacromolecules 7: 288–296.
13. Shikanov A, Ezra A, Domb AJ (2005) Poly(sebacic acid-co-ricinoleic acid) biodegradable carrier for paclitaxel-effect of additives. Journal of Controlled Release 105: 52–67.
14. Slager J, Tyler B, Shikanov A, Domb AJ, Shogen K, et al. (2009) Local Controlled Delivery of Anti-Neoplastic RNAse to the Brain. Pharmaceutical Research 26: 1838–1846.
15. Vaisman B, Motiei M, Nyska A, Domb AJ (2010) Biocompatibility and safety evaluation of a ricinoleic acid-based poly(ester-anhydride) copolymer after implantation in rats. Journal of Biomedical Materials Research Part A 92A: 419–431.
16. Makovitzki A, Avrahami D, Shai Y (2006) Ultrashort antibacterial and antifungal lipopeptides. Proc Natl Acad Sci U S A 103: 15997–16002.
17. Papo N, Oren Z, Pag U, Sahl HG, Shai Y (2002) The consequence of sequence alteration of an amphipathic alpha-helical antimicrobial peptide and its diastereomers. J Biol Chem 277: 33913–33921.
18. Krasko MY, Shikanov A, Ezra A, Domb AJ (2003) Poly(ester anhydride)s prepared by the insertion of ricinoleic acid into poly(sebacic acid). Journal of Polymer Science Part a-Polymer Chemistry 41: 1059–1069.
19. Soren L, Nilsson M, Nilsson LE (1995) Quantitation of Antibiotic Effects on Bacteria by Bioluminescence, Viable Counting and Quantal Analysis. Journal of Antimicrobial Chemotherapy 35: 669–674.
20. Beyth N, Yudovin-Farber I, Perez-Davidi M, Domb AJ, Weiss EI (2010) Polyethyleneimine nanoparticles incorporated into resin composite cause cell death and trigger biofilm stress in vivo. Proc Natl Acad Sci U S A 107: 22038–22043.
21. Stuart CH, Schwartz SA, Beeson TJ, Owatz CB (2006) *Enterococcus faecalis*: its role in root canal treatment failure and current concepts in retreatment. J Endod 32: 93–98.
22. Lee JK, Chang SW, Perinpanayagam H, Lim SM, Park YJ, et al. (2013) Antibacterial Efficacy of a Human beta-Defensin-3 Peptide on Multispecies Biofilms. Journal of Endodontics 39: 1625–1629.
23. Lee JK, Park YJ, Kum KY, Han SH, Chang SW, et al. (2013) Antimicrobial efficacy of a human -defensin-3 peptide using an *Enterococcus faecalis* dentine infection model. International Endodontic Journal 46: 406–412.
24. Lee SH, Baek DH (2012) Antibacterial and Neutralizing Effect of Human beta-Defensins on *Enterococcus faecalis* and *Enterococcus faecalis* Lipoteichoic Add. Journal of Endodontics 38: 351–356.
25. Swenson JM, Clark NC, Sahm DF, Ferraro ML, Doern G, et al. (1995) Molecular Characterization and Multilaboratory Evaluation of *Enterococcus-Faecalis* Atcc-51299 for Quality-Control of Screening-Tests for Vancomycin and High-Level Aminoglycoside Resistance in *Enterococci*. J Clin Microbiol 33: 3019–3021.
26. Jain JP, Sokolsky M, Kumar N, Domb AJ (2008) Fatty acid based biodegradable polymer. Polymer Reviews 48: 156–191.
27. Krishna OD, Kiick KL (2010) Protein- and peptide-modified synthetic polymeric biomaterials. Biopolymers 94: 32–48.

Evaluation of Gene Expression and Alginate Production in Response to Oxygen Transfer in Continuous Culture of *Azotobacter vinelandii*

Alvaro Díaz-Barrera*, Fabiola Martínez, Felipe Guevara Pezoa, Fernando Acevedo

Escuela de Ingeniería Bioquímica, Pontificia Universidad Catolica de Valparaiso, Valparaiso, Chile

Abstract

Alginates are polysaccharides used as food additives and encapsulation agents in biotechnology, and their functional properties depend on its molecular weight. In this study, different steady-states in continuous cultures of *A. vinelandii* were established to determine the effect of the dilution rate (D) and the agitation rate on alginate production and expression of genes involved in alginate polymerization and depolymerization. Both, the agitation and dilution rates, determined the partitioning of the carbon utilization from sucrose into alginate and CO_2 under oxygen-limiting conditions. A low D (0.07 h^{-1}) and 500 rpm resulted in the highest carbon utilization into alginate (25%). Quantitative real-time polymerase chain reaction was used to determine the transcription level of six genes involved in alginate polymerization and depolymerization. In chemostat cultures at 0.07 h^{-1}, the gene expression was affected by changes in the agitation rate. By increasing the agitation rate from 400 to 600 rpm, the *algE7* gene expression decreased tenfold, whereas *alyA1*, *algL* and *alyA2* gene expression increased between 1.5 and 2.8 times under similar conditions evaluated. Chemostat at 0.07 h^{-1} showed a highest alginate molecular weight (580 kDa) at 500 rpm whereas similar molecular weights (480 kDa) were obtained at 400 and 600 rpm. The highest molecular weight was not explained by changes in the expression of *alg8* and *alg44* (genes involved in alginate polymerization). Nonetheless, a different expression pattern observed for lyases could explain the highest alginate molecular weight obtained. Overall, the results suggest that the control of alginate molecular weight in *A. vinelandii* cells growing in continuous mode is determined by a balance between the gene expression of intracellular and extracellular lyases in response to oxygen availability. These findings better our understanding of the biosynthesis of bacterial alginate and help us progress toward obtain tailor-made alginates.

Editor: Baochuan Lin, Naval Research Laboratory, United States of America

Funding: The authors acknowledge FONDECYT Grants 11110311 and DI-PUCV 203.784 for financial support. The funders had no role in study design, data collection and analysis, decision to publish, or preparation of the manuscript.

Competing Interests: The authors have declared that no competing interests exist.

* Email: alvaro.diaz@ucv.cl

Introduction

Alginate is a copolymer of (1–4)-linked residues of β-D-mannuronic acid (M) and α-L-guluronic acid (G) and produced by *Pseudomonas* and *Azotobacter* species [1]. These polymers have a wide range of applications: they are used as food additives and encapsulation agents in biotechnology and have a promising potential for the biomedical field [2]. *Azotobacter vinelandii* is an aerobic bacterium that produces two polymers of industrial interest: poly-β-hydroxybutyrate (PHB) and alginate. Their functional properties depend on their monomer composition and molecular weight [3].

Regarding alginate biosynthesis, it is well known that alginate is synthesized as a polymannuronate from its cytosolic precursor (GDP-M). Nevertheless, the mechanisms involved in the polymerization, modification (acetylation and depolymerization) and translocation steps are poorly elucidated [4]. The alginate depolymerization steps in *A. vinelandii* are complex because this microorganism possesses three intracellular lyases (AlgL, AlyA1 and AlyA2) and two extracellular lyases (AlgE7 and AlyA3) [5]. The polymerization process is catalyzed by Alg8, which is a bottleneck in the biosynthesis of alginate [6]. Alg44 has been postulated to play an indirect role in alginate polymerization, facilitating the transport, modification and secretion of alginate [4]. However, the specific role of Alg44 remains unclear.

In cultures operated in continuous, it has been demonstrated that the oxygen transfer rate (OTR), and the dilution rate (D) affect alginate production, particularly the molecular weight of the final product [7–9]. By varying the agitation rate and increasing the specific oxygen uptake rate from 2.2 to 4.8 mmol g^{-1} h^{-1}, alginate production can be improved. This improvement can be attributed to changes in carbon flux [9]. A previous study [9] reported that a lower alginate molecular weight can be obtained by increasing the specific oxygen uptake rate; increased expression of *algL* (approximately 8-fold) was hypothesized to result in a decreased molecular weight. Similarly, Díaz-Barrera *et al.* [10] showed that increased alginate molecular weight produced by *A. vinelandii* continuous cultures can be linked to higher *alg8* gene expression.

Recently, Flores *et al.* [11] evaluated the expression of genes involved in alginate polymerization and depolymerization, as well as lyase activity in batch cultures of *A. vinelandii* under dissolved oxygen tension (DOT) controlled conditions. These authors showed that in batch cultures at 1% DOT, low lyase activity and high expression levels of *alg8* and *alg44* might be mechanisms by which oxygen regulates the synthesis of alginates. Those experiments were conducted in batch-mode, in which the specific growth rate changes continuously with time, particularly under oxygen-limited conditions. Because the specific growth rate plays an important role in determining the molecular weight of the alginate [8,12] in this work, different steady-states in continuous cultures were established to evaluate the effect of oxygen supply conditions on alginate production and gene expression at a constant specific growth rate. To extend the knowledge about the effects of oxygen availability and D on alginate production and expression of genes involved in alginate polymerization and depolymerization, the objective of this study was to evaluate how alginate production (particularly its molecular weight) and the expression of genes involved in polymerization (*alg8, alg44*) and depolymerization (*algL, algE7, alyA1*, and *alyA2*) are influenced by oxygen transfer in continuous cultures of *A. vinelandii*.

Materials and Methods

Strain and culture medium

Strain ATCC 9046 of *Azotobacter vinelandii* was used. The bacterium was grown under nitrogen fixation conditions. The culture medium used for chemostat cultures was (in g l^{-1}): 10 sucrose, 0.66 K_2HPO_4, 0.16 KH_2PO_4, 0.056 $CaSO_4 \cdot 2H_2O$, 0.2 NaCl, 0.2 $MgSO_4 \cdot 7H_2O$, 0.0029 $Na_2MoO_4 \cdot 2H_2O$, and 0.027 $FeSO_4 \cdot 7H_2O$. The sucrose, K_2HPO_4 and KH_2PO_4 were dissolved in bioreactor and autoclaving at 121°C during 20 min. $CaSO_4 \cdot 2H_2O$ was sterilized separately in the autoclave (121°C, 20 min). To avoid precipitation, the solutions of NaCl, $MgSO_4 \cdot 7H_2O$, $Na_2MoO_4 \cdot 2H_2O$, and $FeSO_4 \cdot 7H_2O$ were separated from the other medium components during sterilization (autoclave at 121°C for 20 min).

Inoculum preparation

The inoculum for the bioreactor was prepared in 500 ml Erlenmeyer flasks with 100 ml of culture medium, as described previously [10]. The initial pH was adjusted to 7.0 using 2 M NaOH. The microorganism was incubated at 200 rpm and 30°C in an orbital incubator shaker (New Brunswick, USA). After 15 h, the cells were transferred (10% v/v) to a bioreactor operated in batch mode.

Fermentation conditions

Chemostat experiments operated at steady-state were conducted in a 3 l bioreactor (Applikon, Schiedam, Netherlands) with a working volume of 2 l at 30°C and pH 7.0 controlled by automatic addition of 2 M NaOH. The stirred fermenter was aerated at 2 l min^{-1} (1.0 vvm) and agitated at 400, 500, and 600 rpm. The DOT was measured with a polarographic oxygen probe (Ingold, Mettler-Toledo) and was not controlled. The bioreactor was operated in batch mode for the first 15 h, then in continuous culture mode with D values of 0.07, 0.09 and 0.11 h^{-1}. The working volume was kept constant by withdrawing culture broth through a continuously operated peristaltic pump (Cole-Parmer, USA). The continuous culture reached steady state conditions when the optical density at 540 nm (OD_{540}) and the sucrose concentration remained constant (<10% variation) after at least 4 residence times.

Samples of cultures (20 ml) were withdrawn from the bioreactor for analytical measurements. All analyses were carried out in triplicate. The results shown are the mean value of two independent chemostat runs, and the standard deviations among replicates are given.

Analytical methods

Cell growth was estimated gravimetrically, as described previously by Díaz-Barrera *et al.* [10]. Sucrose concentration was determined by hydrolysis with β-fructofuranosidase, followed by the determination of reducing sugars with dinitrosalicylic acid (DNS) reagent [13]. PHB was extracted from cells and quantified by HPLC as crotonic acid. PHB was hydrolyzed to crotonic acid using concentrated H_2SO_4: 3 mg of biomass (previously dried) was weighed in an Eppendorf tube; 1 ml of H_2SO_4 was added, and the

Table 1. Primers used for gene expression analysis by quantitative real-time PCR assay.

Gene	Primers
alg8	5′-TGTTGAACCAGCTCTGGAAG-3′
	5′-CCTACCCGCTGATCCTCTAC-3′
alg44	5′-CGACAACTTCACCGAAGGG-3′
	5′-TGACGAAGTAGAGGTCGTAGAG-3′
algE7	5′-AGATAGGTGCGGTTGGTTTC-3′
	5′-CTCCGACCTGATTCTCGATT-3′
algL	5′-GCCCAGTAGGAGTGGTTGTT-3′
	5′-CTGAAATTCTCCAGTTCGCA-3′
alyA1	5′-CGGTCGGTATTGCACATAGA-3′
	5′-CAAGATCCACCGTTTGAGTG-3′
alyA2	5′-AACTGAGTGCAACCTTGACG-3′
	5′-GCTGCACGTTGATCTTGAAT-3′
gyrA	5′-ACCTGATCACCGAGGAAGAG-3′
	5′-AGGTGCTCGACGTAATCCTC-3′

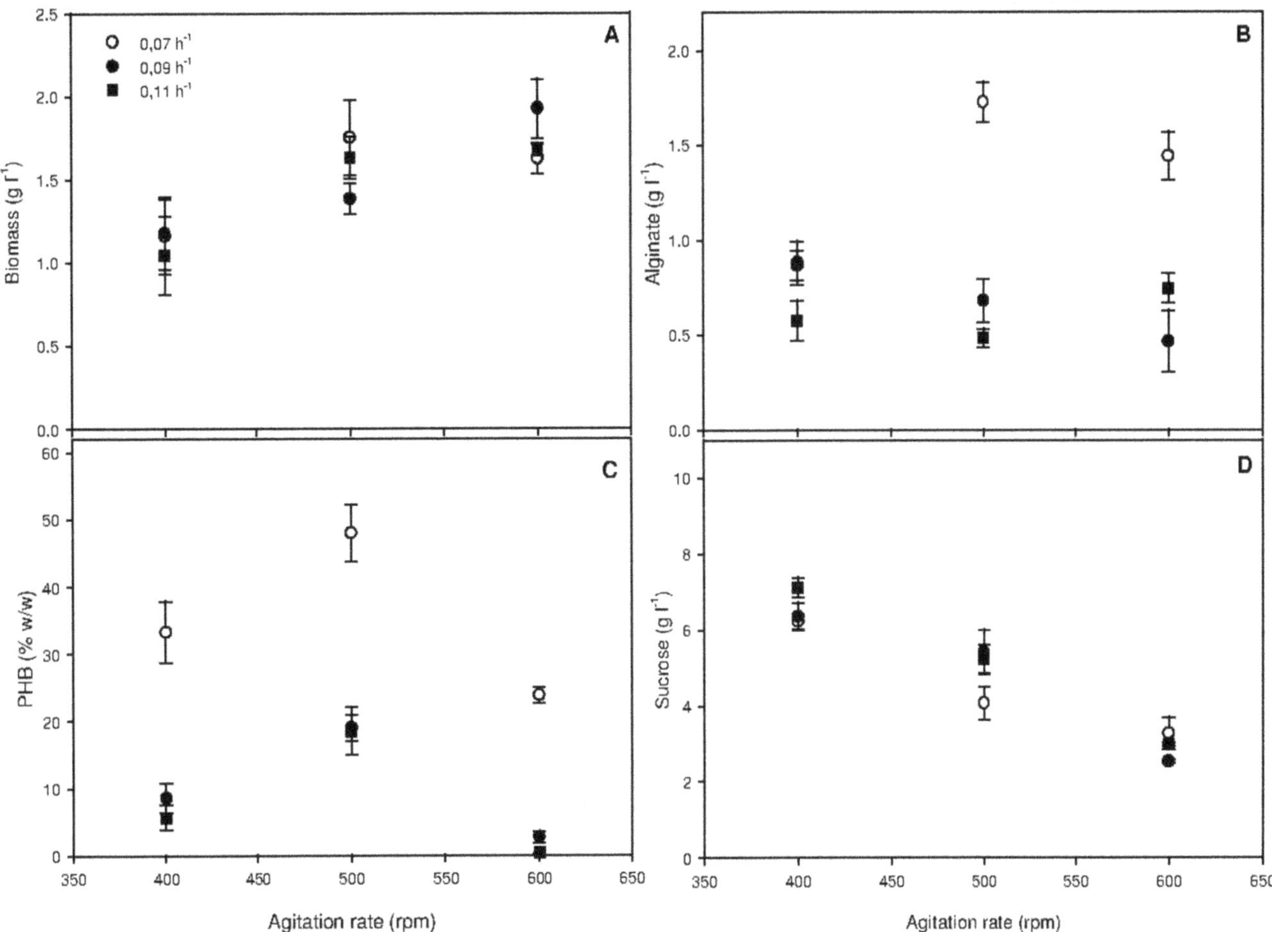

Figure 1. Effect of the agitation rate on biomass, alginate, PHB and sucrose levels in chemostat cultures of *A. vinelandii* conducted at different dilution rates. Data are reported as the mean value ± standard deviation.

tube heated at 90°C and agitated for 1 h. The sample was diluted with Milli-Q water and was assayed using an HPLC-UV system with an Aminex HPX-87H ion-exclusion organic acid column. Elution was performed with 0.005 M H_2SO_4 at 0.6 ml min^{-1} and 35°C [14]. The alginate concentration was quantified by the metahydroxidiphenyl method for the measurement of uronic acid (monomers of alginate) [15]. The molecular weight of alginate was determined by gel permeation chromatography (GPC) using a serial set of Ultrahydrogel columns (UG 500 and Linear Waters) in an HPLC system with a differential refractometer detector (PerkinElmer, USA). Elution was performed with 0.1 M $NaNO_3$ at 35°C at a flow rate of 0.8 ml min^{-1} using pullulans from *Aureobasidium pullulans* as molecular weight standards [16].

Gas analysis and respiratory measurements

Gas analysis was performed by online measurements of O_2 and CO_2 in the exit gas and compared with measurements taken of the inlet gas with a gas analyzer (Teledyne Instruments, model 7500). The OTR and carbon dioxide transfer rate (CTR) were determined by gas analysis and calculated by carrying out gas mass balances [17] as follows:

$$OTR = \frac{M_{O_2} F_G^{in}}{V_R V_M}\left(X_{O_2}^{in} - X_{O_2}^{out}\left(\frac{1 - X_{O_2}^{in} - X_{CO_2}^{in}}{1 - X_{O_2}^{out} - X_{CO_2}^{out}}\right)\right) \quad (1)$$

$$CTR = \frac{M_{CO_2} F_G^{in}}{V_R V_M}\left(X_{CO_2}^{out}\left(\frac{1 - X_{O_2}^{in} - X_{CO_2}^{in}}{1 - X_{O_2}^{out} - X_{CO_2}^{out}}\right) - X_{CO_2}^{in}\right) \quad (2)$$

where M_{o2}, M_{co2} are the molecular mass of oxygen and carbon dioxide (g $mmol^{-1}$), respectively, F_G^{in} the volumetric inlet air flow at standard conditions (l h^{-1}), V_R the working volume (l), V_M the mol volume of the ideal gas at standard conditions (l $mmol^{-1}$), X_{o2}^{in} and X_{co2}^{in} the molar fractions of oxygen and carbon dioxide in the inlet air, respectively (mol mol^{-1}), X_{o2}^{out} and X_{co2}^{out} the molar fractions of oxygen and carbon dioxide in the outlet air of the fermenter, respectively (mol mol^{-1}).

Quantitative real-time PCR assay

Expression of genes *algL, alyA1, alyA2, algE7, alg8 and alg44* were analyzed by quantitative RT-PCR. Measurements were carried out in a Light Cycler 2.0 thermocycler (Roche), using commercial Maxima Sybr Green qPCR Master Mix (Thermo Scientific). Fifty micrograms of RNA purified with an RNA Isolation Kit (Roche) and treated with DNAse (Thermo Scientific) was mixed with Revert Aid H minus Reverse Transcriptase (Thermo Scientific) according to the manufacturer's protocol to obtain the cDNA. The samples were subjected to an initial denaturation at 95°C for 10 min, followed by 40 cycles of 15 s at 95°C and 60 s at 60°C. As an internal standard and control, the

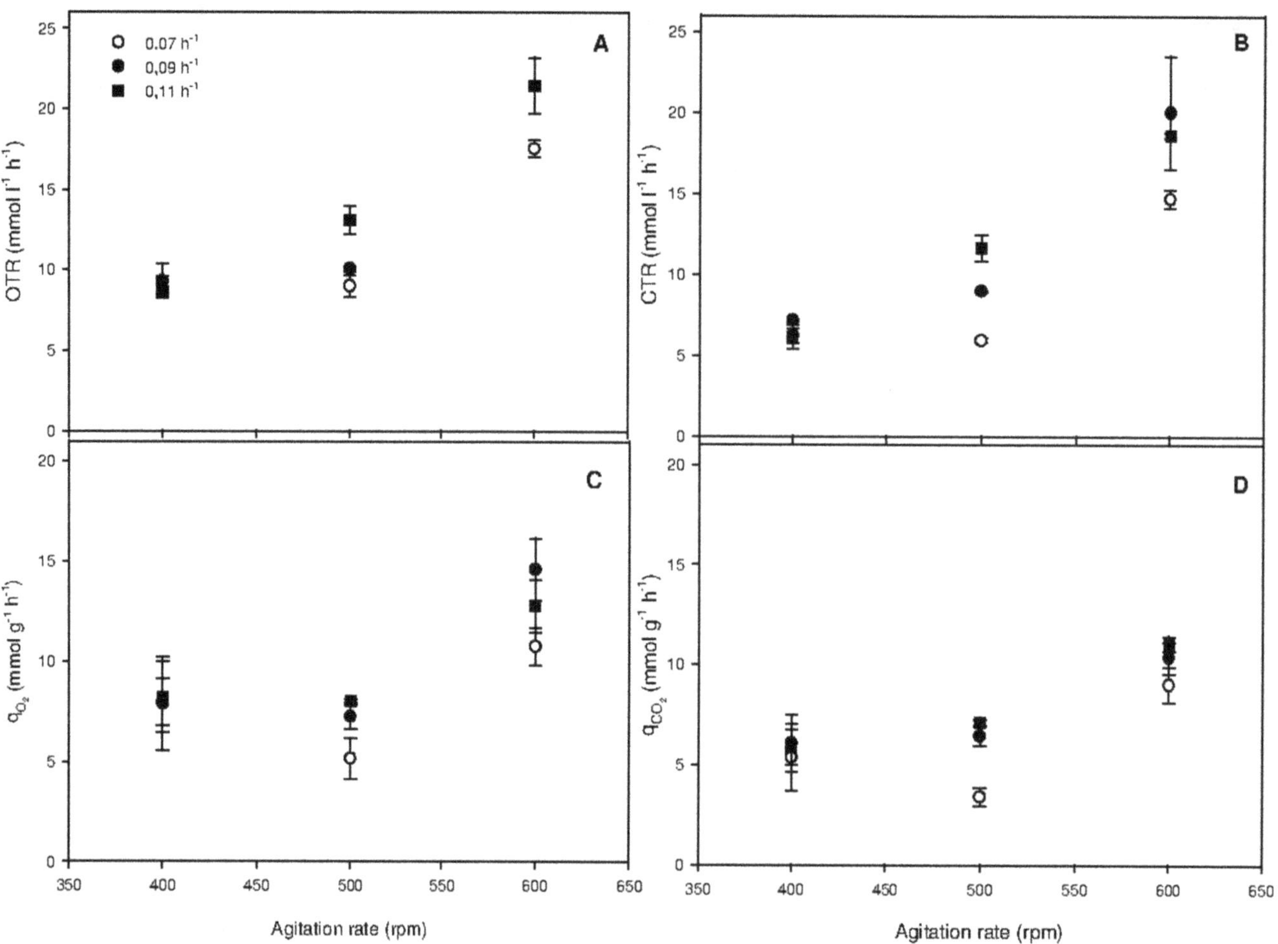

Figure 2. Respiratory activities of *A. vinelandii* as a function of the agitation rate and the dilution rate in continuous cultures. Data are reported as the mean value ± standard deviation.

expression level of *gyrA* was also determined. Relative gene expression values were obtained using the ΔΔCt method. Table 1 shows the primers used for these analyses. The level of genes expression was normalized according to the level of the *gyrA* mRNA and the data are presented as fold changes of mRNA levels respect of calibrator value (500 rpm).

Calculation of the specific uptake/production rates and carbon recovery

The specific alginate production rate (q_p), specific sucrose uptake rate (q_s), specific oxygen uptake rate (q_{o2}), and specific carbon dioxide rate (q_{co2}) were calculated at steady state conditions, considering D, alginate concentration (P), OTR and

Table 2. Sucrose yields and specific sucrose uptake rate for chemostat cultures of *A. vinelandii* at different D and agitation rates.

D (h^{-1})	Agitation rate (rpm)	$Y_{x/s}$ ($g\ g^{-1}$)	$Y_{p/s}$ ($g\ g^{-1}$)	q_s ($g\ g^{-1}\ h^{-1}$)
0.07	400	0.28±0.05	0.18±0.01	0.255±0.045
	500	0.27±0.02	0.27±0.01	0.257±0.014
	600	0.23±0.01	0.17±0.01	0.303±0.010
0.09	400	0.29±0.03	0.19±0.01	0.312±0.040
	500	0.27±0.03	0.13±0.01	0.333±0.041
	600	0.25±0.02	0.06±0.02	0.365±0.026
0.11	400	0.32±0.05	0.15±0.03	0.350±0.062
	500	0.31±0.01	0.09±0.01	0.353±0.002
	600	0.23±0.01	0.09±0.01	0.472±0.013

Values are means of the measurements ± SD.

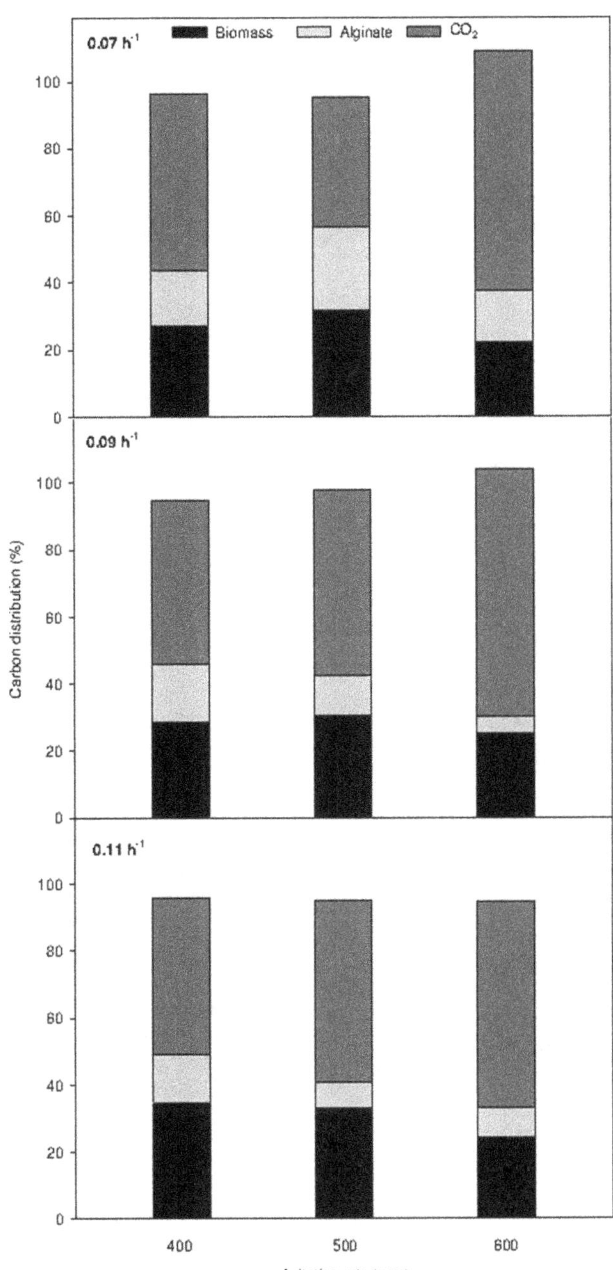

Figure 3. Carbon distribution for biomass, alginate, and carbon dioxide at different agitation rates in chemostat cultures of *A. vinelandii* conducted at 0.07, 0.09 and 0.11 h^{-1}.

CTR values, biomass concentration (X), residual sucrose concentration (S) in the bioreactor and sucrose concentration in the feed medium (S_o) using the following equations:

$$q_p = \frac{DP}{X} \tag{3}$$

$$q_s = \frac{D(So - S)}{X} \tag{4}$$

$$q_{O_2} = \frac{OTR}{X} \tag{5}$$

$$q_{CO_2} = \frac{CTR}{X} \tag{6}$$

Carbon recovery under steady state conditions was determined from reactor mass balances according to Díaz-Barrera *et al.* [9], as described previously.

Results and Discussion

Continuous cultures at different agitation and dilution rates

Figure 1 shows the effect of agitation rate and D on biomass, sucrose, alginate and PHB concentration at steady state. In a chemostat agitated at 400 rpm, the biomass concentration was similar for the different dilution rates evaluated, reaching approximately 1.0 g l^{-1} (Fig. 1a). An increase in the agitation rate from 500 to 600 rpm with D = 0.09 h^{-1} improved biomass production from 1.3 to 1.9 g l^{-1}. It is interesting to note that in chemostat cultures with D = 0.07 h^{-1} and 0.11 h^{-1}, increasing the agitation rate from 500 to 600 rpm did not enhance the biomass concentration. In light of these results, the specific growth rate (i.e., D in chemostats) has more influence than the agitation rate (at least between 500 and 600 rpm) on the diversion of carbon into biomass production.

As shown in Fig. 1b, the alginate production at steady state is strongly affected by D. At the lowest D tested (0.07 h^{-1}) and 500 rpm, the alginate concentration was 2.5 and 3.5 times higher compared to alginate concentrations obtained at 0.09 and 0.11 h^{-1}, respectively. In agreement with this evidence, Díaz-Barrera *et al.* [8] demonstrated that a decrease in the D from 0.08 to 0.05 h^{-1} increased alginate production, whereas biomass and sucrose concentrations remained unchanged.

At a D of 0.09 h^{-1}, the alginate concentration decreased from 0.9 to 0.5 g l^{-1} when the agitation rate was increased from 400 to 600 rpm whereas, at a D of 0.11 h^{-1}, the alginate concentration was only slightly affected by the agitation rate (Fig. 1b). These data indicate that both the specific growth rate and agitation rate influence alginate production. Under the conditions evaluated, lower D and 500 rpm are the optimal operating conditions for the continuous production of alginate.

Behavior similar to that of alginate production was observed for PHB accumulation. The data show that a lower D (i.e., 0.07 h^{-1}) improved PHB production, reaching 48% (w/w) of the dry cell weight for an agitation rate of 500 rpm (Fig. 1c). Regardless of the agitation rate, chemostat cultures conducted at higher D (0.09 and 0.11 h^{-1}) showed a decrease in PHB accumulation (less than 19%), which suggests that D affects intracellular PHB content.

Regardless of the dilution rate, an increase in agitation rate decreased the sucrose concentration at steady-state (Fig. 1d), reaching between 2.5 g l^{-1} and 3.3 g l^{-1} depending on the agitation rate of the culture. Considering that K_s (saturation constant) values are lower than 0.1 g l^{-1} in sucrose-limited chemostat cultures of *A. vinelandii* [18], our chemostat cultures were not limited by sucrose. Under all the conditions evaluated, the DOT at steady-state was nearly zero, which indicates that the cultures were oxygen-limited. Similar behavior has been observed previously [9].

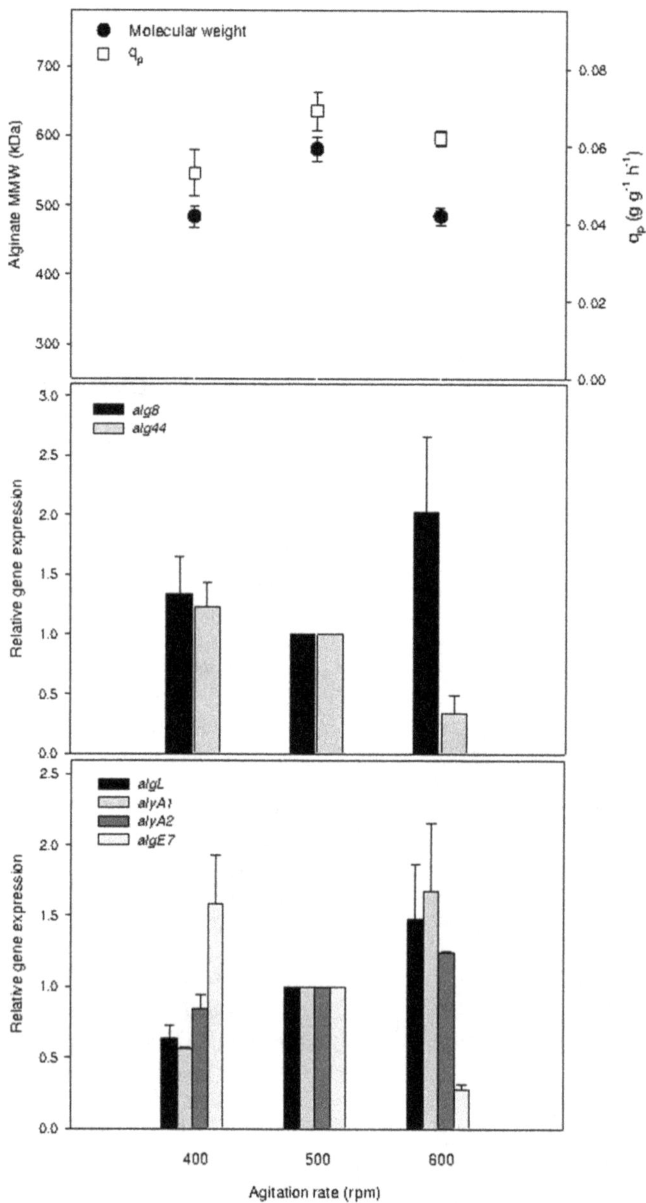

Figure 4. Effect of the agitation rate on alginate molecular weight, specific alginate production rate and expression of genes involved in alginate polymerization and depolymerization in chemostat cultures of *A. vinelandii* performed at a dilution rate of 0.07 h^{-1}. The level of the *algL, alyA1, alyA2, algE7, alg8 and alg44* transcripts was normalized according to the level of the *gyrA* mRNA as described in Section Materials and Methods. The data are presented as fold changes respect of calibrator value (500 rpm). Each point is the mean ± standard deviation.

In the chemostat agitated at 600 rpm, a 1.4 to 1.7-fold increase (depending on D) in sucrose consumption (i.e., inlet sucrose minus sucrose in steady-state) was identified compared to sucrose consumption at 500 rpm. As has been mentioned, in chemostats with D = 0.07 h^{-1} and 0.11 h^{-1}, a change in the agitation rate from 500 to 600 rpm did not increase the biomass and alginate concentrations obtained at steady-state; therefore, this increase in sucrose consumption could be related to increased CO_2 production. To validate this hypothesis, the respiratory activities, i.e., the OTR, CTR, q_{o2} and q_{co2} values of *A. vinelandii* cultures were characterized in all of the chemostat cultures.

Respiratory activities in continuous cultures

Figure 2 shows the OTR, CTR, q_{o2} and q_{co2} obtained for the different chemostat cultures. At the lower agitation rate (400 rpm), a change in D did not affect the respiratory activity, as similar values of OTR, CTR, q_{o2} and q_{co2} were observed at the different dilution rates evaluated. However, at higher agitation rates (500 and 600 rpm), the respiratory activity was affected by D, demonstrating that the influence of D on respiratory activities is dependent on the imposed agitation rate.

As expected, an increase in agitation rate, particularly from 500 to 600 rpm, affected the OTR and q_{o2} level between 2 and 3-fold, depending on the D (Fig. 2a). OTR is a dynamic parameter influenced by operational conditions and the oxygen uptake rate

(OUR) [19]. Under oxygen-limited conditions and at steady-state, the OTR equals the OUR; thus, these results could be interpreted as the effect of oxygen availability on alginate production.

The higher respiratory activity (in particular, the higher OTR and CTR values) obtained with the chemostat at 600 rpm could be related to higher sucrose consumption, suggesting that the carbon flux to the tricarboxylic acid (TCA) cycle is affected by changes in agitation rate or increased oxygen availability. In agreement with this observation, Castillo *et al.* [7] evaluated carbon flux using metabolic flux analysis and demonstrated increased flux through the TCA cycle in cultures of *A. vinelandii* grown under high aeration.

Table 2 shows the sucrose yields and sucrose consumption obtained for chemostat cultures at different D and agitation rates. Depending on the D and agitation rate, the sucrose yield on biomass ($Y_{x/s}$) varied between 0.23 and 0.32 g g^{-1}. Comparing the different dilution rates that were tested, the lowest $Y_{x/s}$ (0.23 or 0.25 g g^{-1}) was obtained in the chemostat agitated at 600 rpm. At 600 rpm, a higher CTR and q_{co2} were obtained (Fig. 2b, d), which may explain the lower $Y_{x/s}$ obtained under these conditions. A decrease in the sucrose yield on alginate ($Y_{p/s}$) was observed when D was increased. For example, in the chemostat agitated at 500 rpm, $Y_{p/s}$ varied between 0.27 to 0.09 g g^{-1} when D was increased from D from 0.07 to 0.11 h^{-1}. It is known that metabolism and physiology in a nutrient-limited state depend on the growth rate [17]; hence, the effect of D on $Y_{p/s}$ could be related changes in carbon flux that allow a higher proportion of sucrose to be diverted to alginate.

As shown in Table 2, q_s varied between 0.26 to 0.47 g g^{-1} h^{-1} at the different steady-states evaluated; these values are lower than the values previously reported by Díaz-Barrera *et al.* [8]. Those authors found values of q_s from 0.42 to 3.19 g g^{-1} h in chemostat cultures conducted using an inlet sucrose concentration of 20 g l^{-1}. In our work, an inlet sucrose concentration of 10 g l^{-1} was used, and it is possible that the differences in q_s could be caused by changes in cell metabolism due the differences in the sucrose consumption. Regardless of the D, an increase in agitation rate increased q_s, indicating a variation in carbon source assimilation. To evaluate how carbon is distributed during continuous cultures, a carbon balance at each steady-state was performed.

Carbon balances at steady-state

Carbon distribution, defined as the percentage of carbon atoms from sucrose converted to alginate, biomass (with PHB), and CO_2 at each steady state condition, is presented as a function of agitation rate and D in Figure 3. The effect of D on carbon distribution to alginate depended on the agitation rate: in the chemostat operated at 500 rpm, an increase in D decreased the carbon diverted to alginate (from 25 to 7.9%). On the other hand, when the agitation rate was 400 rpm, carbon distribution to alginate was not significantly affected by changing D, remaining between 15 and 17%.

A higher carbon utilization for alginate synthesis was observed in the chemostat agitated at 500 rpm and 0.07 h^{-1} (25%), whereas a lower carbon utilization to alginate (5%) was obtained in cultures operated at 600 rpm and 0.09 h^{-1}. This finding confirms that, to divert more carbon to alginate, it is necessary to operate chemostat cultures at an intermediate agitation rate and low D. In contrast, the carbon diverted to CO_2 was greatly influenced by agitation rate in comparison to D. In chemostats with dilution rates of 0.07 and 0.09 h^{-1}, increasing the agitation rate from 400 to 600 rpm increased the carbon diverted to CO_2 from 50% to over 71%. In disagreement with this evidence, Díaz-Barrera *et al.* [9] demonstrated that when the agitation rate increased from 300 to 500 rpm, the percentage of carbon to CO_2 remained almost unchanged, reaching approximately 50%. The different trend in carbon diverted to alginate and CO_2 as a function of agitation rate observed in our study compared to Díaz-Barrera *et al.* [9] could be explained by the different carbon sources used. Considering these results, it is clear that both D and agitation rate affect the partitioning of the carbon flux into alginate and CO_2 under oxygen-limited conditions.

At the different steady-states, the carbon balances closed to within 95–104%, suggesting that *A. vinelandii* cells utilized sucrose efficiently to produce alginate, CO_2 and biomass (including PHB). Other carbon products were not produced, which was confirmed by measurements of organic acids using HPLC (data not shown). In disagreement with these results, Díaz-Barrera *et al.* [10] reported that in chemostat cultures fed sucrose (15 g l^{-1}), a percentage of carbon (20 and 35%) was diverted to acetate and malate, which were released into the culture medium. This different behavior could be attributed to the non-nitrogen-fixation conditions used by Díaz-Barrera *et al.* [10], compared to the nitrogen-fixing conditions used in this work. In light of this observation, further research must be carried out to evaluate how nitrogen fixation conditions affect metabolic fluxes in *A. vinelandii* cultures.

Given that a higher proportion of carbon was diverted to alginate in the chemostat operated at 0.07 h^{-1} (Fig. 3), gene expression (*alg8*, *alg44*, *algL*, *alyA1*, *alyA2*, and *algE7*) and alginate production (in terms of molecular weight and production rate) were evaluated at the different agitation rates explored.

Gene expression and alginate molecular weight at different agitation rates

In previous reports, *algL* and *alg8* gene expression under different oxygen transfer rates (manipulated by agitation rate) in chemostat cultures of *A. vinelandii* has been studied [9,10]. Recently, the expression of genes involved in polymerization and depolymerization of alginate under different DOT in batch cultures was also evaluated [11]. To complement the previous works performed at specific growth rate (i.e., in chemostat cultures), the alginate mean molecular weight (MMW), specific alginate production rate (q_p), and gene expression of *algL*, *alyA1*, *alyA2*, *algE7*, *alg8* and *alg44* were evaluated at different agitation rates under nitrogen-fixation conditions in a continuous culture operated at 0.07 h^{-1} (Fig. 4).

A higher molecular weight was obtained in chemostat cultures agitated at 500 rpm (580 kDa), while a similar molecular weight (480 kDa) was observed for 400 and 600 rpm (Fig. 4a). A similar behavior was observed for q_p due to that values highest (0.069 g g^{-1} h^{-1}) were observed at 500 rpm, such as has been previously described [10]. As shown in Fig. 4b, it is clear that the expression of both *alg44* and *alg8* were affected by changes in agitation rate. The *alg44* expression decreased by a factor of approximately 3.6 when the agitation rate was increased from 400 to 600 rpm, while the expression of *alg8* was highest at 600 rpm compared to the other conditions evaluated. These data suggest that both *alg44* and *alg8* gene expression could be modulated by oxygen availability.

The highest alginate MMW was produced at 500 rpm (Fig. 4a); however, this result cannot be explained by the expression patterns of *alg8* or *alg44*. In disagreement with these results, Díaz-Barrera *et al.* [10] reported that a higher alginate MMW in chemostat cultures can be related to higher *alg8* gene expression, suggesting that in *A. vinelandii* cells, *alg8* encodes the proposed catalytic subunit of alginate polymerase. The difference in behaviors observed in this study and by Díaz-Barrera *et al.* [10] may be explained by physiological differences that result from the different culture conditions used: while Díaz-Barrera *et al.* [10] used non-nitrogen-fixation conditions, nitrogen fixation conditions were used in this work.

Previous studies of in-vitro polymerization showed that the entire cell envelope was required for alginate polymerization, suggesting that Alg8 requires other proteins to function [20]. Although the specific role that Alg44 plays in polymerization remains unclear, Alg44 could play an indirect role. It has been suggested that there is a mutual stability relationship between Alg8 and Alg44 [3]. The findings obtained in our study, which indicate that the expression of *alg8* and *alg44* was not related to higher alginate MMW (obtained at 500 rpm), could be explained by a mutual stability relationship between Alg8 and Alg44. Further studies related with enzymatic activity should be carried out to evaluate this possibility.

Gimmestad *et al.* [5] suggested that AlyA1 and AlyA2 are intracellular enzymes, and it is known that the periplasmic AlgL is intracellular whereas the alginate lyase AlgE7 is extracellular in *A. vinelandii* [21]. Interestingly, different expression patterns were observed for lyases (*algL, alyA1* and *alyA2*) and *algE7* in response to changes in agitation rate (and hence oxygen availability) at steady-state. A 10-fold decrease in *algE7* expression was observed when the agitation rate was increased from 400 to 600 rpm was observed, while an increase in lyase expression (2.8, 2.5 and 1.5 times for *alyA1*, *algL* and *alyA2*, respectively) was observed under the same conditions (Fig. 4c). The results of our work suggest that higher oxygen availability (determined by a higher agitation rate) affect lyase expression levels, increasing the expression of intracellular lyases when OTR level is increased.

Given that increased lyase (*algL, alyA1* and *alyA2*) expression was observed when the agitation rate was increased, the similar alginate MMW values obtained (480 kDa) at 400 and 600 rpm cannot be explained by changes in the expression levels of intracellular lyases. It is possible that at the lower agitation rate (400 rpm), higher expression of *algE7* and, possibly, increased lyase activity could explain the similar alginate MMW obtained at 400 rpm compared to 600 rpm. Similarly, Flores *et al.* [11] recently observed higher expression of *algL* and *alyA2* as well as a high alginate MMW (1200 kDa) in batch cultures at lower DOT (1% compared to 5%). This contradictory behavior was explained by an analysis of alginase activity. Flores *et al.* [11] found a basal level of extracellular lyase activity in the batch cultures at 1% DOT, which is consistent with the high alginate MMW obtained. In light of the evidence obtained, it is possible that alginate molecular weight results from a balance between gene expression of intracellular and extracellular lyases in response to the agitation rate of the culture. To our knowledge, our findings for the first time show lyase expression as a function of oxygen availability at a constant growth rate. Furthermore, these evidences demonstrate how different expression patterns could determine the molecular weight of the alginate synthetized at steady-state.

Conclusion

In this work, the effect of agitation rate on alginate production and the expression of genes involved in alginate polymerization and depolymerization were evaluated in continuous *A. vinelandii* cultures. Under oxygen-limited conditions, the agitation rate influenced the partitioning of carbon into alginate and CO_2. In chemostat cultures performed at 0.07 h^{-1} and 500 rpm, a highest alginate molecular weight (580 kDa) as compared to 400 and 600 rpm was obtained. Different expression levels of lyase genes (intra and extracellular), modulated by the oxygen supply conditions, could explain the changes in the molecular weight, particularly the highest alginate molecular weight obtained at 500 rpm. The findings provide knowledge about alginate polymerization process in *A. vinelandii* and open up new possibilities of synthesizing polymers with particular molecular weight.

Acknowledgments

The technical assistance of M. Luisa Vasquez in HPLC analysis is gratefully acknowledged.

Author Contributions

Conceived and designed the experiments: AD FM FA. Performed the experiments: FM FG. Analyzed the data: FM AD FA FG. Contributed to the writing of the manuscript: AD. Reviewed the manuscript: FA.

References

1. Díaz-Barrera A, Soto E (2010) Biotechnological uses of *Azotobacter vinelandii*: Current state, limits and prospects. Afr J Biotechnol 9: 5240–5250.
2. Capone SH, Dufresne M, Rechel M, Fleury MJ, Salsac AV (2013) Impact of alginate composition: from bead mechanical properties to encapsulated HepG2/C3A cell activities for in vivo implantation. PLoS ONE 8(4): e62032. doi:10.1371/journal.pone.0062032.
3. Galindo E, Peña C, Núñez C, Segura D, Espin G (2007) Molecular and bioengineering strategies to improve alginate and polydydroxyalkanoate production by *Azotobacter vinelandii*. Microb Cell Fact 6: 1–16.
4. Hay ID, Rehman ZU, Moradali MF, Wang Y, Rehm BH (2013) Microbial alginate production, modification and its applications. Microb Biotechnol 6: 637–650.
5. Gimmestad M, Ertesvåg H, Heggeset T, Aarstad O, Svanem B, et al. (2009) Characterization of three new *Azotobacter vinelandii* alginate lyases, one of which is involved in cyst germination. J Bacteriol 191: 4845–4853.
6. Remminghorst U, Hay ID, Rehm B (2009) Molecular characterization of Alg8, a putative glycosyltransferase, involved in alginate polymerization. J Biotechnol 140: 176–183.
7. Castillo T, Galindo E, Peña C (2013) The acetylation degree of alginates in *Azotobacter vinelandii* ATCC9046 is determined by dissolved oxygen and specific growth rate: studies in glucose-limited chemostat cultivations. J Ind Microbiol Biotechnol 40: 715–723.
8. Díaz-Barrera A, Silva P, Berrios J, Acevedo F (2010) Manipulating the molecular weight of alginate produced by *Azotobacter vinelandii* in continuous cultures. Bioresour Technol 101: 9405–9408.
9. Díaz-Barrera A, Aguirre A, Berrios J, Acevedo F (2011) Continuous cultures for alginate production by *Azotobacter vinelandii* growing at different oxygen uptake rates. Process Biochem 46: 1879–1883.
10. Díaz-Barrera A, Soto E, Altamirano C (2012) Alginate production and *alg8* gene expression by *Azotobacter vinelandii* in continuous cultures. J Ind Microbiol Biotechnol 39: 613–621.
11. Flores C, Moreno S, Espín G, Peña C, Galindo E (2013) Expression of alginases and alginate polymerase genes in response to oxygen, and their relationship with the alginate molecular weight in *Azotobacter vinelandii*. Enzyme Microb Technol 53: 85–91.
12. Priego-Jimenéz R, Peña C, Ramírez O, Galindo E (2005) Specific growth rate determines the molecular weight of the alginate produced by *Azotobacter vinelandii*. Biochem Eng J 25: 187–193.
13. Miller G (1959) Use of dinitrosalicylic acid reagent for determination of reducing sugars. Anal Chem 31: 426–428.
14. Karr D, Waters J, Emerich D (1983) Analysis of poly-β-hydroxybutyrate in *Rhizobium japonicum* bacteroids by ion-exclusion high-pressure liquid chromatography and UV detection. Appl Environ Microbiol 46: 1339–1344.
15. Blumenkratz N, Asboe-Hansen G (1973) New method for quantitative determination of uronic acids. Anal Biochem 54: 484–489.
16. Díaz-Barrera A, Peña C, Galindo E (2007) The oxygen transfer rate influences the molecular mass of the alginate produced by *Azotobacter vinelandii*. Appl. Microbiol Biotechnol 76: 903–910.
17. Kayser A, Weber J, Hecht V, Rinas U (2005) Metabolic flux analysis of *Escherichia coli* in glucose-limited continuous culture. I. Growth-rate dependent metabolic efficiency at steady state. Microbiol 151: 693–706.
18. Kuhla J, Oelze J (1988) Dependency of growth yield, maintenance and Ks-values on the dissolved oxygen concentration in continuous cultures of *Azotobacter vinelandii*. Arch Microbiol 149: 509–514.
19. García-Ochoa F, Gomez E (2009) Bioreactor scale-up and oxygen transfer rate in microbial processes: an overview. Biotechnol Adv 27: 153–176.

20. Oglesby L, Jain S, Ohman DE (2008) Membrane topology and roles of *Pseudomonas aeruginosa* Alg8 and Alg44 in alginate polymerization. Microbiol 154: 1605–1615.

21. Svanem BI, Strand WI, Ertesvåg H, Skjåk-Braek G, Hartmann M, et al. (2001) The catalytic activities of the bifunctional *Azotobacter vinelandii* mannuronan C-5-epimerase and alginate lyase AlgE7 probably originate from the same active site in the enzyme. J Biol Chem 276: 31542–31550.

Impact of Extraction Parameters on the Recovery of Lipolytic Activity from Fermented Babassu Cake

Jaqueline N. Silva[1], Mateus G. Godoy[2], Melissa L. E. Gutarra[3]*, Denise M. G. Freire[2]

1 Departamento de Engenharia Bioquímica, Escola de Química, Universidade Federal do Rio de Janeiro, Rio de Janeiro, Brazil, **2** Departamento de Bioquímica, Instituto de Química, Universidade Federal do Rio de Janeiro, Rio de Janeiro, Brazil, **3** Departamento de Engenharia Bioquímica, Escola de Química/Polo Xerém, Universidade Federal do Rio de Janeiro, Rio de Janeiro, Brazil

Abstract

Enzyme extraction from solid matrix is as important step in solid-state fermentation to obtain soluble enzymes for further immobilization and application in biocatalysis. A method for the recovery of a pool of lipases from *Penicillium simplicissimum* produced by solid-state fermentation was developed. For lipase recovery different extraction solution was used and phosphate buffer containing Tween 80 and NaCl showed the best results, yielding lipase activity of 85.7 U/g and 65.7 U/g, respectively. The parameters with great impacts on enzyme extraction detected by the Plackett-Burman analysis were studied by Central Composite Rotatable experimental designs where a quadratic model was built showing maximum predicted lipase activity (160 U/g) at 25°C, Tween 80 0.5% (w/v), pH 8.0 and extraction solution 7 mL/g, maintaining constant buffer molarity of 0.1 M and 200 rpm. After the optimization process a 2.5 fold increase in lipase activity in the crude extract was obtained, comparing the intial value (64 U/g) with the experimental design (160 U/g), thus improving the overall productivity of the process.

Editor: Chenyu Du, University of Nottingham, United Kingdom

Funding: This work was financially supported by the PETROBRAS and the Brazilian governmental funding agencies CNPq (Conselho Nacional de Desenvolvimento Científico e Tecnológico) and FAPERJ (Fundação de Amparo à Pesquisa do Estado do Rio de Janeiro). The funders had no role in study design, data collection and analysis, decision to publish, or preparation of the manuscript.

Competing Interests: This study was funded in part by PETROBRAS. There are no patents, products in development or marketed products to declare.

* Email: gutarra@gmail.com

Introduction

The market for lipases is growing steadily, mostly because of their capacity to catalyze hydrolytic and synthesis reactions using a large range of substrates, and also because they sometimes demonstrate regioselectivity and enantioselectivity [13]. As such, lipases have been widely used in several fields, including the detergent, pharmaceutical and fine chemical industries [23].

Enzymes of industrial interest have traditionally been produced by submerged fermentation (SmF). However, solid-state fermentation (SSF) is being used for the production of enzymes due to its advantages over SmF, such as its high productivity, generation of high-quality products, and use of agro-industrial waste, an abundant and low-cost raw material [26]. The use of agro-industrial waste/complex matrix as a culture medium also induces the production of different hydrolytic enzymes such as amylases and proteases in the same fermentation batch [26], or even pools of lipases with different catalytic properties [6].

In the SSF of agro-industrial waste such as babassu cake, the fungi grow on the surface or inside the solid particles, breaking down their complex nutrients by secreting enzymes and consuming their monomers [11]. In SSF, extracellular enzymes remain absorbed in the fermented solids as the liquid phase within the particles, and mass transfer in SSF is reduced [19]. Lipases produced by SSF can be employed in the liquid form or as fermented solids for the production of biodiesel [22], as described by Salum (2010). However, for almost all industrial applications, lipases are used in liquid solutions and can be concentrated, purified and immobilized in polymeric supports before its use [16]. Then, in SSF the step of enzyme extraction is as important as the step of enzyme production, requiring attention. The production of lipases by SSF has been studied by many authors using different culture media, such as babassu cake [11], wheat bran [17], soybean meal [20] and castor bean waste [8], but only a few studies have been done into the extraction of enzymes from fermented solids, and all of these show that the optimization of this step significantly enhances the overall productivity of the process. The type of extraction solution, temperature, stirring rate, solid/liquid ratio and pH are the main parameters studied for the extraction of xylanase [12], pectinases [3] and proteases [1]. The extraction of lipases produced by the solid-state fermentation of soybean meal was studied and temperature was found to be the most important factor [26].

The fungus *Penicillium simplicissimum* produces high hydrolytic lipase activity in SSF systems using babassu cake, and the lipases present in the crude extractant have attractive catalytic properties such as thermostability and thermophilic characteristics [9;11]. Its crude extract has been successfully applied to the biological/enzymatic treatment of liquid fish waste [25]. Moreover, lipases from this fungal strain produced by the SSF of babassu cake show enantioselectivity when immobilized on supports with different degrees of hydrophobicity, indicating the presence of at least three lipases [4].

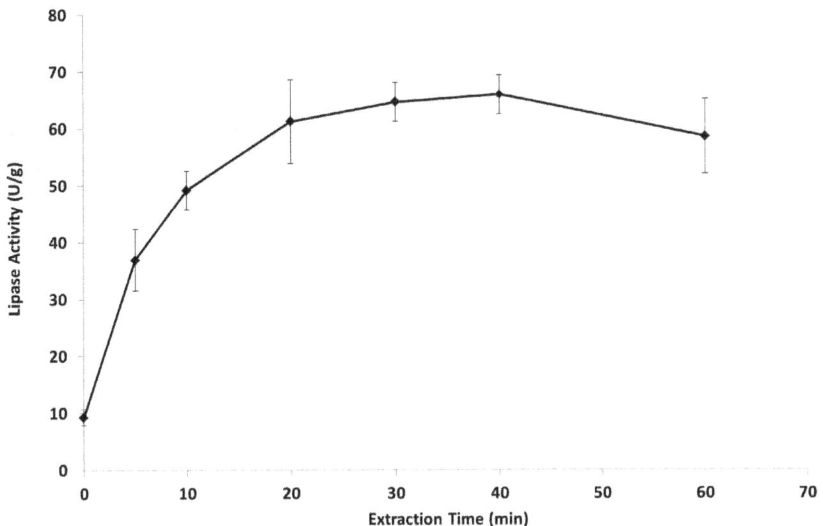

Figure 1. Kinetic of lipase extraction in sodium phosphate buffer (0.1 M, pH 7.0) with agitation of 200 rpm.

The aim of this work was to select suitable solutions for the extraction of a pool of lipases from *P. simplicissimum* produced by the solid-state fermentation of babassu cake; the identification of parameters that statistically influence the process of lipase extraction, using Plackett & Burman experimental design (PB); and process optimization using Central Composite Rotatable Design (CCRD), identifying the extract conditions that yield the highest lipase activity. The optimization of the extraction step could enhance productivity and thus reduce the cost of lipase production by this fungus.

Materials and Methods

Microorganism

Penicillium simplicissimum was selected for use as it is a good lipase producer by SSF [9]. To obtain the spores, the fungus was propagated in an incubator for 7 days at 30°C, in a medium containing (% w/v): 0.5 $(NH_4)_2SO_4$; 2.0 soluble starch; 0.025 $MgSO_4 \cdot 7H_2O$; 0.05 KH_2PO_4; 0.5 $CaCO_3$; 0.1 yeast extract; 1.0 olive oil; and 2.0 agar. The spores were suspended in phosphate buffer (50 mM, pH 7.0). The spore suspension was further used during the SSF. The spore concentration was determined by counting using an optical microscope and a Neubauer chamber.

Solid-state fermentation

Solid waste from the extraction of babassu oil, called babassu cake (TOBASA S.A), was used as the culture medium. The solids were supplemented with 6.25% (w/w) molasses (Usina Santa Elisa, Campinas, SP) and 65% moisture in wet basis(w/w). The cake was ground and sieved to obtain different particle sizes (up to 3.0 mm). Tray-type reactors (Polipropylene, diameter 10 cm, not perforated) were used with 15 g dry solids, which were inoculated with 10^7 spores/gram of dry solid and incubated in a chamber with

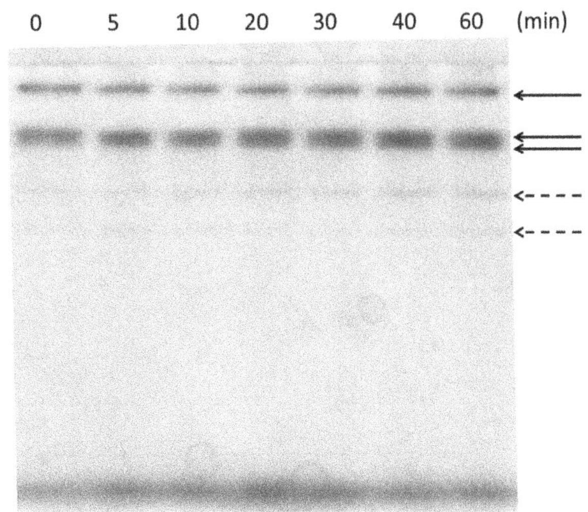

Figure 2. Zymography of a non-denaturing PAGE using α-naphtyl acetate as substrate showing lipase extraction profile major esterase/lipase bands (non-dashed arrows); weak esterase/lipase bands (dashed arrows).

Table 1. Activity of crude extract after extraction with different solvents and Residual activity (%) after storage of crude extract contain lipases at the temperature of −20°C.

Solvent	Activity of crude extract* (U/g)	Residual activity (%)	
		15 days	30 days
Phosphate buffer (100 mM. pH 7.0)	64.7±0.7	105.7	69.0
Buffer + Tween 80 1%	85.7±1.2	101	69.2
Buffer + Triton X-100 0.5%	48±1.6	101.7	64.7
Buffer + Glycerol 20%	63.4±0.4	91.9	65.6
Buffer + NaCl 0.6%	50.7±6.0	141.4	100.1

The values of lipase activity achieved afterwards the extraction were considered as 100%.
*The analyses were done in triplicate.

temperature and relative humidity set to 30°C and 95%, respectively, for 72 h [9].

Enzyme recovery

Determination of extraction time of the pool of lipases. The enzyme was recovered using standard conditions (5 mL/g sodium phosphate buffer (0.1 M, pH 7.0), agitation at 200 rpm, 35°C) for 0, 10, 20, 30, 40 and 60 minutes. Next, the cake was pressed to obtain the crude enzyme extract, which was centrifuged at 3000x*g* for 5 minutes. The supernatant was used to determine lipase activity. The supernatants obtained from 10, 20, 30 40 and 60 min. extraction time were lyophilized and resuspended in a sample buffer, without SDS and b-mercapto-ethanol, at the same protein concentration, then submitted to non-denaturing PAGE and subsequently zymography [15].

Selection of extraction solution. 5 mL sodium phosphate buffer (0.1 M, pH 7.0) was added per gram of babassu cake, followed by shaking at 200 rpm and 35°C (standard condition) for 40 minutes to guarantee total lipase extraction. Sodium phosphate buffer with NaCl (0.6% (w/v)), Tween 80 (0.1% (w/v)), Triton X-100 (0.5% (w/v)) and glycerol (20% (w/v)) was also used as an extraction solution. Lipase activity was determined using the supernatant and then the lipase was stored at −20°C.

Determination of lipase recovery conditions. The most important factors for lipase extraction were screened using Plackett & Burman (PB) experimental design. The factors studied were: stirring rate (100–300 rpm), volume of extraction solution (5–9 mL/g), temperature (25–45°C), buffer molarity (0.01–0.1 M), NaCl concentration (0–0.2%w/v) and Tween 80 concentration (0–0.2%w/v).

Subsequently a Central Composite Rotatable Design (CCRD) was used to optimize the extraction conditions as a function of the important factors identified in the PB design. The factors studied were: volume of extraction solution (3–7 mL/g), temperature (25–

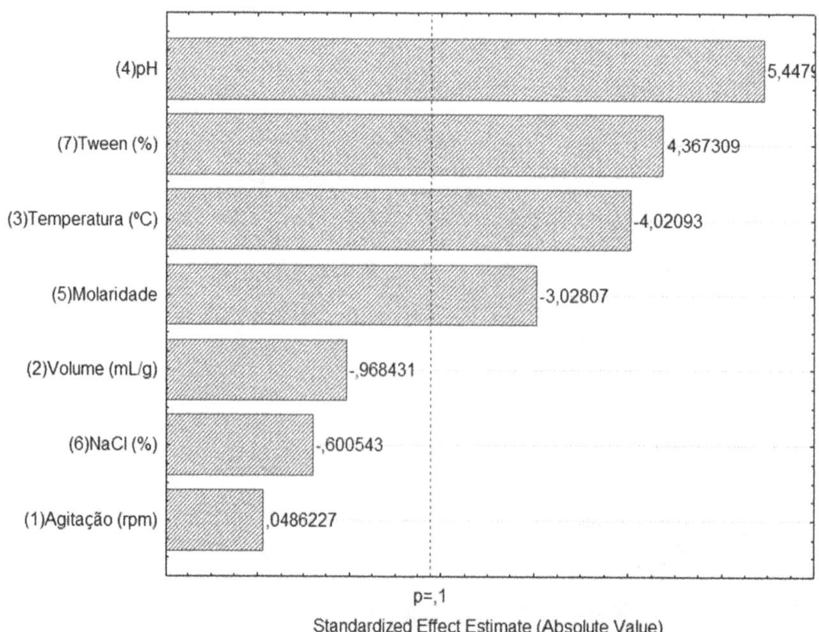

Figure 3. Standardized effects of studied variables and interaction on the lipase extraction, where *p* represents the significance level of the test. Statistically significant terms: p-value >0.1 (90% of confidence).

Table 2. Values for lipase activity achieved after the extraction process for the different experimental conditions (real variable values and coded levels in parenthesis) of the Plackett & Burman design.

Assays	Agitation (rpm)	Volume (mL/g)	Temp. (°C)	pH	Molarity	NaCl conc.(%)	Tween conc.(%)	Lipase activity (U/g)
1	100 (1)	5 (−1)	45 (1)	5 (−1)	0.01 (−1)	0 (−1)	0.2 (1)	58.7
2	100 (1)	9 (1)	25 (−1)	7 (1)	0.01 (−1)	0 (−1)	0 (−1)	85.9
3	300 (−1)	9 (1)	45 (1)	5 (−1)	0.1 (1)	0 (−1)	0 (−1)	11.1
4	100 (1)	5 (−1)	45 (1)	7 (1)	0.01 (−1)	0.2 (1)	0 (−1)	66.0
5	100 (1)	9 (1)	25 (−1)	7 (1)	0.1 (1)	0 (−1)	0.2 (1)	99.3
6	100 (1)	9 (1)	45 (1)	5 (−1)	0.1 (1)	0.2 (1)	0 (−1)	7.3
7	300 (−1)	9 (1)	45 (1)	7 (1)	0.01 (−1)	0.2 (1)	0.2 (1)	93.1
8	300 (−1)	5 (−1)	45 (1)	7 (1)	0.1 (1)	0 (−1)	0.2 (1)	74.9
9	300 (−1)	5 (−1)	25 (−1)	7 (1)	0.1 (1)	0.2 (1)	0 (−1)	67.5
10	100 (1)	5 (−1)	25 (−1)	5 (−1)	0.1 (1)	0.2 (1)	0.2 (1)	69.5
11	300 (−1)	9 (1)	25 (−1)	5 (−1)	0.01 (−1)	0.2 (1)	0.2 (1)	71.2
12	300 (−1)	5 (−1)	25 (−1)	5 (−1)	0.01 (−1)	0 (−1)	0 (−1)	67.0
13	200 (0)	7 (0)	35 (0)	6 (0)	0.055 (0)	0.1 (0)	0.1 (0)	84.8
14	200 (0)	7 (0)	35 (0)	6 (0)	0.055 (0)	0.1 (0)	0.1 (0)	81.0

Table 3. Values of lipase activity achieved after the extraction process and residual activity after 45 days of storage at −20°C for different experimental conditions (real variable values and coded levels in parenthesis) of Central Composite Rotatable Design.

Assays	Temperature (°C)	pH	Tween (% w/v)	Volume (mL)	Lipase activity (U/g)	Residual activity (%) 45 days
1	27.5 (−1)	6.5 (−1)	0.2 (−1)	4 (−1)	68.11	97.4
2	27.5 (−1)	6.5 (−1)	0.2 (−1)	6 (1)	82.58	81.2
3	27.5 (−1)	6.5 (−1)	0.4 (1)	4 (−1)	82.78	75.3
4	27.5 (−1)	6.5 (−1)	0.4 (1)	6 (1)	86.17	81.2
5	27.5 (−1)	7.5 (1)	0.2 (−1)	4 (−1)	92.93	80.6
6	27.5 (−1)	7.5 (1)	0.2 (−1)	6 (1)	109.81	75.8
7	27.5 (−1)	7.5 (1)	0.4 (1)	4 (−1)	99.72	74.5
8	27.5 (−1)	7.5 (1)	0.4 (1)	6 (1)	106.19	80.6
9	32.5 (1)	6.5 (−1)	0.2 (−1)	4 (−1)	69.29	115.8
10	32.5 (1)	6.5 (−1)	0.2 (−1)	6 (1)	90.23	101.7
11	32.5 (1)	6.5 (−1)	0.4 (1)	4 (−1)	99.42	77.1
12	32.5 (1)	6.5 (−1)	0.4 (1)	6 (1)	96.64	87.4
13	32.5 (1)	7.5 (1)	0.2 (−1)	4 (−1)	83.46	86.1
14	32.5 (1)	7.5 (1)	0.2 (−1)	6 (1)	96.94	91.4
15	32.5 (1)	7.5 (1)	0.4 (1)	4 (−1)	90.78	81.8
16	32.5 (1)	7.5 (1)	0.4 (1)	6 (1)	111.15	78.3
17	25 (−2)	7 (0)	0.3 (0)	5 (0)	106.31	78.1
18	35 (2)	7 (0)	0.3 (0)	5 (0)	101.97	78.9
19	30 (0)	6 (−2)	0.3 (0)	5 (0)	93.89	79.8
20	30 (0)	8 (2)	0.3 (0)	5 (0)	109.22	73.4
21	30 (0)	7 (0)	0.1 (−2)	5 (0)	101.16	57.2
22	30 (0)	7 (0)	0.5 (2)	5 (0)	133.74	48.8
23	30 (0)	7 (0)	0.3 (0)	3 (−2)	81.49	53.7
24	30 (0)	7 (0)	0.3 (0)	7 (2)	110.17	59.3
25	30 (0)	7 (0)	0.3 (0)	5 (0)	95.86	57.6
26	30 (0)	7 (0)	0.3 (0)	5 (0)	106.30	71.9
27	30 (0)	7 (0)	0.3 (0)	5 (0)	82.46	86.9

35°C), pH (6.0–8.0) and Tween 80 concentration (0.1–0.5%w/v), keeping the buffer molarity constant at 0.1 M, the stirring rate at 200 rpm and with no addition of NaCl.

For all these experiments it took 40 minutes to guarantee that all the lipase had been extracted.

Determination of lipase activity

Lipase activity was determined by the spectrophotometric method described by Godoy *et al* [7] with *p*-nitrophenyl laurate as a substrate. One unit of lipase activity (U) was defined as the amount of enzyme necessary to hydrolyze 1.0 μmol *p*-nitrophenyl laureate per minute under assay conditions. Lipase activity was expressed in U/g of initial dry weight.

Zymography

The enzymes present in the crude extract with lipase/esterase activity were detected by zymography following non-denaturing PAGE [15]. After electrophoresis, the gel was treated with α-naphthyl acetate 0.02% (w/v) and Fast Blue RR salt 0.05% (w/v) in 0.1 M sodium phosphate buffer, pH 6.2, revealing bands with lipase/esterase activity [14].

Results and Discussion

Determination of extraction time

There was an increase in lipase extraction up to 20 minutes, after which no time difference was observed in the recovery of the enzymes, considering the standard error (Figure 1). This indicates that 20 minutes contact time is long enough for the complete extraction of lipases from *P. simplicissimum*. Aikat *et al* [1] studied the protease recovery profile from fermented wheat bran and found that 90% of protease recovery was obtained in 2 h extraction and that extraction was complete after 10 h. In this study it was possible to recover 100% of lipase in just 40 minutes. For lipases from fermented soybean meal, almost all lipase extraction occurs in the first two minutes of the extraction process, a fact the authors attribute to the growth and secretion of enzymes by the fungus at the surface of the solid particles [25]. In our case, extraction took approximately 10 times longer than reported by Vardanega *et al.* (2008) [24], because *P. simplicissimum* not only grew on the surface of the babassu cake, but also effectively penetrated the solid particles [11], making it harder to extract the lipases secreted within them. Another hypothesis is that the lipases from *P. simplicissimum* bind or interact more strongly with some

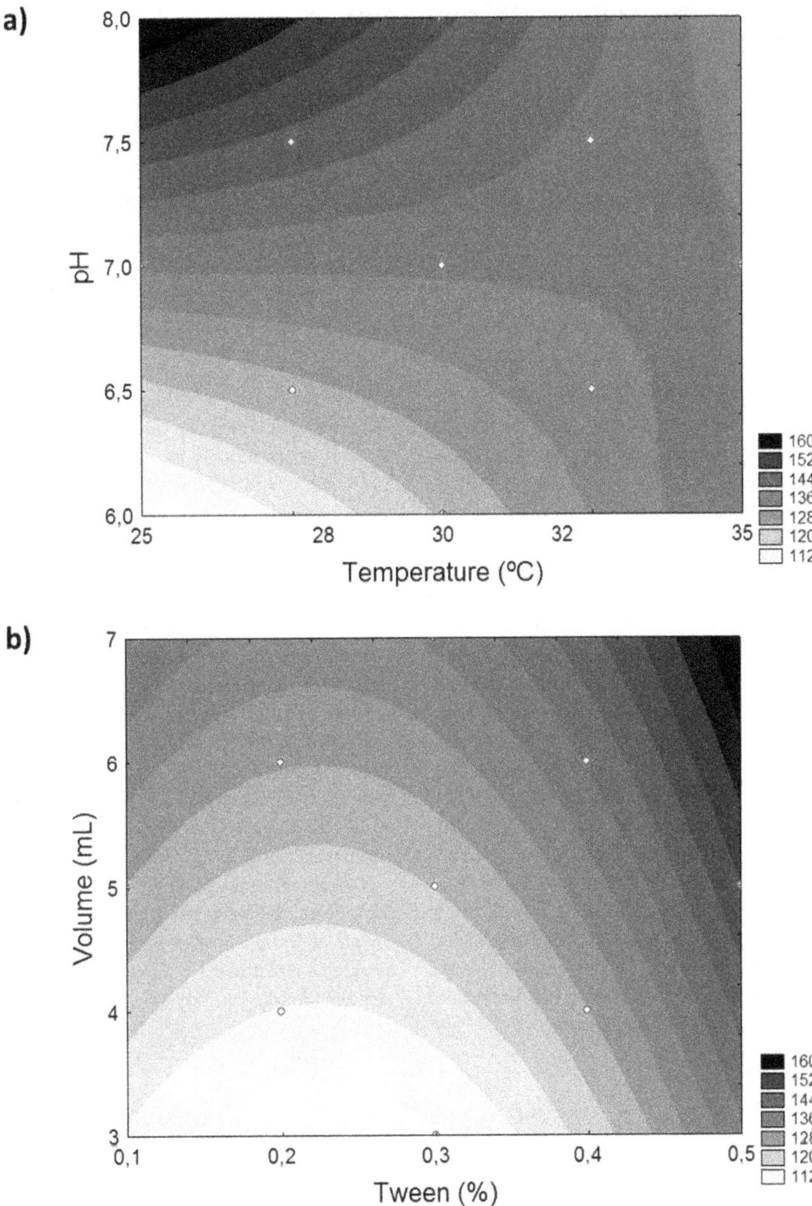

Figure 4. Contour plots for lipase activity in function of the variables temperature and pH (a) and volume and Tween concentration (b).

components at the matrix surface and/or the fungal wall, based on the hydrophobic nature of lipases and fungal surface in SSF (MUNOZ 1995), making it take longer to extract them.

To determine the number of different lipases produced by *P. simplicissimum* in the SSF of babassu cake and the extraction profile of each lipase, non-denaturing PAGE was performed by adding 30 μg protein per lane, followed by a zymogram using α-naphthyl acetate as a substrate. Three major bands with lipase/esterase activity were revealed (Figure 2 – non-dashed arrows). When the SSF of castor bean waste was undertaken, nine lipase/esterase bands were observed [7] with the same *Penicillium* strain, indicating that the type of matrix and fermentation can induce different enzymes or proportions thereof, producing extracts with different catalytic properties. In this work, the three major bands were extracted together for all the extraction times (Figure 2), demonstrating that they are distributed randomly on and in the matrix and probably with no differences in their interaction with the hydrophobic portion of the fungal biomass or the matrix, as they all detach at the same time. These results show that maximum lipase activity could be recovered after 20 minutes of extraction, including all the five bands with esterase/lipase activity.

Effect of extraction solutions on recovery and enzyme storage

Four different solutions were tested, in triplicate: sodium phosphate buffer (100 mM, pH 7.0) and the same buffer with 0.6% NaCl (w/v), 0.1% Tween 80 (w/v), 0.5% Triton X-100 (w/v) and 20.0% glycerol (w/v). The samples were extracted and lipase activity was determined in fresh crude extract and after its storage at −20°C for a period of 30 days. The effect of each

Table 4. Analysis of variance (ANOVA) for the CCRD of enzyme extraction.

Source of variation	Sum of squares	Degrees of freedom	Means squares	F-test
Regression	3363.4	5	672.7	7.71
Residues	1830.2	21	87.2	-
Lack of fit	1544.8	19	81.3	-
Pure error	285.4	2	142.7	-
Total	5193.6	26	-	-

Regression coefficient: $R^2 = 0.65$ ($F_{0.1;5;21}$) = 2.14.

solution on lipase activity was tested by the addition of 0.6% NaCl (w/v), 0.1% Tween 80 (w/v), 0.5% Triton X-100 (w/v) or 20.0% glycerol (w/v) to a crude extract previously obtained by extraction with only sodium phosphate buffer and a comparison of their lipase activity were done. Our results showed that none of the solutions used interfered with lipase activity determination, guaranteeing that the effect of each solution observed later was exclusively on lipase extraction (not shown).

Tween 80, a non-ionic surfactant, proved to be an excellent enzyme recovery agent, yielding the highest extracted lipase activity (85.7 U/g). Lipases have a unique capacity to activate at the water/lipid interface due to a surface conformational change that occurs under this condition. At the water/lipid interface, the structure called the lid appears in its open form, revealing the active site. This occurs when there is adsorption of the hydrophobic pocket around the active center and the internal face of the lid onto a hydrophobic surface [18]. At the end of fermentation, the residual lipid and hydrophobin present at the fungal wall may adsorb the lipases produced [21]. As Tween 80 is a surfactant, it is possible that its hydrophobic portion attracted the lipases, promoting their desorption from the fermented solids, resulting in better extraction. Tween 80 also showed good results for the extraction of other proteins, probably because the micelles it forms surround the proteins [5]. The enzyme recovered with Tween 80 remained stable for 15 days when stored at −20°C in the crude extract form, and had an activity loss of nearly 30% after 30 days (Table 1).

The stability of the lipase extracted and stored at −20°C with sodium phosphate buffer, Triton X-100 and glycerol were similar to the stability observed for Tween 80. However, when the lipase was extracted with NaCl, stability increased by about 41% (comparing with the initial activity of crude extract −50.7 U/g) over a storage period of 20 days and it was not observed any loss of activity after 30 days (Table 1).

Triton X-100 is also a surfactant, but it presented the lowest value for extracted enzyme activity (Table 1). As Triton X-100 is very viscous (240 cP at 25°C), which interferes with mass transfer, it has to be used at higher temperatures. The use of an assay temperature of 35°C could explain the low recovery efficiency.

Despite the protective effect provided by glycerol during the storage of the proteins [25], it was not found to stabilize the *P. simplicissimum* lipase at the used concentrations during the storage period, presenting the lowest residual activity after 30 days of storage (Table 1). Like Triton X-100, the high viscosity of glycerol

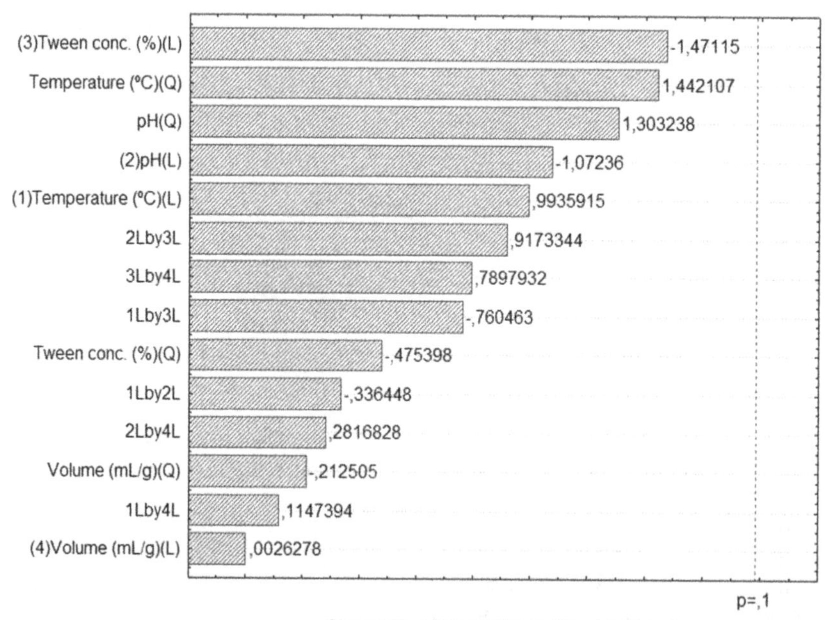

Figure 5. Standardized effects of studied variables and interaction on the residual lipase activity of crude extract after storage at −20°C, where *p* represents the significance level of the test. Statistically significant terms: *p*-value >0.1 (90% of confidence)

could explain its low efficiency during lipase recovery, as shown by Castilho *et al* [2] for pectinase extraction.

The Tween 80 and NaCl solutions were selected due to their highest efficacy in the extraction and stability of *P. simplicissimum* lipases, respectively.

Optimization of lipase extraction using experimental design

In the first stage of the experimental design the following variables were studied: agitation (100 to 300 rpm), volume of solvent (5 to 9 mL solvent/g dry weight of cake), temperature (25 to 45°C), pH (5.0 to 7.0), molarity of solvent (0.01 to 0.1) and concentration of NaCl and Tween 80 (0 to 0.2% (w/v)), to determine the statistically significant variables for the process. The assay conditions and values for extracted lipase activity are shown in Table 2. A Pareto chart (Figure 3) was created, which shows the standardized effects and permits the identification of the statistically significant variables for the process, with a significance level (*p*-value) of 0.1.

The pH and Tween 80 concentration variables had a positive effect, indicating that a higher extraction of lipase could be obtained at the highest values used (pH 7.0 and 0.2% of Tween 80) or even higher values than those studied. On the other hand, temperature presented a negative effect, indicating that lower values should be used to improve the extraction of the enzyme. At higher temperatures, some enzymes could be denaturated, but the lipase from *P. simplicissimum* shows high stability at 50°C [10]. Another hypothesis is that high temperatures enhance the extraction of other compounds from fermented solids, which could hamper the extraction of the lipase, as reported by Diaz *et al.* [5] for the extraction of hydrolytic enzymes from fermented grape pulp.

Molarity demonstrated a significant negative effect, probably because a higher ionic strength helps the adsorption of proteins with hydrophobic surfaces, making it harder to extract them [18]. Even so, buffer molarity was set at 0.1 M, as a lower molarity value reduced the effectiveness of the buffer, affecting the pH of the final crude extract and thereby the stability of the enzyme [10]. Volume was found to have no statistical effect on lipase extraction in the tested range, and could be fixed at any value from 5 to 9 mL per gram. However, the volume/mass ratio is directly related to the product concentration, making the study of this factor in lower ranges necessary.

The second part of the experimental design consisted of optimizing the process using a Central Composite Rotatable Design (CCRD), enabling the analysis of the interaction between the variables, according to Table 3. The effects of temperature, pH, Tween 80 concentration and volume were evaluated on lipase recovery and stability when stored at -20°C for 45 days.

A statistical analysis was carried out to estimate the effects of the variables and the interaction between them. The effects of standardized variables (*t*-values) and the significance probability test (*p*-value) were used to assess the effects of pH, Tween 80 concentration (Tw), volume (V) and temperature (T) on lipase activity. Using a 10% level of significance ($p<0.1$), it was observed that the pH (linear term), Tween 80 concentration (linear and quadratic term), volume (linear term) and pH-temperature interaction had significant effects on lipase activity.

With these experiments, it was possible to construct (through effects table – data not shown) an empirical model for lipase recovery as a function of temperature, pH, Tween 80 concentration and volume (Equation 3.1), including only the statistically significant variables with $p<0.1$.

$$L = 92.4 + 6.1 \times pH + 6.0 \times Tw + 3.9 \times Tw^2 + 6.3 \times V - 3.9 \times T \times pH \quad (3.1)$$

Where:
L = Lipase recovery (U/g)
pH = pH (codified values)
Tw = Concentration of Tween 80 (codified values)
T = Temperature (codified values)
V = Volume (codified values)

The model generated was considered predictive by analysis of variance (ANOVA), since it showed a satisfactory coefficient of determination ($R^2 = 0.7$) and the F test value (7.71) was 3.6 times higher than the critical value (2.14) (Table 4). This model was used to construct contour plots, showing the values of predicted lipase activity for each condition under study (Figure 4).

Effect and coefficient regression analysis indicated that the volume, pH and Tween 80 concentration were the most important factors influencing lipase extraction.

The results were better when the pH was higher than 7.3. Varying the pH of the extraction solution changes the protein surface charge, which probably interferes in the interaction of the enzyme with the solid matrix, making its interaction stronger or weaker. It was observed that the optimal pH for lipase extraction indicated by CCRD did not correspond to the best pH for lipase activity and stability, which were better at pH 5 and 6 [10]. As such, it is important to choose a condition which fosters high extraction without affecting enzyme stability. Similar results were observed for the extraction of lipase from soybean meal, where higher lipase extraction was obtained with pH 8.5, but the authors opted to use pH 7.0 as the enzyme was more stable under this condition [24].

Volume had a greater significance in the final enzyme concentration than in the crude extract, since one of the advantages of SSF is that it yields more concentrated products [19]. Higher extraction was attained at volumes greater than 5 mL, indicating that at lower volumes the solvent was quickly impregnated by proteins and products from the hydrolysis of this agro industrial waste, preventing the lipases from being extracted. In this case the solution is saturated with solutes that diminish mass transfer. Aikat *et al* [16] observed this effect for protease extraction, finding that 5 mL/g was the optimal condition. Vardanega et al. [24] obtained better results for the lowest volume of extraction solution tested (4 mL/g), yielding a more concentrated crude extract.

The addition of Tween 80 to the extraction solution had a positive effect, indicating that higher concentrations of this surfactant would promote greater lipase extraction. This is probably due to the hydrophobicity of lipases, enabling the lipase to be desorbed from the matrix by affinity with Tween 80, and/or the surfactant property of Tween 80, by which it surrounds the proteins extracted during micelle formation, enabling extraction. A similar positive effect was also observed by Díaz *et al.* [5] for the extraction of xylanase. The effect of Tween 80 on lipase extraction had not been previously studied.

The best results were obtained at temperatures lower than 30°C. One hypothesis is that at higher temperatures, other compounds could be extracted from the babassu cake, saturating the solution more quickly and hampering the extraction of lipases. The highest temperature tested in the experimental design (35°C) probable do not cause lipase denaturation effects because lipase from *P. simplicissimum* on in the crude extract is highly stable up to 50°C (half-life of 5 hours at pH 5.0) [10]. Vardanega *et al* [24] also achieved higher lipase extraction at a lower temperature (25°C), but in this case the *Penicillium* sp. lipase was more stable at the same temperature.

Through the analysis of the results of this experimental design, it was possible to obtain maximum values of lipase activity at temperatures from 25 to 30°C, pH from 7.3 to 8.0, volume from 5 to 7 mL/g dry weight, and Tween 80 concentration of 0.5% (w/v). At optimal conditions no changes in the three major lipase extraction bands were observed in the zymography (data not shown).

The effect of the temperature, pH, Tween 80 concentration and volume on lipase stability after the storage of the crude extract at −20°C for 45 days was also evaluated in the CCRD. The assay conditions and the residual activity (%) values after storage are shown in Table 3. The variables studied and their interactions showed no statistical effect on lipase stability (Figure 5). This indicates that the change in variable levels (in the studied range) causes no effect on lipase stability and that for any of the tested conditions approximately 78% of the residual activity can be obtained after storage at −20°C for 45 days.

The conditions for lipase extraction from soybean meal and babassu cake showed similarities at lower temperature (25°C), alkaline pH (about 8.0) and no agitation [25]. However differences in the volume/mass ratio and extraction time were observed, showing that for *P. simplicissimum* lipase extraction from babassu cake, a longer time and higher volume/mass ratio were needed.

Conclusions

The use of 0.1% of Tween 80 in the solvent yielded higher enzyme recovery than sodium phosphate buffer (100 mM, pH 7.0), 0.6% of NaCl, 0.5% Triton X-100 and 20% glycerol. The fungus *Penicillium simplicissimum* grown in the SSF of babassu cake secretes five proteins with lipase/esterase activity, which can be extracted together from fermented solids by the proposed method. These lipases present in the crude extract proved to be stable when extracted with 0.6% NaCl and stored at −20°C for 30 days. The results after the optimization process yielded a 2.5 fold increase in the lipase activity of the crude extract (160 U/g) at 25°C, 0.5% (w/v) Tween 80, pH 8.0 and 7 mL extraction solution per gram of fermented solid, shown in the three major esterase/lipase activity bands. The use of experimental design techniques allowed a greater understanding of the effect of these variables on the extraction and stability of lipases from *P. simplicissimum*, and a significant increase in the recovery of lipases with great biotechnological potential, mainly due to this pool of enzymes high potential for hydrolysis and synthesis, and their enantioselectivity and thermostability.

Author Contributions

Conceived and designed the experiments: DMGF MLEG. Performed the experiments: JNS MGG. Analyzed the data: JNS MGG MLEG DMGF. Wrote the paper: JNS MGG MLEG DMGF.

References

1. Aikat K, Bhattacharyya BC (2000) Protease extraction in solid state fermentation of wheat bran by a local strain of Rhizopus oryzae and growth studies by the soft gel technique. Process Biochem 35: 907–914.
2. Castilho LR, Medronho RA, Alves TLM (2000) Production and extraction of pectinases obtained by solid state fermentation of agroindustrial residues with *Aspergillus niger*. Biores Technol 71: 45–50.
3. Cunha AG, Fernandez-Lorente G, Gutarra MLE, Bevilaqua JV, Almeida RV, et al. (2009) Separation and immobilization of lipase from *Penicillium simplicissimum* by selective adsorption on hydrophobic supports. Appl Biochem Biotechnol 156: 563–575.
4. Díaz AB, Caro I, Ory I, Blandino A (2007) Evaluation of the conditions for the extraction of hydrolitic enzymes obtained by solid state fermentation from grape pomace. Enzyme Microb Technol 41: 302–306.
5. Gekko K, Timasheff SN (1981) Mechanism of Protein Stabilization by Glycerol: Preferential Hydration in Glycerol-Water Mixtures. Biochemistry 20: 4667–4676.
6. Godoy MG, Gutarra MLE, Castro AM, Machado OLT, Freire DMG (2011) Adding value to a toxic residue from the biodiesel industry: production of two distinct pool of lipases from *Penicillium simplicissimum* in castor bean waste. J Ind Microbiol Biotechnol 38: 945–953.
7. Godoy MG, Gutarra MLE, Maciel FM, Felix SP, Bevilaqua JV, et al. (2009) Low cost methodology for biodetoxification of castor bean waste and lipase production. Enzyme Microb Technol 44: 317–322.
8. Gutarra MLE, Cavalcanti EDC, Freire DMG, Castilho LR, Sant'anna GL Jr (2005) Lipase production by solid state fermentation: cultivation conditions and operation of a packed-bed bioreactor. Appl Biochem Biotech 121: 105–116.
9. Gutarra MLE, Godoy MG, Maugeri F, Rodrigues MI, Freire DMG, et al. (2009) Production of an acidic and thermostable lipase of the mesophilic fungus *Penicillium simplicissimum* by solid-state fermentation. Bioresour Technol 100: 5249–5254.
10. Gutarra MLE, Godoy MG, Silva JN, Guedes IA, Lins U, et al. (2009) Lipase production and *Penicillium simplicissimum* morphology in solid-state and submerged fermentations. Biotechnol Journal 4: 1450–1459.
11. Heck JX, Hertz PF, Ayub MAZ (2005) Extraction optimization of xylanases obtained by solid-state cultivation of Bacillus circulans BL53. Process Biochem 40: 2891–2895.
12. Jaeger KE, Dijkstra BW, Reetz MT (1999) Bacterial biocatalysts: molecular biology, three-dimensional structures and biotechnological applications of lipases. Annu Rev Microbiol 53: 315–351.
13. Johri S, Verma V, Parshad R, Koul S, Taneja SC, et al. (2001) Purification and characterisation of an ester hydrolase from a strain of *Arthrobacter* species: its application in asymmetrisation of 2-benzyl-1,3-propanediol acylates. Bioorg Med Chem 9: 269–273.
14. Laemmli UK (1970) Cleavage of structural proteins during assembly of head of bacteriophage-T4. Nature 227: 680–685.
15. Li N, Zong M (2010) Lipases from the genus *Penicillium*: Production, purification, characterization and applications. J Mol Catal 66: 43–54.
16. Mahadik ND, Puntambekar US, Bastawde KB, Khire JM, Gokhale DV (2002) Production of acidic lipase by *Aspergillus niger* in solid state fermentation. Process Biochem 38: 715–721.
17. Mateo C, Palomo JM, Fernandez-Lorente G, Guisan JM, Fernandez-Lafuente R (2007) Improvement of enzyme activity, stability and selectivity via immobilization techniques. Enzyme Microb Technol 40: 1451–1463.
18. Mitchell DA, Berovic M, Krieger N (2002) Overview of solid state bioprocessing. Biotechnol Annu Rev 8: 183–225.
19. Oliveira D, Di Luccio M, Faccio C, Rosa CD, Bender JP, et al. (2004) Optimization of enzymatic production of biodiesel from castor oil in organic solvent medium. Appl Biochem Biotech 115: 771–780.
20. Palomo JM, Peñas MM, Fernández-Lorente G, Mateo C, Pisabarro AG, et al. (2003) Solid-Phase Handling of Hydrophobins: Immobilized Hydrophobins as a New Tool To Study Lipases. Biomacromolecules 4: 204–210.
21. Salum TFC, Villeneuve P, Barea B, Yamamoto CI, Côcco LC, et al. (2010) Synthesis of biodiesel in column fixed-bed bioreactor using the fermented solid produced by *Burkholderia cepacia*LTEB11. Process Biochem 45: 1348–1354.
22. Sharma R, Chisti Y, Banerjee UC (2001) Production, purification, characterization, and applications of lipases. Biotechol Adv 19: 627–662.
23. Valente AM, Alexandre VM, Cammarota MC, Freire DMG (2010) Enzymatic hydrolysis of fat from fish industry effluents aimed at increasing methane production. Ciênc Tecnol Aliment 30: 483–488.
24. Vardanega R, Remonatto D, Arbter F, Polloni A, Rigo E, et al. (2010) Systematic study on extraction of lipase obtained by solid state fermentation of soybean meal by a newly isolated strain of *Penicillium sp*. Food Bioprocess Technol 3: 461–465.
25. Viniegra-González G, Favela-Torres E, Aguilar CN (2003) Advantages of fungal enzyme production in solid state over liquid fermentation systems. Biochem Eng J 13: 157–167.
26. Zeng GM, Shi JG, Yuan XZ, Liu J, Zhang ZB (2006) Effects of Tween 80 and rhamnolipid on the extracellular enzymes of *Penicillium simplicissimum* isolated from compost. Enzym Microb Tech 39: 1451–1456.

12

Hybrid Formulations of Liposomes and Bioadhesive Polymers Improve the Hypotensive Effect of the Melatonin Analogue 5-MCA-NAT in Rabbit Eyes

Daniela Quinteros[1]◔, Marta Vicario-de-la-Torre[2]◔, Vanessa Andrés-Guerrero[2]◔, Santiago Palma[1], Daniel Allemandi[1], Rocío Herrero-Vanrell[2], Irene T. Molina-Martínez[2]*

1 Department of Pharmacy, Facultad de Ciencias Químicas, Universidad Nacional de Córdoba, CONICET, Edificio de Ciencias II, Ciudad Universitaria, Córdoba, Argentina, 2 Department of Pharmacy and Pharmaceutical Technology, Faculty of Pharmacy, Complutense University of Madrid, Pharmaceutical Innovation in Ophthalmology Research Group, Sanitary Research Institute of the San Carlos Clinical Hospital (IdISSC) and the Ocular Pathology National Net (OFTARED) of the Institute of Health Carlos III, Madrid, Spain

Abstract

For the treatment of chronic ocular diseases such as glaucoma, continuous instillations of eye drops are needed. However, frequent administrations of hypotensive topical formulations can produce adverse ocular surface effects due to the active substance or other components of the formulation, such as preservatives or other excipients. Thus the development of unpreserved formulations that are well tolerated after frequent instillations is an important challenge to improve ophthalmic chronic topical therapies. Furthermore, several components can improve the properties of the formulation in terms of efficacy. In order to achieve the mentioned objectives, we have developed formulations of liposomes (150–200 nm) containing components similar to those in the tear film and loaded with the hypotensive melatonin analog 5-methoxycarbonylamino-N-acetyltryptamine (5-MCA-NAT, 100 μM). These formulations were combined with mucoadhesive (sodium hyaluronate or carboxymethylcellulose) or amphiphilic block thermosensitive (poloxamer) polymers to prolong the hypotensive efficacy of the drug. In rabbit eyes, the decrease of intraocular pressure with 5-MCA-NAT-loaded liposomes that were dispersed with 0.2% sodium hyaluronate, 39.1±2.2%, was remarkably higher compared to other liposomes formulated without or with other bioadhesive polymers, and the effect lasted more than 8 hours. According to the results obtained in the present work, these technological strategies could provide an improved modality for delivering therapeutic agents in patients with glaucoma.

Editor: Ted S. Acott, Casey Eye Institute, United States of America

Funding: Research Group UCM 920415, FIS PI10/00645 and FIS PI10/00993, Spanish Ministry of Science and Innovation (MAT 2010-18242), RETICS RD 12/0034/0003. Financial support from Consejo Nacional de Investigaciones Científicas y Técnicas (CONICET), FONCyT PICT 2010-0380 (Préstamo BID N° 2437/OC-AR) and SECyT-UNC is greatly acknowledged. The funders had no role in study design, data collection and analysis, decision to publish, or preparation of the manuscript.

Competing Interests: The authors have declared that no competing interests exist.

* Email: iremm@farm.ucm.es

◔ These authors contributed equally to this work.

Introduction

Ophthalmic drug delivery is one of the most interesting and challenging endeavors facing the pharmaceutical sciences. The anatomy, physiology and biochemistry of the eye render this organ exquisitely impervious to foreign substances [1]. So to get the desired therapeutic effect, repeated and frequent applications of topical ophthalmic formulations are usually required.

Glaucoma is a multifactorial, progressive and neurodegenerative disease characterized by atrophy of the optic nerve and loss of retinal ganglion cells that can eventually lead to loss of visual acuity and visual field [2,3]. High intraocular pressure (IOP) is considered the greatest risk factor for the development of glaucoma, so most treatments consist of the chronic application of eye drops containing hypotensive agents. Despite the usefulness of topical administration, glaucoma treatments are usually associated with adverse reactions generated by the frequent exposure to drugs and excipients undergone by the eye. Among the excipients, preservatives can induce ocular surface alterations [4,5,6,7] that contribute to the development of secondary ophthalmic diseases such as the dry eye syndrome [8,9]. This in turn can compromise patient compliance. However, the elimination of preservatives from ophthalmic formulations is not always enough to avoid side effects on the ocular surface. For that reason, the incorporation of new components in formulations with beneficial properties for the eye and able, at the same time, to increase the bioavailability of the drug, results of great interest in this field.

Melatonin receptors have been identified in the cornea, ciliary body, lens, choroid, and sclera and play a role in aqueous humor dynamics [10,11]. In recent years, the role of melatonin and its analogues in the control of IOP has been investigated. The melatonin analogue 5-methoxycarbonylamino-N-acetyltryptamine

(5-MCA-NAT) induces an IOP reduction after topical administration in normotensive rabbits [12,13] and glaucomatous monkeys [14]. In a previous work, the hypotensive effect of 5-MCA-NAT was enhanced up to 30% (maximum IOP reduction) with the use of mucoadhesive polymers, although the effect could not be prolonged longer than 6 hours [15].

Ophthalmic formulations of liposomes have been proposed to increase the efficacy of topical ophthalmic formulations and, depending on composition, to resemble the tear film lipid layer. Liposomes are spherical vesicles composed of an aqueous core enclosed by concentric phospholipids bilayers [16]. On the ocular surface, these vesicles establish a very tight contact with tissues, thus increasing the residence time of the formulation and, subsequently, the corneal penetration of the drug [17]. When active substances are included in these systems they are protected from the medium and the half-life is usually increased.

Our research group recently developed a novel artificial tear based on liposomes that improve the quality and stability of an unstable tear film [18]. In this formulation, liposomes were dispersed in aqueous solutions of polymers with mucoadhesive properties. The addition of these polymers had positive effects attributed to their rheological and biophysical properties, which are similar to those of mucins involved in the maintenance of the tear film on the ocular surface. Bioadhesive polymers also provide long-lasting hydration and lubrication of the ocular surface while minimizing the loss of formulation through the nasolacrimal drainage pathway [19].

Other materials, such as thermosensitive amphiphilic block copolymers, namely poly (ethylene oxide)–poly (propylene oxide)–poly (ethylene oxide) (PEO–PPO–PEO, poloxamers), have been recently investigated in the design of novel drug delivery systems. When poloxamers are in solution, they form micelles and, depending on the concentration and temperature, are able to self-organize and transform into a viscous gel, allowing a controlled release of the drug [20].

In the present study we have used hybrid combinations of liposomes, mucoadhesive or thermosensitive polymers, and the intraocular hypotensive agent 5-MCA-NAT. These pharmaceutical formulations were characterized in terms of physicochemical properties, *in vitro* release of 5-MCA-NAT, *in vivo* tolerance and *in vivo* intraocular pressure reduction, after topical ophthalmic application in nonsedated normotensive rabbits.

Materials and Methods

Materials

The melatonin agonist 5-MCA-NAT was purchased from Tocris Bioscience (Bristol, UK) and 1.2-propylene glycol (PG) from Guinama (Valencia, Spain). Sodium hyaluronate (SH, Mw 400,000–800,000 g/mol) and sodium carboxymethylcellulose (CMC, 400–800 cps 2% solution at 20°C) were purchased from Abarán Materias Primas S.L. (Madrid, Spain). Phospholipon 90G purified from soy lecithin (>95% phosphatidylcholine, PC) was purchased from Phospholipid GmbH (Cologne, Germany). Cholesterol (Cht) and α-tocopherol (vitE) were acquired from Sigma Chemical Co. (St. Louis, MO, USA). Isotonic NaCl solution was prepared with ultrapure Milli-Q water (EMD Milllipore, Darmstadt, Germany). Poloxamer 407 (PX407) and Poloxamer 188 (PX188) were gifts from BASF S.A. (Buenos Aires, Argentina).

Methods

1. Quantitation of 5-MCA-NAT by high pressure liquid chromatography (HPLC). Quantitative analyses of 5-MCA-NAT were performed with a HPLC instrument (Gilson, Middleton, WI, USA) composed of a solvent delivery pump (305 model), UV-visible detector (118 model), and controller software (UniPoint, all by Gilson). The injector was equipped with a 20 µl loop (7125 Rheodyne, Berkeley, CA, USA). The chromatographic separation was achieved as described previously [12] by a reversed phase protocol with a Mediterranea Sea18 column (25 cm×4 mm, 5 µm particle size; Teknokroma, Barcelona, Spain). The mobile phase was a mixture of methanol (Panreac, Barcelona, Spain) and ultrapure milliQ water (40:60 v/v). The flow rate was set at 0.8 ml/min and the eluent was monitored at 244 nm.

2. Sample preparation. Five formulations (Table 1) containing 100 µM 5-MCA-NAT were prepared: F1, aqueous solution of 5-MCA-NAT; F2, 5-MCA-NAT-loaded liposomes; F3, 5-MCA-NAT-loaded liposomes with 0.2% SH; F4, 5-MCA-NAT-loaded liposomes with 0.5% CMC; and F5, 5-MCA-NAT-loaded liposomes with PX (PX407 and PX188; 12/8, w/w).

Formulation F1, the aqueous non-liposomal solution of 5-MCA-NAT, was prepared by dissolution of 5-MCA-NAT in PG (10 mg/mL) and further dilution with an aqueous solution of NaCl. The final formulation contained 100 µM 5-MCA-NAT, 0.275% PG, and 0.788% NaCl. 5-MCA-NAT-loaded liposomes (F2, F3, F4, and F5) were prepared following the procedure described by Bangham [21] and modified by Vicario-de-la-Torre [22]. According to the procedure PC, Cht and vitE (8:1:0.08) were dissolved in chloroform in a round-bottom flask. Then, the organic solvent was slowly removed at 33°C with a rotator evaporator to produce a thin film of dry lipids on the inner surface of the flask. The dry film was then hydrated with an aqueous solution of 200 µM 5-MCA-NAT to produce the 5-MCA-NAT-loaded liposomes. The liposomes were extruded ten times through 0.22 µm pore polycarbonate membranes (Nucleopore Lipex Biomembrane, Vancouver, Canada). To ensure full lipid hydration, vesicles were allowed to mature overnight under refrigeration. After that, stock formulations were diluted (1:2) with different aqueous solutions to achieve the final liposomal formulations (Table 1): F2, 0.788% NaCl and 0.275% PG; F3, 0.4% SH in 0.788% NaCl and 0.275% PG; F4, 1% CMC in 0.788% NaCl and 0.275% PG); F5, PX407 and PX188 (24/16, w/w) in 0.17% NaCl and 0.275% PG. The concentration of NaCl was adjusted to attain osmolarity values in the final formulations within an acceptable range for ophthalmic administration [23]. Due to the high influence of PX in the osmolarity of the formulations [24], a lower concentration of NaCl was used to adjust F4. Final liposomal formulations were sterilized by filtration through 0.22 µm pore membranes.

3. Determination of the encapsulation efficiency of liposomes. The encapsulation efficiency of 5-MCA-NAT in the liposomes was determined upon separation by centrifugation (18,000 rpm, 4°C, 20 min) from the dispersing medium containing non-encapsulated 5-MCA-NAT. The amount of free 5-MCA-NAT in the supernatant was determined in triplicate by HPLC. The determination of the percentage of drug loading (PDL) in liposomes, expressed as mean ± standard deviation, was calculated using the following equation [25]:

$$PDL = 100\ x\ [total\ drug\ added\ (\mu g) - free\ drug\ (\mu g)] / total\ drug\ added\ (\mu g)$$

The entire amount of 5-MCA-NAT in the liposomal formulations was measured by dissolving liposomes with the appropriate

Table 1. Nomenclature and composition of the 5-MCA-NAT formulations.

Formulation	Composition	Polymer
F1	100 μM 5-MCA-NAT, 0.275% PG and 0.788% NaCl	-
F2	100 μM 5-MCA-NAT, 0.275% PG, 10 mg/mL PC, and 0.788% NaCl	-
F3	100 μM 5-MCA-NAT, 0.275% PG, 10 mg/mL PC, 0.2% SH and 0.788% NaCl	Sodium hyaluronate
F4	100 μM 5-MCA-NAT, 0.275% PG, 10 mg/mL PC, 0.5% CMC and 0.788% NaCl	Carboxymethylcellulose
F5	100 μM 5-MCA-NAT, 0.275% PG, 10 mg/mL PC, PX407 and PX188 (12/8 w/w), and 0.479% NaCl	Poloxamer

5-MCA-NAT: 5-methoxycarbonylamino-N-acetyltryptamine, PG: propylene glycol, PC: phosphatidylcholine, SH: sodium hyaluronate, CMC: carboxymethylcellulose, PX: poloxamer (PX407 and PX188 12/8 w/w).

amount of acetonitrile. Solutions were then centrifuged (15,000 rpm, 4°C, 20 min) and the supernatant analyzed in triplicate by HPLC.

4. Determination of pH, osmolarity, and viscosity. The pH of the formulations was measured with a calibrated pH meter (model 230; Mettler, Barcelona, Spain) equipped with an InLab Microelectrode (Mettler). Measurements were performed in triplicate at room temperature (25°C).

Osmolarity was analyzed by a vapor pressure osmometer (model K-7000 Knauer, Berlin, Germany). Before performing the analyses, the osmometer was calibrated with a solution of NaCl (400 mOsm). The determinations were made in triplicate at 33°C (ocular surface temperature) [26].

Viscosity of liposomal samples was assessed in triplicate at 33°C with a thermostatically controlled rheometer (HaakeRheostress R1, Düsseldorf, Germany) using a parallel plate geometry (diameter 60 mm, gap 0.5 mm). The viscosity was calculated at a shear rate of 100 s^{-1}, which was within the linear viscoelastic range for all of the formulations.

5. Determination of particle size and zeta potential. Measurements of liposome particle size and zeta potential were carried out by photon correlation spectroscopy (PCS, Zetatrac, Largo, FL, USA). For the analyses, formulations were diluted 1/20 (v/v) in an aqueous medium. All determinations were performed in triplicate at room temperature (25°C).

6. Hypotensive efficacy studies *in vivo*: IOP determinations. In this work, experiments were performed in both eyes of nonsedated normotensive male New Zealand white rabbits (2–2.5 kg). Each formulation was evaluated in 10 animals (n = 20 eyes) and each control in 5 (n = 10 eyes). In order to reduce the number of rabbits, the same animals were used multiple times as experimental and control. The animals were kept in individual cages with free access to food and water and maintained in a controlled 12/12 h light/dark cycle. All of the protocols herein were approved by the Ethics Committee for Animal Research of Complutense University of Madrid. Also, animal manipulations followed institutional guidelines, European Union regulations for the use of animals in research, and the ARVO (Association for Research in Vision and Ophthalmology) statement for the use of animals in ophthalmic and vision research.

IOP was measured with a Tonovet rebound tonometer (Tiolat, Helsinki, Finland). With this technique IOP is assessed without the need of topical anesthesia. For each eye, IOP was set at 100% with two basal readings taken 30 min before and immediately before the instillation. Then a single dose of the formulation (25 μL) was applied to both eyes. IOP determinations were performed once every hour over the next 8 hours. As a control, rabbits received formulations without the hypotensive agent. The administration protocol included a washout period of at least 48 hours between experiments.

7. *In vitro* drug release. The release of 5-MCA-NAT from the different formulations was studied using a dialysis method. To do this, a dialisys membrane (Spectra/Por Float-A-Lyzer G2; 20,000 MW cut off; Iberlabo, Madrid, Spain) was employed (0.8 mL of sample). Three membranes were prepared for each formulation. A conventional solution of 100 μM 5-MCA-NAT without liposomes was used as reference. Dialysis membranes were placed inside a flask with 50 mL of a phosphate buffered solution isotonized with NaCl (pH 7.4). The flask was kept on a magnetic stirrer and stirring was maintained at 100 rpm at 33°C. 1 mL of release sample was withdrawn at pre-set times (5 min, 15 min, 30 min, once every hour over a period of 8 and 24 hours). The release of 5MCANAT was analyzed by HPLC following the above mentioned method.

8. *In vivo* tear osmolarity measurements. This study was restricted to formulations F3 (5-MCA-NAT-loaded liposomes dispersed in 0.2% sodium hyaluronate) and F4 (5-MCA-NAT-loaded liposomes dispersed in 0.5% carboxymethylcellulose), that provided the highest *in vivo* hypotensive effect among all the tested formulations. An isotonic saline solution was used as control. Each formulation (control included) was tested in both eyes of three nonsedated male New Zealand white rabbits (n = 6 eyes).

Tear osmolarity was measured with the TearLab Osmolarity System (TearLab Corporation, San Diego, CA, USA). The device, stored in a temperature and humidity controlled room, was re-calibrated using electronic check cards at the beginning of each test session following the manufacturer's guidelines. Tear samples of 50 nL were collected from the lower lateral meniscus of the eyes. To avoid fluctuation of tear osmolarity, all the measurements were made between 9 and 11 AM. Both eyes of rabbits were used for the evaluations. Measurements were taken before the administration of 25 μL of the formulation and 5, 15, 30 and 60 min after instillation.

9. *In vivo* short-term tolerance study. This study was restricted to formulations F3 (5-MCA-NAT-loaded liposomes dispersed in 0.2% sodium hyaluronate) and F4 (5-MCA-NAT-loaded liposomes dispersed in 0.5% carboxymethylcellulose), that provided the highest *in vivo* hypotensive effect among all the tested formulations.

The experiment was carried out on both eyes of six male New Zealand rabbits. A single administration of 25 μL of the formulation were administered in the right eye of the animal.

The contralateral eye was used as control and being instilled with the same volume of an isotonic solution of NaCl (pH 7.4).

Clinical symptoms and signs were evaluated in accordance with a protocol described previously [27]. Ocular surface evaluation was made before the instillation of the formulation or control, and then 1, 8 and 24 h after their administration. The short-term tolerance was evaluated by macroscopic examination of the eye surface of the animal and graded from 0–2, indicating the absence or presence of the following clinical signs: loss of corneal transparency, conjunctival signs (hyperemia, edema), eyelid swelling, and intense blinking (what would show a lack of tolerance).

10. Statistical analysis. Intraocular hypotensive reduction was expressed as means ± standard error of the means (SEM). Other parameters as means ± standard deviation (SD) were also evaluated. Statistical differences between two mean values were evaluated by two-tailed Students t-test. If necessary, an analysis of variance (ANOVA) was employed. Results were taken as significantly different at p-values less than 0.05.

Results

1. 5-MCA-NAT dose determination

The HPLC method to quantify 5-MCA-NAT was validated with respect to linearity, accuracy, and reliability in the range of concentrations between 5 and 50 μg/mL. In all cases, the method allowed sufficient separation of the drug from the vehicle. The retention time of 5-MCA-NAT was 10.7±0.5 minutes. There were non-significant differences ($p = 0.57$) between the theoretical concentration of 5-MCA-NAT (100 μM, ~27.53 μg/mL) and the experimental values obtained (28.5±1.07 μg/mL). The percentage of 5-MCA-NAT loaded in the liposomes was 7.9±3.5%, relative to the entire amount of drug in the liposomal formulation. Taking into consideration that the volume of the instilled formulation in rabbit eyes was 25 μL, the dose of 5-MCA-NAT administered was 0.71±0.04 μg.

2. pH, osmolarity, and viscosity

The pH and osmolarity were measured for all formulations, and the viscosity was measured for liposomal formulations F2–F5 (Table 2). Formulation F1, without liposomes, had an acidic pH of 5.6. The incorporation of lipid vesicles in formulation F2 produced an increase of pH to 6.9, a nearly neutral value. The incorporation of polymer in formulations F3, F4, and F5 produced a minor decrease in pH, with values of 6.4, 6.5, and 6.5 respectively. The osmolarity of formulations F1, F2, F3, and F4 were within the range of isotonicity (Table 2). Formulation F5, with an osmolarity value of 237±3.1 mOsm, was hypotonic. The lowest viscosity, 1.2 mPa·s, was obtained in F2 (Table 2), the liposomal formulation without polymers. The addition of sodium hyaluronate 0.2% (F3) and carboxymethylcellulose 0.5% (F4) increased the viscosity of the formulations to 2.0 mPa·s and 7.3 mPa·s, respectively. The liposomal formulation with poloxamer (F5) had the highest viscosity, 28.3 mPa·s.

3. Mean particle size and zeta potential of liposomal formulations

All of the liposomal formulations showed unimodal size distributions (Figure 1), and the average particle sizes were less than 200 nm in all cases (Table 3). The average zeta potential was neutral (range between −10 to 10 mV) in all cases (Table 3).

4. Effect of 5-MCA-NAT formulations on IOP in rabbits

All formulations reduced IOP in normotensive rabbits though the maximal effects were different for each formulation (Figure 2). The maximum percentage of IOP reduction, the area under the ΔIOP (%) time curve from 0 to 8 hours (estimated by the trapezoidal rule) and the duration of the hypotensive effect (h), were calculated for all the formulations (Table 4). In addition, several other IOP parameters were considered: mean of the minimum and maximum IOP, and mean difference between maximum and minimum IOP (Table 5). The IOP values measured were within the range of 8 to 16 mm Hg (Table 5). Formulation F3, composed of 5-MCA-NAT-loaded liposomes with 0.2% SH, produced the maximum hypotensive effect (Figure 2, Tables 4 and 5). This formulation provided an IOP reduction of 39.13±2.21%. The second highest value, 36.72±2.77%, was achieved by F4, composed of 5-MCA-NAT-loaded liposomes with 0.5% CMC. There were no significant differences between F3 and F4 with regard to the lowering of IOP ($p = 0.55$). The rest of the formulations provided significantly lower values ($p < 0.01$ in all cases). For formulations F2, F3 and F4, the hypotensive effect was maintained for longer than 8 hours ($p < 0.05$ for each formulation compared with the corresponding vehicle) (Figure 2 and Table 4). For F5, composed of 5-MCA-NAT-loaded liposomes dispersed in PX, the hypotensive effect lasted only 5 hours. Differences in the IOP can be observed among the formulations at 4 time points: 3, 6, 7 and 8 hours ($p < 0.05$) although these differences are not systematically generated by the same formulation. Regarding the area under the ΔIOP (%) time curve from 0 to 8 hours, $AUC_0^{8\,h}$, the highest value was shown with F3 (156.15±14.71), while the lowest with F5 (85.20±16.83).

5. *In vitro* release studies

The *in vitro* release behavior of the solution of 5-MCA-NAT (F1), 5-MCA-NAT-loaded liposomes dispersed in NaCl (F2), 5-MCA-NAT-loaded liposomes dispersed in 0.2% SH (F3) and 5-MCA-NAT-loaded liposomes dispersed in 0.5% CMC (F4), is summarized in the cumulative percentage release shown in figure 3.

The release rate of the drug was higher for the solution of 5-MCA-NAT (F1), while the rate of release from 5-MCA-NAT-loaded liposomes dispersed in SH (F3) or CMC (F4) was lower than that of conventional liposomes (F2). Statistically significant differences between the solution of 5-MCA-NAT (F1) and liposomal formulations (F2–F4) were found after 1 hour of the beginning of the study ($p < 0.05$ at all the time points studied). Comparing liposomal formulations, non-significant differences were found between conventional liposomes (F2) and liposomes dispersed in SH ($p > 0.25$ in all cases). In the case of F4, 5-MCA-NAT released from liposomes resulted significantly lower to the ones that were found for the solution of 5-MCA-NAT (F1) or for conventional liposomes (F2) after 1 hour of the beginning of the study and up to 8 hours ($p < 0.003$ in all cases). From 6 to 8 hours, a significant different release rate can be observed between F3 and F4, being F4 (5-MCA-NAT-liposomes dispersed with CMC) the formulation that showed a slower release profile ($p < 0.05$).

6. *In vivo* tear osmolarity

Following the guidelines for the use of experimental animals, the number of rabbits was diminished by selecting the formulations that showed the best *in vivo* hypotensive effect: 5-MCA-NAT-loaded liposomes dispersed in 0.2% SH (F3) and 5-MCA-NAT-loaded liposomes dispersed in 0.5% CMC (F4). Results are collected in table 6.

Table 2. pH, osmolarity, and viscosity data of the 5-MCA-NAT formulations.

Formulation	Polymer	pH	Osmolarity (mOsm)	Viscosity (mPa·s)
F1	-	5.6±0.02	295.8±0.5	-
F2	-	6.9±0.03	303.8±0.2	1.2±0.1
F3	Sodium hyaluronate	6.4±0.02	307.1±0.3	2.0±0.04*
F4	Carboxymethylcellulose	6.5±0.01	294.2±0.5	7.3±0.1*
F5	Poloxamer	6.5±0.01	237.2±3.1*	28.3±1.4*

Data are expressed as means ± SD (n = 3).
*Significant differences with formulation F2 (p-value<0.05).

Instillation of a single drop of saline solution (Control Group) decreased the tear osmolarity for 5 min. However, 30 and 60 min later, osmolarity values were not significantly different to basal ($p = 0.12$). When animals were treated with liposomes combined with bioadhesive polymers (F3 and F4) the tear osmolarity diminished significantly at any time in comparison to basal ($p = 0.04$ and $p = 0.0026$, respectively).

7. *In vivo* tolerance study

Before testing, all animals had a normal ocular surface and corneal transparency. None had any conjunctival disorders including hyperemia or edema, eyelid swelling, or intense blinking (grade 0). Animals showed no discomfort or irritation during the test or within 24 h after the administration of formulations F3 and F4, which contained 5-MCA-NAT-loaded liposomes dispersed in 0.2% SH and 0.5% CMC, respectively. The cornea remained transparent (no vessels) throughout the assay (grade 0) and the coloration of the conjunctiva remained normal (grade 0). No animal presented signs of mucus secretion.

Discussion

In general, topical ophthalmic anti-glaucoma therapies combine the use of one or more hypotensive drugs with several instillations per day in chronic treatments. In these patients, the eye surface is continuously exposed to drugs and preservatives that produce, in most cases, ocular surface alterations and generally lead to a therapeutic failure [28]. Eye drops for chronic treatments should include components that are well tolerated for the ocular surface and, at the same time, enable enhanced drug bioavailability

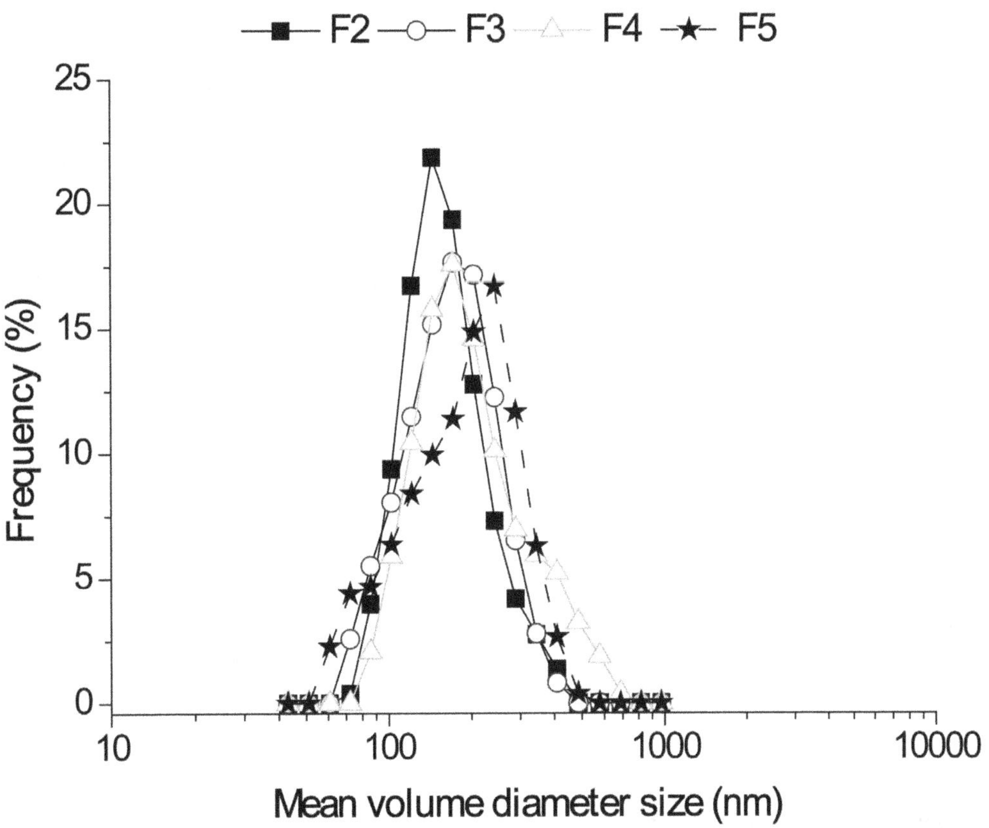

Figure 1. Size distribution of 5-MCA-NAT-loaded liposomes dispersed in NaCl (F2), 0.2% sodium hyaluronate (F3), 0.5% carboxymethylcellulose (F4) and 12/8 w/w poloxamer (F5).

Table 3. Mean diameter and zeta potential of the 5-MCA-NAT liposomal formulations.

Formulation	Polymer	Size (nm)	Zeta potential (mV)
F1	-	-	-
F2	-	155.1±4.0	10.2±2.7
F3	Sodium hyaluronate	162.8±7.7	8.7±2.7
F4	Carboxymethylcellulose	188.5±16.9*	7.2±4.5
F5	Poloxamer	181.8±12.8*	6.4±0.9

Data are expressed as means ± SD (n = 3).
*Significant differences with formulation F2 (p-value<0.05).

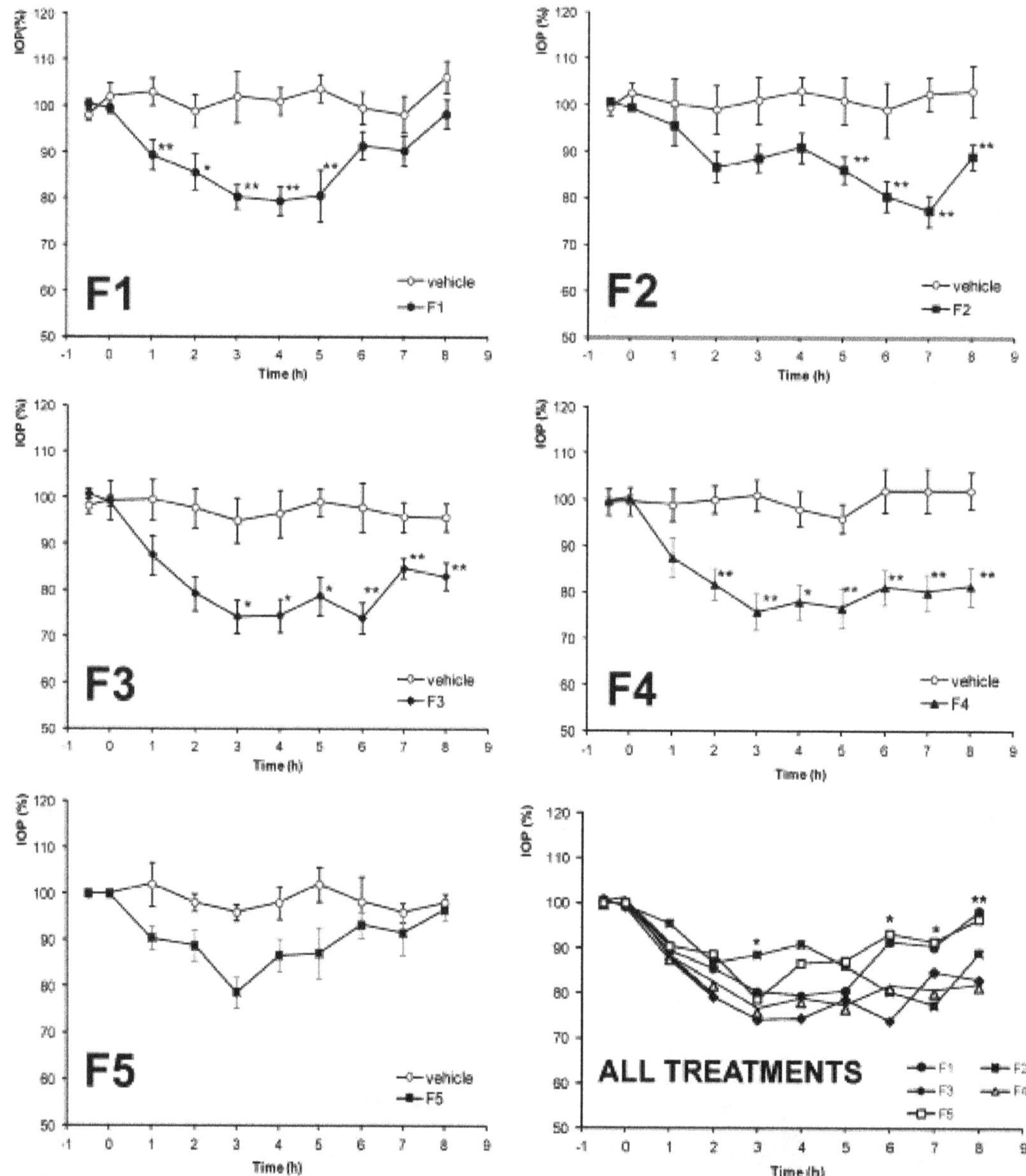

Figure 2. Ocular hypotensive effect of 5-MCA-NAT formulations. F1, solution of 5-MCA-NAT. F2, 5-MCA-NAT-loaded liposomes dispersed in NaCl; F3, 5-MCA-NAT-loaded liposomes dispersed in 0.2% sodium hyaluronate; F4, 5-MCA-NAT-loaded liposomes dispersed in 0.5% carboxymethylcellulose; F5, 5-MCA-NAT-loaded liposomes dispersed in 12/8 w/w poloxamer. Vehicles without 5-MCA-NAT were used as control in each case. Data are expressed as the mean ± SEM (n = 20).

Table 4. Maximum percent IOP reduction over 8 hours.

Formulation	Polymer	Maximum IOP reduction (%)	$AUC_0^{8\ h}$ (%·h)	Duration (h)
F1	-	29.27±2.42	118.02±14.62	7
F2	-	29.44±2.40	112.07±15.93	+8
F3	Sodium hyaluronate	39.13±2.21*	156.15±14.71*	+8
F4	Carboxymethylcellulose	36.72±2.77*	131.84±16.81*	+8
F5	Poloxamer	29.12±2.11	85.20±16.83	5

IOP, intraocular pressure; Maximum IOP reduction (%), percentage of reduction ± SEM; $AUC_0^{8\ h}$, ΔIOP (%) versus time (h) from 0 to 8 hours. Data are expressed as means ± SEM (n = 20 eyes).
*Significant differences with formulation F2 (p-value<0.05).

[29,30]. In general, toxicity of formulations can be reduced by removing preservatives from the composition. However, in some cases the toxic effect of the drug still remains [31,32,33]. The inclusion of bioadhesive polymers have demonstrated to increase the tolerance of formulations while extending the effect of hypotensive drugs such as timolol maleate [34]. Nevertheless, the benefits produced by these polymers in terms of efficacy are limited, so it becomes necessary to use other technological approaches, such as combination of the drugs with other colloidal systems. Liposomes have gained considerable attention for ocular drug delivery. They have been primarily investigated as a modality to enhance corneal drug absorption through the ability to come into intimate contact with the corneal and conjunctival surfaces, thereby increasing the ocular drug penetration [17].

In this study, we designed novel liposomal formulations loaded with 5-MCA-NAT, a melatonin derivative that reduces IOP [12,15]. The liposomes that we employed (150–200 nm, 6–10 mV zeta potential) simulate the lipid composition of the pre-ocular tear film and have demonstrated to be effective in the treatment of the dry eye syndrome [18]. The anti-glaucoma formulations of the current study were designed to (a) control the delivery of 5-MCA-NAT, thus prolonging the hypotensive effect, and (b) replenish the tear film, which may improve an ocular surface damaged by chronic exposure to topical treatments. To achieve the objectives, we prepared liposomes with the biocompatible components PC, Cht, and vitE – all of which are present in the natural tear film – that were loaded with 100 µM 5-MCA-NAT. Once prepared, the liposomes were dispersed in an isotonic aqueous solution alone or with the bioadhesive biopolymers SH or CMC, or a thermo-sensitive polymer, PX, to determine whether or not the use of these combinations would enhance the hypotensive effect of 5-MCA-NAT. The bioadhesive polymers and concentrations were selected among the available existing commercial artificial tear products [35]. A conventional solution of 100 µM 5-MCA-NAT without liposomes was used as the reference.

All the formulations developed satisfied the given requirements of pH and osmolarity for ophthalmic solutions [36]. Regarding pH values, the presence of lipid vesicles in formulation F2 produced a slight increase of pH to a nearly neutral value, probably due to the presence of PC in the formulation. The incorporation of polymers in formulations F3, F4, and F5 produced minor decreases in pH, presumably due to the acidic nature of SH, CMC and PX. For the formulations F2–F5, the pH values were nearly neutral, thus helping to maintain the optical properties of the eye surface, epithelial cell functions, and cellular homeostasis. With respect to the osmolarity, all of the formulations were within an acceptable range for ophthalmic administration [23].

To obtain a homogeneous film after instillation, the viscosity of a formulation should be similar to that of natural tears, 1–9 mPa·s [37,38]. Previous studies showed that the most significant relative improvement in ocular bioavailability occurred for vehicles in the viscosity range from 1 to 15 mPa·s [39,40]. In our study, the viscosities of formulations with bioadhesive polymers were within this range, and the most effective formulations for reducing IOP, F3 and F4, had viscosity values of 2.0 mPa·s and 7.3 mPa·s, respectively. The formulation F5 had the highest viscosity, 28.3 mPa·s, but it was no more effective in reducing IOP than were the non-liposomal F1 or the liposomal F2 without a bioadhesive polymer. This anomalous behavior of F5 (liposomes dispersed with Poloxamer) could be associated with its high viscosity and sol–gel phase transition phenomena [41].

The bioadhesive properties and rheological behavior of SH and CMC are widely described [42,43]. Mucins present in the pre-ocular tear film can interact with the bioadhesive polymers and

Table 5. IOP parameters over 8 hours.

Formulation	IOP_{max}	IOP_{min}	ΔIOP
F1	15.5±1.8	10.1±1.8	5.4±1.7
F2	14.7±1.8	9.7±1.4	5.0±1.6
F3	16.2±1.9	9.3±1.5	6.9±1.8
F4	14.7±3.0	8.6±1.5	6.1±2.5
F5	13.2±0.5	8.9±1.1	4.3±1.3

IOP, intraocular pressure in mmHg, IOP_{max}, mean maximum IOP; IOP_{min}, mean minimum IOP; ΔIOP, mean difference between IOP_{max} and IOP_{min}. Data are expressed as means ± SEM (n = 20 eyes).

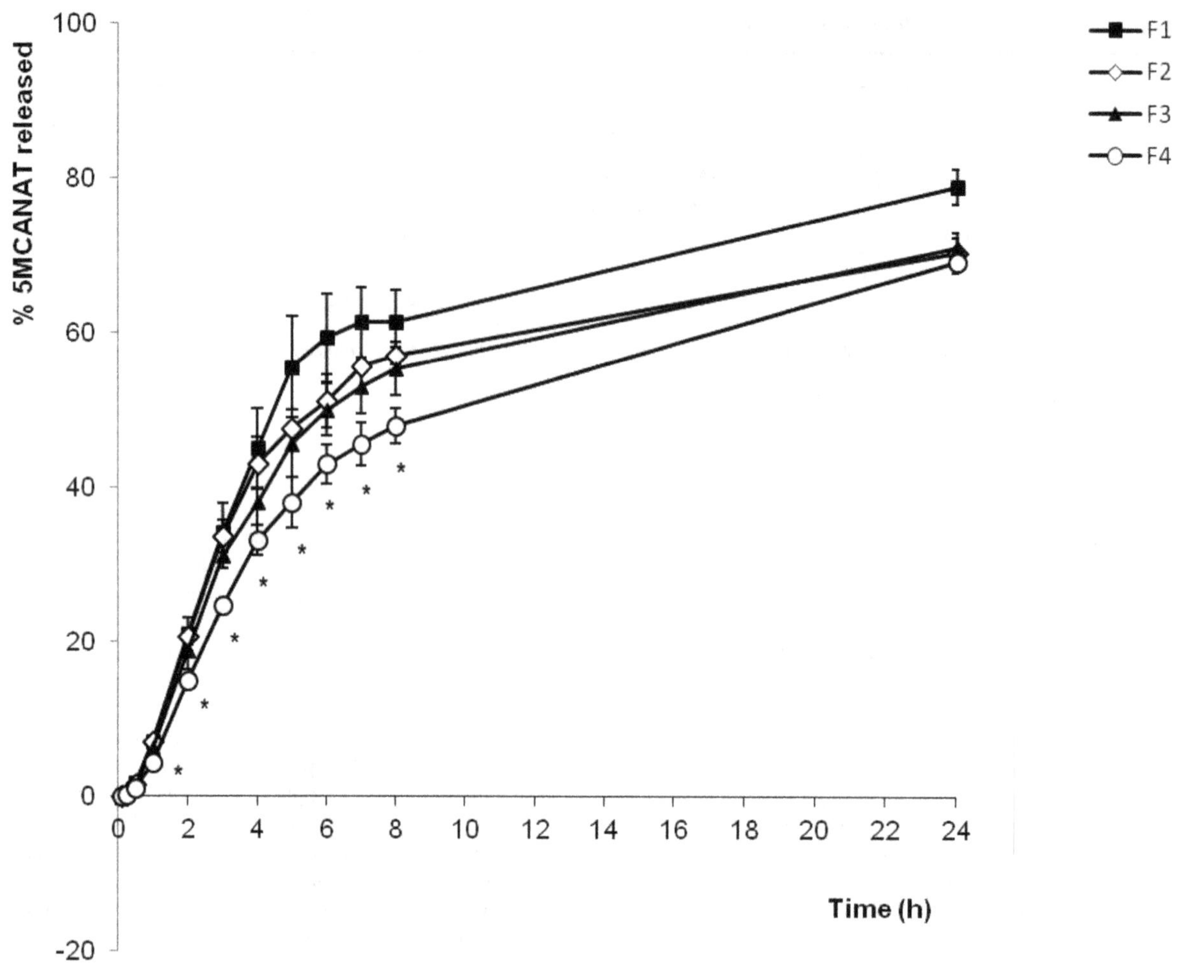

Figure 3. ***In vitro*** **release curve of 5-MCA-NAT formulations in phosphate buffered solution at 33°C.** F1: solution of 5-MCA-NAT; F2: 5-MCA-NAT-loaded liposomes dispersed in NaCl; F3: 5-MCA-NAT-loaded liposomes dispersed in 0.2% sodium hyaluronate; F4: 5-MCA-NAT-loaded liposomes dispersed in 0.5% carboxymethylcellulose. * Significant differences between formulations F2–F4 and formulation F1 (reference, p-value< 0.05).

increase the residence time of the formulation on the eye surface and improve drug bioavailability. Formulations F2 (5-MCA-NAT-liposomes), F3 (5-MCA-NAT-liposomes dispersed with SH) and F4 (5-MCA-NAT-liposomes dispersed with CMC) present hypotensive effect up to 8 hours after a single administration unlike formulation F1 (5-MCA-NAT dissolved in 0.788% NaCl and 0.275% PG). This effect can be attributed to possible interactions between components of the formulations (liposomes and bioadhesive polymers) with the ocular surface (mucins and/or tear film). *In vitro* release studies also corroborate these results. Formulation F4 (5-MCA-NAT-liposomes dispersed with CMC) shows significant differences in the release rate from 6 h to 8 h with the rest of formulations and a tendency to extend the hypotensive effect in animals. Taking into account that hypertensive animals show higher IOP decrease after the use of hypertensive drugs (in comparison to normotensive animals), an induced hypertensive

Table 6. Tear osmolarity in the 3 Study Groups.

t (min)	Control Group	F3 Group	F4 Group
0	325.2±3.8	324.3±2.8	323.7±1.7
5	311.3±3.0	314.9±3.3*	316.7±1.0**
30	326.7±3.6	314.0±2.9*	313.2±3.4*
60	324.6±4.1	315.4±3.6*	311.8±1.9**

Control Group: animals were administered 25 μL of saline solution in both eyes; F3 Group animals were instilled 25 μL of 5-MCA-NAT-loaded liposomes dispersed in 0.2% SH (F3) in both eyes; F4 Group: animals received 25 μL of 5-MCA-NAT-loaded liposomes dispersed in 0.5% CMC (F4). Data are expressed as means ± SD (n = 6).
*Significant differences with Control Group (p-value<0.05).
**Significant differences with Control Group (p-value<0.01).

animal models might be useful to appreciate differences in the extension of the hypotensive effect [44,45].

In addition, these polymers hydrate and protect the ocular surface by forming a film that covers the eye. So, the liposomes dispersed with bioadhesive polymers contribute to the modulation of drug delivery by adequately mixing with the natural tears that cover the ocular surface. The liposomes also reduce water evaporation from tears and help improve symptoms such as dryness, redness, and visual impairment associated with dry eyes. The mucoadhesive behavior of these hybrid formulations was exhibited when rabbit eyes' tear osmolarity decreased to reach osmolarity values similar to the formulations F3 and F4 that were maintained for at least 60 minutes after a single instillation. These results must be confirmed in humans due to the physiological differences between both species that can affect the residence time of the formulations on the eye surface (i.e. lower blink rate than humans). In any case, further studies are necessary to demonstrate the efficacy of these formulations, in terms of ocular surface damage, that require experimental animal models.

The goal of the study was to enhance the effect of the hypotensive agent 5-MCA-NAT by using an unpreserved drug delivery system that could increase its bioavailability while being soft with the ocular surface. The novel hybrid formulations simulate tear film composition and include biocompatible components that might relieve symptoms of secondary diseases affecting the ocular surface derived from the chronic use of anti-glaucoma topical medications. With these improvements, patient adherence to topical glaucoma therapy could be enhanced.

Conclusions

A number of applications of liposomes in ophthalmic drug delivery have been extensively studied [46,47,48,49,50]. The improvement in the precorneal retention, transcorneal permeation, and therapeutic efficacy has been explored in detail, providing information about the interaction between liposomes and ocular tissues. In this work we prepared novel liposomal formulations dispersed in aqueous solutions of bioadhesive or amphiphilic block copolymers for the delivery of the ocular hypotensive agent 5-MCA-NAT. *In vivo* efficacy studies in rabbits showed that the hypotensive effect of the drug was remarkably increased with the combination of liposomes and bioadhesive polymers and also tear osmolarity decreased significantly for 60 min. This improvement could be attributed to the influence of liposomes as drug carriers, the increased residence time of the formulation on the eye surface derived from the mucoadhesive properties of the polymers and the ability of these hybrid formulations to simulate and replenish the tear film. Addition of the thermosensitive polymer poloxamer to the drug-loaded liposomes was no more effective than the non-liposomal 5-MCA-NAT. In this work, the application of liposomes combined with bioadhesive polymers for the improvement of the precorneal retention of hypotensive drugs has shown potential for further investigation. In conclusion, hybrid nanosystems composed by liposomes combined with bioadhesive polymers might serve as potential ocular drug carriers that prolong drug retention and improve biocompatibility of the formulations on the ocular surface.

Author Contributions

Conceived and designed the experiments: SP DA RH IM. Performed the experiments: DQ MV VA. Analyzed the data: MV VA RH IM. Contributed reagents/materials/analysis tools: MV VA SP RH IM. Contributed to the writing of the manuscript: DQ MV VA.

References

1. Urtti A (2006) Challenges and obstacles of ocular pharmacokinetics and drug delivery. Adv Drug Deliv Rev 58: 1131–1135.
2. Schuman JS (2000) Antiglaucoma medications: a review of safety and tolerability issues related to their use. Clin Ther 22: 167–208.
3. Flanagan JG (1998) Glaucoma update: epidemiology and new approaches to medical management. Ophthalmic Physiol Opt 18: 126–132.
4. Baudouin C (2008) Detrimental effect of preservatives in eyedrops: implications for the treatment of glaucoma. Acta Ophthalmol 86: 716–726.
5. Baudouin C, Labbe A, Liang H, Pauly A, Brignole-Baudouin F (2010) Preservatives in eyedrops: the good, the bad and the ugly. Prog Retin Eye Res 29: 312–334.
6. Debbasch C, Pisella PJ, Warnet JM, Rat P, Baudouin C (2001) Quaternary ammoniums and other preservatives' contribution in oxidative stress and apoptosis on Chang conjunctival cells. Invest Ophthalmol Vis Sci 42: 642–652.
7. Brasnu E, Brignole-Baudouin F, Riancho L, Warnet JM, Baudouin C (2008) Comparative study on the cytotoxic effects of benzalkonium chloride on the Wong-Kilbourne derivative of Chang conjunctival and IOBA-NHC cell lines. Mol Vis 14: 394–402.
8. Leung EW, Medeiros FA, Weinreb RN (2008) Prevalence of ocular surface disease in glaucoma patients. J Glaucoma 17: 350–355.
9. Moss SE, Klein R, Klein BE (2000) Prevalence of and risk factors for dry eye syndrome. Arch Ophthalmol 118: 1264–1268.
10. Lundmark PO, Pandi-Perumal SR, Srinivasan V, Cardinali DP, Rosenstein RE (2007) Melatonin in the eye: implications for glaucoma. Exp Eye Res 84: 1021–1030.
11. Wiechmann AF, Wirsig-Wiechmann CR (2001) Multiple cell targets for melatonin action in Xenopus laevis retina: distribution of melatonin receptor immunoreactivity. Vis Neurosci 18: 695–702.
12. Andres-Guerrero V, Alarma-Estrany P, Molina-Martinez IT, Peral A, Herrero-Vanrell R (2009) Ophthalmic formulations of the intraocular hypotensive melatonin agent 5-MCA-NAT. Exp Eye Res 88: 504–511.
13. Andres-Guerrero V, Herrero-Vanrell R (2008) [Ocular drug absorption by topical route. Role of conjunctiva]. Arch Soc Esp Oftalmol 83: 683–685.
14. Serle JB, Wang RF, Peterson WM, Plourde R, Yerxa BR (2004) Effect of 5-MCA-NAT, a putative melatonin MT3 receptor agonist, on intraocular pressure in glaucomatous monkey eyes. J Glaucoma 13: 385–388.
15. Andrés-Guerrero V, Molina-Martínez IT, Peral A, de las Heras B, Pintor J (2011) The Use of Mucoadhesive Polymers to Enhance the Hypotensive Effect of a Melatonin Analogue, 5-MCA-NAT, in Rabbit Eyes. Invest Ophthalmol Vis Sci 52: 1507–1515.
16. Bangham AD, Standish MM, Watkins JC (1965) Diffusion of univalent ions across the lamellae of swollen phospholipids. Journal of Molecular Biology 13: 238–252.
17. Law SL, Huang KJ, Chiang CH (2000) Acyclovir-containing liposomes for potential ocular delivery: Corneal penetration and absorption. Journal of Controlled Release 63: 135–140.
18. Vicario-de-la-Torre M, Benitez del Castillo J, Vico E, de las Heras B, Andres V, et al. (2010) Formulation based on liposomes and bioadhesive polymers for dry eye treatment. Tolerance studies. Invest Ophthalmol Vis Sci 51: E- Abstract 6275.
19. Ludwig A (2005) The use of mucoadhesive polymers in ocular drug delivery. Advanced Drug Delivery Reviews 57: 1595–1639.
20. Cao F, Zhang X, Ping Q (2010) New method for ophthalmic delivery of azithromycin by poloxamer/carbopol-based in situ gelling system. Drug Deliv 17: 500–507.
21. Bangham AD, Standish MM, Watkins JC (1965) Diffusion of univalent ions across the lamellae of swollen phospholipids. J Mol Biol 13: 238–252.
22. Vicario-de-la-Torre M, Herrero-Vanrell R, Benitez-del-Castillo JM, Molina-Martinez IT (2007) New formulations for dry eye treatment. Arch Soc Esp Oftalmol 82: 395–396.
23. Burstein N (1997) Clinical Ocular Pharmacology. Boston: Butterworth-Heinemann. 21–45.
24. Zhao M, Thuret G, Piselli S, Pipparelli A, Acquart S (2008) Use of poloxamers for deswelling of organ-cultured corneas. Invest Ophthalmol Vis Sci 49: 550–559.
25. Bhatia A, Kumar R, Katare OP (2004) Tamoxifen in topical liposomes: development, characterization and in-vitro evaluation. J Pharm Pharm Sci 7: 252–259.
26. Purslow C, Wolffsohn JS (2005) Ocular surface temperature: a review. Eye Contact Lens 31: 117–123.
27. Enríquez de Salamanca A, Calonge M, García-Vazquez C, Callejo S, Vila A, et al. (2006) Chitosan nanoparticles as a potential drug delivery system for the ocular surface: toxicity, uptake mechanism and in vivo tolerance. 1416–1425.
28. Camras CB, Toris CB, Tamesis RR (1999) Efficacy and adverse effects of medications used in the treatment of glaucoma. Drugs Aging 15: 377–388.

29. Andres-Guerrero V, Vicario-de-la-Torre M, Molina-Martinez IT, Benitez-del-Castillo JM, Garcia-Feijoo J, et al. (2011) Comparison of the in vitro tolerance and in vivo efficacy of traditional timolol maleate eye drops versus new formulations with bioadhesive polymers. Invest Ophthalmol Vis Sci 52: 3548–3556.
30. Kaur IP, Smitha R (2002) Penetration enhancers and ocular bioadhesives: two new avenues for ophthalmic drug delivery. Drug Dev Ind Pharm 28: 353–369.
31. Ishibashi T, Yokoi N, Kinoshita S (2003) Comparison of the short-term effects on the human corneal surface of topical timolol maleate with and without benzalkonium chloride. J Glaucoma 12: 486–490.
32. Whitson JT, Ochsner KI, Moster MR, Sullivan EK, Andrew RM, et al. (2006) The safety and intraocular pressure-lowering efficacy of brimonidine tartrate 0.15% preserved with polyquaternium-1. Ophthalmology 113: 1333–1339.
33. Ayaki M, Yaguchi S, Iwasawa A, Koide R (2008) Cytotoxicity of ophthalmic solutions with and without preservatives to human corneal endothelial cells, epithelial cells and conjunctival epithelial cells. Clin Experiment Ophthalmol 36: 553–559.
34. Andres-Guerrero V, Vicario-de-la-Torre M, Molina-Martinez IT, Benitez-del-Castillo JM, Garcia-Feijoo J, et al. (2011) Comparison of the in vitro tolerance and in vivo efficacy of traditional timolol maleate eye drops versus new formulations with bioadhesive polymers. Invest Ophthalmol Vis Sci 52: 3548–3556.
35. Bartlett JD (2012) Ophthalmic Drug Facts: Facts and Comparisons; (Ed.) iH, editor: Lippincott Williams & Wilkins.
36. Burstein N (1997) Clinical Ocular Pharmacology. Boston: Butterworth-Heinemann.
37. Tiffany JM (1991) The viscosity of human tears. Int Ophthalmol 15: 371–376.
38. Tiffany JM (1994) Viscoelastic properties of human tears and polymer solutions. Adv Exp Med Biol 350: 267–270.
39. Chrai SS, Robinson JR (1974) Ocular evaluation of methylcellulose vehicle in albino rabbits. J Pharm Sci 63: 1218–1223.
40. Patton TF, Robinson JR (1975) Ocular evaluation of polyvinyl alcohol vehicle in rabbits. J Pharm Sci 64: 1312–1316.
41. Dumortier G, Grossiord JL, Agnely F, Chaumeil JC (2006) A review of poloxamer 407 pharmaceutical and pharmacological characteristics. Pharm Res 23: 2709–2728.
42. Krause WE, Bellomo EG, Colby RH (2001) Rheology of sodium hyaluronate under physiological conditions. Biomacromolecules 2: 65–69.
43. Le Bourlais C, Acar L, Zia H, Sado PA, Needham T, et al. (1998) Ophthalmic drug delivery systems–Recent advances. Progress in Retinal and Eye Research 17: 33–58.
44. Stein PJ, Clack JW (1994) Topical application of a cyclic GMP analog lowers IOP in normal and ocular hypertensive rabbits. Invest Ophthalmol Vis Sci 35: 2765–2768.
45. Martinez-Aguila A, Fonseca B, Bergua A, Pintor J (2013) Melatonin analogue agomelatine reduces rabbit's intraocular pressure in normotensive and hypertensive conditions. Eur J Pharmacol 701: 213–217.
46. Lin HH, Ko SM, Hsu LR (1996) The preparation of norfloxacin-loaded liposomes and their in-vitro evaluation in pig's eye. J Pharm Pharmacol 48: 801–805.
47. Moon JW, Song YK, Jee JP, Kim CK, Choung HK, et al. (2006) Effect of subconjunctivally injected, liposome-bound, low-molecular-weight heparin on the absorption rate of subconjunctival hemorrhage in rabbits. Invest Ophthalmol Vis Sci 47: 3968–3974.
48. Nii T, Ishii F (2005) Encapsulation efficiency of water-soluble and insoluble drugs in liposomes prepared by the microencapsulation vesicle method. Int J Pharm 298: 198–205.
49. Simmons ST, Sherwood MB, Nichols DA, Penne RB, Sery T, et al. (1988) Pharmacokinetics of a 5-fluorouracil liposomal delivery system. Br J Ophthalmol 72: 688–691.
50. Vemuri S, Rhodes CT (1995) Preparation and characterization of liposomes as therapeutic delivery systems: a review. Pharm Acta Helv Jul; 70: 95–111.

13

PCB-153 Shows Different Dynamics of Mobilisation from Differentiated Rat Adipocytes during Lipolysis in Comparison with PCB-28 and PCB-118

Caroline Louis[1], Gilles Tinant[1], Eric Mignolet[1], Jean-Pierre Thomé[2], Cathy Debier[1]*

1 Institut des Sciences de la Vie, Université catholique de Louvain, Louvain-la-Neuve, Belgium, **2** Laboratoire d'Ecologie animale et d'Ecotoxicologie, Université de Liège, Liège, Belgium

Abstract

Background: Polychlorinated biphenyls (PCBs) are persistent organic pollutants. Due to their lipophilic character, they are preferentially stored within the adipose tissue. During the mobilisation of lipids, PCBs might be released from adipocytes into the bloodstream. However, the mechanisms associated with the release of PCBs have been poorly studied. Several *in vivo* studies followed their dynamics of release but the complexity of the *in vivo* situation, which is characterised by a large range of pollutants, does not allow understanding precisely the behaviour of individual congeners. The present *in vitro* experiment studied the impact of (*i*) the number and position of chlorine atoms of PCBs on their release from adipocytes and (*ii*) the presence of other PCB congeners on the mobilisation rate of such molecules.

Methodology/Principal Findings: Differentiated rat adipocytes were used to compare the behaviour of PCB-28, -118 and -153. Cells were contaminated with the three congeners, alone or in cocktail, and a lipolysis was then induced with isoproterenol during 12 hours. Our data indicate that the three congeners were efficiently released from adipocytes and accumulated in the medium during the lipolysis. Interestingly, for a same level of cell lipids, PCB-153, a hexa-CB with two chlorine atoms in *ortho*-position, was mobilised slower than PCB-28, a tri-CB, and PCB-118, a penta-CB, which are both characterised by one chlorine atom in *ortho*-position. It suggests an impact of the chemical properties of pollutants on their mobilisation during periods of negative energy balance. Moreover, the mobilisation of PCB congeners, taken individually, did not seem to be influenced by the presence of other congeners within adipocytes.

Conclusion/Significance: These results not only highlight the obvious mobilisation of PCBs from adipocytes during lipolysis, in parallel to lipids, but also demonstrate that the structure of congeners defines their rate of release from adipocytes.

Editor: Arun Rishi, Wayne State University, United States of America

Funding: The authors have no support or funding to report.

Competing Interests: The authors have declared that no competing interests exist.

* Email: cathy.debier@uclouvain.be

Introduction

Polychlorinated biphenyls (PCBs) are a class of environmentally persistent pollutants that biomagnify throughout food chains. Ingestion of contaminated food, and especially fat-rich animal products, represents 90% of the mean uptake of humans [1]. Adipose tissue is then the main reservoir for the storage of these highly lipophilic molecules [2]. The cytoplasm of adipocytes is almost exclusively composed of lipid droplets (LDs) [3], which appear to be the principal targets for PCBs [4]. These cells therefore have an enormous capacity to accumulate lipophilic pollutants [5].

During periods of weight loss in humans, lipids from adipose tissue are mobilised, leading to an increase of PCB concentrations in this tissue [6,7]. Evidence from wildlife indicates same trends [8–14]. This phenomenon suggests that PCBs are less efficiently mobilised from adipocytes than fatty acids. However, a release of PCBs in the blood circulation does occur during such periods of weight loss and appears to become more important when the adipose stores are already significantly reduced [2,6,15–18]. In addition to being a reservoir, the adipose tissue can thus also be an internal source of lipophilic pollutants for the rest of the body [2]. Once in the bloodstream, pollutants are able to contaminate other tissues or be transferred in maternal milk. The exposure to PCBs is associated to adverse effects on human and animal health [19,20]. Among others, PCBs are involved in endocrine disruption, immuno- and neuro-toxicity as well as in the development of cardiovascular diseases and type-2 diabetes [21–25]. A correlation between the rise of persistent organic pollutants (POPs), such as PCBs, in the serum and the alterations of skeletal muscle oxidative capacity has been suggested in humans [16]. Furthermore, individuals who underwent bariatric surgery exhibited a positive association between POP serum levels and a diminished improvement of lipid values and liver markers [6].

Even if several *in vivo* studies report a release of PCBs from adipose tissue during lipolytic process [6,10,13–15], little is known concerning the chemical and biochemical factors that govern their mobilisation and transfer into the circulation. *In vivo* studies on long-term fasting wild animals report an impact of the fasting stage as well as the degree of lipophilicity of PCB congeners on their dynamics of mobilisation from the adipose tissue [8,10,13,14]. The transfer from adipose tissue into the blood circulation appears to be selective and strongly dependent on the log K_{ow} value of the compounds, with less lipophilic PCBs being more efficiently released. *In vivo* models being usually complex, the *in vitro* cultures of adipocytes would be useful to precisely understand the mobilisation of PCB congeners as a function of their chemical structure. A recent study from our group investigated the dynamics of accumulation of three PCB congeners, differing in the position and number of their chlorine atoms (PCB-28, log $K_{ow} = 5.71$; PCB-118, log $K_{ow} = 6.57$ and PCB-153, log $K_{ow} = 6.80$) in cultured adipocytes [26]. The accumulation profile revealed significant differences between PCB congeners. Their release during lipolysis was however not investigated.

In this study, we followed and compared the dynamics of mobilisation of PCB-28, PCB-118 and PCB-153 from *in vitro* differentiated rat adipocytes. Cells were contaminated by the three congeners, added individually or in cocktail, at the same concentrations in the culture medium. Lipolysis was then triggered over 12 hours with a lipolytic medium supplemented with isoproterenol, a well-known synthetic catecholamine [27,28]. The levels of PCBs in the extracellular medium and adipocytes were regularly assessed. The present experiment allowed (*i*) to estimate the impact of the number and position of chlorine atoms of PCBs on their release from adipocytes and (*ii*) to assess the impact of the presence of other PCB congeners on the mobilisation dynamics of such molecules.

Experimental Procedures

Primary cultures of rat adipocytes

Differentiated rat adipocytes were obtained and cultured as described previously [5,28]. Experimental procedures in animals were approved by the Animal Care and Use Committee of the Université catholique de Louvain (#103201) and were performed in accordance with the "Principles of Laboratory Animal Care" (NIH Publication 85–23). Two-month-old male Wistar rats (Centre d'Elevage Janvier, Le Genest Saint Isle, France) were sacrificed by decapitation. The fat tissue of the stromal-vascular fraction was sampled and then digested in a solution of collagenase (1250 U/ml type II; Sigma-Aldrich, Bornem, Belgium). The digested tissue underwent three filtrations and three centrifugations in order to obtain a final pellet of stromal-vascular cells, which was suspended in a medium composed of Dulbecco's Modified Eagle Medium (DMEM, 4.5 g/l glucose, Gibco-Invitrogen, Merelbeke, Belgium), 10% (*v:v*) heat-inactivated foetal bovine serum (FBS, PAA, A&E Scientific, Marcq, Belgium) and an antibiotic and antifungal mixture. Cells were seeded at a mean density of 18,000 cells per cm^2 on 6-well plates (Corning CellBIND Surface, Corning, Elscolab, Kruibeke, Belgium) (day 0) and incubated at 37°C in a humidified atmosphere containing 10% CO_2 in air for 24 hours to allow cell sedimentation and adhesion. Twenty-four hours after the isolation of progenitor cells (day 1), the medium was replaced by a differentiation medium composed of DMEM (4.5 g/l glucose), 10% (*v:v*) heat-inactivated FBS, 100 U/ml penicillin – 100 U/ml streptomycin – 250 ng/ml amphotericin B mixture (Lonza, Verviers, Belgium), 10 nM dexamethasone (Sigma-Aldrich), 10 μM ciglitizone (Sigma-Aldrich) and 5 μg/ml insulin (Sigma-Aldrich). This medium was renewed every 48 hours until day 10 in order to obtain differentiated adipocytes.

Cell treatment

At day 10, adipocytes were incubated with a medium supplemented with PCB congeners (Dr. Ehrenstorfer GmbH, Ausburg, DE) during 12 hours (37°C – 10% CO_2 in air). The PCBs were added to the culture medium as ethanolic solution and four conditions of PCB contamination were tested: (*i*) 2,4,4′-trichlorobiphenyl (PCB-28); (*ii*) 2,3′,4,4′,5-pentachlorobiphenyl (PCB-118); (*iii*) 2,2′,4,4′,5,5′-hexachlorobiphenyl (PCB-153); (*iv*) an equimolar mixture of three PCB congeners (PCB-28, -118 and -153), also called cocktail of PCBs. In all conditions of contamination, each PCB was added to the culture medium at a concentration of 300 nM, which is within the range of concentrations found in *in vivo* and *in vitro* studies [4,29,30]. Impact of the ethanol vehicle was tested earlier [4].

Lipolysis experiment

At day 11, lipolytic process was induced to differentiated adipocytes as previously described in Louis et al. [28]. The differentiation medium in contact with adipocytes was removed and replaced by a lipolytic medium composed of DMEM (1.0 g/l glucose, Gibco–Invitrogen), 5% (*v:v*) heat-inactivated FBS, 2% (*w:v*) bovine albumin (Sigma-Aldrich) and 1 μM isoproterenol (Sigma-Aldrich). The lipolytic medium was renewed every 3 hours and the process was carried out for 12 hours. In the same way as in Louis et al. [28], cells from one plate (*i.e.* cells coming from the same PCB contamination) were collected every 3 hours and pooled in order to assess the cellular PCB and protein contents as well as the levels of fatty acids of cellular neutral lipids (NLs). Likewise, free fatty acids (FFAs), glycerol and PCBs released in the extracellular medium were quantified every 3 hours in all conditions.

Cellular protein assessment

Every 3 hours, cells were washed with phosphate-buffer saline (Sigma-Aldrich) at 37°C and then collected in an aqueous solution composed of 35 mM sodium dodecyl sulfate (Merck, Darmstadt, Germany), 60 mM Tris buffer (Merck) and 10 mM ethylenediaminetetraacetic acid (Sigma-Aldrich). After homogenisation, the cellular protein content was quantified by using the Bicinchoninic Acid Protein Assay kit (Sigma-Aldrich) with bovine serum albumin (Sigma-Aldrich) as calibration curve [28].

Cellular neutral lipid assessment

Cells were collected as described for the determination of protein content. The method used for the extraction and the isolation of the NL fraction (i.e. triglycerides (TGs), diglycerides, monoglycerides and cholesterol esters) from cell lysates is described in details in Louis et al. [28]. Briefly, the lipids were extracted with a mixture of chloroform/methanol/water (2:2:1, *v:v:v*) (Biosolve, Valkenswaard, The Netherlands) containing triheptadecanoin (Larodan, Malmö, Sweden) used as internal standard. After centrifugation, the supernatant was discarded; the chloroform phase was evaporated and samples were then suspended into 200 μl chloroform. In order to isolate NLs, samples were loaded on solid-phase extraction columns (Bond Elut NH_2, 200 mg, Varian, Middelburg, The Netherlands). The NL fraction collected with chloroform/2-propanol (2:1, *v:v*) (Biosolve) was evaporated and a methylation step was performed by adding 0.1 M KOH (Sigma-Aldrich) in methanol at 70°C for 1 hour and then, by adding 1.2 M HCl in methanol at 70°C for 15 min. The addition of hexane followed by deionized water allowed extracting the fatty acid methyl esters by a centrifugation step. Those were separated by gas chromatography [31]. Each peak was then identified and quantified by comparison with pure methyl ester standards (Larodan and Nu-Check Prep, Elysian, MN, USA). Data were processed with the ChromQuest 4.2 software (ThermoFinnigan, Milan, Italy). Thereafter, results were expressed by moles of fatty acids in cellular NLs. For the sake of simplicity, we refer to µmol NLs/mg protein in the text [28].

Extracellular free fatty acid assessment

FFA contents in the lipolytic medium were quantified with the Wako NEFA HR kit (Sopachem, Eke, Belgium) following the manufacturer's instructions [28].

Extracellular glycerol assessment

Glycerol released in the extracellular medium was measured with an *in vitro* enzymatic colorimetric test using glycerol-3-phosphate-oxidase (Diasys Free Glycerol FS kit, Sopachem) according to the manufacturer's instructions.

PCB assessment

At each studied time of the lipolysis, cells and extracellular medium were collected in EPA vials (Alltech, Lokeren, BE) with 5 ml of *n*-hexane (Biosolve) in order to perform a liquid-liquid extraction by a 10-min shaking. The hexane phase was transferred into a tube and PCB-112 (Dr. Ehrenstorfer GmbH) was added as internal standard. All samples were then purified by acid and Florisil clean-up steps as described in Debier et al. [8]. Purified samples were collected in *n*-hexane. Five µl of anhydrous nonane (Sigma-Aldrich) were added to samples and solvent was then evaporated under a gentle stream of nitrogen. The purified extracts were suspended into a hexane solution of Mirex (200 pg/µl) (Dr. Ehrenstorfer GmbH) used as internal standard for the correction of the extract volume injected in GC/MS. PCB congeners were separated and quantified with a gas chromatograph (GC Trace, ThermoFinnigan) equipped with an automatic split/splitless type injector (CTC Analytics, Zwingen, Switzerland), a fused silica capillary column (30 m × 0.25 mm internal diameter; 0.25 µm film) (Rxi-5ms, Restek, Bellefonte, PA, USA) and a mass spectrometer (Trace DSQ, ThermoFinnigan). The system used helium as carrier gas at a constant flow rate of 1.1 ml per minute. The temperature of injector was 230°C. The oven temperature program was as follows: 2 min at 60°C, gradual heating from 60 to 140°C at the rate of 20°C per minute, 1 minute at 140°C, gradual heating from 140 to 290°C at the rate of 2.5°C per minute, 10 minutes at 290°C and gradual cooling from 290°C to 60°C at the rate of 10°C per minute. Molecules were sent to mass spectrometer by the line transfer at 290°C. The ion source of the detector was kept at 230°C. PCBs were identified according to their retention time. Data were recorded using XCalibur 1.3 software (ThermoFinnigan). Quantification was performed by comparison to an external standard composed of 28 congeners (IUPAC numbers: PCB-8, -18, -28, -44, -52, -66, -77, -81, -101, -105, -114, -118, -123, -126, -128, -138, -153, -156, -157, -167, -169, -170, -180, -187, -189, -195, -206 and -209) in a certified calibration mixture (AccuStandard, New Haven, CT, USA). Five dilutions (concentration ranging from 25 to 500 pg/µl) were used in order to draw a linear calibration curve for each PCB.

Quality control

Blanks were run with sample series to control extraction and clean-up steps. The PCB recovery was calculated on the basis of the internal standard, PCB-112. Results were accepted only if the recoveries were between 70 and 130% according to EC [32]. All results were corrected to obtain 100% recovery [8]. The quality control was assessed through an interlaboratory comparison.

Cytotoxicity assessment

The potential cytotoxicities of the PCBs and the lipolytic treatment were assessed by measuring the release of lactate dehydrogenase (LDH) in the extracellular medium. The activity of LDH was determined using the cytotoxicity detection kit (Roche Diagnostics, Vilvoorde, Belgium) according to the manufacturer's instructions in (*i*) the differentiation medium after 12 hours of PCB exposure and (*ii*) the lipolytic media collected every 3 hours during the lipolytic process. Before the PCB exposure and the lipolysis, some cells were lysed with 1% Triton x-100 (Sigma-Aldrich) and were used as full toxicity control [28]. No treatments appeared toxic (< 5% of control) as compared to the full toxicity control (results not shown).

Statistical analysis

Data are presented as means ± SEM of three independent experiments for the conditions with PCB congeners alone and means ± SEM of five independent experiments for the condition with the cocktail of PCBs. The statistical analysis was performed by SAS 9.3 software (SAS Institute Inc., Cary, USA). Differences between treatments were assessed with mixed linear models and a Tukey's test [28]. Differences were deemed significant at p-values<0.05.

Results

1. Incorporation of PCBs in adipocytes

At day 10, differentiated rat adipocytes were exposed to four different treatments of PCBs during a 12-hour period. Three PCB congeners (PCB-28, -118 and -153) were added to the culture medium at a concentration of 300 nM, alone or in cocktail. Assessment of PCBs within adipocytes was carried out before the induction of lipolysis. The concentration of each congener in adipocytes, expressed as nmol per unit of cellular protein, was statistically similar, whatever the kind of contamination (alone or in cocktail) ($0.328<p<1.000$) (Table 1). Same conclusions were drawn when the results were expressed per unit of NLs ($0.207<p<1.000$). Since the dynamics of accumulation of PCBs in cells vary with cellular lipid content [26], the levels of cellular NLs were quantified and no statistical difference was noted ($0.457<p<0.978$) (Table 1).

2. The time-course of lipolysis

At day 11, adipocytes contaminated with PCBs (alone or in cocktail) were incubated with a lipolytic medium supplemented with 1 µM isoproterenol, which was replaced every 3 hours for 12 hours. Since the aim of this work was to study the kinetics of PCB release during the mobilisation of lipids in adipocytes, we firstly ensured of the efficiency of lipolysis. Cellular NLs as well as FFAs and glycerol released in the culture medium were quantified over 12 hours (Figure 1). The FFA and glycerol initially present in the lipolytic medium, before being in contact with adipocytes, were subtracted from each result, leading to a value of 0 at 0 hour (Figure 1B–C). The total FFAs and glycerol released by the adipocytes throughout a given period were calculated by adding the quantities measured at each period of 3 hours. For example,

Table 1. Efficient accumulation of PCBs in adipocytes during a 12-hour period.

	Contamination by a PCB congener alone:			Contamination by a cocktail of three PCB congeners:		
	PCB-28	**PCB-118**	**PCB-153**	**PCB-28**	**PCB-118**	**PCB-153**
PCBs (nmol/mg protein)	1.7±0.1	2.0±0.1	1.8±0.1	1.8±0.1	2.1±0.1	1.8±0.1
PCBs (nmol/mg NLs)	0.47±0.08	0.62±0.09	0.48±0.08	0.61±0.08	0.70±0.08	0.60±0.08
NLs (μmol/mg protein)	13.6±1.1	13.1±1.3	15.6±1.1	11.4±1.1	11.4±1.1	11.4±1.1

Quantities of PCBs accumulated in differentiated rat adipocytes, expressed in nmol per unit of cellular protein and per unit of neutral lipids (NLs), after a dose of PCBs was added during 12 hours in the culture medium. Quantities of NLs, expressed in μmol per unit of cellular protein are also presented. Data represent the means of (*i*) three independent experiments ± SEM for conditions with one PCB congener alone, (*ii*) five independent experiments ± SEM for conditions with the cocktail of PCBs. There was no significant difference of PCB and NL concentrations between the three PCB congeners, whatever the kind of contamination, alone or in cocktail ($p>0.05$).

the amount of total FFAs or glycerol released after 6 hours of lipolysis corresponds to the sum of FFAs or glycerol released between 0 and 3 hours and between 3 and 6 hours [28].

A significant decrease of cellular NLs was observed between early and late lipolysis for the conditions with the PCB-28 ($p = 0.005$) and cocktail of PCBs ($p = 0.029$). The cell NLs slightly decreased over 12 hours within adipocytes contaminated with PCB-118 ($p = 0.298$) and PCB-153 ($p = 0.097$). For all conditions, the greatest reduction occurred during the first 3 hours of the experiment (Figure 1A). Indeed, a mean loss of 57±3% initial NLs was noted between 0 and 12 hours, whereas after 3 hours of lipolysis, adipocytes already lost 33±6% of NLs on average, compared to the initial levels ($0.372<p<0.721$ for all conditions of contamination between 0 and 3 hours). The comparison of NL levels between the four PCB treatments at a given period did not highlight any difference ($0.255<p<1.000$).

The decrease of cellular lipid contents was accompanied by a significant increase of total FFAs in the lipolytic medium over the 12-hour period ($p<0.001$) (Figure 1B). The increase of total FFAs was more pronounced during the first 3 hours of lipolysis, during which an average release of 63±4% of total FFAs occurred ($p<0.001$). FFA concentrations then continued to increase slightly, but not significantly, between the subsequent consecutive periods ($0.491<p<0.985$ for all conditions of contamination between 3 and 6 hours, 6 and 9 hours, 9 and 12 hours). No difference in the release of FFAs was noted between the four conditions of contamination at a given time of the lipolytic process ($0.273<p<1.000$). As for total FFAs, contaminated adipocytes released a significant amount of total glycerol throughout the lipolytic process ($p<0.001$ between 0 and 12 hours) (Figure 1C). The major part of the release (49±3%) also occurred during the first 3 hours of lipolysis for all conditions ($p<0.001$) (Figure 1C). Here again, no difference in the release of glycerol was noted between the four conditions of PCB contamination over the 12-hour period ($0.119<p<0.999$). Taking all conditions of PCB contamination together, the FFA/glycerol ratios lied between 1.4 and 1.8.

3. Comparative dynamics of mobilisation of PCB-28, -118 and -153 present in cocktail in adipocytes

In the present section, we compared the dynamics of mobilisation of PCB-28, -118 and -153 (present in cocktail within the cells) from the same adipocytes undergoing a lipolytic process over a 12-hour period. In order to strictly compare the different mobilisations between congeners, we expressed the results as percentages of the amounts initially present in adipocytes (i.e. at 0 hour). The potential presence of PCBs in the lipolytic medium, before being in contact with the cells, has been tested and was negligible (result not shown). The proportions of PCBs in the lipolytic medium at 0 hour were thus set at 0%. Regarding the subsequent studied times of the lipolysis (3, 6, 9 and 12 hours), the PCB levels in the medium were calculated by adding the quantities evaluated at each period of 3 hours. For example, the amounts of each PCB congener released after 6 hours of lipolytic treatment correspond to the sum of PCB released between 0 and 3 hours and between 3 and 6 hours.

In parallel to the lipid mobilisation, there was a release of PCBs from adipocytes into the culture medium (Figure 2). An important drop of the PCB cellular content occurred during the first half of the lipolytic process ($p<0.001$ between 0 and 6 hours for the three PCBs). It was followed by a slight, but still significant reduction of PCB-28, -118 and -153 in adipocytes during the second part of the lipolysis ($p<0.001$ between 6 and 12 hours for the three PCBs). Accordingly, a reversed trend was noted in the lipolytic medium: the proportions of PCB-28, -118 and -153 accrued during the first 6 hours ($p<0.001$ between 0 and 6 hours for the three PCBs). The PCB accumulation in the culture medium was more moderate, but still significant ($0.001<p<0.006$), during the second part of the lipolytic process.

Even if all congeners dropped in the cells and increased in the culture medium, the dynamics of release of PCB-153 somewhat differed from those of PCB-28 and -118. Indeed, despite the fact that the amounts of the three congeners within adipocytes were statistically similar before the beginning of the lipolysis (Table 1), the percentages of PCB-153 in adipocytes remained higher than the ones of PCB-28 and PCB-118 at 3, 6 and 9 hours of lipolysis ($0.001<p<0.032$ for PCB-28; $0.031<p<0.037$ for PCB-118). The difference disappeared at 12 hours between PCB-153 and PCB-28 ($p = 0.139$), but remained significant between PCB-153 and PCB-118 ($p = 0.027$). On the other hand, the cellular percentage of PCB-28 and PCB-118 did not differ between each other throughout the lipolytic process ($0.373<p<1.000$). The slower mobilisation of PCB-153 from adipocytes was reflected by a slower increase in the culture medium as compared to PCB-28 and PCB-118 after 3 hours of lipolysis ($p<0.001$ for the both comparisons). After 6 hours, the accumulation in the medium was weaker for PCB-153 than PCB-28 ($p = 0.027$) and similar between PCB-153 and PCB-118 ($p = 0.178$). The proportions of PCB-153 were then similar to those of PCB-28 and -118 for the subsequent hours ($0.275<p<0.986$ at 9 and 12 hours). On the other hand, the percentages of PCB-28 and -118 in the extracellular medium were similar to each other over the lipolysis ($0.429<p<1.000$).

4. Impact of the presence of other PCB congeners on the dynamics of mobilisation of PCB-28, -118 and -153

We investigated if the dynamics of mobilisation of one PCB congener was influenced by the presence of the other congeners.

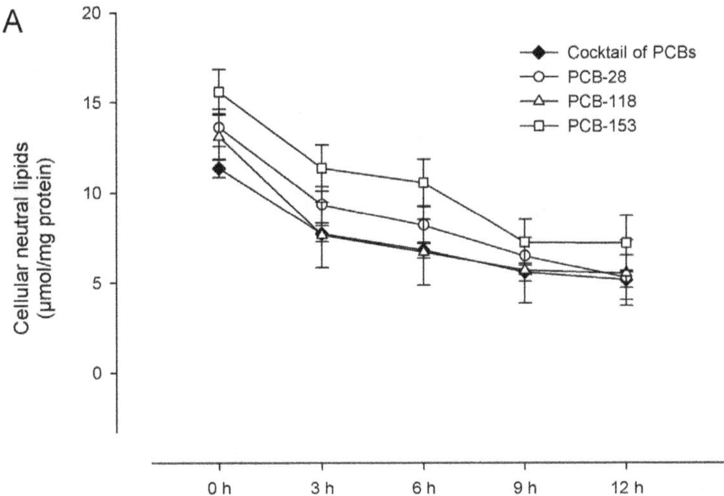

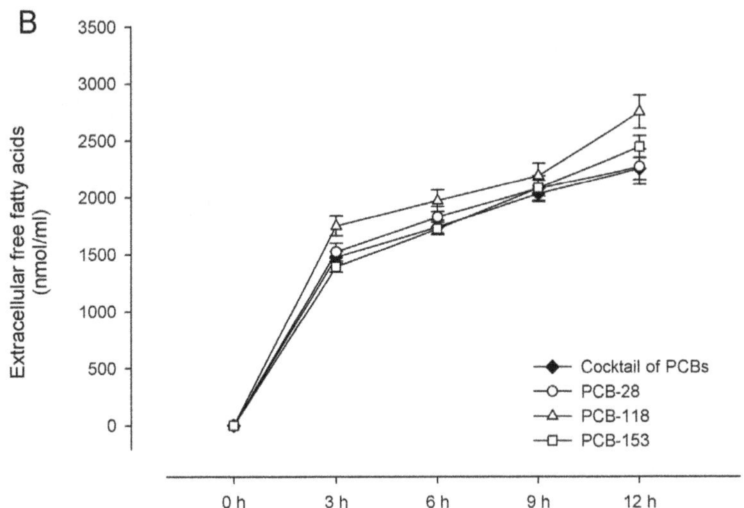

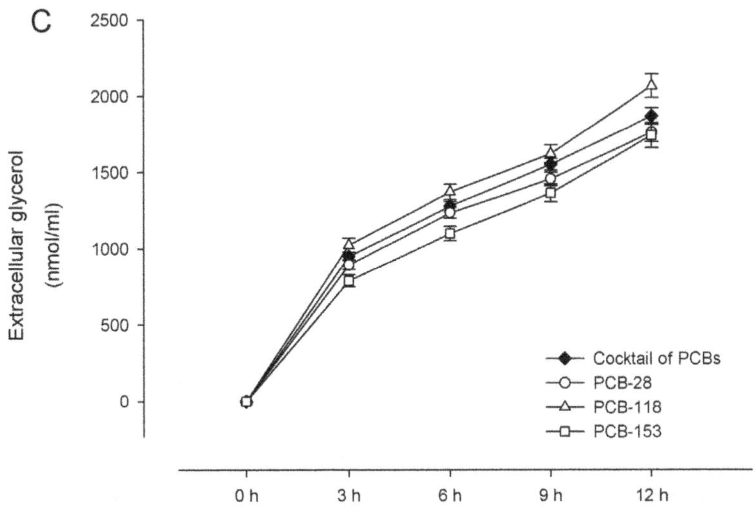

Time of lipolysis

Figure 1. Lipolytic treatment decreased cellular neutral lipids and increased extracellular fatty acids and glycerol. At day 11, differentiated rat adipocytes, which were previously contaminated with PCBs, were incubated with a lipolytic medium supplemented with 1 μM isoproterenol. We renewed the lipolytic medium every 3 hours for 12 hours. Cellular neutral lipids (corresponding to μmol of fatty acids in cellular neutral lipids) were expressed per mg of total cell protein (A). A significant decrease of cellular lipid contents was noted throughout the lipolytic process for the conditions with PCB-28 ($p = 0.005$) and the cocktail of PCBs ($p = 0.029$). The decrease was slighter for the conditions with PCB-118 ($p = 0.298$) and PCB-153 ($p = 0.097$). Total extracellular free fatty acids (B) and total extracellular glycerol (C) were expressed per ml of medium. Quantities of total free fatty acids and glycerol in the medium were obtained by adding the quantities released during the periods of 3 hours (e.g. total free fatty acids at 6 hours correspond to the sum of total free fatty acids released between 0 and 3 hours and between 3 and 6 hours). A significant increase of total free FAs and total glycerol was observed over the 12-hour lipolytic treatment ($p < 0.001$ for all conditions). Data represent the means of (*i*) three independent experiments ± SEM for conditions with one PCB congener alone, (*ii*) five independent experiments ± SEM for conditions with the cocktail of PCBs.

To achieve this goal, we contaminated adipocytes either with one of the three PCB congeners (PCB-28, -118 or -153) or with a cocktail of the three congeners. Here also, the results are expressed as percentages of the amounts initially present in adipocytes. As previously described, the proportions of PCBs just before the lipolytic process (i.e. 0 hour) was set to 0% since no PCB congener was quantified in the initial lipolytic medium. Here also, the total PCB congeners released by the adipocytes throughout a given period were calculated by adding the quantities measured at each period of 3 hours.

At each studied time throughout the 12-hour period, similar proportions of PCB-28 (Figure 3A), PCB-118 (Figure 3B) and PCB-153 (Figure 3C) within adipocytes were quantified between both conditions of contamination (i.e. congeners alone and in cocktail) ($0.134 < p < 0.934$). Accordingly, no difference was noted between the percentages of PCBs released in the lipolytic medium in contact with adipocytes in both conditions of contamination ($0.122 < p < 0.916$), except after 3 hours of lipolysis, where the proportion of PCB-153 measured in the medium were higher when it was added in cocktail than when it was added alone ($p = 0.046$).

Discussion

Considerable accumulation of PCBs within adipocytes

Differentiated rat adipocytes were exposed to three targeted PCB congeners (PCB-28, -118 and -153), which differ by the number and the position of the chlorine atoms [19,33]. After 12 hours, similar concentrations of PCBs were stored within adipocytes. Same observations were already drawn after 4 hours of exposure for the same congeners [26]. The different molecular structures of PCBs did thus not seem to influence their accretion within adipocytes on long term. Previous studies from our group highlighted the importance of the amount of cellular NLs, acting as a trap for the accumulation of PCBs within adipocytes [5,26]. The fact that cellular NL levels were similar between our

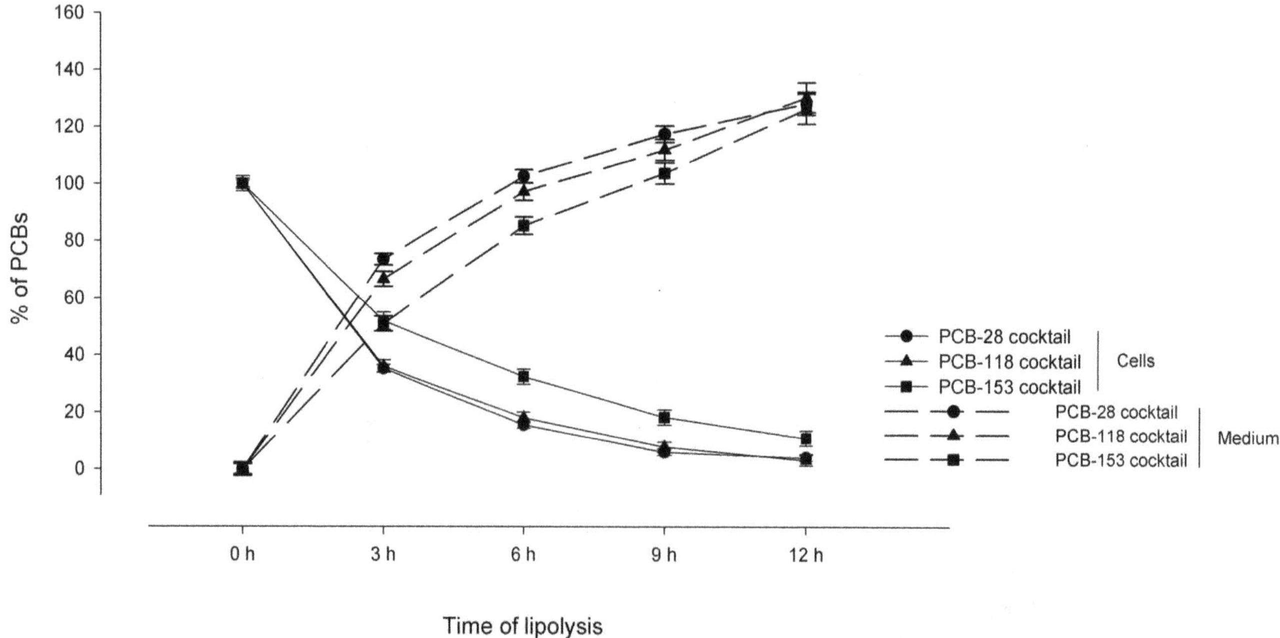

Figure 2. Lipolytic treatment decontaminated the adipocytes, inducing an accumulation of PCB congeners in the extracellular medium. At day 11, contaminated adipocytes underwent a lipolytic process. The PCB contents as well as the proportions of PCBs released in the extracellular medium were assessed every 3 hours. For each PCB congener, all results were expressed in percentage of the amounts initially present in the cells. Proportions of one PCB congener in the medium were obtained by adding the amounts released during periods of 3 hours (e.g. proportion of one congener at 6 hours corresponds to the sum of this congener released between 0 and 3 hours and between 3 and 6 hours). An important drop of cell PCB proportions was observed during the first 6 hours of lipolysis ($p < 0.001$) and was followed by a more moderate decrease of the cell PCB percentages during the last 6 hours ($p < 0.001$). In parallel, an important increase of PCBs was noted in the extracellular medium during the first half of lipolysis ($p < 0.001$) and was followed by a slower accumulation during the second half of lipolysis ($0.001 < p < 0.006$). Data represent the means of five independent experiments ± SEM.

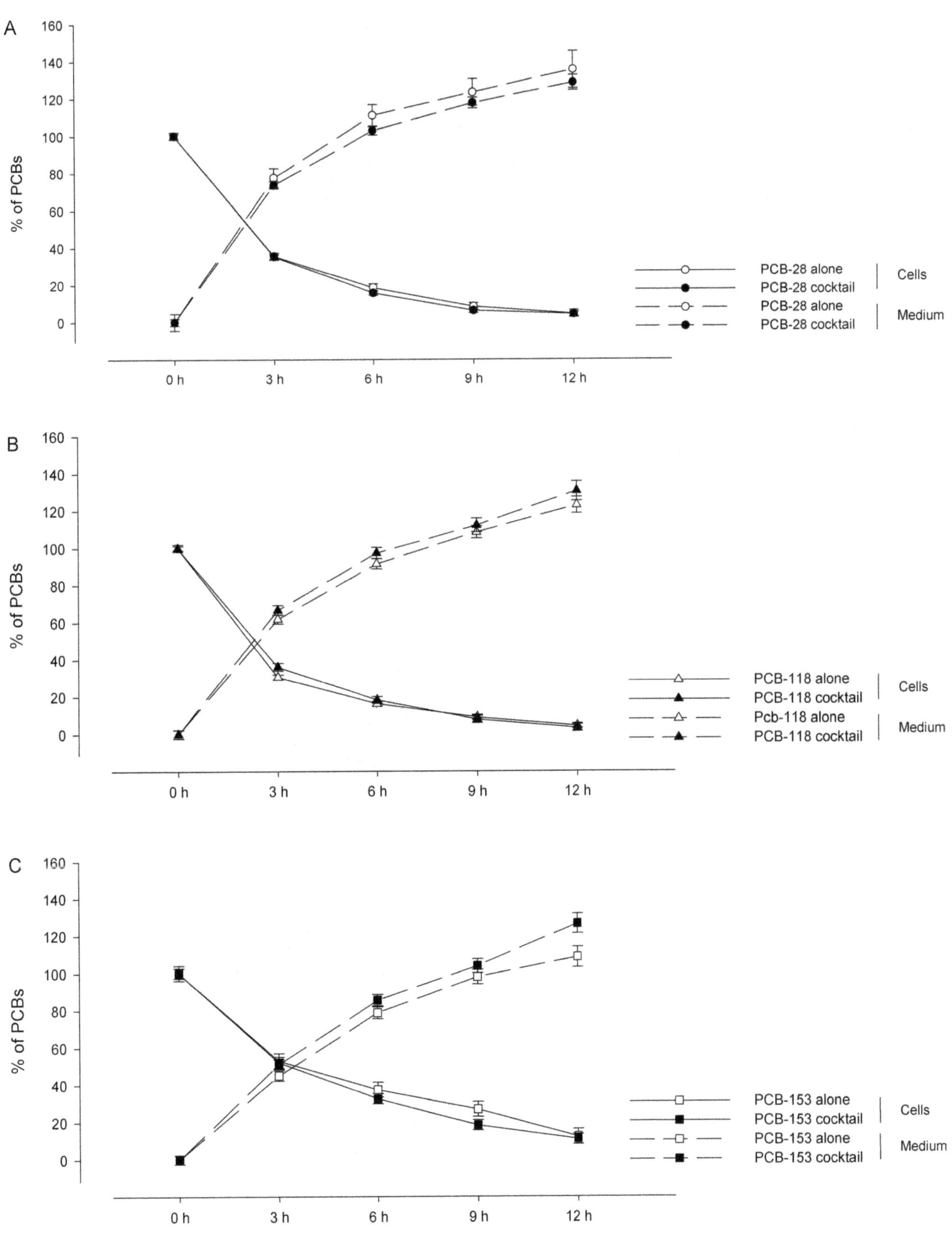
A
B
C
% of PCBs
0 h
3 h
6 h
9 h
12 h
PCB-28 alone
PCB-28 cocktail
PCB-28 alone
PCB-28 cocktail
Cells
Medium
PCB-118 alone
PCB-118 cocktail
Pcb-118 alone
PCB-118 cocktail
PCB-153 alone
PCB-153 cocktail
PCB-153 alone
PCB-153 cocktail
Time of lipolysis

Figure 3. Presence of other congeners did not influence the dynamic of PCB mobilisation. At day 11, differentiated rat adipocytes, which were previously contaminated with either individual PCB congeners or with a cocktail of PCBs, underwent a lipolytic process. The cellular levels of PCBs before the lipolytic process were quantified and set at 100%. During 12-hour period of lipolysis, the contents of PCB congeners within adipocytes and in the extracellular medium were assessed every 3 hours. The results for PCB-28 (A), PCB-118 (B) and PCB-153 (C) were expressed by the percentage of initial amounts of each congener. Within a condition of contamination, proportions of one PCB congener in the medium were obtained by adding the quantities released during the periods of 3 hours (e.g. proportion of one congener at 6 hours corresponds to the sum of this congener released between 0 and 3 hours and between 3 and 6 hours). At each given time of lipolytic treatment, no differences were noted between the proportions of each PCB (i.e. PCB-28, -118 and -153) in both conditions of contamination (i.e. congeners alone or in cocktail), either in the cells or in the lipolytic medium ($0.122<p<0.916$). Only the percentage of PCB-153 in the lipolytic medium was lower when taken alone as compared to the condition in cocktail after 3 hours of lipolysis ($p = 0.046$). Data represent the means of (*i*) three independent experiments ± SEM for conditions with one PCB congener alone, (*ii*) five independent experiments ± SEM for conditions with the cocktail of PCBs.

experimental conditions is most probably at the origin of the identical accumulation of PCBs within adipocytes. While PCB concentrations in the culture medium were in the same range than those measured in the human serum [29,30], the PCB levels found in cultured adipocytes after 12 hours of incubation (data from Table 1 are equivalent ~175 ng PCB congeners per mg NLs) were much higher than those measured *in vivo*, in the human adipose tissue (from 0.02 to 0.66 ng total PCBs per mg lipids) [34–38]. It reflects a high propensity of differentiated rat adipocytes to store PCBs [4]. Such differences have already been noticed previously and the reasons are discussed in details elsewhere [5]. Briefly, the higher *in vitro* concentrations of PCBs within adipocytes could result from the extended contact between the cells and the contaminated culture medium (12 hours). On the contrary, in the *in vivo* situation, the PCB congeners, transported in the circulation by lipoproteins and plasma albumin [33,39,40], are in continual movement thanks to the blood flow. In addition, the culture medium contains only a low concentration of serum (10%) [4], and therefore very low levels of lipoproteins and albumin, which could contribute to a smaller retention of PCBs in this hydrophilic compartment and a higher storage in the lipophilic compartment represented by adipocytes. The differentiated rat adipocytes are also organised as a monolayer whereas *in vivo* adipose tissue shows a complex 3D-structure. Furthermore, lipolysis, which occurs regularly *in vivo*, may lead to the mobilisation of PCBs from adipocytes. Finally, the circulating PCBs may be taken up by other tissues such as the liver.

Similar dynamics of mobilisation between cellular lipids and PCBs

Once the lipolytic pathway was induced, adipocytes started to mobilise their lipid content. A decrease of the cellular NLs could be observed throughout the 12-hour experiment, with a more pronounced lipolytic action during the first 3 hours. This sharper decrease of NL content at early lipolysis was in accordance with our previous study [28]. As a result of the mobilisation of cellular NLs, FFAs and glycerol were released in the extracellular medium. The lipolytic treatment also led to the release of PCBs from adipocytes to the extracellular medium. The dynamics of mobilisation of PCBs exhibited some parallelism with those of cellular lipids, as the major part of PCB discharge occurred during the first hours of lipolysis as well. Previously, it was shown that *in vitro* epididymal adipocytes isolated from rats also unloaded PCB-153 during a lipolytic treatment of 50 min with 0.8 µM isoproterenol [41]. The release of PCBs might accompany the mobilisation of cell lipids, which agrees with previous studies on the behaviour of dioxins [42,43]. In addition, cellular TG content is an important parameter governing the accumulation of PCBs in adipocytes [5,26]. PCBs are stored almost exclusively within the LDs [4]. As this lipophilic pool is reduced during lipolysis [28], the capacity of storage is thus also lessened, promoting the release of PCBs in the extracellular medium, where they could be tightly associated with diverse lipoproteins (present in the 5% serum) and bovine albumin (2%).

Although an obvious release of PCBs occurred from adipocytes during the lipolytic experiment, some molecules of PCBs could be taken up again by the cells as previously suggested in *in vivo* studies [13,14]. This phenomenon is well known for FFAs, which are partly reabsorbed by adipocytes and re-esterified into newly synthesized TGs [28,44]. A complete hydrolysis of one mole of TGs leads to the release of three moles of free FAs and one mole of glycerol. This could be translated by a free FA/glycerol ratio equivalent to 3.0. However, free FA/glycerol ratios were lower than 3.0 in our experiments, which likely reflects a reuptake of free FAs by the cells. Nevertheless, this phenomenon might have been somewhat limited in our experimental conditions, because of the renewal of the lipolytic medium every 3 hours [28].

Studies investigating the release of POPs from adipose tissue during periods of weight loss in animals and humans usually report an increase of the concentrations of PCBs and related compounds in adipose tissue, despite their significant discharge in the blood circulation [6,14,15,18]. This increase suggests a less efficient mobilisation of PCBs from this tissue than lipids and a concentration of these lipophilic pollutants in the remaining amount of fat cells. The adipose tissue is a macroscopic structure that is irrigated by blood vessels. During adipose tissue lipolysis, it is possible that PCBs are transferred to adipocytes that still contain significant amounts of lipids in their LDs instead of being all released into the circulation. It is also possible that PCBs are released in the bloodstream together with the lipids and then reabsorbed by the adipose tissue as a result of their higher affinity for the remaining lipids present in the cells [13]. Our *in vitro* model differs from the *in vivo* situation among others by the fact that it is characterised by only one layer of cells, which is in direct contact with the extracellular medium that is regularly renewed. A reuptake of PCBs by the cells and/or a migration of PCBs to deeper adipocyte layers that are still filled with fat are thus not possible. Moreover, the high PCB concentrations, which were found in the cultivated adipocytes before the lipolytic induction, might also promote the massive release of congeners in the extracellular medium.

Differences of release according to the kind of PCB congener

When the three congeners were added in cocktail to the culture medium, we could observe that PCB-153 was less efficiently mobilised from adipocytes than PCB-28 and PCB-118 during the first part of the lipolytic process. This difference however disappeared at 12 hours of lipolysis. The slower mobilisation of PCB-153 from adipocytes reflects the fact that, besides the cellular lipid content, the rate of release is also governed by the physico-chemical properties of the congeners, which are defined by the number and the position of chlorine atoms on the biphenyl core [2]. If we consider the electrostatic potentials of PCBs [26], PCB-

153 exhibits a large electron-deficient zone. This characteristic makes this congener rather lipophilic, which is reflected by the higher partition coefficient *n*-octanol/water (log $K_{ow} = 6.80$). On the other hand, PCB-28 and PCB-118 have a reduced electron-deficient zone, translated by lower log K_{ow} (PCB-28: log $K_{ow} = 5.71$ and PCB-118: log $K_{ow} = 6.57$). PCB-153 could thus be more trapped within LDs than PCB-28 and PCB-118 and as a consequence, be released more slowly.

In addition, it was previously observed that a small proportion of PCB-153 was sequestrated in the cell membranes when isolated primary adipocytes absorbed the PCB congeners present in the culture medium for 2 hours [4]. It was not the case for PCB-28 and -118. Likewise, several studies showed that PCB-52 and -153, two di-*ortho*-substituted PCBs, intercalate between membrane phospholipids similarly to cholesterol and have an impact on the membrane fluidity in fish, rodent and chicken cells [45–49]. In the present study, the release of PCB-153 from adipocytes could thus be slowed down by its association with cell membranes, as compared to PCB-28 and PCB-118, two mono-*ortho*-substituted congeners. The fact that PCB-153 has two chlorine atoms in the *ortho* position on the biphenyl core induces a more perpendicular layout of the phenyl rings. It means that PCB-153 occupies a larger bulk than PCB-28 and PCB-118, which could be involved in the sequestration of PCB-153 within membranes and its slower mobilisation from adipocytes.

Previous findings from our group, investigating the uptake of PCBs by differentiated adipocytes, highlighted that PCB-28 enters the cells more rapidly than PCB-118 and PCB-153 [4]. If, during lipolysis, a reuptake of PCBs by the cells occurs, PCB-28 might thus be taken up more rapidly than the other congeners, which could lead to an underestimation of the differences of mobilisation kinetics during the lipolytic treatment.

Similar rate of release when PCBs are present alone or in cocktail

In the *in vivo* situation, tissues are exposed to a cocktail of contaminants that might interact with each other's, regarding either the toxicokinetics or the toxicodynamics of the molecules. Here, we investigated the effect of a simple combination of PCBs (three congeners) on their release by adipocytes during lipolysis. To do this, the discharge of PCB-28, -118 and -153 from adipocytes was followed either alone, or in cocktail (i.e. with two other congeners). In the two conditions of contamination, the dynamics of PCB release were similar, meaning that the mobilisation of PCB congeners was not influenced by the presence of other congeners within adipocytes in these experimental conditions. As noted above, PCB-153 influences the properties of cell membranes. One could thus have expected that this congener could influence the dynamics of release of other PCBs from the cells.

Conclusion

Our results showed an efficient accumulation of PCB-28, -118 and -153 in adipocytes. Once lipolysis was induced, the congeners were massively mobilised from cells into the culture medium, in parallel with the release of lipids. The dynamics of discharge however differed between the three investigated congeners. The release of PCB-153 was slightly but significantly slower than the ones of PCB-28 and -118. The phenomenon might be explained by the fact that PCB-153 is more lipophilic than the two other congeners and could thus be more trapped in LDs. In addition, PCB-153 being a di-*ortho*-substituted congener, it is more bulky, which could be involved in its partial sequestration within cell membrane [5] and its slower mobilisation from adipocytes. On the other hand, the dynamics of mobilisation was not influenced by the presence of the other two congeners.

Acknowledgments

The authors are very grateful to Coralie Piget and Marie-Thérèse Ahn from "Institut des Sciences de la Vie" (ISV), UCLouvain, for technical assistance. Members of "Support en méthodologie et calcul statistique" (Institut multidisciplinaire pour la modélisation et l'analyse quantitative, UCLouvain, Belgium) are gratefully acknowledged for the collaboration in the statistical analyses. We also greatly appreciated the help and advice of Guillaume Bernard for picture processing.

Author Contributions

Conceived and designed the experiments: CL GT. Performed the experiments: GT. Analyzed the data: CL GT. Contributed reagents/materials/analysis tools: EM JPT CD. Contributed to the writing of the manuscript: CL. Engaged in active discussions: CL GT EM JPT CD.

References

1. Djien Liem AK, Furst P, Rappe C (2000) Exposure of populations to dioxins and related compounds. Food Additives & Contaminants 17: 241–259.
2. La Merrill M, Emond C, Kim MJ, Antignac JP, Le Bizec B, et al. (2013) Toxicological function of adipose tissue: focus on persistent organic pollutants. Environmental Health Perspectives 121: 162–169.
3. Sbarbati A, Accorsi D, Benati D, Marchetti L, Orsini G, et al. (2010) Subcutaneous adipose tissue classification. European Journal of Histochemistry 54: e48.
4. Bourez S, Le Lay S, Van den Daelen C, Louis C, Larondelle Y, et al. (2012) Accumulation of polychlorinated biphenyls in adipocytes: Selective targeting to lipid droplets and role of caveolin-1. PLoS ONE 7: e31834.
5. Bourez S, Joly A, Covaci A, Remacle C, Larondelle Y, et al. (2012) Accumulation capacity of primary cultures of adipocytes for PCB-126: Influence of cell differentiation stage and triglyceride levels. Toxicology Letters 214: 243–250.
6. Kim MJ, Marchand P, Henegar C, Antignac JP, Alili R, et al. (2011) Fate and complex pathogenic effects of dioxins and polychlorinated biphenyls in obese subjects before and after drastic weight loss. Environmental Health Perspectives 119: 377–383.
7. Chevrier J, Dewailly E, Ayotte P, Mauriege P, Despres JP, et al. (2000) Body weight loss increases plasma and adipose tissue concentrations of potentially toxic pollutants in obese individuals. International Journal of Obesity and Related Metabolic Disorders 24: 1272–1278.
8. Debier C, Pomeroy PP, Dupont C, Joiris C, Comblin V, et al. (2003) Quantitative dynamics of PCB transfer from mother to pup during lactation in UK grey seals *Halichoerus grypus*. Marine Ecology Progress Series 247: 237–248.
9. Debier C, Pomeroy PP, Thomé JP, Mignolet E, de Tillesse T, et al. (2004) An unexpected parallelism between Vitamin A and PCBs in seal milk. Aquatic Toxicology 68: 179–183.
10. Debier C, Chalon C, Le Bœuf BJ, de Tillesse T, Larondelle Y, et al. (2006) Mobilization of PCBs from blubber to blood in northern elephant seals (*Mirounga angustirostris*) during the post-weaning fast. Aquatic Toxicology 80: 149–157.
11. Vanden Berghe M, Mat A, Arriola A, Polain S, Stekke V, et al. (2010) Relationships between vitamin A and PCBs in grey seal mothers and pups during lactation. Environmental Pollution 158: 1570–1575.
12. Debier C, Crocker DE, Houser DS, Vanden Berghe M, Fowler M, et al. (2012) Differential changes of fat-soluble vitamins and pollutants during lactation in northern elephant seal mother–pup pairs. Comparative Biochemistry and Physiology Part A 162: 323–330.
13. Vanden Berghe M, Weijs L, Habran S, Das K, Bugli C, et al. (2012) Selective transfer of persistent organic pollutants and their metabolites in grey seals during lactation. Environment International 46: 6–15.
14. Louis C, Dirtu AC, Stas M, Guiot Y, Malarvannan G, et al. (2014) Mobilisation of lipophilic pollutants from blubber in northern elephant seal pups (*Mirounga angustirostris*) during the post-weaning fast. Environmental Research 132: 438–448.
15. Hue O, Marcotte J, Berrigan F, Simoneau M, Doré J, et al. (2006) Increased plasma levels of toxic pollutants accompanying weight loss induced by hypocaloric diet or by bariatric surgery. Obesity Surgery 16: 1145–1154.

16. Imbeault P, Tremblay A, Simoneau J-A, Joanisse DR (2002) Weight loss-induced rise in plasma pollutant is associated with reduced skeletal muscle oxidative capacity. American Journal of Physiology 282: E574–E579.
17. Debier C, Le Boeuf BJ, Ikonomou MG, de Tillesse T, Larondelle Y, et al. (2005) Polychlorinated biphenyls, dioxins, and furans in weaned, free-ranging northern elephant seal pups from central California, USA. Environmental Toxicology and Chemistry 24: 629–633.
18. Dirtu AC, Dirinck E, Malarvannan G, Neels H, Van Gaal L, et al. (2013) Dynamics of organohalogenated contaminants in human serum from obese individuals during one year of weight loss treatment. Environmental Science & Technology 47: 12441–12449.
19. Carpenter DO (2006) Polychlorinated biphenyls (PCBs): routes of exposure and effects on human health. Reviews on Environmental Health 21: 1–23.
20. Robertson L, Hansen L (2001) PCBs: Recent advances in environmental toxicology and health effects. The University Press of Kentucky, Lexington, Kentucky.
21. Kester MH, Bulduk S, Tibboel D, Meinl W, Glatt H, et al. (2000) Potent inhibition of estrogen sulfotransferase by hydroxylated PCB metabolites: A novel pathway explaining the estrogenic activity of PCBs. Endocrinology 141: 1897–1900.
22. Gupta C (2000) Reproductive malformation of the male offspring following maternal exposure to estrogenic chemicals. Proceedings of the Society for Experimental Biology and Medicine 224: 61–68.
23. Schantz SL, Widholm JJ, Rice DC (2003) Effects of PCB exposure on neuropsychological function in children. Environmental Health Perspectives 111: 357–576.
24. Lee DH, Steffes MW, Sjödin A, Jones RS, Needham LL, et al. (2011) Low dose organochlorine pesticides and polychlorinated biphenyls predict obesity, dyslipidemia, and insulin resistance among people free of diabetes. PLoS ONE 6: e15977.
25. Dirinck E, Jorens PG, Covaci A, Geens T, Roosens L, et al. (2011) Obesity and persistent organic pollutants: Possible obesogenic effect of organochlorine pesticides and polychlorinated biphenyls. Obesity 19: 709–714.
26. Bourez S, Van den Daelen C, Le Lay S, Poupaert J, Larondelle Y, et al. (2013) The dynamics of accumulation of PCBs in cultured adipocytes vary with the cell lipid content and the lipophilicity of the congener. Toxicology Letters 216: 40–46.
27. Zhou L, Wang X, Yang Y, Wu L, Li F, et al. (2011) Berberine attenuates cAMP-induced lipolysis via reducing the inhibition of phosphodiesterase in 3T3-L1 adipocytes. Biochimica et Biophysica Acta (BBA) 1812: 527–535.
28. Louis C, Van den Daelen C, Bourez S, Donnay I, Larondelle Y, et al. (2014) Efficient *in vitro* adipocyte model of long-term lipolysis: A tool to study the behaviour of lipophilic compounds. In Vitro Cellular & Developmental Biology - Animal 50: 507–518.
29. Wassermann M, Wassermann D, Cucos S, Miller HJ (1979) World PCBs map: Storage and effects in man and his biologic environment in the 1970s. Annals of the New York Academy of Sciences 320: 69–124.
30. Meeker JD, Maity A, Missmer SA, Williams PL, Mahalingaiah S, et al. (2011) Serum concentrations of polychlorinated biphenyls in relation to *in vitro* fertilization outcomes. Environmental Health Perspectives 119: 1010–1016.
31. Dang Van QC, Focant M, Mignolet E, Turu C, Froidmont E, et al. (2011) Influence of the diet structure on ruminal biohydrogenation and milk fatty acid composition of cows fed extruded linseed. Animal Feed Science and Technology 169: 1–10.
32. EC (2002) European Commission. Council L221/8. Official Journal of the European Communities 262: 8–36.
33. Matthews HB, Surles JR, Carver JG, Anderson MW (1984) Halogenated biphenyl transport by blood components. Fundamental and Applied Toxicology 4: 420–428.
34. Wang N, Kong D, Cai D, Shi L, Cao Y, et al. (2010) Levels of polychlorinated biphenyls in human adipose tissue samples from Southeast China. Environmental Science & Technology 44: 4334–4340.
35. De Saeger S, Sergeant H, Piette M, Bruneel N, Van de Voorde W, et al. (2005) Monitoring of polychlorinated biphenyls in Belgian human adipose tissue samples. Chemosphere 58: 953–960.
36. Moon HB, Lee DH, Lee Y, Choi M, Choi HG, et al. (2012) Polybrominated diphenyl ethers, polychlorinated biphenyls, and organochlorine pesticides in adipose tissues of Korean women. Archives of Environmental Contamination and Toxicology 62: 176–184.
37. Arrebola JP, Cuellar M, Claure E, Quevedo M, Antelo SR, et al. (2012) Concentrations of organochlorine pesticides and polychlorinated biphenyls in human serum and adipose tissue from Bolivia. Environmental Research 112: 40–47.
38. Malarvannan G, Dirinck E, Dirtu AC, Pereira-Fernandes A, Neels H, et al. (2013) Distribution of persistent organic pollutants in two different fat compartments from obese individuals. Environment International 55: 33–42.
39. Becker MM, Gamble W (1982) Determination of the binding of 2,4,5,2',4',5'-hexachlorobiphenyl by low density lipoprotein and bovine serum albumin. Journal of Toxicology and Environmental Health 9: 225–234.
40. Spindler-Vomachka M, Vodicnik MJ, Lech JJ (1984) Transport of 2,4,5,2',4',5'-hexachlorobiphenyl by lipoproteins *in vivo*. Toxicology and Applied Pharmacology 74: 70–77.
41. Gallenberg LA, Ring BJ, Vodicnik MJ (1987) Influence of lipolysis on the mobilization of 2,4,5,2',4',5'-hexachlorobiphenyl from adipocytes *in vitro*. Journal of Toxicology and Environmental Health 20: 163–171.
42. Koppe JG (1995) Nutrition and breast-feeding. European Journal of Obstetrics & Gynecology and Reproductive Biology 61: 73–78.
43. Irigaray P, Mejean L, Laurent F (2005) Behaviour of dioxin in pig adipocytes. Food and Chemical Toxicology 43: 457–460.
44. Edens NK, Leibel RL, Hirsch J (1990) Mechanism of free fatty acid re-esterification in human adipocytes *in vitro*. Journal of Lipid Research 31: 1423–1431.
45. López-Aparicio P, Merino MJ, Sánchez E, Recio MN, Pérez-Albarsanz MA (1997) Effect of Aroclor 1248 and two pure PCB congeners upon the membrane fluidity of rat renal tubular cell cultures. Pesticide Biochemistry and Physiology 57: 54–62.
46. Gonzalez A, Odjélé A, Weber JM (2013) PCB-153 and temperature cause restructuring of goldfish membranes: Homeoviscous response to a chemical fluidiser. Aquatic Toxicology 144–145: 11–18.
47. Yilmaz B, Sandal S, Chen CH, Carpenter DO (2006) Effects of PCB 52 and PCB 77 on cell viability, $[Ca^{2+}]_i$ levels and membrane fluidity in mouse thymocytes. Toxicology 217: 184–193.
48. Campbell AS, Yu Y, Granick S, Gewirth AA (2008) PCB association with model phospholipid bilayers. Environmental Science & Technology 42: 7496–7501.
49. Katynski AL, Vijayan MM, Kennedy SW, Moon TW (2004) 3,3',4,4',5-pentachlorobiphenyl (PCB 126) impacts hepatic lipid peroxidation, membrane fluidity and β-adrenoceptor kinetics in chick embryos. Comparative Biochemistry and Physiology Part C 137: 81–93.

14

Medium Effects on Minimum Inhibitory Concentrations of Nylon-3 Polymers against *E. coli*

Heejun Choi[1], Saswata Chakraborty[1], Runhui Liu[1], Samuel H. Gellman[1,2]*, James C. Weisshaar[1,2]*

1 Department of Chemistry, University of Wisconsin-Madison, Madison, Wisconsin, United States of America, **2** Molecular Biophysics Program, University of Wisconsin-Madison, Madison, Wisconsin, United States of America

Abstract

Minimum inhibitory concentrations (MICs) against *E. coli* were measured for three nylon-3 polymers using Luria-Bertani broth (LB), brain-heart infusion broth (BHI), and a chemically defined complete medium (EZRDM). The polymers differ in the ratio of hydrophobic to cationic subunits. The cationic homopolymer is inert against *E. coli* in BHI and LB, but becomes highly potent in EZRDM. A mixed hydrophobic/cationic polymer with a hydrophobic *t*-butylbenzoyl group at its N-terminus is effective in BHI, but becomes more effective in EZRDM. Supplementation of EZRDM with the tryptic digest of casein (often found in LB) recapitulates the LB and BHI behavior. Additional evidence suggests that polyanionic peptides present in LB and BHI may form electrostatic complexes with cationic polymers, decreasing activity by diminishing binding to the anionic lipopolysaccharide layer of *E. coli*. In contrast, two natural antimicrobial peptides show no medium effects. Thus, the use of a chemically defined medium helps to reveal factors that influence antimicrobial potency of cationic polymers and functional differences between these polymers and evolved antimicrobial peptides.

Editor: Jürgen Harder, University Hospital Schleswig-Holstein, Campus Kiel, Germany

Funding: This work was supported by the National Institute of General Medical Sciences of the National Institutes of Health, Grant # R01GM094510 (JCW as PI) and R01GM093265 (JCW and SHG as co-PIs). The funders had no role in study design, data collection and analysis, decision to publish, or preparation of the manuscript.

Competing Interests: The authors have declared that no competing interests exist.

* Email: gellman@chem.wisc.edu (SHG); weisshaar@chem.wisc.edu (JCW)

Introduction

There is a pressing need to develop new ways to kill harmful bacteria while causing minimal damage to eukaryotic cells [1,2]. An important component of the innate response to invasive bacteria is the release of antimicrobial peptides (AMPs) that permeabilize bacterial membranes and ultimately kill the invaders [3–5]. These peptides target bacterial membranes selectively relative to eukaryotic membranes. Thousands of natural AMPs are now known [6]. Substantial effort has been devoted to development of synthetic analogues containing α-amino acid residues and/or unnatural subunits that mimic the selective antibacterial action of AMPs. Examples include discrete oligomers generated from L-α-amino acids [7,8], N-alkyl glycines ("peptoids") [9–11], β-amino acids ("β-peptides") [12,13], or combinations of these building blocks. Many of these oligomers have been designed to adopt an amphipathic helical conformation, because this structural motif is common among natural AMPs.

The synthesis of sequence-specific oligomers requires a step-by-step approach, typically involving solid-phase methods, which is time-consuming and expensive. This synthetic problem has encouraged several research groups to explore polymerization-based methods to generate antimicrobial materials [14–22]. In many cases, a pair of precursors is copolymerized, with one precursor giving rise to a hydrophobic subunit and the other giving rise to a cationic subunit in the polymer chains. The resulting materials are heterogeneous, containing chains that vary in length, composition, subunit sequence and, frequently, subunit stereochemistry. Antibacterial activity in such cases cannot depend on adoption of a single amphipathic conformation. Nevertheless, careful tuning of the proportion and identities of the cationic and hydrophobic subunits can provide polymeric materials that exhibit strong bacteriostatic action against both Gram negative and Gram positive bacteria at concentrations that do not cause destruction of red blood cells ("hemolysis").

Our recent structure-function study of binary nylon-3 copolymers (β-peptide backbone) showed that a specific proportion of hydrophobic and cationic subunits plus inclusion of a hydrophobic group such as p-*t*-butylbenzoyl at the N-terminus provided a favorable balance of bacteriostatic and hemolytic properties [19]. Bacteriostatic potency was evaluated in terms of the minimum inhibitory concentration (MIC), the lowest polymer concentration that halted bacterial growth. Four bacterial species were evaluated, among which *Escherichia coli* was the only Gram negative organism. *E. coli* MIC measurements were carried out using brain-heart infusion (BHI) growth medium. Hemolytic activity was measured as the minimum hemolytic concentration (MHC), the smallest polymer concentration that caused detectable release of hemoglobin from human red blood cells.

We have been developing fluorescence microscopy methods that monitor membrane disruption induced by antimicrobial peptides acting on single bacterial cells in real time [23–25]. The broths typically used for rapid bacterial growth, including brain-heart infusion (BHI) and Luria-Bertani (LB), are unsuitable for sensitive fluorescence work due to their strong background fluorescence on excitation with visible light. Instead, we use a low-fluorescence, chemically defined medium called "EZ rich, defined medium" (EZRDM) [26]. As a prelude to studies of the mechanism of nylon-3 action against *E. coli*, we measured MIC values for a panel of nylon-3 polymers in the EZRDM medium. Surprisingly, we observed a dramatic reduction of MIC values (greater polymer efficacy) in EZRDM as compared with either BHI or LB.

In particular, cationic homopolymers (lacking hydrophobic subunits) were much more effective against *E. coli* in EZRDM than in BHI or LB media.

We report studies intended to elucidate the effect of bacterial growth medium on MIC values measured for *E. coli*. By adding tryptone, the tryptic digest of bovine casein, to EZRDM, we recover the pattern of activity observed in BHI. We also present evidence indicating that anionic peptide components of tryptone (and, by extension, those in BHI and LB) diminish the ability of the highly cationic polymers to attack *E. coli* relative to the effects observed in EZRDM. This functional attenuation presumably results from formation of relatively inert electrostatically bound complexes between the cationic polymers and the anionic peptide components. In sharp contrast to the behavior of the cationic polymers, two natural AMPs (LL-37 and Cecropin A) showed consistent MIC values in all media. Another natural AMP, Magainin 2, was inactive against *E. coli* in EZRDM and had low activity in BHI as well. We suggest that the MIC measurements in EZRDM reveal the "intrinsic activity" of a polymer or AMP.

These results expand our understanding of structure-activity relationships among the nylon-3 polymers and suggest new design strategies for the future. Moreover, these observations highlight a previously undocumented feature of natural AMPs, which can apparently be fine-tuned by evolutionary selection to avoid the polyanion-based neutralization mechanism suggested by our polymer findings.

Materials and Methods

Materials

The nylon-3 polymers used in this study (Fig. 1) were synthesized as previously described [19]. Polymers were prepared from racemic β-lactams, and each polymer was therefore heterochiral. Polymer samples had mean chain lengths of 27 (polymer **A**), 27 (**B**), and 24 (**C**). Polydispersity index (PDI) values ranged from 1.02–1.15. To check for possible batch-to-batch variation in polymer properties, each polymer was synthesized twice. The MIC values measured for different batches of the same polymer were indistinguishable in each case. Human cathelicidin LL-37 was purchased from Bachem, and moth Cecropin A and Magainin 2 were purchased from Anaspec. All three peptides were used without purification. EZ rich defined medium (EZRDM, Teknova), brain heart infusion broth (BHI, Difco), and Luria-Bertani broth (LB, Sigma Aldrich) were purchased as powders and dissolved in water. Tryptone powder was purchased from BD Sciences. We compared the effects of supplements made from tryptone powder as received vs tryptone powder that was dissolved, dialyzed to remove small ions and solutes (1 kDa cutoff), and then lyophilized. No differences were observed. We use "tryptone" or "dialyzed tryptone" to mean a solution of the dialyzed tryptone powder dissolved in EZRDM. We use "raw tryptone" to mean a solution of the tryptone powder as received. The free base ion-exchange resin Amberlite IRA67 was purchased from Sigma Aldrich.

We devised two tests of the effects of anionic species within tryptone on MIC values. First, the large polyanions in raw tryptone were removed from dialyzed tryptone solution using a washed, free-base anion-exchange resin (Amberlite IRA67, Sigma Aldrich). At pH = 7, the resin removes H^+ and polyanions from solution. A solution of dialyzed tryptone was incubated with 1.2X (wt/wt) resin for 2 hr and then filtered to remove the beads. We designate the product of this operation "anion-exchanged tryptone". The filtered product was lyophilized and added to EZRDM for MIC assays. Second, we used solid-phase methods to synthesize a single, specific anionic peptide, FQSEEQQTE-DELQDK, and tested its effects at 400 μM in EZRDM.

This peptide is a putative component of raw tryptone based on the predicted products of digestion of bovine beta-casein by trypsin. The peptide was used without purification (estimated purity ~85%).

Minimum Inhibitory Concentration (MIC) Assay

E. coli strains JM109 and MG1655 were studied. MIC values were determined using a standard serial microdilution method. Serial two-fold dilutions of each nylon-3 polymer, and of LL-37, Cecropin A, and Magainin 2, were performed in separate rows of a polystyrene 96-well plate in the chosen medium containing an inoculum of either JM109 or MG1655. Polymer concentrations between 6.3 μg/mL and 200 μg/mL were evaluated. Each assay plate contained a dilution series with ampicillin as a positive control. To test for possible effects of the MIC procedural details, the measurements in BHI medium were carried out with cells initially sampled either from stationary phase (as in the earlier study) [19] or from mid-log phase. In the stationary phase procedure, a culture grown overnight at 37°C to stationary phase was sampled and diluted to $OD_{600} = 0.05$ with medium at the same temperature. In the mid-log phase procedure, the stationary culture was diluted in fresh medium (1:100) and grown until it reached $OD_{600} = 0.5$. For the MIC measurements, the plate was incubated at 37°C for 6 hr. For both the stationary phase procedure and the mid-log phase procedure, we tested for effects of stationary incubation in a VWR 1525 digital incubator vs increased aeration due to shaking at 200 rpm in a Lab-Line Orbital Environ Shaker (Model 3527). For experiments augmenting EZRDM medium with tryptone solution, the appropriate amount of raw or dialyzed tryptone powder was dissolved in EZRDM and used in the dilution steps.

The MIC results were not affected by the choice of *E. coli* strain or any of the other variations in experimental procedure. Following these initial tests, our standard procedure used stationary phase cultures with no shaking during outgrowth and tryptone solution made from dialyzed tryptone powder. The overnight culture of *E. coli* was grown in EZRDM prior to the addition of tryptone supplemented EZRDM.

The MIC is reported as the lowest concentration for which no cell growth could be detected after 6 hr ($OD = 0.00 \pm 0.05$), as determined by measurements at 595 nm using an EnVision 2100 Multilabel Reader (Perkin-Elmer). Examples of OD vs time curves are provided in Figs. S1, S2, and S3. With care, the resulting MIC measurements are more accurate than a factor of two, as evidenced by exact reproduction over repeated trials of the particular concentration step that halted growth over 6 hr. When growth inhibition did not occur at the highest concentration of polymer studied (200 μg/mL), we report 200 μg/mL as a lower bound on the MIC, denoted by an upward arrow in the MIC bar graphs (Fig. 2).

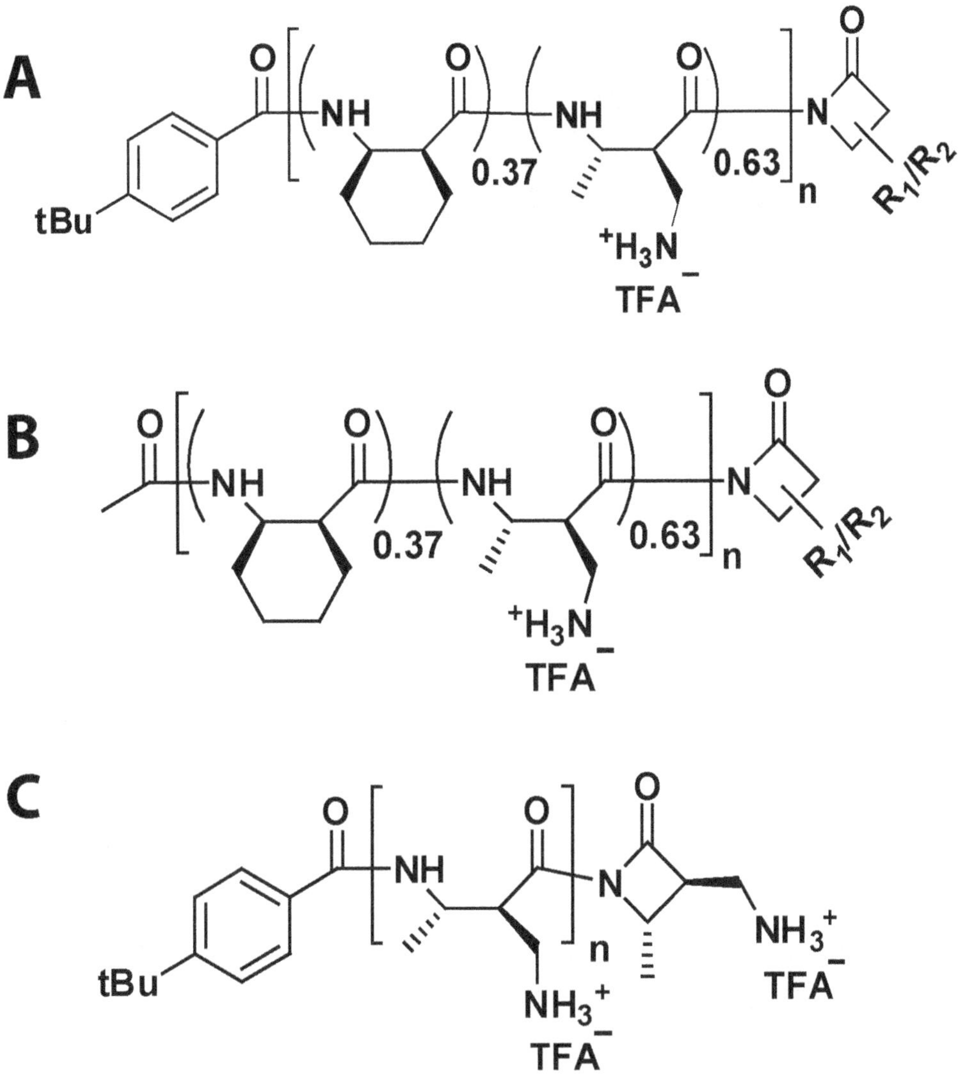

Figure 1. Structures of the random, heterochiral nylon-3 polymers used in this study. **A** bears a hydrophobic p-*t*-butylbenzoyl group at the N-terminus. **B** shares the 37:63 CH:MM ratio of **A**, but lacks a hydrophobic group at the N-terminus. **C** is a homopolymer of the cationic MM subunit.

Results

Experimental Design

This study focuses on three nylon-3 polymers that have been previously described [19] (Fig. 1; **A–C**). **A** is a copolymer generated from a ring-opening polymerization reaction mixture containing two β-lactams, CHβ and MMβ, in 37:63 molar ratio. "CH" denotes a hydrophobic cyclohexyl side chain within the monomer and "MM" (monomethyl) denotes a single methyl group at the C_α site of the cationic monomer. Since the β-lactams are racemic, the resulting polymer sample contains chains with many different stereochemistries, as well as many different sequences. After polymerization, deprotection provides a side chain amino group in the MM subunit. These amino groups are protonated when the polymer is dissolved in aqueous solution near or below neutral pH, which confers a positive charge on the polymer.

The 37:63 ratio of CH and MM subunits was previously shown [19] to provide a favorable compromise between antibacterial activity in BHI medium (lower MIC preferred) and hemolytic activity (higher minimum hemolytic concentration, MHC, preferred). Antibacterial activity in BHI medium was enhanced by the presence of a hydrophobic group at the N-terminus of the polymer chains; **A** bears a p-*t*-butylbenzoyl group at this position. In BHI medium, nylon-3 polymers **B** and **C** showed diminished antibacterial activity relative to **A**. Copolymer **B** shares the 37:63 CH:MM ratio of **A**, but **B** lacks a hydrophobic group at the N-terminus. **C** is a homopolymer of the cationic MM subunit.

We were originally motivated to explore cationic-hydrophobic copolymers such as **A** because AMPs are commonly rich in both cationic and hydrophobic amino acid residues. A net positive charge is believed to be necessary to attract AMPs to the lipopolysaccharide (LPS) surface of a bacterial cell, which bears a net negative charge. Hydrophobic side chains are thought to be essential for interaction with the nonpolar interior of a lipid bilayer and the resulting membrane barrier disruption. From this perspective, it is sensible that a cationic homopolymer such as **C** would not display strong antibacterial effects, as observed in previous studies of *E. coli* MICs conducted in BHI medium [19]. We were therefore surprised to discover that cationic homopolymer **C** is highly active against *E. coli* in the chemically defined

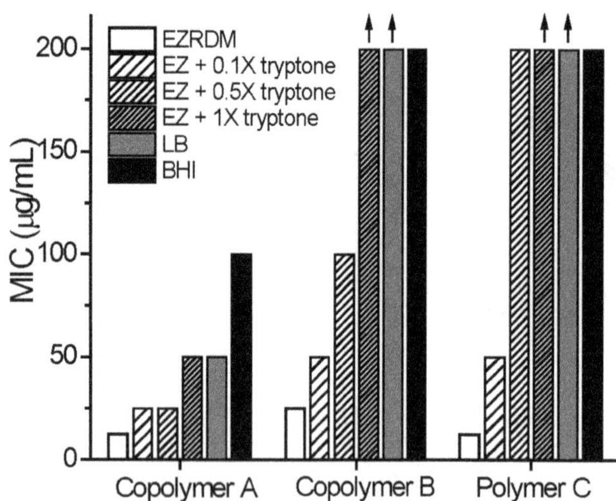

Figure 2. Minimum inhibitory concentrations (MICs) of three nylon-3 polymers against *E. coli* for different media. See Fig. 1 for structures of **A**, **B**, and **C**. Vertical arrows mark bars that are lower limits only. EZRDM, LB, and BHI as described in main text. The designation "EZ + trypt" refers to EZRDM supplemented with 10 g/L of dialyzed tryptone powder (1X tryptone). "EZ + trypt - PA" refers to EZRDM supplemented with anion-exchanged tryptone at the equivalent of 10 g/L. "EZ + FQS..." refers to EZRDM supplemented with 400 µM of the single anionic peptide FQSEEQQTEDELQDK (net −5 charge).

EZRDM. The experiments described here were undertaken to try to determine why MIC results differ between a traditional but chemically undefined medium such as BHI and chemically defined EZRDM.

EZRDM is a MOPS-buffered solution that contains glucose (0.2%), supplemental amino acids and vitamins, nucleotides, 1.32 mM K_2HPO_4, and 76 mM NaCl. This is our preferred medium for optical imaging experiments on bacteria because of its low autofluorescence [26].

In contrast, BHI medium is not chemically defined, since it is generated from calf brain and heart. BHI medium is supplemented with peptone, a digest of an undefined mixture of proteins from cow and pig. The enzyme used in the digest is proprietary, and the peptide mixture in peptone is therefore uncharacterized. As shown below, we found that *E. coli* MIC studies conducted in LB provide results similar to those conducted in BHI. LB is a common bacterial growth medium that, like BHI medium, is chemically undefined. LB medium contains tryptone along with yeast extract and sodium chloride. Tryptone is generated via tryptic digestion of casein.

We considered the hypothesis that peptides from a supplement such as tryptone (an additive in LB) or peptone (an additive in BHI medium) might interact with cationic nylon-3 polymers in a way that affects antibacterial potency. Since EZRDM contains amino acids rather than enzymatically generated peptides, the peptide-polymer interactions that we proposed to inhibit antibacterial activity of nylon-3 polymers would not be possible in EZRDM. Thus, the absence of enzymatically generated peptides in EZRDM might explain why polymer MIC values were lower in this medium than in LB or BHI medium.

This hypothesis was tested by supplementing EZRDM with dialyzed tryptone solution.

The supplement was generated by dialyzing raw tryptone solution against pure water for 48 hr to remove small ions and molecules below 1000 Da. The retained material was assumed to be composed largely of peptides generated via the cleavage of casein by trypsin. The retained solution was lyophilized, and the resulting dialyzed tryptone powder was added in varying proportions to EZRDM. As shown by the data presented below, this simple supplementation strategy caused a shift in nylon-3 MIC values for *E. coli* from the EZRDM profile to the profile observed in complex media such as BHI or LB.

Two additional sets of measurements tested whether or not anionic peptides were responsible for the larger MIC values observed on addition of dialyzed tryptone to EZRDM.

We measured MICs against *E. coli* for copolymers **A**, **B**, and **C** in EZRDM supplemented by the single anionic peptide FQSEEQQTEDELQDK. We also measured MICs for **A**, **B**, and **C** in EZRDM supplemented with anion-exchanged tryptone, which should lack anionic polypeptides. Both tests support the hypothesis that complexation of the highly cationic copolymer with polyanionic peptides diminishes antimicrobial activity.

MIC measurements for polymers

Results obtained for the nylon-3 polymers in BHI medium (Fig. 2) agree with previous reports: cationic-hydrophobic copolymer **A** displays moderate activity against *E. coli* in this medium, but absence of the hydrophobic N-terminal group (copolymer **B**) or absence of the hydrophobic CH subunits (homopolymer **C**) leads to a profound loss of activity. MIC values measured in LB medium show the same pattern as those measured in BHI medium (Fig. 2).

In contrast, the MIC pattern in EZRDM is quite different: all three polymers are more active in EZRDM than in BHI or LB, and there is little distinction among the three polymers in EZRDM.

We supplemented EZRDM with dialyzed tryptone solution at three concentrations, 1 g/L, 5 g/L or 10 g/L, to generate media designated "EZ + 0.1X tryptone", "EZ + 0.5X tryptone" and "EZ + 1X tryptone". These designations are based on the fact that LB medium typically contains 10 g of tryptone powder per liter. Detailed MIC curves and MIC values vs tryptone concentrations are included in Figs. S3A and B. The antimicrobial activity of each polymer decreases (i.e., MIC value increases) as the concentration of dialyzed tryptone in EZRDM increases. The effects of 1X tryptone supplement on the MIC values for the three polymers are included in Fig. 2. The factor by which the MIC increases depends on polymer composition: polymer **A**, which is the most active of the three in complex media, is less strongly affected than polymers **B** and **C**, both of which are inactive in complex media but highly active (low MIC values) in EZRDM. Supplementation of EZRDM with 1X dialyzed tryptone recapitulates the effects of BHI and LB on MIC values. Polymer **A**, the most active polymer in BHI and LB, is the only polymer that retains reasonably good activity in EZRDM supplemented with 1X tryptone, in BHI, and in LB.

To test the ability of polyanions within dialyzed tryptone to alter MICs, we supplemented EZRDM with anion-exchanged tryptone, from which polyanionic components have been removed. The concentration of anion-exchanged tryptone was 10 g/L (as in 1X tryptone) *minus* the mass of the polyanions removed by the resin. The effects on MIC values for the three polymers are summarized in Fig. 2. Removal of polyanions from tryptone drastically reduces the MICs of the three polymers to the point where they are comparable to the MICs in EZRDM.

In addition, the pattern of relative MICs in EZRDM is recovered by removal of anionic molecules from the tryptone additive.

To investigate further the effect of polyanionic peptides in tryptone, the anionic peptide FQSEEQQTEDELQDK was added to EZRDM without tryptone supplementation. The anionic peptide concentration was chosen as 200 μM or 400 μM. The value 400 μM matches the estimated concentration of this particular peptide in 1X tryptone (based on 10 g/L of the digest products of the 23 kDa protein bovine beta-casein). The MICs of all three polymers increased with added concentration of the anionic peptide, as shown in Fig. S5. Addition of 400 μM of added anionic peptide increases the MICs of all three polymers by a factor of four to eight compared with EZRDM (Fig. 2).

MIC measurements for antimicrobial peptides

For comparison, we measured the MIC values against *E. coli* of two natural antimicrobial peptides, LL-37 and Cecropin A, in four media: BHI, EZRDM, and EZRDM supplemented with 0.5X or 1X tryptone. As shown in Fig. 3, LL-37 and Cecropin A showed strong antimicrobial activity in all four media. Detailed OD data are provided in Fig. S4. In fact, MIC values for these two AMPs were slightly smaller in EZRDM plus 1X tryptone than in unsupplemented EZRDM. The natural AMP Magainin 2 was not active against *E. coli* in EZRDM (MIC>100 μM) and had an MIC of 40 μM in BHI. Unlike polymers **A–C**, Magainin 2 does not gain activity when the medium changes from BHI to EZRDM (Fig. 3).

Discussion

Medium Effects on polymer MIC values

The susceptibility of bacteria to antibiotics, including antimicrobial peptides, can depend on environmental conditions such as temperature, aeration, and pH, as well as the concentrations of ionic species [4]. A high concentration of Ca^{2+} [27] or Mg^{2+} [28] in the growth medium increases MIC values for AMPs, presumably due to competition between the divalent cation and the cationic peptide for binding sites within the bacterial cell surface. Addition of polyanions such as DNA [29] also increases MIC values, presumably by binding to cationic AMPs and decreasing the concentration of free AMP available for disruptive interactions with the bacterial cell surface. To avoid such complications, most laboratories screen for antimicrobial peptide activity without varying the growth medium. The few studies that compare MIC values between two different media [30,31] have not provided clear conclusions because of the inherent complexity of the broths employed. The present work compares cationic polymer and AMP activities in two widely used complex media (BHI and LB) with activities in a chemically defined medium (EZRDM). This experimental design enables us to draw some tentative conclusions about the underlying causes of the strong variation in MIC values across media.

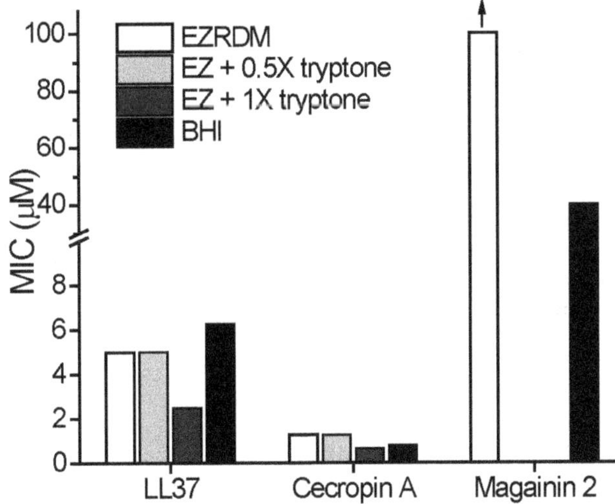

Figure 3. Minimum inhibitory concentrations (MICs) of three natural antimicrobial peptides, LL-37, Cecropin A, and Magainin 2, in different media. EZRDM and BHI as described in text. See Fig. 5 for sequences of the antimicrobial peptides. Note break in vertical scale for Magainin 2. The designations "EZ + 0.5X tryptone" and "EZ + 1X tryptone" refer to EZRDM supplemented with 5 g/L, and 10 g/L of dialyzed tryptone powder, respectively. Vertical arrow marks a bar that is a lower limit only.

Our primary finding is that BHI (which usually contains peptone supplement) and LB (which contains tryptone supplement) suppress polymer activity against *E. coli* as compared with EZRDM (which contains only small molecules). By adding 1X dialyzed tryptone solution to EZRDM, we were able to recapitulate polymer performance in BHI and LB. To avoid possibly confounding effects of small molecules such as salts, sugar, vitamins, and divalent cations, we used material derived from tryptone powder via dialysis vs water for several days, followed by lyophilization. We therefore attribute the effects of adding tryptone to EZRDM to the polypeptides that are generated via tryptic digestion of casein.

What are these polypeptides? Trypsin cleaves peptide bonds specifically at the C-terminal side of lysine and arginine residues. Complete digestion necessarily produces peptides with only one positive charge (at the C-terminus). Digest products are thus intrinsically biased to be negatively charged or hydrophobic or both. The sequence of bovine casein and the predicted products of its complete digestion by trypsin are shown in Fig. 4. This predicted mixture includes six peptides containing 1–7 residues, which have a net charge of +1 or are neutral; such small peptides are presumably depleted from raw tryptone solution during dialysis. In addition, there are five longer peptides, including three that contain 16–24 residues (net charge ranging from −1 to −6), one with 48 residues (neutral) and one with 56 residues (net charge −3). Highly anionic components include 16-mer FQSEEQQTEDELQDK (net charge −5), and 24-mer ELE..., with net charge −6. The two longest peptides are highly hydrophobic.

Supplementation of EZRDM with 1X tryptone causes a profound diminution in the ability of polymers **A–C** to inhibit growth of *E. coli* (Fig. 2). We propose that all three polymers bind to the polyanionic peptides generated via tryptic digestion of casein. The resulting polymer-anionic peptide complexes presumably bind less strongly to the anionic lipopolysaccharide (LPS) layer found on the outer surface of *E. coli* than do uncomplexed polymers. Even for polymer **A**, which displays significant activity in BHI, the MIC decreases fourfold from EZRDM plus 1X tryptone to EZRDM or from LB to EZRDM, and eightfold from EZRDM to BHI from EZRDM.

This hypothesis is further supported by the MIC pattern of the three polymers in EZRDM supplemented with anion-exchanged tryptone, which is quite similar to the MIC pattern in EZRDM alone (Fig. 2). Addition of the anionic peptide FQSEEQQTEDELQDK, which is predicted to be generated by digestion of beta-casein with trypsin, increased the MICs of the three polymers by a factor of four to eight (Fig. 2). Addition of this peptide did not, however, match the full tryptone-induced increases in MICs for polymers **B** and **C**. This finding with a specific peptide supports our hypothesis regarding the MIC-suppressing effects of polyanionic components on a polymer's antibacterial activity. How-

Sequence of Bovine Beta-Casein

MKVLILACLV ALALARELEE LNVPGEIVES LSSSEESITR INKKIEKFQS EEQQQTEDEL

QDKIHPFAQT QSLVYPFPGP IPNSLPQNIP PLTQTPVVVP PFLQPEVMGV SKVKEAMAPK

HKEMPFPKYP VEPFTESQSL TLTDVENLHL PLPLLQSWMH QPHQPLPPTV MFPPQSVLSL

SQSKVLPVPQ KAVPYPQRDM PIQAFLLYQE PVLGPVRGPF PIIV

Predicted Digestion of Bovine Beta-Casein with Trypsin

Sequence	Charge
K	+1
R	+1
VK	+1
HK	+1
INK	+1
IEK	0
EAMAPK	0
GPFPIIV	0
EMPFPK	0
VLPVPQK	+1
AVPYPQR	+1
FQSEEQQQTEDELQDK	-5
DMPIQAFLLYQEPVLGPVR	-1
ELEELNVPGEIVESLSSSEESITR	-6
IHPFAQTQSLVYPFPGPIPNSLPQNIPPLTQTPVVVPPFLQPEVMGVSK	0
YPVEPFTESQSLTLTDVENLHLPLPLLQSWMHQPHQPLPPTVMFPPQSVLSLSQSK	-3

Figure 4. Sequence of bovine beta-casein and of the predicted components of complete digestion of bovine β-casein with trypsin. Red, blue, and bold-black letters denote anionic, cationic, and hydrophobic residues, respectively. Net charges as shown.

ever, this observation also suggests that the interactions between constituents of the polyanion mixture in tryptone and a heterogeneous cationic polymer sample may be difficult to understand in detail.

There is precedent for our observation that the antibacterial effect of highly cationic polymers can be sensitive to the growth medium. The cationic homopolymer ε-polylysine is used as a food preservative based on its antibacterial effects [32]. The reported MIC for ε-polylysine against *E. coli* decreases from 50 μg/mL [33] to 1 μg/mL [34] (50-fold) when the growth medium is changed from nutrient buffer (which includes peptone) to Davis medium (which is a chemically defined medium containing "casamino acids", a mixture of *monomeric* α-amino acids). Since the ingredients of Davis medium do not include a *peptide*-rich component generated via enzymatic degradation of proteins, such as peptone or tryptone, the pronounced effect of medium on the antibacterial activity of ε -polylysine is consistent with our conclusions about the antibacterial activity of cationic nylon-3 polymers. Our results may also be relevant to studies using cationic peptides to enhance the permeability of the outer membrane to hydrophobic drugs [35].

Implications for the design of antimicrobial polymers

The three polymers evaluated here were selected based on a previous study of structure-function relationships within the nylon-3 family [19]. That study measured MIC (in BHI medium) and minimum hemolytic concentration (MHC) values for polymers that varied in the nature of the hydrophobic and cationic subunits, the ratio of hydrophobic to cationic subunits, the N-terminal group, and the mean chain length. The composition embodied in polymer **A**, at ~25-mer average chain length, exhibited the best performance overall. This nylon-3 polymer displayed moderate MIC values against both Gram negative and Gram positive bacteria along with a high MHC value. Analogous polymer **B**, which lacks the hydrophobic unit at the N-terminus, was far less active than **A** against *E. coli*. Homopolymer **C** also was far less active than **A** against *E. coli*, which indicated that hydrophobic CH subunits are critical for conferring antibacterial activity on polymer **A**. The trends previously observed among nylon-3 polymers **A**–**C** are consistent with conclusions drawn from studies of natural AMPs and analogous synthetic peptides, which suggest that optimizing antibacterial activity while simultaneously minimizing eukaryotic cell toxicity (e.g., hemolytic activity) requires the

LL-37

LLGDFFRKSKEKIGKEFKRIVQRIKDFLRNLVPRTES-NH2

Cecropin A

KWKLFKKIEKVGQNIRDGIIKAGPAVAVVGQATQIAK-NH2

Magainin 2

GIGKFLHSAKKFGKAFVGEIMNS

Figure 5. Sequences of the natural AMPs LL-37, Cecropin A, and Magainin 2. Red, blue, and bold-black letters denote anionic, cationic, and hydrophobic residues, respectively.

presence of both hydrophobic and cationic subunits, with a proper balance between net charge and net lipophilicity [36].

The previous nylon-3 studies were conducted in BHI medium. The present results show that the impact of introducing hydrophobic subunits or cationic N-terminal groups can be dramatically altered by changing the medium. We observe that all three nylon-3 polymers are more active against *E. coli* (lower MIC values) in EZRDM than in BHI medium. Most striking is the observation that cationic homopolymer **C** is highly active in EZRDM, while this polymer has very little activity against *E. coli* in BHI medium.

Our study raises the general question of which media are most appropriate for evaluating the activity of new antimicrobial candidates. The ε-polylysine precedent [32] and the sensitivity to medium we document for highly cationic polymers suggest that investigations in chemically defined media have substantial merit for such evaluations. In addition, many pathogenic bacteria are more virulent in minimal growth conditions than in rich growth conditions [37,38], a trend that argues for using a chemically defined *minimal* medium to evaluate the antimicrobial activities of new polymers.

The natural cationic AMPs LL-37 and Cecropin A are remarkably impervious to the medium effects that we find to be so strong for the highly cationic polymers (Fig. 3). AMPs may have evolved so as to maintain their efficacy in a variety of environments. The sequences of LL-37 and Cecropin A (Fig. 5) are perhaps instructive. The positive charge density is smaller, on a per-subunit basis, in the natural AMPs than in nylon-3 polymers **A-C** because the peptides contain residues that are neither positively charged nor hydrophobic. The lower positive charge density and the inclusion of some negatively charged residues may diminish binding of the natural AMPs to polyanionic species in BHI and in tryptone-supplemented EZRDM. This speculation suggests that it may prove fruitful to study random polymers comprising three or even four components, with proportions of cationic, anionic, and hydrophobic monomers chosen to mimic those found in natural AMPs. Unlike the nylon-3 polymers, which gained activity when moved from BHI to EZRDM, Magainin 2 showed low activity (MIC = 40 μM) in BHI and no activity (MIC>100 μM) in EZRDM (Fig. 3). Magainin 2, with 23 residues and a net charge of +3 at neutral pH, may not be sufficiently cationic to bind effectively to the LPS layer of Gram negative species.

It is possible that AMPs and highly positive nylon-3 polymers attack bacterial cells by different mechanisms, and that the mechanism of action varies among the nylon-3 polymers. Many AMPs fold into amphipathic helices and may form membrane pores of reasonably well defined structure. Defined conformations are not available to the sequence-random copolymers or the purely cationic polymers studied here. It is not clear how a highly cationic polymer such as **C** could cause membrane disruption, which evidently underlies the antibacterial effects of peptides such as LL-37 and Cecropin A [23–25]. The observation that diverse nylon-3 polymers display strong growth-inhibitory activity toward *E. coli* in EZRDM will enable detailed investigations of the mechanism(s) of action via optical imaging methods.

Supporting Information

Detailed data behind MIC measurements for the polymers and the antimicrobial peptides for different media and conditions are provided as supporting information.

Figure S1 Optical density (O.D.) vs concentrations of two different batches of polymers in BHI medium.

Figure S2 Optical density (O.D.) vs concentration compared for two different media and two different *E. coli* strains.

Figure S3 MIC measurements in different media as indicated. (A) Optical density (O.D.) vs concentration of polymers **A**, **B**, and **C** in different media. (B) Bar graph of MIC values.

Figure S4 Optical density (O.D.) vs concentration of natural antimicrobial peptides Magainin 2, Cecropin A, and LL-37 for different media.

Figure S5 Minimum inhibitory concentrations (MICs) of nylon-3 polymers A, B, and C in EZRDM supplement-

ed with 200 μM and 400 μM of the anionic peptide FQSEEQQTEDELQDK.

Acknowledgments

Research reported in this publication was supported by the National Institute of General Medical Sciences of the National Institutes of Health. The content is solely the responsibility of the authors and does not necessarily represent the official views of the National Institutes of Health.

Author Contributions

Conceived and designed the experiments: HC SC RL SHG JCW. Performed the experiments: HC SC RL. Analyzed the data: HC JCW. Contributed reagents/materials/analysis tools: HC SC RL SHG JCW. Contributed to the writing of the manuscript: HC SC RL SHG JCW.

References

1. Hancock REW, Sahl H-G (2006) Antimicrobial and host-defense peptides as new anti-infective therapeutic strategies. Nat Biotech 24: 1551–1557.
2. Coates A, Hu YM, Bax R, Page C (2002) The future challenges facing the development of new antimicrobial drugs. Nat Rev Drug Discovery 1: 895–910.
3. Brogden KA (2005) Antimicrobial peptides: Pore formers or metabolic inhibitors in bacteria? Nat Rev Micro 3: 238–250.
4. Yeaman MR, Yount NY (2003) Mechanisms of antimicrobial peptide action and resistance. Pharmacological Rev 55: 27–55.
5. Zasloff M (2002) Antimicrobial peptides of multicellular organisms. Nature 415: 389–395.
6. Zhao X, Wu H, Lu H, Li G, Huang Q (2013) Lamp: A database linking antimicrobial peptides. PLOS One 8.
7. Spindler EC, Hale JDF, Giddings TH Jr, Hancock REW, Gill RT (2011) Deciphering the mode of action of the synthetic antimicrobial peptide Bac8c. Antimic Agents Chemo 55: 1706–1716.
8. Munk JK, Uggerhoj LE, Poulsen TJ, Frimodt-Moller N, Wimmer R, et al. (2013) Synthetic analogs of anoplin show improved antimicrobial activities. J Peptide Science 19: 669–675.
9. Goodson B, Ehrhardt A, Ng S, Nuss J, Johnson K, et al. (1999) Characterization of novel antimicrobial peptoids. Antimic Agents Chemo 43: 1429–1434.
10. Patch JA, Barron AE (2003) Helical peptoid mimics of magainin-2 amide. JACS 125: 12092–12093.
11. Chongsiriwatana NP, Patch JA, Czyzewski AM, Dohm MT, Ivankin A, et al. (2008) Peptoids that mimic the structure, function, and mechanism of helical antimicrobial peptides. PNAS USA 105: 2794–2799.
12. Porter EA, Wang X, Lee HS, Weisblum B, Gellman SH (2000) Non-hemolytic beta-amino-acid oligomers. Nature 404: 565.
13. Godballe T, Nilsson LL, Petersen PD, Jenssen H (2011) Antimicrobial beta-peptides and alpha-peptoids. Chem Bio & Drug Design 77: 107–116.
14. Tiller JC, Liao CJ, Lewis K, Klibanov AM (2001) Designing surfaces that kill bacteria on contact. PNAS USA 98: 5981–5985.
15. Lewis K, Klibanov AM (2005) Surpassing nature: Rational design of sterile-surface materials. Trends Biotech 23: 343–348.
16. Sellenet PH, Allison B, Applegate BM, Youngblood JP (2007) Synergistic activity of hydrophilic modification in antibiotic polymers. Biomacromolecules 8: 19–23.
17. Allison BC, Applegate BM, Youngblood JP (2007) Hemocompatibility of hydrophilic antimicrobial copolymers of alkylated 4-vinylpyridine. Biomacromolecules 8: 2995–2999.
18. Sambhy V, Peterson BR, Sen A (2008) Antibacterial and hemolytic activities of pyridinium polymers as a function of the spatial relationship between the positive charge and the pendant alkyl tail. Angewandte Chemie 47: 1250–1254.
19. Mowery BP, Lindner AH, Weisblum B, Stahl SS, Gellman SH (2009) Structure-activity relationships among random nylon-3 copolymers that mimic antibacterial host-defense peptides. JACS 131: 9735–45.
20. Chakraborty S, Liu R, Lemke JJ, Hayouka Z, Welch RA, et al. (2013) Effects of cyclic vs acyclic hydrophobic subunits on the chemical structure and biological properties of nylon-3 copolymers. ACS Macro Lett 2: 753–756.
21. Takahashi H, Palermo EF, Yasuhara K, Caputo GA, Kuroda K (2013) Molecular design, structures, and activity of antimicrobial peptide-mimetic polymers. Macromol Biosci 13: 1285–1299.
22. Kuroda K, Caputo GA (2013) Antimicrobial polymers as synthetic mimics of host-defense peptides. Wiley Interdiscip Rev Nanomed Nanobiotech 5: 49–66.
23. Sochacki KA, Barns KJ, Bucki R, Weisshaar JC (2011) Real-time attack on single *Escherichia coli* cells by the human antimicrobial peptide LL-37. PNAS USA 108: E77–E81.
24. Barns KJ, Weisshaar JC (2013) Real-time attack of ll-37 on single *Bacillus subtilis* cells. Biochim Biophys Acta-Biomembranes 1828: 1511–1520.
25. Rangarajan N, Bakshi S, Weisshaar JC (2013) Localized permeabilization of *E. coli* membranes by the antimicrobial peptide Cecropin A. Biochemistry 52: 6584–6594.
26. Bakshi S, Bratton BP, Weisshaar JC (2011) Subdiffraction-limit study of Kaede diffusion and spatial distribution in live *Escherichia coli*. Biophys J 101.
27. Sugimura K, Nishihara T (1988) Purification, characterization, and primary structure of *Escherichia-coli* protease-vii with specificity for paired basic residues - identity of protease-vii and Ompt. J Bacteriology 170: 5625–5632.
28. Bryan LE (1984) Antimicrobial drug resistance. Orlando: Academic Press.
29. Lewenza S (2013) Extracellular DNA-induced antimicrobial peptide resistance mechanisms in pseudomonas aeruginosa. Frontiers Microbiology 4: 21–21.
30. Wesolowski D, Alonso D, Altman S (2013) Combined effect of a peptide-morpholino oligonucleotide conjugate and a cell-penetrating peptide as an antibiotic. PNAS USA 110: 8686–8689.
31. Schwab U, Gilligan P, Jaynes J, Henke D (1999) *In vitro* activities of designed antimicrobial peptides against multidrug-resistant cystic fibrosis pathogens. Antimic Agents Chemo 43: 1435–1440.
32. Hiraki J, Ichikawa T, Ninomiya S, Seki H, Uohama K, et al. (2003) Use of adme studies to confirm the safety of epsilon-polylysine as a preservative in food. Regulatory Toxicology and Pharmacology 37: 328–340.
33. Yoshida T, Nagasawa T (2003) Epsilon-poly-l-lysine: Microbial production, biodegradation and application potential. Appl Microbiology Biotech 62: 21–26.
34. Shima S, Matsuoka H, Iwamoto T, Sakai H (1984) Antimicrobial action of epsilon-poly-L-lysine. J Antibiotics 37: 1449–1455.
35. Vaara M (1992) Agents that increase the permeability of the outer-membrane. Microbiological Rev 56: 395–411.
36. Wimley WC (2010) Describing the mechanism of antimicrobial peptide action with the interfacial activity model. ACS Chem Biol 5: 905–917.
37. Yoon H, McDermott JE, Porwollik S, McClelland M, Heffron F (2009) Coordinated regulation of virulence during systemic infection of salmonella enterica serovar typhimurium. PLOS Pathogens 5.
38. Kim KS, Rao NN, Fraley CD, Kornberg A (2002) Inorganic polyphosphate is essential for long-term survival and virulence factors in shigella and salmonella spp. PNAS USA 99: 7675–7680.

Biomass Enzymatic Saccharification Is Determined by the Non-KOH-Extractable Wall Polymer Features That Predominately Affect Cellulose Crystallinity in Corn

Jun Jia[1,2,3], Bin Yu[1,2,4], Leiming Wu[1,2,3], Hongwu Wang[5], Zhiliang Wu[1,2,3], Ming Li[1,2,4], Pengyan Huang[1,2,3], Shengqiu Feng[1,2,3], Peng Chen[1,2,3], Yonglian Zheng[1,4], Liangcai Peng[1,2,3]*

1 National Key Laboratory of Crop Genetic Improvement and National Centre of Plant Gene Research (Wuhan), Huazhong Agricultural University, Wuhan, P.R. China, 2 Biomass and Bioenergy Research Centre, Huazhong Agricultural University, Wuhan, P.R. China, 3 College of Plant Science and Technology, Huazhong Agricultural University, Wuhan, P.R. China, 4 College of Life Science and Technology, Huazhong Agricultural University, Wuhan, P.R. China, 5 Institute of Crop Sciences, Chinese Academy of Agricultural Sciences, Beijing, P.R. China

Abstract

Corn is a major food crop with enormous biomass residues for biofuel production. Due to cell wall recalcitrance, it becomes essential to identify the key factors of lignocellulose on biomass saccharification. In this study, we examined total 40 corn accessions that displayed a diverse cell wall composition. Correlation analysis showed that cellulose and lignin levels negatively affected biomass digestibility after NaOH pretreatments at $p<0.05$ & 0.01, but hemicelluloses did not show any significant impact on hexoses yields. Comparative analysis of five standard pairs of corn samples indicated that cellulose and lignin should not be the major factors on biomass saccharification after pretreatments with NaOH and H_2SO_4 at three concentrations. Notably, despite that the non-KOH-extractable residues covered 12%–23% hemicelluloses and lignin of total biomass, their wall polymer features exhibited the predominant effects on biomass enzymatic hydrolysis including Ara substitution degree of xylan (reverse Xyl/Ara) and S/G ratio of lignin. Furthermore, the non-KOH-extractable polymer features could significantly affect lignocellulose crystallinity at $p<0.05$, leading to a high biomass digestibility. Hence, this study could suggest an optimal approach for genetic modification of plant cell walls in bioenergy corn.

Editor: Ying Xu, University of Georgia, United States of America

Funding: This work was supported in part by grants from the 111 Project of MOE (B08032), the National Natural Science Foundation of China (31200911), the China Postdoctoral Science Foundation (20100471197, 201104475), the Transgenic Plant and Animal Project of MOA (2009ZX08009-119B), the 973 Pre-project of MOST (2010CB134401), and HZAU Changjiang Scholar Promoting Project (52204-07022). The funders had no role in study design, data collection and analysis, decision to publish, or preparation of the manuscript.

Competing Interests: The authors have declared that no competing interests exist.

* Email: lpeng@mail.hzau.edu.cn

These authors contributed equally to this work.

Introduction

Lignocellulose has been regarded as a major biomass resource for biofuels and chemicals [1,2]. Traditional field crops constitute the bulk of lignocellulosic resources, and thus the application of such materials complements that of food supplies. Lignocellulosic biomass process involves three major steps: physical and chemical pretreatments to disrupt the cell wall; enzymatic hydrolysis to release soluble sugar; and yeast fermentation to produce ethanol. However, plant cell wall recalcitrance entails a costly biomass process to produce biofuels [3,4]. Principally, recalcitrance is characterized by cell wall composition and wall polymer features [5–8]. To reduce recalcitrance, genetic modification of plant cell walls has been proposed as a promising solution in bioenergy crops [9–12]. Hence, the key factors of plant cell walls affecting biomass enzymatic saccharification should be identified in various pretreatment conditions.

Plant cell walls are mainly composed of cellulose, hemicelluloses, and lignin. Cellulose is a crystalline linear polymer of β-(1,4)-linked glucose moieties, accounting for approximately 30% of the dry mass of primary cell walls and a maximum of 40% of secondary cell walls [13,14]. It has been characterized that cellulose crystallinity is the key factor that negatively affects biomass enzymatic digestions in plants [5,7,15].

Hemicelluloses are the polysaccharides accounting for approximately 20% to 35% of lignocellulosic biomass [16]. Hemicelluloses can be effectively extracted using different concentrations of alkali that dissociates the hydrogen bonds of wall polymers [17]. For example, 4 M KOH has been used to remove hemicelluloses and other associated wall polymers in plants [6,8,17]. Hemicelluloses have been considered as the positive factor affecting biomass digestibility in *Miscanthus* [5], but they are not the main factors on biomass digestibility in rice, wheat and sweet sorghum [8,33]. Although xylan is the major hemicellulose in grasses, the degree of arabinose (Ara) substitution is reported as the main factor positively affecting biomass enzymatic saccharification in plants [6,8,33].

Lignin is a very stable phenolic polymer composed of *p*-coumaryl alcohol (H), coniferyl alcohol (G), and sinapyl alcohol (S). Lignin has been considered as the major contributor to lignocellulosic recalcitrance because of its structural diversity and heterogeneity. However, recent reports have indicated that lignin could play dual roles in biomass enzymatic digestions due to the distinctive lignin compositions and monolignol ratios in different plant species [5,8,18,19,34].

As a highly photosynthetic-efficient C4 grass, corn is one of the major food crops with large amounts of lignocellulosic residues that can be used for biofuels [20]. Despite various pretreatment technologies have been applied in corn lignocellulose digestions [21–26], limited information is available regarding cell wall characteristics that affect biomass digestibility in corn. However, due to the complicated structures and diverse biological functions of plant cell walls, it becomes technically difficult to identify the main factors on biomass digestions. In the present study, we determined total 40 natural corn accessions that displayed a diverse cell wall composition. We then selected five standard pairs of corn samples that exhibited characteristic cell wall composition and features. Hence, the current study focused on identification of the main factors of the three major wall polymers that affect biomass enzymatic digestibility under various chemical pretreatments in corn stalk.

Materials and Methods

Plant materials

Total 539 corn accessions collected from China, American and International Maize and Wheat Improvement Center (CIMMYT) were planted in Jianshui, Yunnan, China in 2010. Among them, 40 accessions were used in this study. Five matured plants of each accession were harvested, and the stem tissues without leaves were dried at 50°C. The dried tissues were ground through a 40 mesh sieve and stored in a dry container until use. In addition, the field study was carried out on private exprimental land and did not involve endangered or protected species, and no specific permissions were required.

Plant cell wall fractionation

The plant cell wall fractionation method was described by Peng et al. [17] and Xu et al. [5] with minor modification. The well-mixed biomass powder samples were used for cell wall fractionation. The soluble sugar, lipids, starch and pectin of the samples were consecutively removed by potassium phosphate buffer (pH 7.0), chloroform-methanol (1:1, v/v), DMSO-water (9:1, v/v) and 0.5% (w/v) ammonium oxalate. The remaining pellet was extracted with 4 M KOH with 1.0 mg/mL sodium borohydride for 1 h at 25°C, and the combined supernatant with two parallels, one parallel was neutralized, dialyzed and lyophilized as KOH-extractable hemicelluloses monosaccharides; and one parallel was collected for determination of free pentoses and hexoses as the KOH-extractable hemicelluloses. The non-KOH-extractable residues were sequentially extracted with trifluoroacetic acid (TFA) at 120°C for 1 h as non-KOH-extractable hemicelluloses, and the remaining pellet was used as crystalline cellulose. All experiments were performed in independent triplicate.

Colorimetric assay of hexoses and pentoses

UV-VIS Spectrometer (V-1100D, Shanghai MAPADA Instruments Co., Ltd. Shanghai, China) was used for total hexoses and pentoses assay as described by Huang et al. [27] and Wu et al. [8]. Hexoses were detected using the anthrone/H_2SO_4 method [14] and pentoses were tested using the orcinol/HCl method [28]. Anthrone was purchased from Sigma-Aldrich Co. LLC., and ferric chloride and orcinol were obtained from Sinopharm Chemical Reagent Co., Ltd. The standard curves for hexoses and pentoses were drawn using D-glucose and D-xylose as standards (purchased from Sinopharm Chemical Reagent Co., Ltd.) respectively. Total sugar yield from pretreatment and enzymatic hydrolysis was subject to the sum total of hexoses and pentoses. Considering the high pentoses level can affect the absorbance reading at 620 nm for hexoses content by anthrone/H_2SO_4 method, the deduction from pentoses reading at 660 nm was carried out for final hexoses calculation. A series of xylose concentrations were analyzed for plotting the standard curve referred for the deduction, which was verified by gas chromatography-mass spectrometry (GC-MS) analysis. All experiments were carried out in triplicate.

GC-MS determination of hemicelluloses monosaccharides

Determination of monosaccharide composition of hemicelluloses by GC-MS was described by Li et al. [6] and Wu et al. [8] with minor modification. The combined supernatants from 4 M KOH fraction were dialyzed for 36 h after neutralization with acetic acid. The sample from the dialyzed KOH-extractable supernatant or the non-KOH-extractable residue was hydrolyzed by 2 M TFA for free monosaccharide release in a sealed tube at 121°C in an autoclave for 1 h. *Myo*-inositol (200 μg) was added as the internal standard for GC-MS (SHIMADZU GCMS-QP2010 Plus) analysis.

GC-MS analytical conditions: Restek Rxi-5 ms, 30 m× 0.25 mm ID×0.25 μm df column; carrier gas: helium; injection method: split; injection port: 250°C; interface: 250°C; injection volume: 1.0 μL; the temperature program: from 155°C (held for 23 min) to 200°C (held for 5 min) at 3.8°C/min, and then from 200°C to 300°C (held for 2 min) at 20°C/min; ion source temperature: 200°C; ACQ Mode: SIM. The mass spectrometer was operated in the EI mode with ionization energy of 70 ev. Mass spectra were acquired with full scans based on the temperature program from 50 to 500 m/z in 0.45 s. Calibration curves of all analytes routinely yielded correlation coefficients of 0.999 or better.

Total lignin measurement and high performance liquid chromatography (HPLC) detection of lignin monomers

Total lignin was determined by two-step acid hydrolysis method according to Laboratory Analytical Procedure of the National Renewable Energy Laboratory [29], as described by Wu et al. [8]. All samples were carried out in triplicate.

Lignin monomer determination was described by Wu et al. [8]. Standard chemicals: *p*-Hydroxybenzaldehyde (H), vanillin (G) and syringaldehyde (S) were purchased from Sinopharm Chemical Reagent Co., Ltd. The sample was extracted with benzene-ethanol (2:1, v/v) in a Soxhlet for 4 h, and the remaining pellet was collected as cell wall residue (CWR). The procedure of nitrobenzene oxidation of lignin was conducted as follows; 0.05 g CWR was added with 5 mL 2 M NaOH and 0.5 mL nitrobenzene, and a stir bar was put into a 25 mL Teflon gasket in a stainless steel bomb. The bomb was sealed tightly and heated at 170°C (oil bath) for 3.5 h and stirred at 20 rpm. Then, the bomb was cooled with

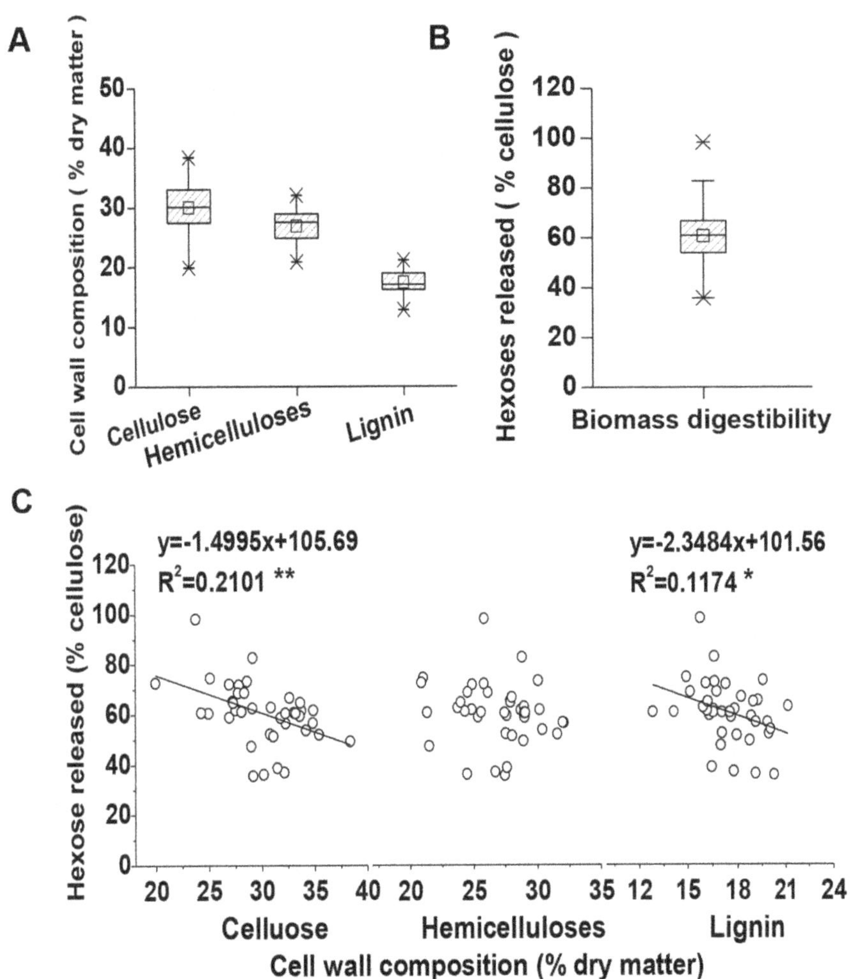

Figure 1. Effects of major wall polymer levels on biomass digestibility in total 40 representative corn accessions. (**A**) Variations of three major wall polymers (cellulose, hemicelluloses and lignin); (**B**) Diversity of hexoses yields from enzymatic hydrolysis after 1% NaOH pretreatment; (**C**) Correlations between wall polymers and hexoses yields after 1% NaOH pretreatment. * and ** Indicated as significant correlations at $p<0.05$ and 0.01 (n = 40), respectively.

cold water. The chromatographic internal standard (ethyl vanillin) was added to the oxidation mixture. This alkaline oxidation mixture was washed 3 times with 30 mL CH_2Cl_2/ethyl acetate mixture (1:1, v/v) to remove nitrobenzene and its reduction by-products. The alkaline solution was acidified to pH 3.0–4.0 with 6 M HCl, and then extracted with CH_2Cl_2/ethyl acetate (3×30 mL) to obtain the lignin oxidation products which were in the organic phase. The organic extracts were evaporated to dryness under reduced pressure at 40°C. The oxidation products were dissolved in 10 mL chromatographic pure methanol.

For HPLC analysis the solution was filtered with a membrane filter (0.22 μm). Then 20 μL solution was injected into the HPLC (Waters 1525) column Kromat Universil C18 (4.6 mm×250 mm, 5 μm) operating at 28°C with CH_3OH:H_2O:HAc (16:63:1, v/v/v) carrier liquid (flow rate: 1.1 mL/min). Calibration curves routinely yielded correlation coefficients 0.999 or better, and the components were detected with a UV-detector at 280 nm.

Measurement of lignocellulose crystallinity

The X-ray diffraction (XRD) method was applied for detection of cellulose crystallinity index (CrI) as described by Zhang et al. [7] and Wu et al. [8]. The biomass samples were examined by means of wide-angle X-ray diffraction on a Rigaku-D/MAX instrument (Uitima III, Japan) with 0.0197°/s from 10° to 45°. The crystallinity index (CrI) was estimated using the intensity of the 200 peak (I_{200}, $\theta = 22.5°$) and intensity at the minimum between the 200 and 110 peaks (I_{am}, $\theta = 18.5°$), based on the equation: $CrI = 100 \times (I_{200}-I_{am})/I_{200}$. I_{200} represents both crystalline and amorphous materials while I_{am} represents amorphous material. The standard error of the CrI method was detected at ±0.05 to approximately 0.15 using five representative samples in triplicates.

Scanning electron microscopy (SEM) observations

Scanning electron microscopy (SEM) was used to examine the biomass residue, as described by Li et al. [6]. The well-mixed biomass powder samples were pretreated with 1% NaOH or 1% H_2SO_4, and hydrolyzed with the mixed-cellulases. Then, the lignocellulose samples were rinsed with distilled water until the pH was 7.0, dried under air, and sputter-coated with gold in a JFC-1600 ion sputter (Mito City, Japan). The surface morphology of the treated samples was observed by SEM (JSM-6390/LV, Hitachi, Tokyo, Japan), and the representative imagines of each sample were photographed from 5–10 views.

Table 1. Cell wall composition (% dry matter) of biomass residues in typical pairs of corn samples.

Pair	Sample	Cell wall composition (% dry matter)					
		Cellulose		Hemicelluloses		Lignin	
I-1	Zm23(H)[a]	**24.94±0.43****	**-26.02%**[b]	27.46±0.43	-0.59%	16.52±0.33	0.50%
	Zm15(L)	**31.43±0.55**		27.62±0.30		16.44±1.00	
I-2	Zm01(H)	30.74±0.84	5.57%	27.54±0.55	0.27%	**17.05±0.46****	**-18.72%**
	Zm10(L)	29.12±0.29		27.46±0.20		**20.25±0.14**	
I-3	Zm27(E1)	24.26±0.31	-2.82%	**21.32±0.56****	**-28.80%**	16.42±0.85	-0.65%
	Zm23(E2)	24.94±0.43		**27.46±0.43**		16.52±0.33	
II-1	Zm18(H)	30.82±0.74	5.84%	28.96±0.69	5.46%	21.16±0.60	4.49%
	Zm10(L)	29.12±0.29		27.46±0.50		20.25±0.54	
II-2	Zm40(H)	31.05±0.59	-3.38%	28.03±0.45*	4.90%	17.93±0.46	0.96%
	Zm03(L)	32.10±0.61		26.72±0.07		17.76±0.29	

* and ** Indicated significant difference between the two samples of each pair by *t*-test at $p<0.05$ and 0.01, respectively ($n = 3$).
[a] Sample in the pair with high (H) or low (L) or equal (E) biomass digestibility.
[b] Percentage of the increased or decreased level between the two samples of each pair: subtraction of two samples divided by low value.

Chemical pretreatment of biomass samples

The biomass pretreatments were performed as previously described by Huang et al. [27], Li et al. [6] and Wu et al. [8] with minor modifications. All samples were carried out in triplicate. H_2SO_4 pretreatment: the well-mixed powder of biomass sample (0.3 g) was added with 6 mL H_2SO_4 at three concentrations (0.25%, 1%, 4%, v/v). The tube was sealed and heated at 121°C for 20 min in an autoclave (15 psi) after sample was mixed well. Then, the tube was shaken at 150 rpm for 2 h at 50°C, and centrifuged at 3,000 *g* for 5 min. The pellet was washed three times with 10 mL distilled water, and stored at −20°C for enzymatic hydrolysis. All supernatants were collected for determination of total sugars (pentoses and hexoses) released from acid pretreatment, and samples with 6 mL distilled water were shaken for 2 h at 50°C as the control.

NaOH pretreatment: the well-mixed powder of biomass sample (0.3 g) was added with 6 mL NaOH at three concentrations (0.5%, 1%, 4%, w/v). The tube was shaken at 150 rpm for 2 h at 50°C, and centrifuged at 3,000 *g* for 5 min. The pellet was washed three times with 10 mL distilled water, and stored at −20°C for enzymatic hydrolysis. All supernatants were collected for determination of total sugars released from alkali pretreatment, and samples with 6 mL distilled water were shaken for 2 h at 50°C as the control.

Enzymatic hydrolysis of lignocellulose residues

The lignocellulose residues obtained from various pretreatments were washed 2 times with 10 mL distilled water, and once with 10 mL mixed-cellulase reaction buffer (0.2 M acetic acid-sodium acetate, pH 4.8). The washed residues were added with 6 mL (1.6 g/L) of mixed-cellulases containing β-glucanase ($\geq 2.98\times10^4$ U), cellulase (≥298 U) and xylanase ($\geq 4.8\times10^4$ U) from Imperial Jade Bio-technology Co., Ltd). During the enzymatic hydrolysis, the samples were shaken under 150 rpm at 50°C for 48 h. After centrifugation at 3,000 *g* for 10 min, the supernatants were collected for determining amounts of pentoses and hexoses released from enzymatic hydrolysis. The samples with 6 mL reaction buffer were shaken for 48 h at 50°C as the control. All samples were carried out in triplicate.

Statistical calculation of correlation coefficients

The statistical software (SPSS 17.0) was applied for any statistical analysis. Correlation coefficient values were calculated by performing Spearman rank correlation analysis for all pairs of the measured aspects (or traits, factors) across the whole populations. The measured aspects were derived from the average values of duplications. The box plot, histogram and line graph presented in the study were generated by using software (Origin 8.0).

Results and Discussion

Effects of major wall polymers on biomass digestibility in corn

Corn is a typical C4 food crop with large amounts of lignocellulose residues. Although corn stove has been applied in biofuels, little information is available about the characteristics of wall polymer involved in biomass process [8,12]. In the current study, a total of 40 corn samples were selected from hundreds of natural corn accessions collected worldwide, including various ecological types and genetic germplasms (Fig. 1). In general, the selected corn samples exhibited a diverse cell wall composition (Fig. 1A, Table S1). For instance, the cellulose contents vary from 19.94% to 38.35% (% dry matter), hemicelluloses from 20.89% to

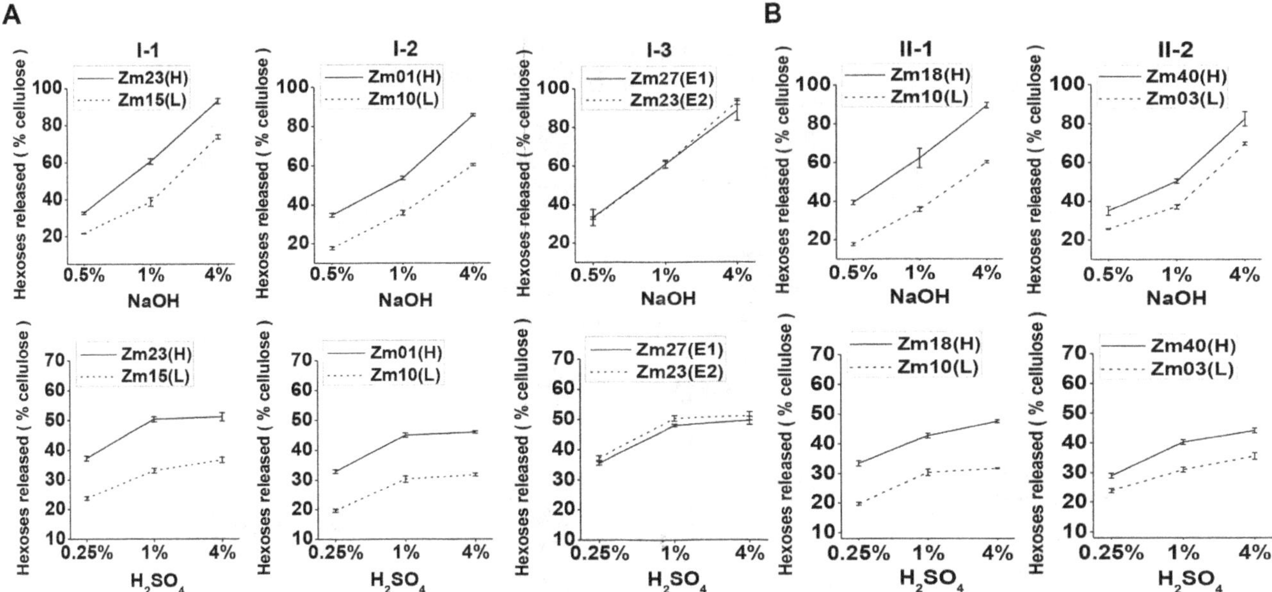

Figure 2. Hexoses yields released from enzymatic hydrolysis after NaOH and H_2SO_4 pretreatments in corn samples. (A) Pairs I-1, I-2 and I-3 samples; **(B)** Pairs II-1 and II-2 samples. Bar indicated as ± SD (n = 3).

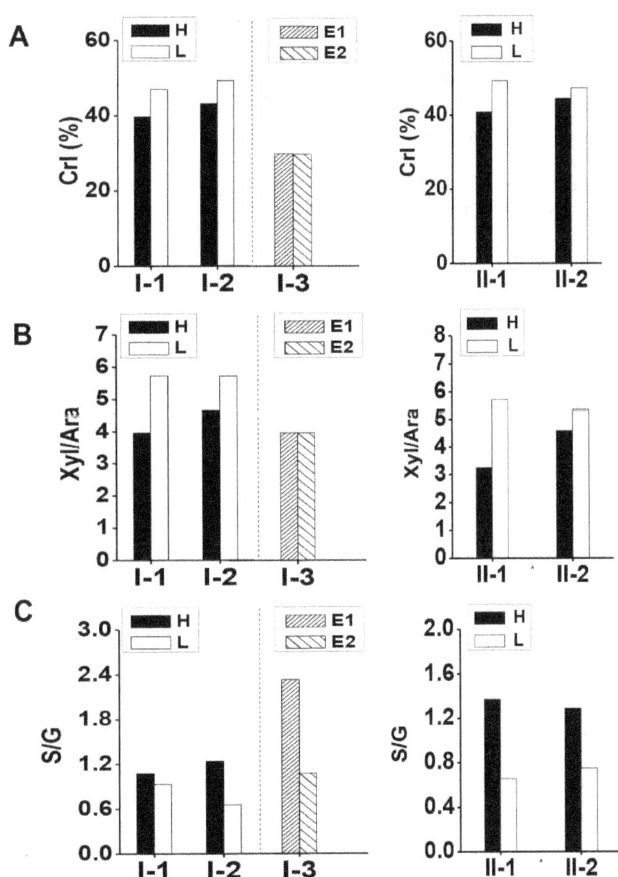

Figure 3. Detection of cell wall features in typical five pairs of corn samples. (A) Lignocellulosic CrI of raw material. **(B)** Xyl/Ara ratio of the non-KOH-extractable hemicelluloses; **(C)** S/G ratio of the non-KOH-extractable lignin. H/L/E Indicated as relatively high/low/equal biomass digestibility at pair.

32.04%, and lignin from 12.83% to 21.16%. Hence, the corn samples showed a relatively low average lignin level compared with *Miscanthus* [5] and wheat [8].

Biomass enzymatic digestibility (saccharification) has been defined by measuring the hexoses yields (% cellulose) released by hydrolysis of lignocellulose using crude cellulase mixture after the samples were exposed to various pretreatment conditions [5,6,8]. The hexoses yields of the 40 corn samples were measured after they were pretreated with 1% NaOH (Fig. 1B, Table S1). As a result, the corn samples displayed diverse biomass digestibility with hexoses yields ranging from 35.76% to 98.16%, and almost half of the samples exhibited high hexoses yields up to 60%. Hence, the 40 corn samples were suitable for the experiments investigating the effects of wall polymers on biomass enzymatic digestibility.

Correlation analysis has been extensively performed to determine the effects of wall polymer on biomass saccharification in plants [7,8,27]. In this study, the correlations were analyzed between the three major wall polymer levels and hexoses yields after the 40 corn samples were pretreated with 1% NaOH (Fig. 1C). As a result, cellulose and lignin showed a significantly negative correlation with hexoses yields at $p<0.05$ and 0.01 levels. By comparison, hemicelluloses were not correlated with hexoses yields ($p>0.05$), different from the previous observations in *Miscanthus* [5], rice, and wheat [8], but similar to the finding in sweet sorghum [33]. Hence, the hemicelluloses level is not the main factor affecting biomass enzymatic saccharification in corn.

Analysis of biomass digestions in five typical pairs of corn samples

To test the effects of major wall polymer levels on biomass digestibility, we selected five standard pairs of corn samples (Table 1), and compared their hexoses yields released from enzymatic hydrolysis after various chemical pretreatments (Fig. 2). Each of the three pairs (I-1, I-2, and I-3) displayed significant changes in single wall polymers (cellulose, lignin, and hemicelluloses; $p<0.01$) by 26.02%, 18.72%, and 28.80%, respectively. The two other major wall polymers of each pair

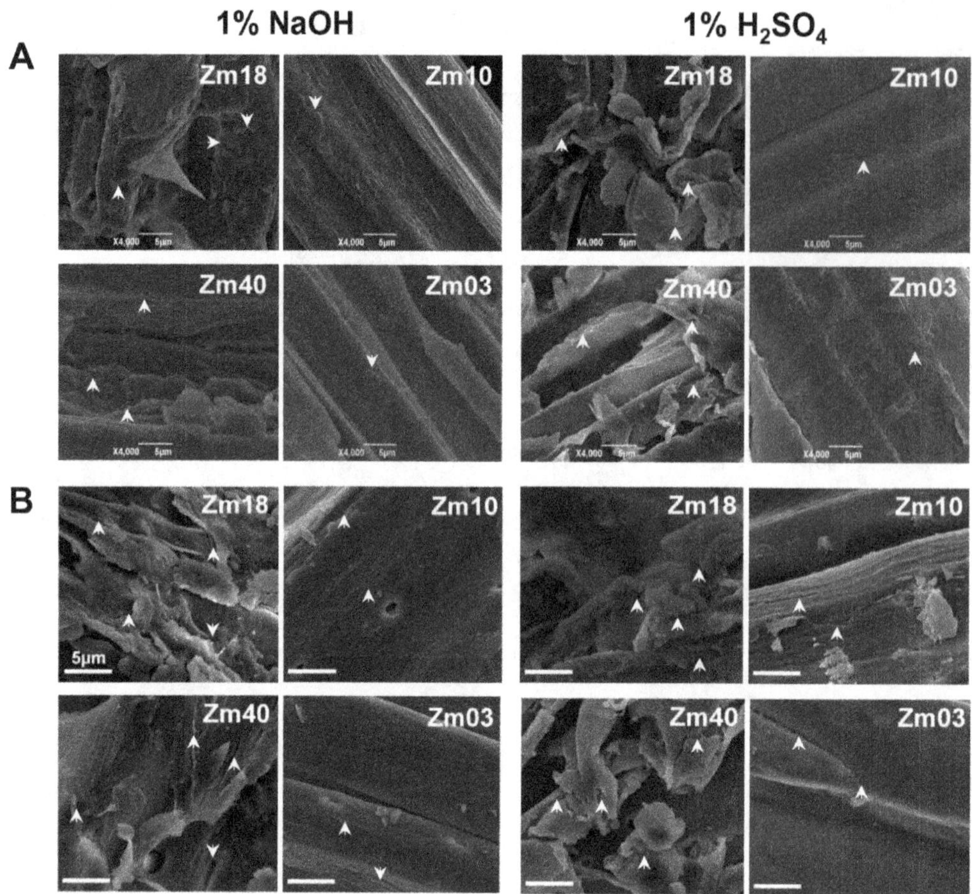

Figure 4. Scanning electron microscopic observation of biomass residues in pairs II-1 and II-2 corn samples. (**A**) Biomass residues released from pretreatments with 1% NaOH and 1% H_2SO_4; (**B**) Biomass residues released from enzymatic hydrolysis after 1% NaOH and 1% H_2SO_4 pretreatments. Arrow indicated the rough face.

did not show significant differences less than 6% (Table 1). Pretreated with three different concentrations (0.25% or 0.5%, 1%, and 4%) of NaOH and H_2SO_4, the Zm23 and Zm01 samples of pairs I-1 and I-2 with less cellulose and lignin contents (Table 1), respectively exhibited significantly 1.6- and 1.9-fold higher hexoses yields than those of their paired samples (Zm15 and Zm10; $p<0.01$; Fig. 2A, Table S2). The Zm27 sample of pair I-3 contained 28.80% less hemicelluloses showed the similar hexoses yields to its paired Zm23 sample from NaOH pretreatments or slightly lower hexoses yields from H_2SO_4 pretreatments. Hence, the cellulose and lignin levels exhibited significantly negative effects on hexoses yields from various chemical pretreatments, whereas the hemicelluloses did not affect biomass enzymatic digestibility in corn.

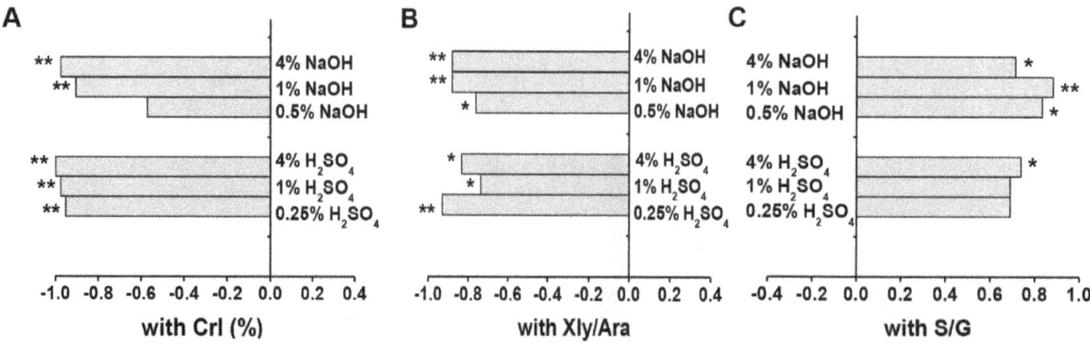

Figure 5. Correlation analysis between wall polymer features and hexoses yields from enzymatic hydrolysis after pretreatments. (**A**) Correlative coefficients between lignocellulose CrI (%) of raw material and the hexoses yields (% cellulose) released from enzymatic hydrolysis after NaOH and H_2SO_4 pretreatments at three concentrations; (**B**) Coefficients with Xyl/Ara of the non-KOH-extractable hemicelluloses; (**C**) Coefficients with S/G of the non-KOH-extractable lignin. * and ** Indicated as significant correlations at $p<0.05$ and 0.01 levels (n = 8), respectively.

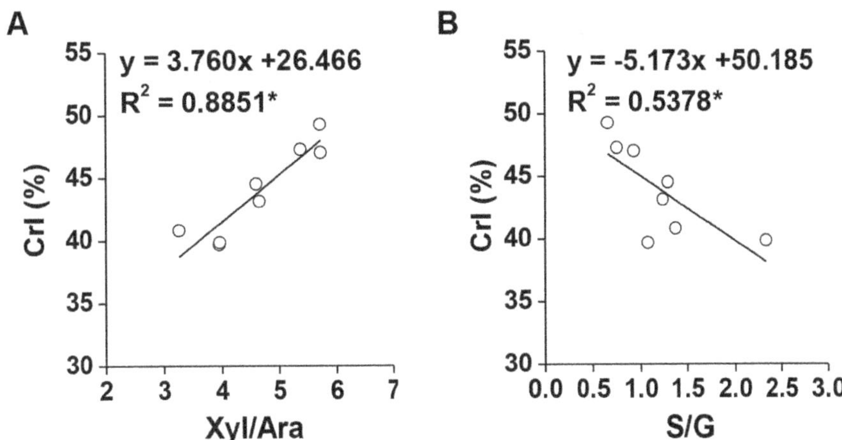

Figure 6. Correlation analysis between lignocellulose CrI and the non-KOH-extractable wall polymer features. (**A**) Xyl/Ara of non-KOH-extractable hemicelluloses; (**B**) S/G of the non-KOH-extractable lignin.*Indicated as significant correlations at $p<0.05$ levels (n = 8).

Furthermore, we detected other two typical pairs of corn sample (II-1, II-2) that each displayed a similar cell wall composition with wall polymer alterations by less than 6% (Table 1). Although the three major wall polymer levels of paired samples were slightly different, both pairs displayed remarkable changes in hexoses yields after these samples were pretreated with NaOH and H_2SO_4 (Fig. 2B, Table S2). In particular, pair II-1 exhibited the highest increase in hexoses yields, reaching a maximum of 2.2-fold increase after pretreated with 0.5% NaOH, which was even higher than that of Pair I-2 (1.9-fold increase). Thus, the data indicated that the wall polymer content should not be the major factor on biomass enzymatic digestibility, but the wall polymer features may play a dominant role as reported in wheat and rice species [8].

Detection of wall polymer features in five pairs of samples

The crystalline index (CrI) of lignocellulose has been used to determine cellulose crystallinity in plants [5,7,8]. In the present study, the CrI of lignocellulose in the five standard pairs of corn samples was examined (Fig. 3A, Table S3). Corn samples with relatively high biomass digestibility (Fig. 2) in the four sample pairs (I-1, I-2, II-1, and II-2) exhibited lower CrI values than that of their paired samples (Fig. 3A). The reduced CrI ratios ranged from 6% to 17% in the four pairs of corn samples (Table S3). By comparison, pair I-3 samples did not show different CrI values, consistent with their similar biomass digestibility (Fig. 2). Hence, the results indicated that lignocellulose CrI was also a key factor on biomass enzymatic saccahrification in corn.

With regard to the hemicelluloses feature, we determined monosaccharide composition of hemicelluloses in the five standard pairs of corn samples (Table S4). Similar to *Miscanthus*, rice, wheat and sweet sorghum [5,8,33], all corn samples contained two major pentoses: xylose (Xyl) and arabinose (Ara), indicating that xylan was one of major hemicelluloses. Significant amount glucose was also found in hemicelluloses, suggesting that a rich β-1,3; 1,4-glucans could also be present in corn [16,30]. Considering that the substitution degree of Ara in xylan has been reported as a major factor affecting biomass enzymatic digestibility in grasses [6,8], we calculated Xyl/Ara values as the reverse indicator of the substitution degree of Ara in xylan in two types of hemicelluloses (KOH-extractable and non-KOH-extractable). However, only the non-KOH-extractable hemicelluloses, neither the KOH-extractable nor total hemicelluloses, displayed consistently reduced Xyl/Ara ratios in the samples of four pairs (Pairs I-1, I-2, II-1, II-2) with relatively higher hexoses yields (Table S4 and Fig. 3B). By comparison, pair I-3 samples with similar hexoses yields (Fig. 2), showed close Xyl/Ara values in the non-KOH-extractable hemicelluloses. Hence, the substitution degree (reverse Xyl/Ara) of Ara in the non-KOH-extractable xylan was a positive factor affecting biomass saccharification in corn. However, corn is distinct from *Miscanthus* that displays the positive effects of the substitution degrees of Ara in both KOH-extractable and non-KOH-extractable xylans [6].

Lignin has been recently characterized with dual effects on biomass enzymatic digestibility due to monolignin constitution distinctive in different plant species [5,8,18,19,34]. In the present study, three monolignin ratios (S/G, H/G, and S/H) were calculated from KOH-extractable and non-KOH-extractable lignins in the five pairs of corn samples (Table S5). The biomass samples with relatively high hexoses yields exhibited the higher S/G ratios than that of their paired samples in the non-KOH-extractable lignin, rather than KOH-extractable or total lignin (Fig. 3C, Table S5). This result suggested that the S/G ratios in the non-KOH-extractable lignin could be applied as the positive indicators on biomass saccharification in corn. However, this

Table 2. Correlative coefficients between lignocellulose CrI and two wall polymer features (Xyl/Ara, S/G).

	Xyl/Ara			S/G		
CrI (%)	KOH-extractable	Non-KOH-extractable	Total	KOH-extractable	Non-KOH-extractable	Total
	0.286	**0.833***	0.595	−0.429	**0.738***	−0.429

* Indicated significant difference at $p<0.05$ (n = 8).

Table 3. Proportions of two types of hemicelluloses and lignin in the typical corn samples.

	KOH-extractable	Non-KOH-extractable	Total
Hemicelluloses	1341.4±122.6[a]	409.0±45.2	1750.4±140.1
	(1051.71496.5)[b]	(340.8477.4)	(1392.51874.1)
	76.60%[c]	23.40%	100%
Lignin	1064.4±100.3	144.7±45.5	1209.1±130.3
	(956.21273.5)	(86.4238.3)	(1093.71444.3)
	88.00%	12.00%	100%

[a] Mean value ± SD (n = 8);
[b] Minimum and maximum values;
[c] Percentage of total polymer.

result is contrary to that in a previous study, in which the S/G of *Miscanthus* was a negative indicator on biomass saccharification [5]. In addition, the corn was different from the rice and wheat that displayed a positive impact of H/G in the KOH-extractable lignin, rather than the non-KOH-extractable residue [8]. Although pair I-3 exhibited similar hexoses yields (Fig. 2A), two samples showed a different S/G in the non-KOH-extractable lignin (Fig. 3C).

Observation of biomass residues from pretreatment and enzymatic hydrolysis

Biomass residue surfaces in pairs II-1 and II-2 samples were observed under scanning electron microscopy (Fig. 4). After pretreated with 1% NaOH and 1% H_2SO_4, the Zm18 and Zm40 samples of pairs II-1 and II-2 with relatively high hexoses yields displayed coarse biomass residue surfaces, whereas their paired samples (Zm10 and Zm03) showed smooth surfaces (Fig. 4A). In sequential enzymatic hydrolysis, both Zm18 and Zm40 samples exhibited rougher surfaces than the pretreated samples (Fig. 4B); this result is similar to that in previous studies on *Miscanthus*, rice, and wheat [5,7,8]. Hence, the rough surface of the biomass residue could indicate a relatively effective biomass enzymatic hydrolysis. It could also suggest that biomass residue surface was mainly affected by the characteristics of wall polymers because pairs II-1 and II-2 showed similar cell wall compositions.

Correlation among wall polymer features and biomass saccharification

To confirm the predominant effects of wall polymer features on biomass digestibility, we further performed a correlation analysis by using the five pairs of corn samples (Fig. 5). A negative correlation was found between lignocellulose CrI and hexoses yields released from enzymatic hydrolyses after the samples were pretreated with NaOH and H_2SO_4 except 0.5% NaOH pretreatment ($p<0.01$ or $p<0.05$; Fig. 5A, Table S6). A similar result was not observed in the samples pretreated with 0.5% NaOH possibly because of relatively low hexose yields that caused small variations among the five pairs of corn samples. However, our results showed that the CrI of lignocellulose negatively affected biomass enzymatic saccharification in corn.

Furthermore, correlations were calculated between hexoses yields and Xyl/Ara ratios in the two types of hemicelluloses (Fig. 5B, Table S7). Only non-KOH-extractable hemicelluloses showed significantly negative correlations ($p<0.01$ or $p<0.05$) compared with KOH-extractable or total hemicelluloses. This result indicated that the non-KOH-extractable hemicelluloses predominantly affected biomass enzymatic digestion in corn. A positive correlation was also found between hexoses yields and S/G ratios in the non-KOH-extractable lignin, compared with KOH-extractable or total lignin (Fig. 5C, Table S8). Although the two pretreatment conditions (0.25% and 1% H_2SO_4) did not show significant correlations, correlation coefficients remained high (Table S8). Therefore, the non-KOH-extractable lignocellulosic characteristics could predominantly affect biomass enzymatic digestibility in corn. Considering that the S/G of the total lignin negatively affects biomass enzymatic digestibility in *Miscanthus* and the H/G of KOH-extractable lignin positively affects biomass enzymatic digestibility in wheat and rice [6,8], we found that corn was different from wheat and rice because its non-KOH-extractable S/G elicited a positive effect on biomass enzymatic hydrolysis.

Mechanism on the wall polymer features that affect biomass enzymatic digestion

Lignocellulose crystallinity is the key factor that negatively affects biomass enzymatic digestibility in plants [5–8,31,32]. In general, lignocellulose crystallinity is distinctively affected by two major wall polymer (hemicelluloses and lignin) characteristics [5–8]. To understand the predominant effects of wall polymer characteristics on biomass enzymatic digestion in corn, we further performed a correlation analysis between lignocellulose CrI and two major wall polymer features (Xyl/Ara, S/G; Fig. 6). Our results showed that Xyl/Ara in non-KOH-extractable hemicelluloses exhibited a significantly positive correlation with lignocellulose CrI ($p<0.05$; $R^2 = 0.885$; Fig. 6A, Table 2). Hence, the branched Ara of the non-KOH-extractable xylan may interact with β-1,4-glucan chains via hydrogen bonds that reduce the lignocellulose crystallinity for high biomass digestibility, as observed in other grasses [6–8]. By comparison, the non-KOH-extractable S/G was negatively correlated ($p<0.05$; $R^2 = 0.538$; Fig. 6B, Table 2). This result suggested that G-monomer may be associated with β-1,4-glucan chains or S-monomer may interact with the Ara of xylan that indirectly reduces lignocellulose crystallinity. This result could also explain the different observation in pair I-3 regarding S/G, which did not affect hexoses yields, because the non-KOH-extractable S- and G-monomers may not affect lignocellulose crystallinity. However, the data confirmed that the non-KOH-extractable wall polymer features could play a predominant role in biomass enzymatic hydrolysis in corn.

Potential cell wall modification for high biomass digestibility

Corn is the typical C4 food crop with enormous biomass residues for biofuels. However, desirable cell walls for high biomass digestibility can not be identified easily [1], because plant biomass is composed of many different cell types with diverse wall components. Furthermore, any genetic modification of plant cell walls could consequently lead to plant growth defect and mechanical strength reduction because of the cell wall is involved in diverse biological functions [12]. In the present study, we have screened out natural corn varieties with normal plant growth and high biomass digestibility. More importantly, we have found that the non-KOH-extractable Xyl/Ara predominately affect biomass enzymatic saccharification under various chemical pretreatments. Since the non-KOH-extractable biomass residue respectively covers 23.4% of total hemicelluloses and 12.0% of total lignin (Table 3), its genetic modification should cause less defects on plant growth and development than that of the KOH-extractable biomass. Despite that lignocellulose crystallinity is the key negative factor on biomass digestibility, it could be reduced by enhancing Ara substitution degree (reverse Xyl/Ara) in the non-KOH-extractable biomass. Hence, the current study indicated that mild cell wall modifications for enhancing biomass enzymatic saccharification could be performed by expressing genes involved in branched Ara biosynthesis or by partially silencing genes associated with xylan backbone synthesis in corn.

Conclusion

Correlative analysis of 40 representative corn germplasm accessions and comparative analysis of the five standard pairs of corn samples have demonstrated that either the substitution degree of Ara in xylan or the S/G of lignin in non-KOH-extractable biomass could positively affect biomass enzymatic digestibility under various chemical pretreatments by negatively affecting lignocellulose crystallinity. The results have provided the potential approaches that could be performed to modify plant cell wall for high biofuel production in corn.

Supporting Information

Table S1 Variations of wall polymers and biomass digestibility in total 40 corn accessions.

Table S2 Hexoses yields (% cellulose) released from enzymatic hydrolysis after NaOH and H_2SO_4 pretreatments in five typical pairs of corn samples.

Table S3 Lignocellulose crystaline index (CrI) of raw materials in the five typical pairs of corn samples.

Table S4 Monosaccharide composition of hemicelluloses.

Table S5 Monomer composition of lignin.

Table S6 Correlation coefficients between lignocellulose CrI values and hexoses yields from enzymatic hydrolysis after various chemical pretreatments in the typical corn samples.

Table S7 Correlation coefficients between hemicellulosic Xyl/Ara ratios and hexoses yields from enzymatic hydrolysis after various chemical pretreatments in the typical corn samples.

Table S8 Correlation coefficients between monolignin ratios and hexoses yields from enzymatic hydrolysis after various chemical pretreatments in the typical corn samples.

Acknowledgments

We thank all members of the Biomass and Bioenergy Research Center of Huazhong Agricultural University (HZAU) for reading this manuscript, and the staff of Microscopy Center of HZAU for assisting scanning electron microscopic observation.

Author Contributions

Conceived and designed the experiments: LP. Performed the experiments: JJ BY LW ZW ML PH. Analyzed the data: JJ BY. Contributed reagents/materials/analysis tools: HW SF PC YZ. Contributed to the writing of the manuscript: LP.

References

1. Pauly M, Keegstra K (2010) Plant cell wall polymers as precursors for biofuels. Curr Opin Plant Biol 13: 305–312.
2. Chen P, Peng LC (2013) The diversity of lignocellulosic biomass resources and their evaluation for use as biofuels and chemicals. In: Sun JZ, Ding SY, Peterson JD, editors. Biological Concerstion of Biomass for Fuels and Chemicals: Exploration from Natural Biomass Utilization Systems. Oxfordshire: Royal Society of Chemistry. pp. 83–109.
3. Himmel ME, Ding SY, Johnson DK, Adney WS, Nimlos MR, et al. (2007) Biomass recalcitrance: Engineering plants and enzymes for biofuels production. Science 315: 804–807.
4. Ragauskas AJ, Williams CK, Davison BH, Britovsek G, Cairney J, et al. (2006) The path forward for biofuels and biomaterials. Science 311: 484–489.
5. Xu N, Zhang W, Ren SF, Liu F, Zhao CQ, et al. (2012) Hemicelluloses negatively affect lignocellulose crystallinity for high biomass digestibility under NaOH and H_2SO_4 pretreatments in *Miscanthus*. Biotechnol Biofuels 5: 58.
6. Li FC, Ren SF, Zhang W, Xu ZD, Xie GS, et al. (2013) Arabinose substitution degree in xylan positively affects lignocellulose enzymatic digestibility after various NaOH/H_2SO_4 pretreatments in *Miscanthus*. Bioresour Technol 130: 629–637.
7. Zhang W, Yi ZL, Huang JF, Li FC, Hao B, et al. (2013) Three lignocellulose features that distinctively affect biomass enzymatic digestibility under NaOH and H_2SO_4 pretreatments in *Miscanthus*. Bioresour Technol 130: 30–37.
8. Wu ZL, Zhang ML, Wang LQ, Tu YY, Zhang J, et al. (2013) Biomass digestibility is predominantly affected by three factors of wall polymer features distinctive in wheat accessions and rice mutants. Biotechnol Biofuels 6: 183.
9. Gressel J (2008) Transgenics are imperative for biofuel crops. Plant Science 174: 246–263.
10. McCann MC, Carpita NC (2008) Designing the deconstruction of plant cell walls. Curr. Opin Plant Biol 11: 314–320.
11. Rubin EM (2008) Genomics of cellulosic biofuels. Nature 454: 841–845.
12. Xie GS, Peng LC (2011) Genetic Engineering of Energy Crops: A Strategy for Biofuel Production in ChinaFree Access. J Integr Plant Biol 53: 143–150.
13. Arioli T, Peng LC, Betzner AS, Burn J, Wittke W, et al. (1998) Molecular analysis of cellulose biosynthesis in *Arabidopsis*. Science 279: 717–720.
14. Fry SC (1988) The growing plant cell wall: chemical and metabolic analysis. London: Blackburn Press.
15. Zhu L, O'Dwyer JP, Chang VS, Granda CB, Holtzapple MT (2008) Structural features affecting biomass enzymatic digestibility. Bioresour Technol 99: 3817–3828.
16. Saha BC (2003) Hemicellulose bioconversion. J Ind Microbiol Biotechnol 30: 279–291.
17. Peng LC, Hocart CH, Redmond JW, Williamson RE (2000) Fractionation of carbohydrates in *Arabidopsis* root cell walls shows that three radial swelling loci are specifically involved in cellulose production. Planta 211: 406–414.

18. Davison BH, Drescher SR, Tuskan GA, Davis MF, Nghiem NP (2006) Variation of S/G ratio and lignin content in a Populus family influences the release of xylose by dilute acid. hydrolysis. Appl Biochem Biotechnol 129–132: 427–435.
19. Studer MH, DeMartini JD, Davis MF, Sykes RW, Davison B, et al. (2011) Lignin content in natural Populus variants affects sugar release. Proc Natl Acad Sci U S A 108: 6300–6305.
20. Carpita NC, McCann MC (2008) Maize and sorghum: genetic resources for bioenergy grasses. Trends Plant Sci 13: 415–420.
21. Kim TH, Lee YY (2007) Pretreatment of corn stover by soaking in aqueous ammonia at moderate temperatures. Appl Biochem Biotechnol 137: 81–92.
22. Lau MW, Dale BE (2009) Cellulosic ethanol production from AFEX-treated corn stover using *Saccharomyces cerevisiae* 424A (LNH-ST). Proc Natl Acad Sci U S A 106: 1368–1373.
23. Lei HW, Ren SJ, Julson J (2009) The Effects of Reaction Temperature and Time and Particle Size of Corn Stover on Microwave Pyrolysis. Energy Fuels 23: 3254–3261.
24. Zhu ZG, Sathitsuksanoh N, Vinzant T, Schell DJ, McMillan JD, et al. (2009) Comparative Study of Corn Stover Pretreated by Dilute Acid and Cellulose Solvent-Based Lignocellulose Fractionation: Enzymatic Hydrolysis, Supramolecular Structure, and Substrate Accessibility. Biotechnol Bioeng 103: 715–724.
25. Sun YS, Lu XB, Zhang R, Wang XY, Zhang ST (2011) Pretreatment of Corn Stover Silage with $Fe(NO_3)_3$ for Fermentable Sugar Production. Appl Biochem Biotechnol 164: 918–928.
26. Wang H, Srinivasan R, Yu F, Steele P, Li Q, et al. (2012) Effect of Acid, Steam Explosion, and Size Reduction Pretreatments on Bio-oil Production from Sweetgum, Switchgrass, and Corn Stover. Appl Biochem Biotechnol 167: 285–297.
27. Huang JF, Xia T, Li A, Yu B, Li Q, et al. (2012) A rapid and consistent near infrared spectroscopic assay for biomass enzymatic digestibility upon various physical and chemical pretreatments in Miscanthus. Bioresour Technol 121: 274–281.
28. Dische Z (1962) Color reactions of carbohydrates. In: Whistler RL, Wolfrom ML, editors. Methods in carbohydrate chemistry. New York: Academic Press. pp. 477–512.
29. Sluiter A, Hames B, Ruiz R, Scarlata C, Sluiter J, et al. (2008) Determination of structural carbohydrates and lignin in biomass. Tech. Rep. NREL/TP-510-42618, NREL, Golden, Co.
30. Scheller HV, Ulvskov P (2010) Hemicelluloses. Annu Rev Plant Biol 61: 263–289.
31. Puri VP (1984) Effect of crystallinity and degree of polymerization of cellulose on enzymatic saccharification. Biotechnol Bioeng 26: 1219–1222.
32. Park S, Baker JO, Himmel ME, Parilla PA, Johnson DK (2010) Cellulose crystallinity index: measurement techniques and their impact on interpreting cellulase performance. Biotechnol Biofuels 3: 10.
33. Li M, Feng SQ, Wu LM, Li Y, Fan CF, et al. (2014) Sugar-rich sweet sorghum is distinctively affected by wall polymer features for biomass digestibility and ethanol fermentation in bagasse. Bioresour Technol 167: 14–23.
34. Li M, Si SL, Hao B, Zha Y, Wan C, et al. (2014) Mild alkali-pretreatment effectively extracts guaiacyl-rich lignin for high lignocellulose digestibility coupled with largely diminishing yeast fermentation inhibitors in *Miscanthus*. Bioresour Technol 169: 447–456.

16

Novel Intramedullary-Fixation Technique for Long Bone Fragility Fractures Using Bioresorbable Materials

Takanobu Nishizuka[1]*, Toshikazu Kurahashi[1], Tatsuya Hara[1], Hitoshi Hirata[1], Toshihiro Kasuga[2]

1 Department of Hand Surgery, Nagoya University Graduate School of Medicine, Nagoya, Japan, **2** Department of Frontier Materials, Nagoya Institute of Technology, Nagoya, Japan

Abstract

Almost all of the currently available fracture fixation devices for metaphyseal fragility fractures are made of hard metals, which carry a high risk of implant-related complications such as implant cutout in severely osteoporotic patients. We developed a novel fracture fixation technique (intramedullary-fixation with biodegradable materials; IM-BM) for severely weakened long bones using three different non-metallic biomaterials, a poly(L-lactide) (PLLA) woven tube, a nonwoven polyhydroxyalkanoates (PHA) fiber mat, and an injectable calcium phosphate cement (CPC). The purpose of this work was to evaluate the feasibility of IM-BM with mechanical testing as well as with an animal experiment. To perform mechanical testing, we fixed two longitudinal acrylic pipes with four different methods, and used them for a three-point bending test (N = 5). The three-point bending test revealed that the average fracture energy for the IM-BM group (PLLA + CPC + PHA) was 3 times greater than that of PLLA + CPC group, and 60 to 200 times greater than that of CPC + PHA group and CPC group. Using an osteoporotic rabbit distal femur incomplete fracture model, sixteen rabbits were randomly allocated into four experimental groups (IM-BM group, PLLA + CPC group, CPC group, Kirschner wire (K-wire) group). No rabbit in the IM-BM group suffered fracture displacement even under full weight bearing. In contrast, two rabbits in the PLLA + CPC group, three rabbits in the CPC group, and three rabbits in the K-wire group suffered fracture displacement within the first postoperative week. The present work demonstrated that IM-BM was strong enough to reinforce and stabilize incomplete fractures with both mechanical testing and an animal experiment even in the distal thigh, where bone is exposed to the highest bending and torsional stresses in the body. IM-BM can be one treatment option for those with severe osteoporosis.

Editor: Jie Zheng, University of Akron, United States of America

Funding: This work was supported by Japan Science and Technology Agency (AS2414028P, http://www.jst.go.jp/tt/EN/univ-ip/a-step.html#supportContent). The funders had no role in study design, data collection and analysis, decision to publish, or preparation of the manuscript.

Competing Interests: The authors have declared that no competing interests exist.

* Email: nishizuka1@mail.goo.ne.jp

Introduction

Metallic implants such as locked plating or intramedullary nailing can instantaneously strengthen weakened long bone metaphyses; however, they sometimes cause complications such as cutout because the strength of the bone is much less than that of the metallic implants in severely osteoporotic patients [1]. We therefore developed a new technique, intramedullary-fixation with biodegradable materials (IM-BM), to instantaneously strengthen severely weakened long bone metaphyses, imitating vertebroplasty. Vertebroplasty [2] can immediately make the collapsed vertebrae stronger using calcium phosphate cement (CPC) [3]; however, there is no such procedure with CPC for long bone metaphyses due to the limited torsional and bending strength associated with CPC [3,4]. We therefore combined a poly(L-lactide) (PLLA) woven tube with CPC based on a concept of reinforced concrete in order to strengthen severely weakened long bone.

Reinforced concrete is one of the most widely used modern building materials to strengthen the framework. Typical concrete has high resistance to compressive stresses (about 28 MPa); however, any appreciable tension (*e.g.*, due to bending) will break the microscopic rigid lattice, resulting in cracking and concrete separation. Reinforced concrete is a composite material made by inserting steel bars in concrete, and resists not only compression but also bending and torsional stresses.

We also combined a nonwoven polyhydroxyalkanoates (PHA) fiber mat to prevent cement leakage from the fracture site. PHA are biodegradable materials and show excellent extendibility.

The purpose of this work was to evaluate the feasibility of IM-BM in both mechanical testing and animal experiment models. The present work demonstrated that IM-BM was strong enough to reinforce and stabilize incomplete fractures with both mechanical testing and an animal experiment even in the distal thigh, where bone is exposed to the highest bending and torsional stresses in the body.

Materials and Methods

Ethics statement

The Institutional Committee for Animal Care of Nagoya University approved the experimental protocol (reference number: 25166).

Intramedullary-fixation with Biodegradable Materials (IM-BM)

The procedure for IM-BM begins with reaming the intramedullary cavity and inserting a PLLA woven tube into the cavity, followed by injection of CPC paste both inside and outside the tube using a syringe. However, massive CPC paste leakage occurs from the fracture site, and it appears likely to inhibit bone union. Therefore, we use a nonwoven PHA fiber mat to prevent CPC leakage. The 3-hydroxybutyrate (3HB) and 4-hydroxybutyrate (4HB) copolymers (poly [P](3HB-co-4HB)) used in the PHA fiber mat in this work demonstrate both strength and flexibility (Figure 1). The nonwoven PHA fiber mat can prevent CPC leakage. Therefore, when we injected the CPC paste inside and outside the PLLA woven tube, the nonwoven PHA fiber mat was expanded until it fit the cavity (Figures 2 and 3).

Preparation of PLLA Woven Tube

Nipro Co., Ltd. (Tokyo, Japan) produced the PLLA woven tube used in this study. First, a plain-stitch fabric was knitted using PLLA monofilaments (0.2 mm in diameter). Then, three sheets of fabric were stacked, and formed into a cylindrical shape by painting with dichloromethane (Wako Junyaku Kogyo Co., Ltd., Osaka, Japan), with a resulting external and internal diameter of 10 and 7 mm, respectively. Finally, they were cut and shortened to 65 mm in length for mechanical testing. Smaller size PLLA woven tubes for the animal experiment, with external and internal diameters of 5 and 4 mm, respectively, were also prepared. They were cut and shortened to 38 mm in length. Our previous experiment indicated that the mean modulus of elasticity in bending for these tubes was 46.7 MPa.

Preparation of a Nonwoven PHA Fiber Mat

PHA represent a complex class of biopolymers consisting of various hydroxyalkanoic acids. Microorganisms synthesize PHA as storage compounds for energy [5]. PHA exhibit biodegradable, biocompatible, thermoplastic and elastomeric properties once extracted from the cells [6]. Attempts have been made for many years to develop PHA applications in medical devices [7]. Polymerization of 4HB with other hydroxyl acids such as 3HB can produce elastomeric compositions at moderate 4HB contents (15%–35%), and relatively hard rigid polyesters at lower 4HB contents [7]. 3HB and 4HB copolymer (P(3HB-co-4HB)) at 18% 4HB content (G5 JAPAN Co., Ltd., Osaka, Japan) was used in our work as a tool to prevent CPC leakage.

Electrospinning is a process that can generate a polymer fiber mat with high flexibility and porosity [8,9]. A porous material with continuous pore structure is expected to be useful for implants used in the regeneration of damaged tissue, because the pore would allow penetration of nutriments and/or ingrowth of tissues, blood vessels, and cells [10]. The nonwoven fiber mat consisting of a biodegradable polymer may be one of the best biomaterial candidates, because the interlocking fibers easily form large connective pores [11]. A nonwoven PHA fiber mat with 3HB and 4HB copolymers was prepared using an electrospinning method. First, 2 g of PHA powder consisting of P(3HB-co-4HB) copolymers containing 18% 4HB (G5 JAPAN Co.) were dissolved in chloroform at 6 wt% to prepare the solution for electrospinning. The samples were then spun on the electrospinning unit (NEU, Kato Tech Co., Kyoto, Japan). The solution for electrospinning was loaded into a glass syringe and pushed out at a flow rate of 30 μl/min through a metallic needle (22 gauge) that was connected to a +10 kV electrical field at room temperature and approximately 55% relative humidity. The fibers were collected on an aluminum drum rotating at 2000 mm/min, traversing at 100 mm/min, and positioned 80 mm from the tip of the needle. The nonwoven PHA fiber mat, consisting of microfibers with diameters of approximately 10 μm (Figure 4), was produced with a final thickness of 100 μm after spinning for 120 min. The resulting nonwoven PHA fiber mat showed excellent expandability (Figure 1); it did not fracture easily even when the CPC paste was injected to expand it (Figure 3).

CPC Preparation

The CPC used in the present work was Biopex-R® (advanced type) (BPRad, HOYA Co., Ltd., Tokyo, Japan). The CPC powder consisted of α-tricalcium phosphate (α-$Ca_3(PO_4)_2$; α-TCP), tetracalcium phosphate ($Ca_4(PO_4)_2O$; TECP), dicalcium phosphate dihydrate ($CaHPO_4 \cdot 2H_2O$), hydroxyapatite ($Ca_{10}(PO_4)_6(OH)_2$; HAp) and magnesium phosphate ($Mg_3(PO_4)_2$). The malaxation liquid was composed of sodium chondroitin sulfate, disodium succinate ($(CH_2COONa)_2$), sodium hydrogensulfite ($NaHSO_3$) and water (H_2O). The CPC powder was mixed with the malaxation liquid, turning it into a paste, which hardened over time via hydration to form a hydroxyapatite structure [12]. Twelve grams of powder and 4 ml of malaxation liquid were used in the present mechanical testing and animal experiment. BPRad takes only 1 day to reach maximum compressive strength after kneading of the powder and malaxation liquid mixture.

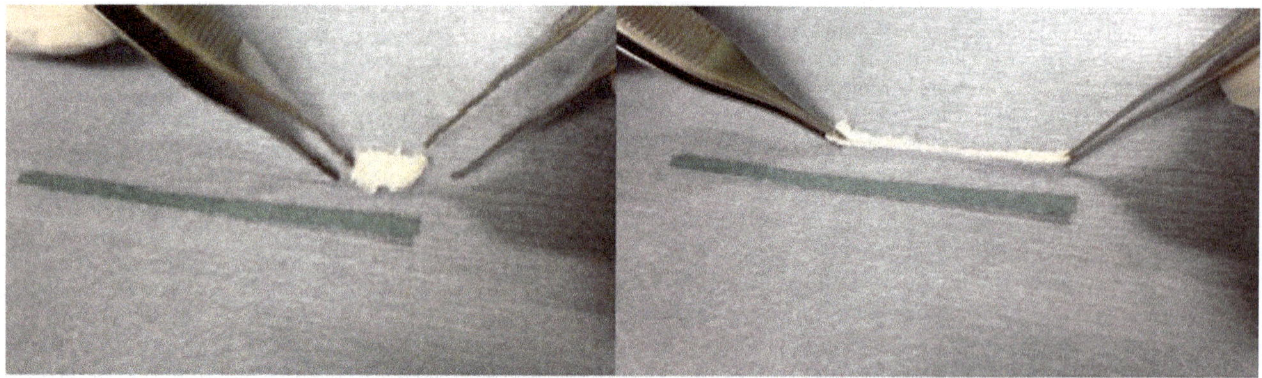

a b

Figure 1. Expandability of PHA fiber mat. Before elongation (a) and after elongation (b).

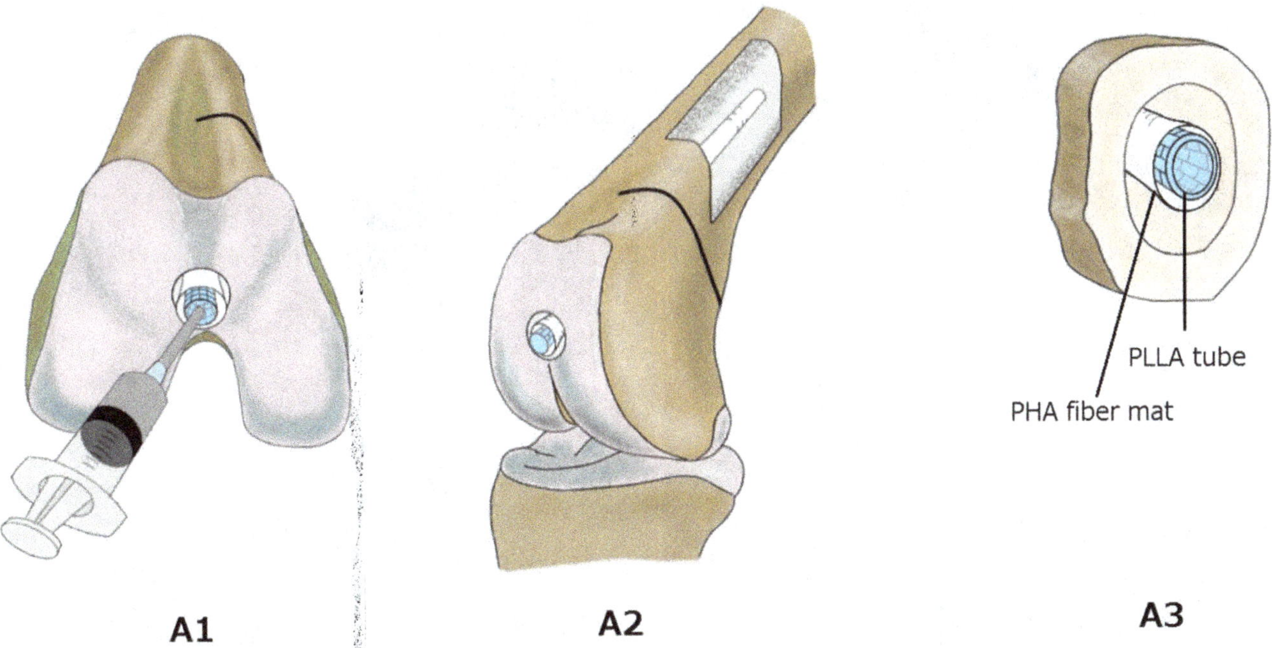

Figure 2. IM-BM procedures. View of the distal femur (A1, A2) and schema of bone cross section (A3) before CPC injection. PLLA: PLLA woven tube; PHA: PHA fiber mat; CPC: calcium phosphate cement.

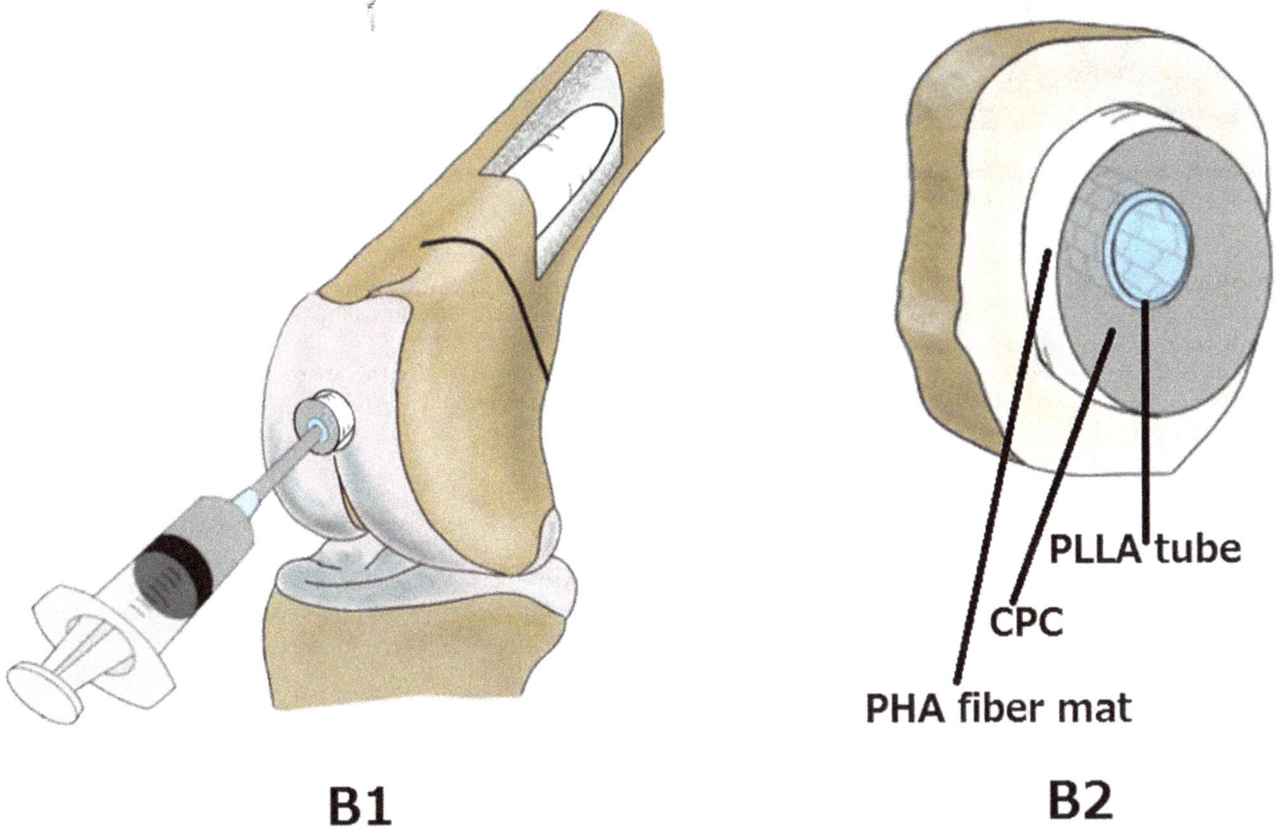

Figure 3. IM-BM procedures. View of the distal femur (B1) and bone cross section schema (B2) after CPC injection. PLLA: PLLA woven tube; PHA: PHA fiber mat; CPC: calcium phosphate cement.

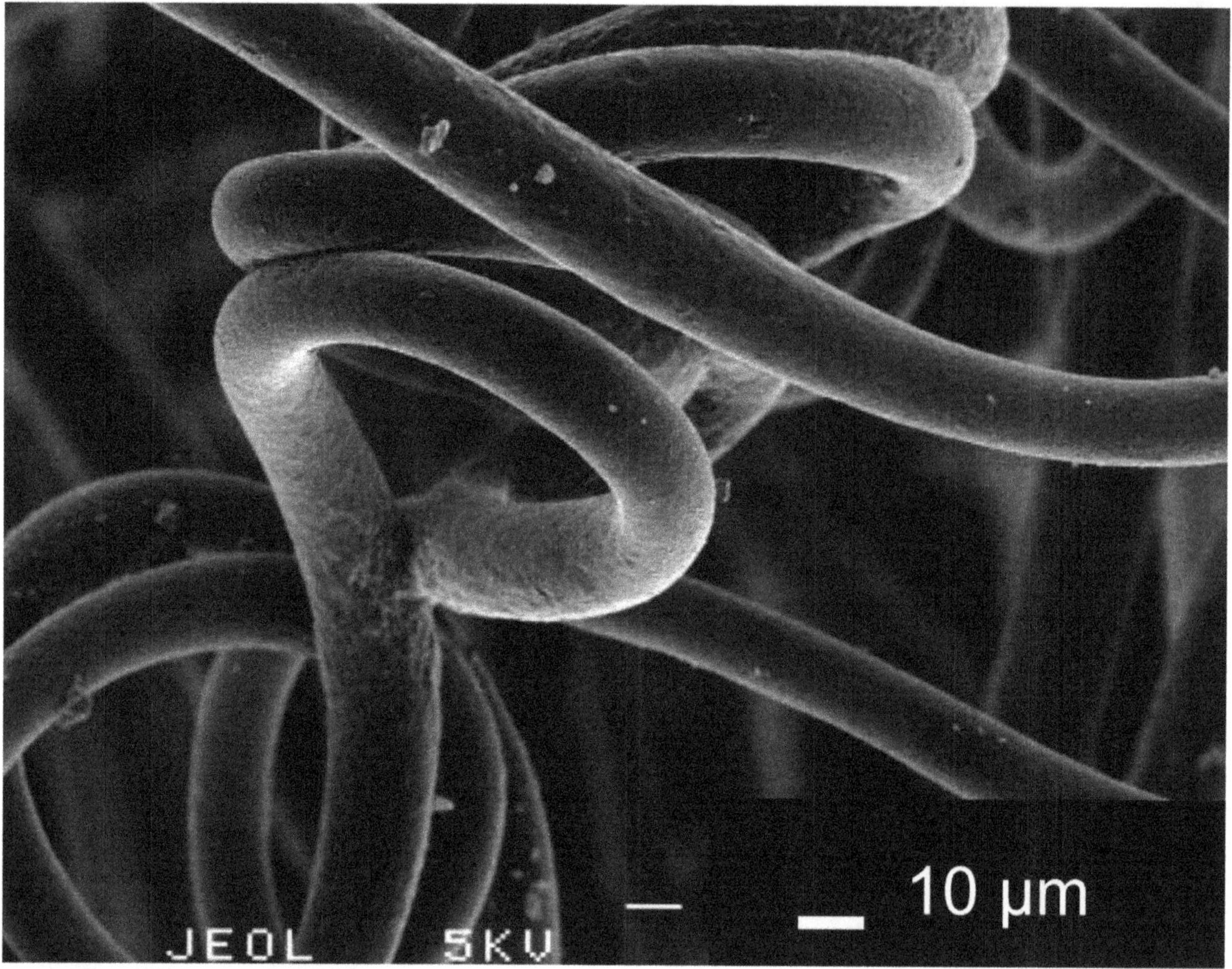

Figure 4. Scanning Electron Microscopy (SEM) photograph of a nonwoven PHA fiber mat. A nonwoven PHA fiber mat, consisting of microfibers with diameters of approximately 10 µm.

Mechanical Testing

Five profile specimens were tested for each sample composition. As shown in Figure 5, we fixed two longitudinal acrylic pipes with four different methods. For group 1, the PLLA woven tube was wrapped by the nonwoven PHA fiber mat, and it was inserted into two acrylic pipes placed longitudinally. The outer diameter of the pipe was 16 mm, and the length of each pipe was 35 mm. The powder and the liquid in the incubator (NTT-2200, EYELA, Tokyo, Japan) were warmed to 30°C before the injection. Just after kneading, the CPC paste was injected inside and outside the PLLA woven tube using a syringe. After the injection, the specimens were submerged in simulated body fluid (Na^+ 142.0, K^+ 5.0, Mg^{2+} 1.5, Ca^{2+} 2.5, Cl^- 148.8, HCO_3^- 4.2, PO_4^{2-} 1.0, and SO_4^{2-} 0.5 mM and buffered at pH 7.40 with trishydroxymethylaminomethane) [13] maintained at 37°C. The CPC started to harden gradually. After 10 min, a three-point bending test was performed using 858 Mini Bionix II (MTS, Eden Prairie, MN) following the Japanese Industrial Standard (JIS) K 7074. The cross-head speed was 0.5 mm/min, and the support span was 60 mm. For group 2, the PLLA woven tube was inserted into two acrylic pipes placed longitudinally. Then, the CPC paste was injected inside and outside the PLLA woven tube. For group 3, the cylindrical shape of the nonwoven PHA fiber mat was inserted into acrylic pipes. Then, the CPC paste was injected into the nonwoven PHA fiber mat. Only the CPC paste was injected into acrylic pipes for group 4.

The maximum flexural strength was determined using the following equation:

$$\delta_{fmax} = 8F_{max}L/\pi d^3,$$

where δ_{fmax} is the maximum flexural strength (MPa), F_{max} is the maximum load (N), L is the support span (mm), and d is the diameter of the specimen (mm).

The modulus of elasticity in bending was determined using the following equation:

$$E_f = (4L^3/3\,\pi d^4) \times (\Delta F/\Delta S),$$

where E_f is the modulus of elasticity in bending (MPa), L is the support span (mm), d is the diameter of the specimen (mm), ΔF is the variation of the load (N), and ΔS is the variation of the central deflection (mm).

Fracture energy (J/m^2) was determined using integral calculus from the beginning to the yield point at the stress–strain curve. A

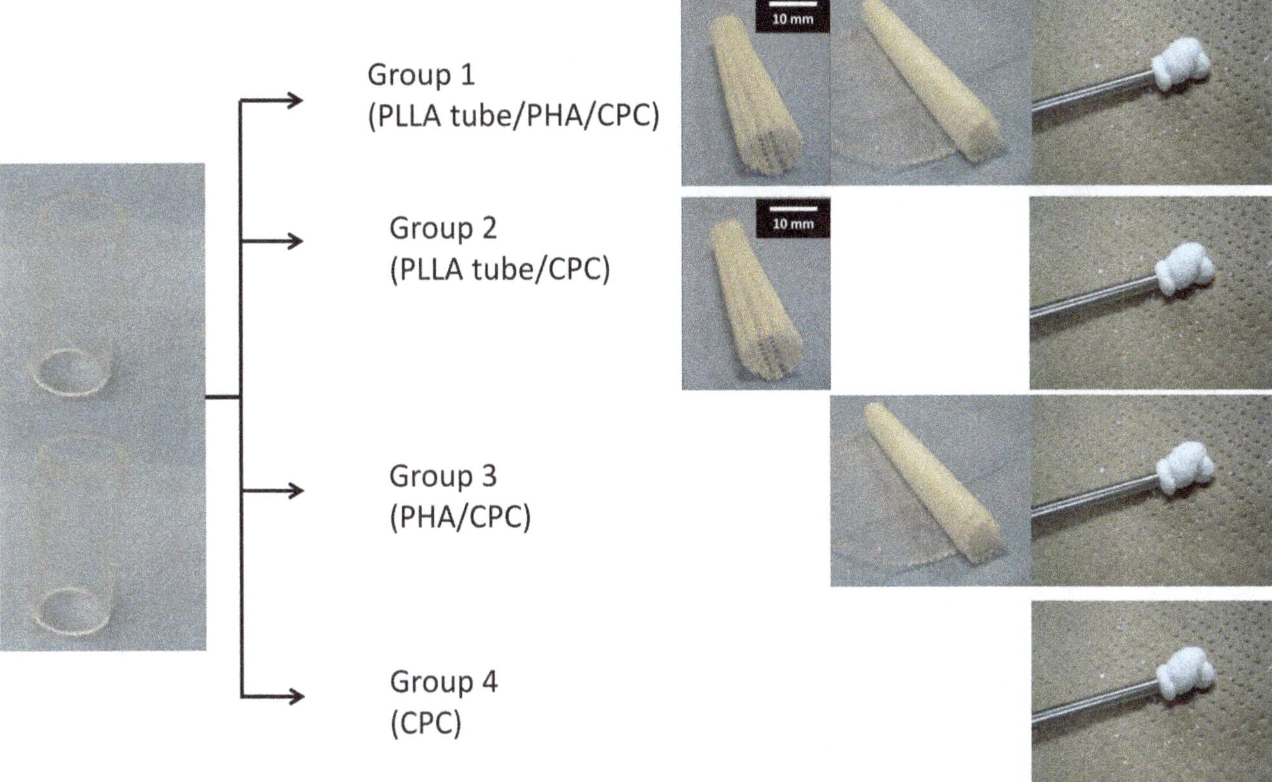

Figure 5. Method of mechanical testing. Two acrylic pipes placed longitudinally were fixed in four groups by the different method. Pictures of group 1 show a PLLA woven tube, a nonwoven PHA fiber mat, and CPC paste sequentially from the left.

mean value and standard deviation were calculated concerning the δ_{fmax}, the modulus of elasticity in bending, and the fracture energy.

All statistical analyses were conducted using SPSS version 19.0 (SPSS, Tokyo, Japan). Data from multiple groups were compared using a one-way analysis of variance (ANOVA). We performed multiple comparisons of the three treatment groups using the Bonferroni test when significant differences were detected. Significance levels of all the tests were established at $p<0.05$.

Experimental Animal Model of Osteoporosis and Surgical Procedure

Sixteen skeletally mature, 8-month-old (3.0–3.5 kg body weight), female New Zealand white rabbits were used. Experimental rabbit osteoporosis models were made by performing bilateral ovariectomy followed by intramuscular injection with methylprednisolone sodium succinate (Solu-Medrol; Pfizer, Tokyo, Japan) at a dosage of 1 mg/kg/day for four consecutive weeks, as previously described by Castaneda et al. [14] The sixteen rabbits were then randomly allocated into four experimental groups. The first was an IM-BM group (PLLA woven tube + CPC + nonwoven PHA fiber mat). The second was a PLLA + CPC group. The third was a CPC group. The fourth was a Kirschner wire (K-wire) group.

All the PLLA woven tubes and the nonwoven PHA fiber mat used in our animal experiments were sterilized with ethylene oxide gas (20% by weight; CO_2 80% by weight) with 50% H_2O at 45°C for 5 h. The rabbits were anesthetized with an intramuscular injection of ketamine hydrochloride (75 mg/kg body weight) and xylazine (10 mg/kg body weight) 8 weeks after the ovariectomy. The surgical site was shaved and prepared with a solution of Betadine (povidone-iodine). A lateral parapatellar incision was made, and the patella was medially dislocated. A half-round fracture was produced 10 mm proximal to the distal end of the femur with an electrical cutter to reproduce an insufficiency fracture. In the K-wire group, two 1.5-mm K-wires were inserted intramedullary from the intercondylar area 5 mm proximal from the edge of the intercondylar notch until penetrating the proximal bone cortex. Finally, the fascia and the skin wounds were closed. In the IM-BM group, PLLA + CPC group, and the CPC group, a drill hole was produced in the intercondylar area 5 mm proximal from the edge of the intercondylar notch using 1.5-mm K-wire. Next, an intramedullary cavity was reamed from 3 mm to 7 mm by the drill reamer, and the cavity was curetted.

In the IM-BM group, the PLLA tube wrapped with nonwoven PHA fiber mat was inserted into the cavity (Figure 2-A1, A2, A3). Next, CPC paste was injected into the lumen of the PLLA tube just after kneading the powder and malaxation liquid, which was preheated to 30°C in the incubator (Figure 3-B1, B2). The paste began to harden about 10 min after kneading. In the PLLA + CPC group, CPC paste was injected after insertion of the PLLA woven tube. In the CPC group, only CPC paste was injected. A small amount of CPC paste leaked from the fracture site in the PLLA + CPC group and the CPC group. In the IM-BM group, the PLLA + CPC group, and the CPC group, the fascia and the skin wounds were closed within 15 min after kneading of the CPC paste, and the rabbits began to bear weight a few hours after the operation.

Postoperative lesions were evaluated with the use of soft X-ray (SOFTEX, Yokohama, Japan) at weeks 1, 4, 8, 12, and 20. Lateral radiographs were made with an exposure of 45 kV, 10 mA, and 15 s. The soft X-ray data at postoperative week 1 from multiple groups were compared using a one-way ANOVA. When significant differences were detected in ANOVA, we performed multiple comparisons of the four treatment groups using the Bonferroni test. A significance level for the test was established at $p<0.05$.

With pentobarbital overdose, two of the four rabbits in the IM-BM group were sacrificed at week 20, and the other two in the IM-BM group were sacrificed at week 52. A 15 mm segment of the metaphyseal portion of the femur was removed for histological examination. The specimens were fixed with 0.2% glutaraldehyde for 24 h and then in 10% phosphate-buffered formalin for 48 h. The specimens were subsequently dehydrated in ethanol and embedded in methyl methacrylate (MMA). The embedded blocks were trimmed with a cutter and ground with abrasive paper. Thereafter, the sections were further ground to a final thickness of approximately 10 μm. Finally, the specimens were stained with hematoxylin and eosin, and examined under the microscope.

Results

Mechanical Testing

Table 1 shows the mean maximum flexural strength values for each group at 10 min immediately following the CPC injection. The group 1 values (PLLA + CPC + PHA; 2.71±0.66 MPa) were significantly higher ($p<0.001$) than the values of group 3 (CPC + PHA; 0.79±0.23 MPa) and group 4 (CPC only; 0.52±0.24 MPa). Although there is no statistically significant difference, the values of group 1were higher than the values of group 2 (PLLA + CPC; 2.37±0.13 MPa) ($p=0.15$). Table 1 also shows the average fracture energy at 10 min following the CPC injection. The group 1 values (PLLA + CPC + PHA; 1210±334 J/m^2) were significantly higher ($p<0.001$) than the group 2 values (PLLA + CPC; 434±63 J/m^2), the group 3 values (CPC + PHA; 19.4±6.4 J/m^2) and the group 4 values (CPC only; 5.5±5.7 J/m^2). Figure 6 shows the representative stress–strain curves for group 1 (PLLA + CPC + PHA) and group 4 (CPC only). In group 1, the curve can be categorized into four zones (zone 1: steep slope zone, zone 2: gentle slope zone, zone 3: almost flat slope zone, and zone 4: negative slope zone). In zones 1 and 2, the stress kept increasing and the curve did not drop until the end of zone 3. In contrast, in groups 3 and 4, the curve immediately reached the yield point. Once the curve reached the yield point, the curve dropped abruptly due to the fragmentation of the materials. The average moduli of elasticity in bending in group 1 were 179±89 MPa in zone 1 and 18.2±8.7 MPa in zone 2. The average moduli of elasticity in bending in groups 2, 3, and 4 were 144±31 MPa, 183±64 MPa and 223±69 MPa, respectively.

Animal Experiments

Soft X-ray photographs revealed that there were no fracture displacements (0/4) over the entire postoperative period in the IM-BM group (Figure 7a), whereas two of four rabbits in PLLA + CPC group, three of four rabbits in the CPC group (Figure 7f), and three (including one cutout) of four rabbits in the K-wire group (Figure 7d) had fracture displacements at postoperative week 1. Although statistical analysis did not indicate a significant difference, IM-BM group had fewer fracture displacements than other groups (PLLA + CPC, CPC, K-wire) ($p=0.35$, 0.143, 0.143, respectively: all Bonferroni multiple comparison test), as shown in Table 2. Figure 7b shows that all rabbits in the IM-BM group achieved bony union in 8 weeks. There were no postoperative infections or clinical signs of implant reaction in the IM-BM group. On the other hand, all rabbits in the PLLA + CPC group showed CPC leakage from the fracture site (Figure 7e), and three of four rabbits in the CPC group had wound dehiscence within the first postoperative week.

Histologic examination at week 20 in the IM-BM group indicated that the PLLA tube was not degraded yet (Figure 8a). Histologic examination at week 52 in the IM-BM group showed that the PLLA tube seemed to have degraded gradually (Figure 8b). A multinucleated giant cell and neovascularization were observed in the PHA fiber mat layer (Figure 8c).

Discussion

The incidence of fragility fractures has been increasing in developed countries. Since Galibert et al. in 1987 introduced percutaneous poly (methyl methacrylate) (PMMA)-assisted vertebroplasty [15], many authors have reported treatment outcomes for vertebroplasty or balloon kyphoplasty. PMMA has excellent compressive, bending, and tensile strength [16]. However, PMMA monomer toxicity produces a risk of hypotension [17], and can damage surrounding cells due to heat evolution during the hardening process. In addition to these problems, PMMA is not osteoconductive and does not enhance bone remodeling.

Alternatively, CPC (Biopex) is one of the most commonly used injectable bone cement pastes and is highly biocompatible, with excellent osteoconductivity [12]. The cement is absorbed progressively from the outer surface and replaced by bone tissue through the normal remodeling process [12]. Moreover, it can set without heat evolution. However, CPC is brittle and has generally low mechanical strength [3,4], except for compressive strength. Thus far, CPC has been mainly limited to vertebroplasty and balloon kyphoplasty, which require a high compressive strength.

We inserted a PLLA woven tube into the intramedullary cavity with injection of CPC paste in order to overcome the above-mentioned problems. We wrapped the PLLA tube with nonwoven PHA fiber mat to prevent both CPC leakage from the fracture site

Table 1. The mean maximum flexural strength values and the mean fracture energy for each group.

	Group 1 (PLLA tube/PHA/CPC)	Group 2 (PLLA tube/CPC)	Group 3 (PHA/CPC)	Group 4 (CPC)
Maximum flexural strength (MPa)	*2.71±0.66	2.37±0.13	0.79±0.23	0.52±0.2
Fracture energy (J/m^2)	*1210±334	434±63	19.4±6.4	5.5±5.7

*Significant difference (p<0.001) in the maximum flexural strength between group 1(*) and group 3/group 4 (Bonferroni test).
*Significant difference (p<0.001) in the fracture energy between group 1(*) and group 2/group 3/group 4 (Bonferroni test).
PLLA: poly(L-lactide); PHA: polyhydroxyalkanoates; CPC: calcium phosphate cement.

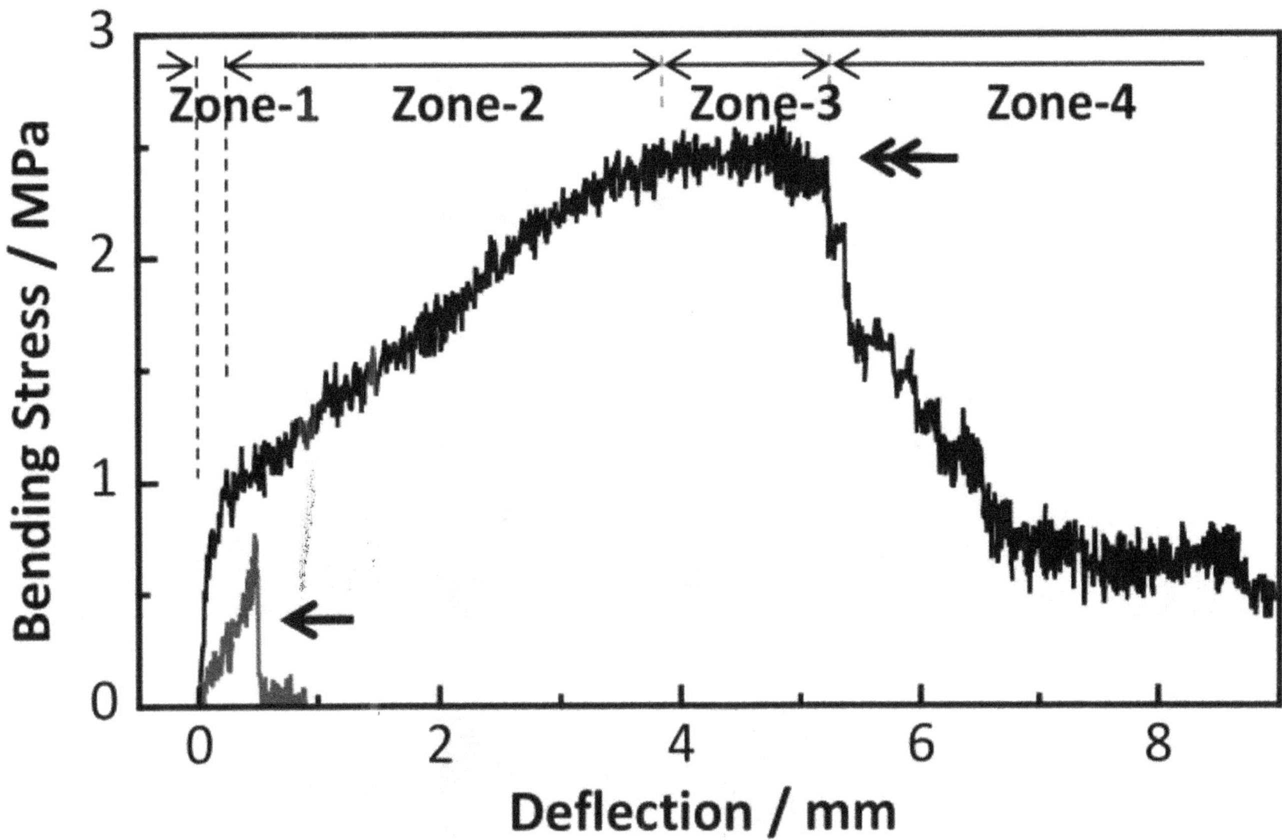

Figure 6. Result of mechanical testing. A representative stress–strain curve 10 min after the CPC injection in group 1 (double arrow) and group 4 (arrow).

and embolism. The PHA used in the present work was comprised of the copolymers 3HB and 18% 4HB. We also attempted to use a nonwoven poly(lactic-co-glycolic acid) (PLGA) fiber mat as a leakage prevention material in a preliminary trial with acrylic tubes. However, the PLGA mat did not extend far enough into the cavity and easily ruptured during injection of CPC paste. In contrast, the present animal experiment demonstrated that a nonwoven PHA fiber mat wrapped around a PLLA tube can efficiently prevent cement leakage from the fracture site and allow almost complete filling of the reamed cavity with CPC, as observed by postoperative soft X-ray analysis as well as histologic examination (Figures 7a and 8a).

This animal experiment also demonstrated that IM-BM markedly improved mechanical strength when compared with CPC only. Rabbits in the IM-BM group had no fracture displacement, whereas three of four rabbits in the CPC group had fracture displacements within 1 week after the surgery. Furthermore, the IM-BM also revealed the better results than other two groups (PLLA + CPC, K-wire). Two of four rabbits in the PLLA + CPC group and three of four rabbits in the K-wire group suffered fracture displacement, and one of four rabbits in the K-wire group suffered cutout at postoperative week 1. Histologic examination at week 52 in the IM-BM group showed partial degradation of the PLLA tube in the CPC. It has been previously reported that it takes 1.5–5 years for the degradation of PLLA to be complete, as shown in Table 3 [7]. The absorption of the nonwoven PHA fiber mat is preferable, because it enables the CPC to make contact with the bone gradually. However, partial absorption of the nonwoven PHA fiber mat was unclear at weeks 20 and 52. A multinucleated giant cell and neovascularization were observed in the PHA fiber mat layer. This suggests that the PHA used in our study have good biocompatibility. We are now conducting a long-term study to evaluate the degradation behavior of PHA.

Mechanical testing has clearly shown that the combination of the PLLA woven tube with CPC and PHA fiber mat improved the

Table 2. Fracture displacement rate in four groups at postoperative week 1.

	IM-BM group	PLLA + CPC group	CPC group	K-wire group
Fracture displacement rate	0/4	2/4	3/4	3/4

IM-BM group had fewer fracture displacements than the PLLA + CPC, the CPC, and the K-wire groups although statistical analysis did not indicate a significant difference ($p = 0.35$, 0.143, 0.143, respectively: all Bonferroni multiple comparison test).
IM-BM: intramedullary-fixation with biodegradable materials; CPC: calcium phosphate cement; K-wire: Kirschner wire.

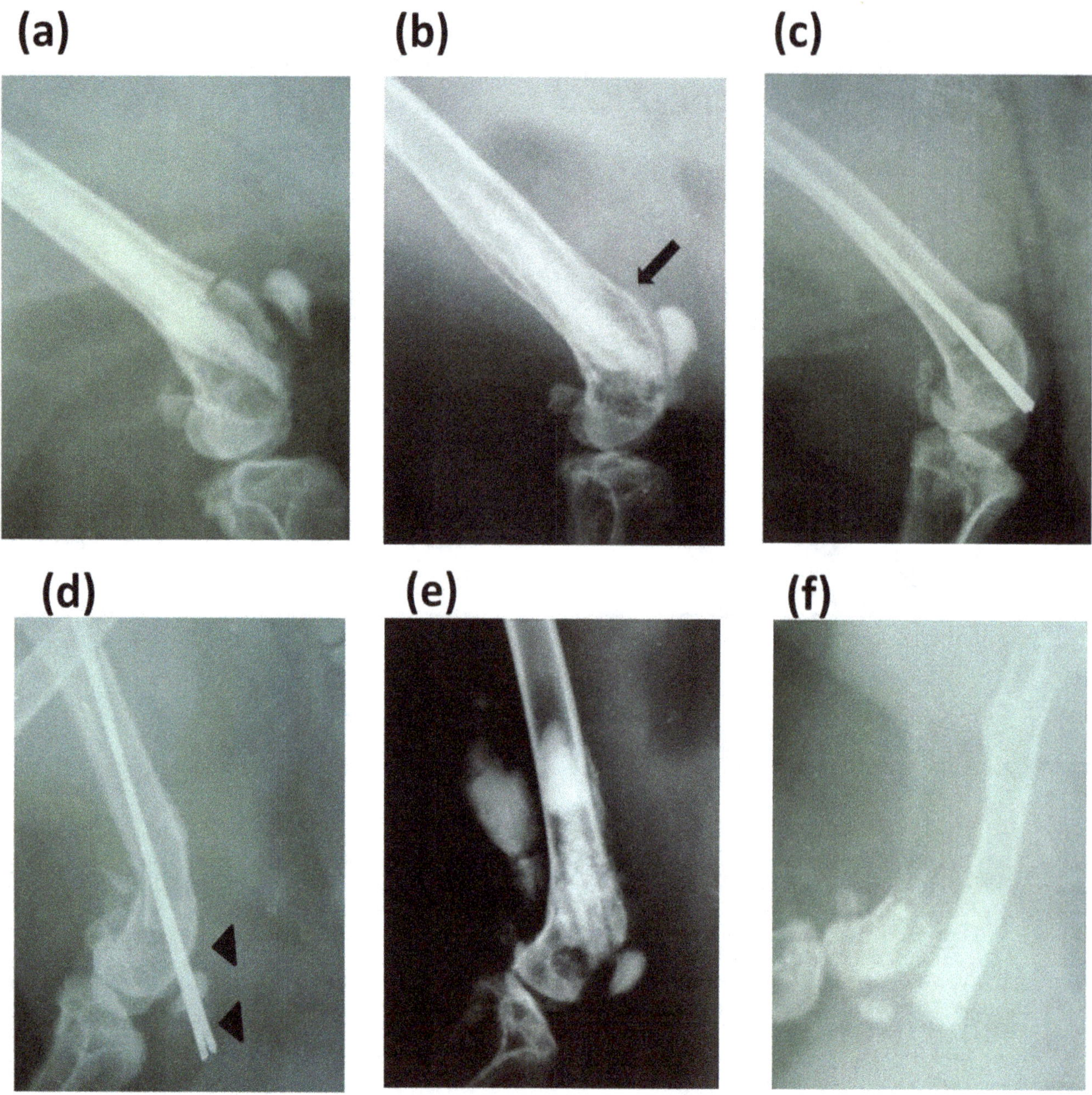

Figure 7. Representative postoperative radiographs. Representative postoperative radiographs in the IM-BM group (PLLA + CPC + PHA) at week 0 (a) and week 8 (b). (b) The fracture site obtained complete bony union (arrow). Representative postoperative radiographs in the Kirschner wire group at week 0 (c) and week 1 (d). (d) Cutout happened at the fracture site (arrowhead). Representative postoperative radiograph in the PLLA + CPC group at week 0 (e). (e) It showed CPC leakage from the fracture site. Representative postoperative radiograph in the CPC group at week 1 (f). (f) It showed fracture displacement.

apparent mechanical properties, including a significant increase in fracture energy and flexural strength. All the stress–strain curves for group 1 (PLLA + CPC + PHA) showed slope changes at the end of zone 1 (Figure 6). The change may be due to fracture of bonding at the interface between the CPC and the PHA fiber mat. Thus, cracks in the CPC followed the fracture of bonding. The slope in zone 2 was smaller than that in zone 1. It is thought that the PLLA woven tube mainly supported the load with the CPC, because the PHA fiber mat prevented the CPC from breaking into pieces. In zone 3, the slope became flat. The PLLA woven tube was believed to be considerably deformed. In zone 4, the slope became negative. The CPC was not able to support the load any further because it was broken into pieces.

It should be emphasized that the average fracture energy for the IM-BM group (PLLA + CPC + PHA) was 60 to 200 times greater than that of CPC + PHA group and CPC group in our mechanical testing. Considering the result that the mean fracture energy of PLLA + CPC + PHA group (1210 ± 334 J/m^2) was 3 times higher ($p<0.001$) than that of the PLLA + CPC group (434 ± 63 J/m^2), the enhanced fracture energy of PLLA + PHA + CPC implant was mainly due to the PLLA tube, but PHA fiber mat also improved the fracture energy significantly.

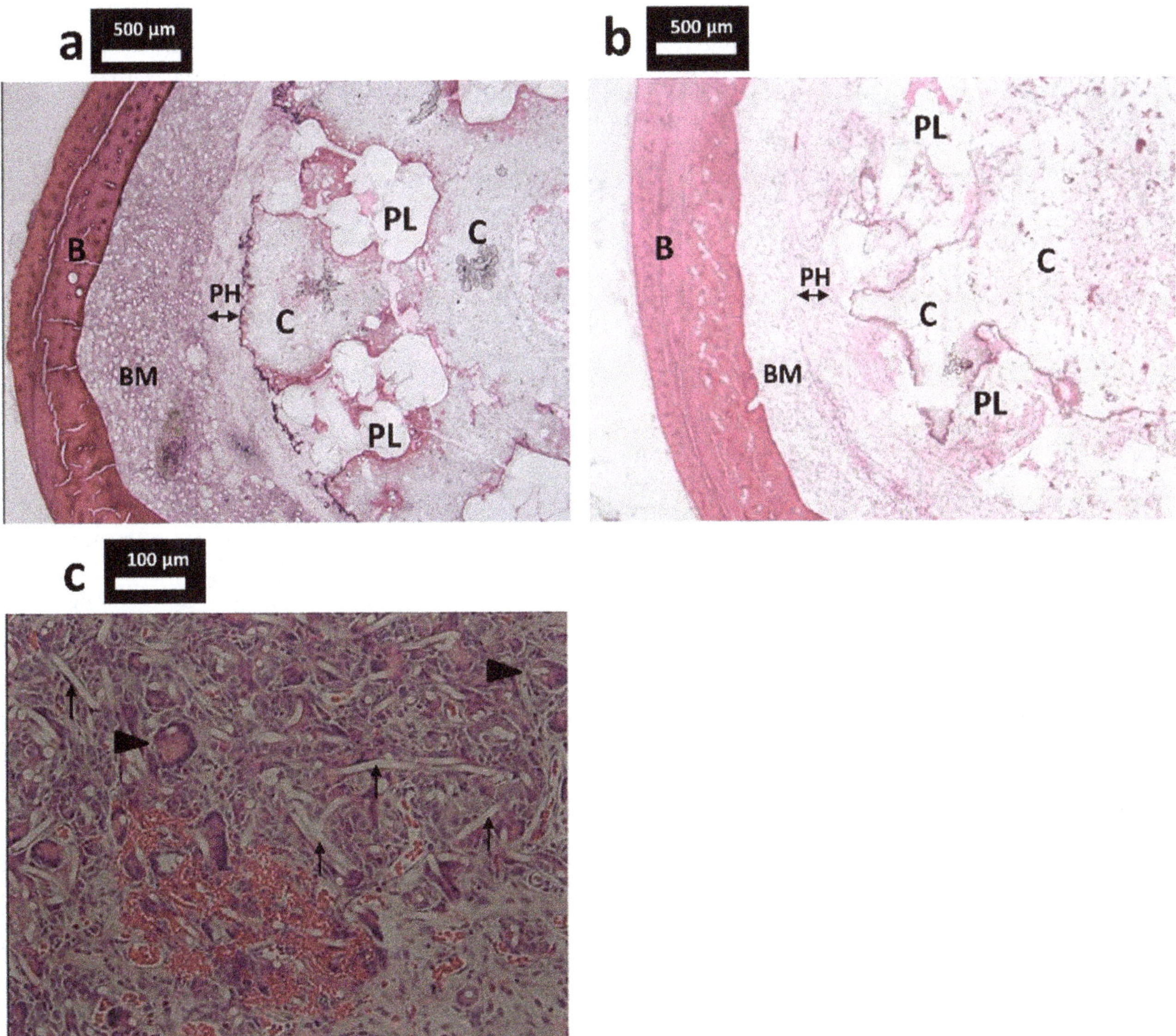

Figure 8. Histologic cross sections of specimens from the IM-BM group. Histologic cross sections of specimens from the IM-BM group (PLLA + CPC + PHA) stained with hematoxylin and eosin. (a) A PHA fiber mat layer surrounded the CPC, and the PLLA tube was not degraded at week 20. (b) At week 52, the PLLA tube seemed to have degraded gradually. (c) Multinucleated giant cell (arrowhead) and neovascularization were observed in the PHA fiber mat layer at week 52. PHA fiber (arrow). PL: PLLA woven tube; PH: PHA fiber mat C: CPC; B: bone cortex; BM: bone marrow.

Table 3. Properties of biodegradable thermoplastic polyesters. (Wu Q. 2009 [7]).

	Tm (°C)	Tg (°C)	Tensile strength (MPa)	Tensile modulus (GPa)	Elongation at break (%)	Absorption rate
PGA	225	35	70	6900	<3	6 weeks
PLLA	175	65	28–50	1200–2700	6	1.5–5 years
PDLLA	Amorphous	50–55	29–35	1900–2400	6	3 months
P(3HB)	180	1	36	2500	3	2 years
P(3HB-co-4HB) (4HB 16%)	152	−8	26	Not measured	444	Unknown

PGA: poly glycolic acid; PLLA: poly(L-lactide); PDLLA: poly($_{D}$ $_{L}$-lactide); P(3HB): poly(3-hydroxybutyrate); P(3HB-co-4HB) (4HB 16%): poly(3-hydroxybutyrate-co-16% 4-hydroxybutyrate); Tm: melting temperature; Tg: glass transition temperature.

The maximum flexural strength of K-wire fixation in our previous mechanical study using two 1.6-mm K-wires was almost same as that of PLLA + PHA + CPC group in our current study (3.0 MPa in K-wire group vs 2.71 MPa in PLLA + PHA + CPC group). However, we think that IM-BM was strong enough to reinforce and stabilize incomplete fractures even in the distal thigh where bone is exposed to the highest bending and torsional stresses. In fact, a benefit of the IM-BM implant is that it is not harder than necessary and the risk of cutout is low, whereas metallic implants sometimes cause cutout in severely osteoporotic patients. The current study indicates that IM-BM can also be applied to insufficiency fractures in other areas in the body such as rib and proximal humerus.

Conclusions

In conclusion, the present work demonstrated that IM-BM was strong enough to reinforce and stabilize incomplete fractures with both mechanical testing and an animal experiment even in the distal thigh, where bone is exposed to the highest bending and torsional stresses in the body. The combination of three biomaterials with different physical and biological properties can be one treatment option for those with severe osteoporosis.

Acknowledgments

We thank Dr. Seiichi Kato of the Pathology Department at Nagoya University Hospital for advice on histology.

Author Contributions

Conceived and designed the experiments: TN HH T.Kasuga. Performed the experiments: TN T.Kurahashi TH. Analyzed the data: TN. Contributed reagents/materials/analysis tools: T.Kasuga. Contributed to the writing of the manuscript: TN HH T.Kasuga.

References

1. Owsley K, Gorczyca JT (2008) Fracture displacement and screw cutout after open reduction and locked plate fixation of proximal humeral fractures [corrected]. J Bone Joint Surg 90: 233–240.
2. Nakano M, Hirano N, Matsuura K, Watanabe H, Kitagawa H, Ishihara H, et al. (2002) Percutaneous transpedicular vertebroplasty with calcium phosphate cement in the treatment of osteoporotic vertebral compression and burst fractures. J Neurosurg 97: 287–293.
3. Ambard AJ, Mueninghoff L (2006) Calcium phosphate cement: review of mechanical and biological properties. J Prosthodont 15: 321–328.
4. Ishikawa K, Asaoka K (1995) Estimation of ideal mechanical strength and critical porosity of calcium phosphate cement. J Biomed Mater Res 29: 1537–1543.
5. Valappil SP, Misra SK, Boccaccini AR, Roy I (2006) Biomedical applications of polyhydroxyalkanoates, an overview of animal testing and in vivo responses. Expert Rev Med Devices 3: 853–868.
6. Steinbüchel A, Hustede E, Liebergesell M, Pieper U, Timm A (1992) Molecular basis for biosynthesis and accumulation of polyhydroxyalkanoic acids in bacteria. FEMS Microbiol Rev 9: 217–230.
7. Wu Q, Wang Y, Chen GQ (2009) Medical application of microbial biopolyesters polyhydroxyalkanoates. Artif Cells Blood Substit Immobil Biotechnol 37: 1–12.
8. Sill TJ, von Recum HA (2008) Electrospinning: applications in drug delivery and tissue engineering. Biomaterials 29: 1989–2006.
9. Li WJ, Laurencin CT, Caterson EJ, Tuan RS, Ko FK (2002) Electrospun nanofibrous structure: a novel scaffold for tissue engineering. J Biomed Mater Res 60: 613–621.
10. Mizutani Y, Hattori M, Okuyama M, Kasuga T, Nogami M (2005) Preparation of porous poly(L-lactic acid) composite containing hydroxyapatite whiskers. Chemistry Letters 34: 1110–1111.
11. Bhattarai SR, Bhattarai N, Yi HK, Hwang PH, Cha DI (2004) Novel biodegradable electrospun membrane: scaffold for tissue engineering. Biomaterials 52: 2595–2602.
12. Kurashina K, Kurita H, Kotani A, Takeuchi H, Hirano M (1997) In vivo study of a calcium phosphate cement consisting of alpha-tricalcium phosphate/dicalcium phosphate dibasic/tetracalcium phosphate monoxide. Biomaterials 18: 147–151.
13. Kokubo T, Kushitani H, Sakka S, Kitsugi T, Yamamuro T (1990) Solutions able to reproduce in vivo surface-structure changes in bioactive glass-ceramic A-W. J Biomed Mater Res 24: 721–734.
14. Castañeda S, Calvo E, Largo R, González-González R (2008) Characterization of a new experimental model of osteoporosis in rabbits. J Bone Miner Metab 26: 53–59.
15. Galibert P, Deramond H, Rosat P, Le Gars D (1987) Preliminary note on the treatment of vertebral angioma by percutaneous acrylic vertebroplasty. Neurochirurgie 33: 166–168.
16. Yamamuro T, Nakamura T, Iida H, Kawanabe K, Matsuda Y, et al. (1998) Development of bioactive bone cement and its clinical applications. Biomaterials 19: 1479–1482.
17. Phillips H, Cole PV, Lettin AW (1971) Cardiovascular effects of implanted acrylic bone cement. Br Med J 3: 460–461.

17

Real-Time UV-Visible Spectroscopy Analysis of Purple Membrane-Polyacrylamide Film Formation Taking into Account Fano Line Shapes and Scattering

María Gomariz, Salvador Blaya*, Pablo Acebal, Luis Carretero

Departamento de Ciencia de Materiales, Óptica y Tecnología Electrónica, Universidad Miguel Hernández, Elx (Alicante), Spain

Abstract

We theoretically and experimentally analyze the formation of thick Purple Membrane (PM) polyacrylamide (PA) films by means of optical spectroscopy by considering the absorption of bacteriorhodopsin and scattering. We have applied semiclassical quantum mechanical techniques for the calculation of absorption spectra by taking into account the Fano effects on the ground state of bacteriorhodopsin. A model of the formation of PM-polyacrylamide films has been proposed based on the growth of polymeric chains around purple membrane. Experimentally, the temporal evolution of the polymerization process of acrylamide has been studied as function of the pH solution, obtaining a good correspondence to the proposed model. Thus, due to the formation of intermediate bacteriorhodopsin-doped nanogel, by controlling the polymerization process, an alternative methodology for the synthesis of bacteriorhodopsin-doped nanogels can be provided.

Editor: Mark G. Kuzyk, Washington State University, United States of America

Funding: The authors have no funding or support to report.

Competing Interests: The authors have declared that no competing interests exist.

* Email: salva@dite.umh.es

Introduction

Bacteriorhodopsin (bR) is a photochromic protein related to the visual pigment rhodopsin contained in the cone cells of the human retinal, and has been widely explored for its use in electronics and photonic applications [1,2]. Among several biological molecules, bR has received most attention because of its outstanding optical properties and excellent stability against chemical, thermal and photochemical degradation [1,2]. In this sense, the use of bR has been proposed and demonstrated for a variety of technological applications in optics such as data storage [1,3,4], real-time holography [5,6], optical display and spatial light modulation [7,8], optical image processing [9], slow light [10].

bR is the simplest natural light energy transducer and the major protein component of the purple membrane (PM) of the archea *Halobacterium salinarium* [11,12]. After the absorption of a photon from the visible range (≈ 570nm) a cyclic sequence of reactions is produced in bR leading to the proton moving from the cytoplasmic side to the intracellular surface and the generation of an electrochemical potential that is used by the archea to maintain its metabolism by driving the synthesis of adenine triphosphate (ATP) [11]. This light-driven photocycle is well-known [13] and the chromophore passes through a sequence of transient optical states, being the sequence of these processes is characterized spectroscopically, defining photocycle intermediates (K, L, M, N, O) that differ in their absorption spectra in the UV-vis (λ_{max} values of 410, 560, and 630 nm for the M, N, and O intermediates respectively) [14,15].

Absorption spectra of bR in suspensions of purple membrane and in polymeric films have been widely studied as a function of different variables such as pH, temperature, environment, etc [16–18]. Asymmetrical Gaussian or Lorentzian bands are found, and analyzed by different phenomenological mathematical expressions [19–22]. Furthermore, due to the availability of high resolution crystal structure a better understanding of the properties of these systems has been reached since the spectral and optical properties of these kinds of biomolecules can be determined by the chemical nature of the chromophore, the electronic interactions between the different chromophores, and the interactions between chromophores and their environment [23–25].

In 1961, Fano proposed a theoretical treatment of the interaction of a discrete state coupled to a degenerate continuum under the condition that both the discrete and the continuum levels are excited by some external perturbation. As a result asymmetric resonant line shape (Fano profile) associated with absorption by the coupled system is obtained which has been widely used for describing phenomenons throughout nuclear, atomic and solid-state physics, photonic devices, nanoestructures, metamaterials as well as molecular spectroscopy [26,27].

The aim of this paper is to theoretically and experimentally analyze the formation of PM-polyacrylamide (PM-PA) films from the real-time variation of UV-visible spectra at different pHs. To do so, a model of the formation of those films is proposed and demonstrated by fitting the UV-Visible spectra. In order to explain the behaviour of the UV-Visible spectra of PM-doped polymeric suspension, in our model we have included the scattering of purple membrane and the absorption of bacteriorhodopsin taking into account the previous mentioned Fano profiles. The main reason for this study is to provide a tool for the development of biosensitized nanofilms engineered from biomembrane components and inorganic nanoparticles that is a promising field of colloid and interface science and technologies [28]. Recent nano-bioengineering approaches employing quantum dots (QDs) permit the enhancement of the purple membrane (PM) "light-harvesting capacity" compared to native PMs. In this sense, it has been reported several advances for the feasibility of bacteriorhodopsin as biophotosensitizer in excitonic solar cells and nanoscale devices [29,30]. Requirements of bR-containing nanofilms and nanoparticles are determined by the absorption spectra and, for this, our study is important because explain the resulting anomalies with the medium. Furthermore, Optogenetics is a technology that allows targeted, fast control of precisely defined events in biological systems as complex as freely moving mammals [31]. By delivering optical control at the speed (millisecond-scale) and with the precision (cell type-specific) required for biological processing, optogenetic approaches have opened new landscapes for the study of biology, both in health and disease. This technique has revolutionized the ability to remotely control neurons and also can be used in muscle, and cardiac and embryonic stem cells [32]. For this purpose, in order to deliver light of sufficient intensity to deep structures, absorption and scattering must be characterized with high precision and the presented technique could be also applied in this field [33].

Theoretical Background

In this paper, we will describe the mechanism of formation of PM-PA film formation from the corresponding PM and acrylamide solutions. We propose the gelification process to be made in four steps as summarized in Figure 1. Basically, when the polymerization initiator system is added to the PM suspension (step 1 in Figure 1) highly reactive radicals are generated which initiate the polymerization reaction of acrylamide. The formation of polyacrylamide chains is located around PM center in a similar manner as the nucleation step in solid state crystal formation (step 2 in Figure 1). As the polyacrylamide chains grow those PM centers increase their size reaching a maximal value, which can be described as PM-PA nanogels (step 3 in Figure 1). Finally, due to the number of PA-centers and their size, these PA spheres collapse to obtain the PM-polyacrylamide film (step 4 in Figure 1).

In order to analyze the process of formation PM-PA film formation we will perform an analysis of the UV-Visible spectra. Thus, the starting point for this study is the equation of the radiative energy transfer given by:

$$\frac{\partial I(\zeta)}{\partial \zeta} = -(\beta+\beta_R)I(\zeta) \tag{1}$$

Where I is the intensity, ζ is the propagation coordinate of the electromagnetic field, while β and β_R are the macroscopic magnitudes related to the microscopic absorption and scattering cross sections respectively. By integrating Equation 1 with respect to ζ, assuming that β and β_R do not depend on ζ and I and d being the thickness of the film, the Optical Density (D) is given by:

$$D = -\log_{10}\left(\frac{I(\zeta=d)}{I(\zeta=0)}\right) = \frac{(\beta+\beta_R)\,d}{Log\,[10]} \tag{2}$$

As it can be deduced from equation 2, the measured optical density depends on two macroscopic quantities, β related to bR absorption and β_R to the PM-scattering. For clarity, we will continue by describing both contributions (absorption and scattering) separately.

Absorption

The macroscopic parameter β can be obtained from the corresponding microscopic properties by means of Statistical Mechanics according to:

$$\beta(\omega) = N_{bR}\sum_{i}^{3}\sigma_{ii}(\omega) \tag{3}$$

Where σ_{ii} corresponds to the components of the microscopic absorption cross section of the ground state of bR and N_{bR} is the population density of bacteriorhodopsin units in the ground state. Due to the complexity of biomolecular systems such as bacteriorhodopsin, it is very difficult to describe the experimental optical spectra. However, quantum mechanical methods are becoming more important for these analyses. Biomolecular spectra is highly complex and, can be understood on the basis of the spectroscopic properties of building blocks, the simplest case being a single chromophore unit which dominates the spectral signature. We propose that, the microscopic absorption cross section of the ground state of bR can be described by two terms (equation 4), the first one corresponding to the chromophore ($\sigma_{ii,0}(\omega)$) and the second one to the Fano line shape ($\sigma_f(\omega)$) produced by the interaction of a discrete state with a background of continuum of states under the condition that both, the discrete and the continuum levels, are excited by some external perturbation [34–36].

$$\sigma_{ii}(\omega) = \sigma_{ii,0}(\omega)\sigma_f(\omega) \tag{4}$$

Thus, when the applied radiation field excite the discrete state and the broad-band system as well, something analogous to the Fano effect [26,34], can be expected, where the characteristic asymmetric line shape is characterized by the Fano factor q. The Fano factor is the ratio of the transition probabilities of the indirect transition and the direct transition into the ground state being the Fano cross section is given by:

$$\sigma_f(\omega) = \left(\frac{(q+\epsilon)^2}{1+\epsilon^2}\right) \tag{5}$$

where $\epsilon = \dfrac{2(\omega-\omega_f)}{\hbar\gamma_f}$, ω_f is the Fano resonance frequency and γ_f the line width. Equation 5 can be interpreted as interference between the transition into the continuum and the discrete state. It has been satisfactorily employed for justifying the complete optical spectrum of graphene by the excitonic resonance that forms near the van Hove singularity at the saddle point of the band structure and couples to the Dirac continuum [37]; spectral line shapes in both plasmonic and all-dielectric symmetric oligomers [38]; in bulk materials or heavily doped semiconductors in terms of electron-phonon interaction [39]; in nanostructures, where the discrete phonons can interfere with continuum of electronic states available in the material as a result of quantum confinement [40]; for coupled molecules that interact with electron-hole pairs or

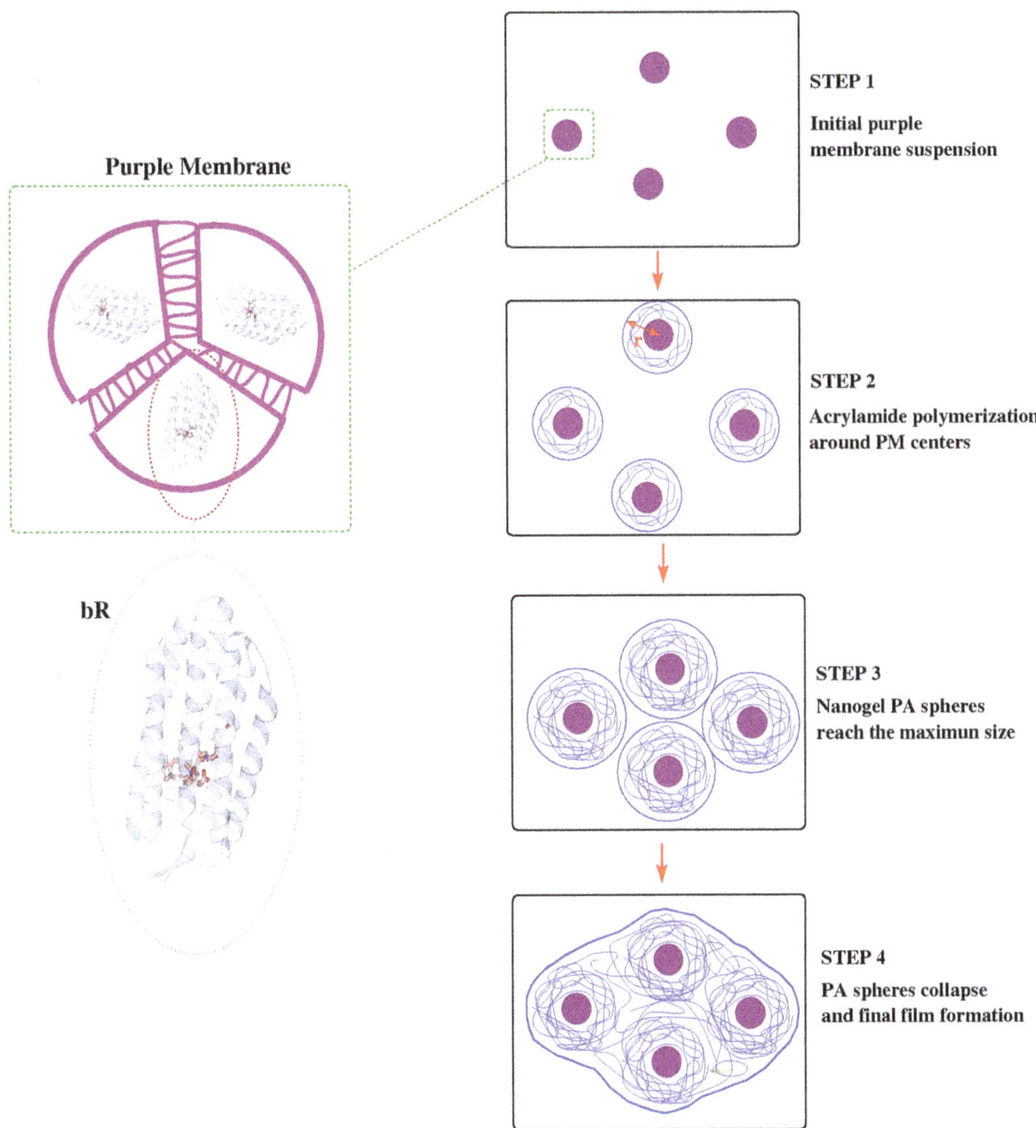

Figure 1. Scheme of the model of formation of PM-polyacrylamide films.

optical phonons in the substrate [35]. In the case of biomolecules such as bR, the Fano formula has been used in several ways such as the interaction between the chromophore (polyene chain) and protein interaction with the surrounding protein environment or the interaction between chromophores [24,41]. In this sense, the origin of electronic energy transfer in several photosynthetic pigment-protein complexes involves quantum coherence, which is a property that is directly connected to the intimate details of the medium surrounding the chromophores, due to the decoherence caused by the coupling of the electronic transitions to the fluctuations of the environment [25,42]. Moreover, experimentally observed and unusually high wavenumber in the IR spectra of the amide I band of bR has been explained by interhelical coupling within a bR monomer [43]. Finally, it is important to add that the visible CD spectrum of bR in purple membrane has a negative CD band at 600 nm and a positive band at 530 nm and has been interpreted by exciton coupling within the bR trimer [44,45], which is an effect that can also justify the Fano profile.

The microscopic absorption cross section of the ground state is given by [46]:

$$\sigma_{ii,0}(\omega) = \frac{\omega(\mu_{1g,i})^2}{c\hbar\epsilon_0} \frac{Exp[-\frac{(\omega-\omega_{1g})^2}{2\gamma^2}]}{\sqrt{2\pi}\gamma} \tag{6}$$

where μ_{1g} is the ground to excited state transition dipole moment of bR, ω_{1g} ($\omega_{1g} = 2\pi c/\lambda_{1g}$) is the frequency of this transition, γ denotes the width of the absorption curve, c is the light speed and ϵ_0 is the dielectric permittivity of the vacuum. In order to assign reliable numerical values to these microscopic parameters we follow two different strategies. On the one hand, we use experimental values present in the literature for μ_{1g} and λ_{1g} [47,48], which allow us to have a complete description of the absorption cross section of the system 6. To be exact, the values of the microscopic optical properties of bR were $\mu_{1g,x} = 3.75 \times 10^{-29}$ (Cm) and $\lambda_{1g} = 568$ nm [49].

By introducing equation 5 and 6 in equation 4, we can study the effect on absorption of the two terms shown on the absorption cross section given by equation 4. Figure 2 shows the effect of Fano parameter (q) on $\sigma_f(\omega)$, also on $\sigma_0(\omega)$ and the resulting $\sigma(\omega)$ obtained from our model. As can be seen, $\sigma_f(\omega)$ presents the typical Fano profile [34,36], the line shape being asymmetric with a dip into the background. As q rises, the line shape becomes symmetric Lorentzian (the external perturbation does not couple to the background state) and when $q \to 0$ the external perturbation does not couple to the discrete state. The inset (a) of Figure 2 shows the typical microscopic absorption cross section of the ground state (Gaussian line shape) which is slightly asymmetrical at long wavelengths due to the ω factor (equation 6). Finally, inset (b) describes the resulting absorption cross section (equation 4) as a function of q parameter. As can be seen, as q rises the asymmetry to short wavelengths increases as well as a bathochromic shift of the maximum.

Scattering

Regarding scattering, β_R, we have assumed that the scatter-particles are spherical with a small radii compared to the wavelength of the scattered light. Thus, β_R is expressed as a function of the Rayleigh cross section according to [50]:

$$\beta_R(\omega) = N_{PM} \frac{8\pi}{3c^4} \omega^4 \kappa^2 \tag{7}$$

κ is a parameter that includes the effect of time-dependent magnitudes during the formation process of the film and it is given by:

$$\kappa = n_{water}^2 \, r^3 \left(\frac{m^2 - 1}{m^2 + 2} \right) \tag{8}$$

where r is the scatter-particle radius, $m = n_{PM-PA}/n_{water}$ is the ratio of the refractive index of the particle (PM-PA) to that of the surrounding medium (water) and N_{PM} is the population density of PM. As can be deduced from equation 7 as the scatter-particle increases the value of β_R rises, this effect being more significant at short wavelengths. Thus, from the analysis of β_R, it is possible to obtain information related to the microscopical structure changes produced during the formation of PM-PA films, i.e. the size variation of the scatter-particle during the film formation or the collapse of those particles. In this sense, taking into account equation 8 and the model described in Figure 1, during the formation of PM-PA films the temporal variation of the scattering coefficient is expected to have two different zones. Initially, there is an increase in the values of the scattering coefficient associated to the size of the scatter-particle (r variable), which grows due to the polymerization process around the PM centers. As a result, a maximum value of the scattering coefficient is obtained when the scatters reach their maximum size, which can be estimated according to:

$$r_{max} = \left(\frac{3\eta}{4\pi N_{PM}} \right)^{(1/3)} \tag{9}$$

where η is the maximum packing fraction of hard spheres randomly distributed (around 0.62 [51]) and N_{PM} is the density of spheres, which can be assumed to be the same as the number of PM centers. Once the spheres reach the maximum-allowed size, the scattering coefficient begins to decrease due to the collapse of spheres, which produces the approach of the m-parameter to the unity and the end of the gelification process. As a result an homogeneous PA medium of PA is obtained instead of water.

Therefore, by using the expressions that describe β and β_R (equations 3–7), the optical density of the system is given by:

$$D = \frac{\left(N_{bR} \sum_i^3 \frac{\omega(\mu_{1g,i})^2}{c\hbar\varepsilon_0} \frac{Exp[-\frac{(\omega-\omega_{1g})^2}{2\gamma^2}]}{\sqrt{2\pi}\gamma} \left(\frac{(q+\frac{2(\omega-\omega_f)}{\hbar\gamma_f})^2}{1+(\frac{2(\omega-\omega_f)}{\hbar\gamma_f})^2} \right) + N_{PM} \frac{8\pi}{3c^4} \omega^4 \kappa^2 \right) d}{Log[10]} \tag{10}$$

Materials and Methods

Purple membrane from Halobacterium salinarum culture were purified following protocol optimized by Oesterhelt and Stoeckenius with minor modifications [52]. Before films processing, PM purity was analyzed by an electrophoresis under denaturing conditions (SDS-PAGE) and absorption spectrum was also measured. We considered the samples with a ratio $Abs_{280nm}/Abs_{568nm} \leq 2.2$ as high-quality. PM-doped films preparation was carried out by using lyophilized PM which were suspended in a solution (10 mg/ml) containing acrylamide-*N,N'*-methylene-bisacrylamide (20%) and Tris(hydroxymethyl)-aminomethane-HCl buffer 0.1 mM at increasing pH (6.0, 7.0, 8.0, 9.0 and 10.0). Each PM-mixture was homogenized using a sonicator (Sonopuls HD 2200) and ammonium persulfate 0.05% (w/v) and *N,N,N',N'*-tetramethylethyldiamine (1 $\mu l/ml$) were added for catalyst and initiation of the polymerization reaction. In all films, gel solution was poured in a 1 mm thick cuvette, where the polymerization process occurred. Finally, the temporal evolution of the polymeric film was carried out by using UV-Visible absorption spectrum (400–700 nm) measured by a spectrophotometer (Agilent Tecnologie) every minute for one hour.

Results and Discussion

In the previous section, we described the mechanism of the formation of PM-PA films from the corresponding PM and acrylamide solutions (Figure 1). In order to prove it, we have performed a real-time analysis of the UV-Visible spectra of the formation of PM-PA film formation at different pHs. As an example, the temporal variation of the UV-Visible spectra at pH = 7 is shown in Figure 3. As can be observed, at the beginning of the process the optical density increases at all wavelengths, this amount being most important at short wavelengths. Furthermore, this effect is observed at all analyzed pHs and depends on PM as demonstrated in Figure 4, where it can be seen that the Optical Density (D) in all the spectra region is in absence of PM is seen to be much lower than when PM is present in the material. Thus, the optical density of UV-Visible spectra at pH = 7 without PM is very low and does not present increase previously described in the initial period.

Taking into account these experimental results, we are going to demonstrate the proposed mechanism of the formation of PM-PA films, by using Equation 10. As stated, the theoretical expression of the optical density, has two terms, the effect of bR absorption and scattering due to the PM. Moreover, bR absorption is described by two terms (equation 4), the one corresponding to the chromophore and the Fano line shape given by the interaction of a discrete state with a background of continuum of states. Thus, in order to characterize the absorption and scattering of PM-PA films, in this

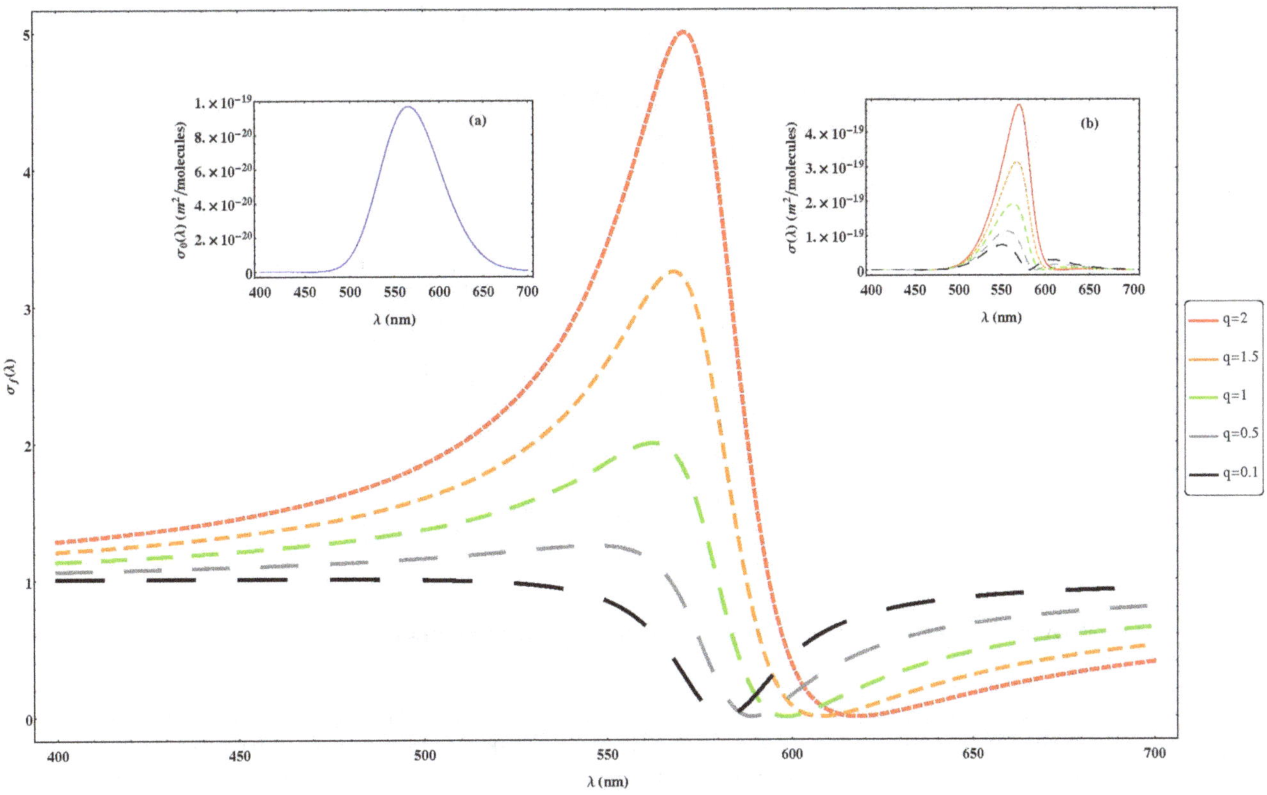

Figure 2. Simulation of $\sigma_f(\lambda)$ (equation 5) for different Fano parameters (q). The inset (a) of the figure represents the microscopic absorption cross section of the ground state ($\sigma_0(\lambda)$) (Equation 6) and the inset (b) the resulting absorption cross section ($\sigma_f(\lambda) \times \sigma_0(\lambda)$) (Equation 4). For these simulations we used the following parameters: $\lambda_f = 580$ nm, $\gamma = \gamma_f = 2 \times 10^{14}$ (s^{-1}) and $\gamma_f = 2 \times 10^{14}$ s^{-1}, $\mu_{1g,x} = 3.75 \times 10^{-29}$ (Cm) and $\lambda_{1g} = 568$ nm.

paper we are going to theoretically analyze the temporal variation of the UV-Visible spectrum (Figure 3) at the pH range of 6.0 to 10.0 by a non-linear fit procedure using equation 10. To do so, λ_f ($\lambda_f = 2\pi c/\omega_f$), q, γ_f, γ and κ are used as free parameters. Furthermore, regarding to the concentration shown in equation 10, it has been assumed that 75% of the PM weight is bR [53] and the concentrations of bR and PM (N_{bR} and N_{PM}) have also been taken as free parameters as $N_{bR} = N_{bR}^0 \times N_{eff}$ and $N_{PM} = N_{PM}^0 \times N_{eff}$ where (N_{bR}^0 and N_{PM}^0) correspond to the initial concentration (experimental) and N_{eff} is a factor that takes into account the active concentration (diminution of the initial one) due to possible

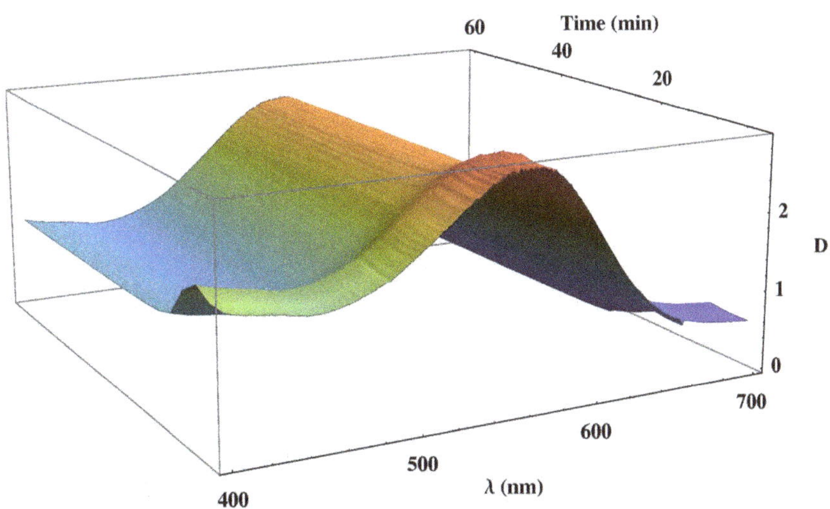

Figure 3. Temporal variation of UV-Visible spectra of the formation of PM-PA films at pH = 7.

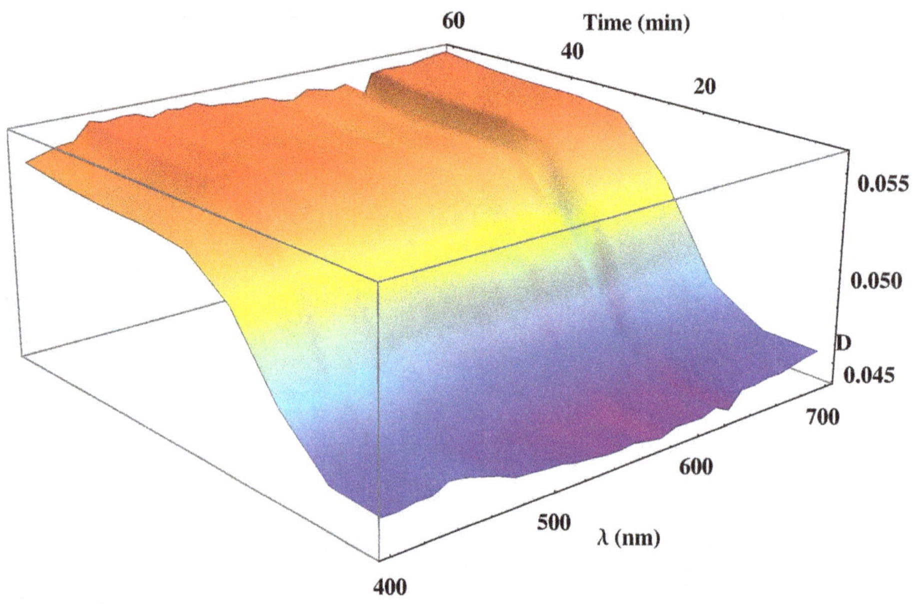

Figure 4. Temporal variation of UV-Visible spectra of the formation of PA films at pH = 7.

irreversible changes that occur during the preparation by thermal decomposition, bleaching, polymerization, pH, etc [17].

In Figure 5, two examples of the experimental and fitted data at pH 6 and 10 are shown for a fixed time of 40 minutes, where the good concordance between theory and experiment ($r^2 > 0.999$) can be seen. In order to measure the effect of pH, by this methodology, the temporal range of between 10 to 60 minutes has been analyzed, and no significant variations are observed in the spectra (Figure 3). Thus, more than 50 experiments at each pH have been fitted, observing that all the free parameters are nearly constant for a fixed pH, the corresponding values obtained at steady-state are shown at table 1. As can be seen the Fano resonance wavelength (λ_f) varies between 592 to 602 *nm*, the lower value reached being at pH = 7.0. Regarding to the Fano parameter (q), which characterizes the line shape and the asymmetric response, increase from 1.7 to 2.8, where the lowest is given at pH 10 and the highest at pH 7. However, in relation to the Fano effect, the bandwidth is quite similar for all cases, ranging between 6.3×10^{14} to 6.6×10^{14}. With respect to the bandwidth of the chromophore absorption, this parameter is similar for all cases except for pH = 7, since, in order to show the effect of all these parameters, the corresponding variation of the term associated to the absorption ($\beta(\omega)$) (equation 3) is analyzed for different pH in Figure 6. As it can be seen all the curves are slightly asymmetric and the effect of pH is not important. At pH = 7 a broader absorption response is obtained whereas at higher pH it is sharper. Regarding the maximum value of $\beta(\omega)$ it is observed that a small shift (1 nm) to shorter wavelengths is obtained for pH 6 and 7, 2 nm for pH 9 and 10, meanwhile for pH 8 the shift is 1 nm to longer wavelengths and the highest value. As previously pointed

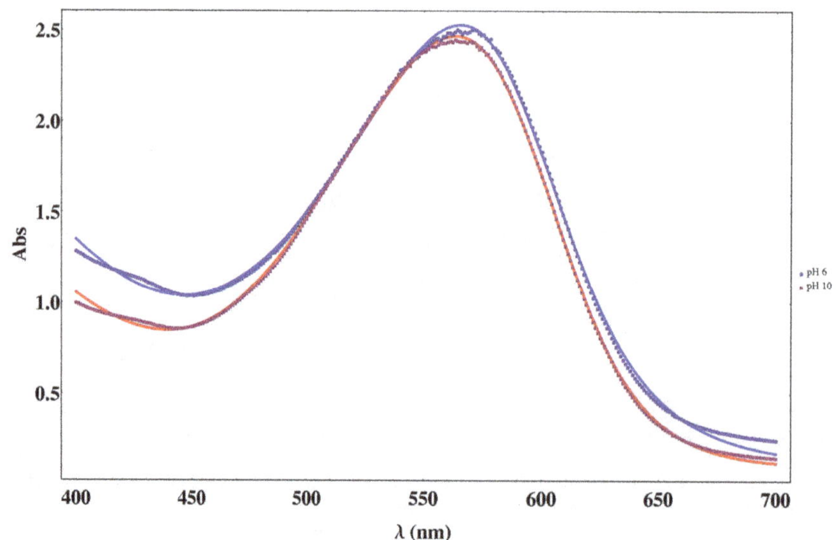

Figure 5. Theoretical (lines) and experimental (points) spectra of PM solutions at pH 6 and pH 10 for a fixed time of 40 minutes.

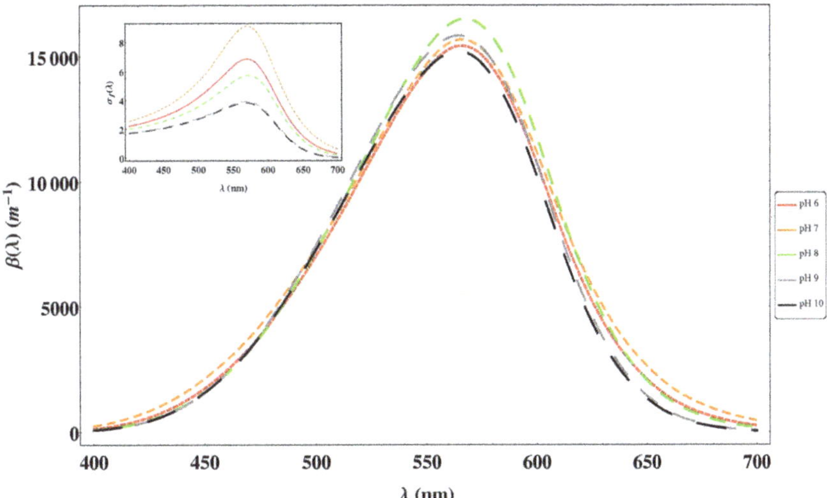

Figure 6. Variation of $\beta(\omega)$ (equation 3) taking into account the parameters obtained as described in Table 1. The inset of the figure represents the contribution of the Fano effect ($\sigma_f(\omega)$) (Equation 5).

out, $\beta(\omega)$ considers the effect of the chromophore and the Fano line shape ($\sigma_f(\omega)$) (equation 5) is produced by the interaction of a discrete state with a background of continuum of states. In order to analyze the Fano effect, the inset of Figure 6 shows the variation of $\sigma_f(\omega)$ obtained by using the parameters given in table 1. At pH 7 a larger and broader asymmetric shape is obtained which could be produced by the protonation effects of several residues or conformation changes of proteinic structure. Regarding the shift of the maxima, a red shift is observed in all cases with respect to the reference value of 568 nm, the larger one (5 nm) being at pH = 8. Finally, the parameter related to scattering (κ) at steady-state is similar at pH 6 and 7 and increases at higher pH values.

In order to analyze the effects observed in the initial period of polymerization (Figure 3) (non steady-state case) and validate the proposed model of film formation, we are going to perform a complete temporal study of the UV-Visible spectrum. Following a similar methodology to the one given above, by a non-linear fitting procedure using equation 10 the temporal UV-Visible spectra has been analyzed by taking only a time variable parameter (κ) related to the scattering losses of the purple membrane and by using the parameters described in table 1 for each pH (except the κ value at the steady-state). As a result, in Figure 7, the temporal variation of the κ values obtained from the non-linear fit ($r^2 \approx 0.999$) as a function of the pH is analyzed. As can be seen, a similar temporal response is obtained for different pH-solutions; this term associated to scattering is nearly constant at the beginning of the process (a few minutes) and after that a quick amount of scattering is produced until finally a weak decrease is produced reaching a saturation value. The differences among the pH values studied are given by the different value of the maximum scattering reached and the rate of reaching it. Moreover, at pH≤7 the saturation value reached is lower than the initial one, and at pH = 7 scattering increases at the beginning (there is no constant period).

These results can be justified by the proposed model of PM-PA formation (Figure 1). According to Figure 7 the amount of scattering (size of scatter-center) is given by the growth of the polyacrylamide chains around the PM centers similar to the nucleation step in solid state crystal formation. All the scatter-centers increase their size and reorient in order to homogeneously distribute in a minimal energetic configuration. When the size of these PM-doped nuclei are sufficient important, a transition from scatter-center to a network is produced, therefore the scattered-center reaches a maximum size. Thus, as the spatial extension of polyacrylamide increases the scattering term diminishes to a saturation value. Finally, the different temporal behaviour obtained as a function of pH is related to the polymerization rate and the length of the formed polymeric chains which depends on the pH [54]. Note at this point, that taking into account these results, by controlling the polymerization process it is possible to obtain bR-doped nano-gels which could be useful for biomedical and technological applications.

Table 1. Values of the parameters λ_f, q, γ_f and γ obtained from the non-linear fitting of PM solutions at different pH.

pH	λ_f(nm)	q	γ_f $(10^{14}s^{-1})$	γ $(10^{14}s^{-1})$	κ $(10^{-50}m^6/\text{molecule})$
6.0	594±4	2.4±0.4	6.6±0.6	4.9±0.2	31±5
7.0	592±3	2.8±0.6	6.8±0.1	5.5±0.8	32±5
8.0	599±3	2.2±0.3	6.3±0.2	4.6±0.3	26±4
9.0	602±2	2.2±0.3	6.3±0.1	4.4±0.2	16±2
10.0	602±2	1.7±0.2	6.3±0.1	4.5±0.2	16±2

The values correspond to the mean at the stationary state and the error to the standard deviation. N_{eff} varies between 0.3 to 0.5 for the different cases studied.

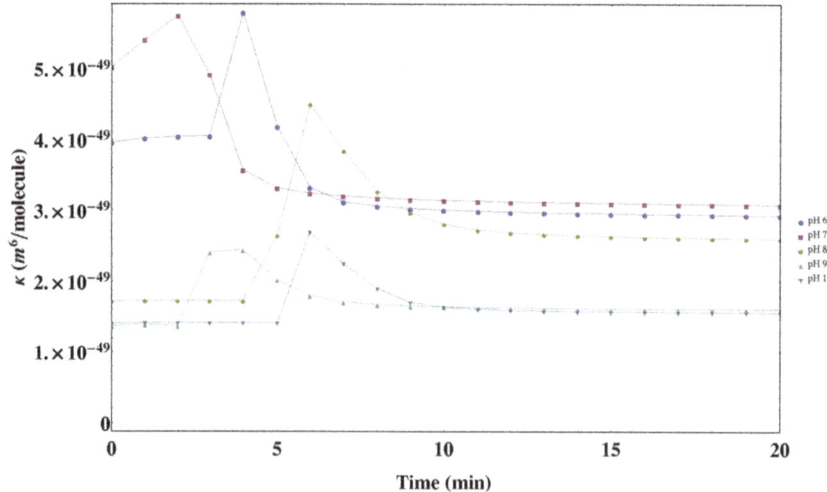

Figure 7. Temporal variation of the κ-parameter during the polymerization of acrylamide as a function of the pH of the PM solution.

Finally, according to equation 8, the range of radius (r) and the refractive index of PM-PA compatibles to the experimental value of κ obtained from the fittings are shown in Figure 8. Due to the polymerization process, it is difficult to ensure the refractive index of PM-PA, which could oscillate between (1.35 and 1.55) [55,56], this figure, therefore, presents an estimation of the radius of PM-PA. As can be seen, a wide range of particle refractive index can be given between 15 to 22 nm. It is important to note that these values are in accordance with the radius sizes estimated by equation 9 for $\eta = 0.62$ and the used PM concentration used, resulting in scatter particle sizes around 17 to 20 nm. Finally, on

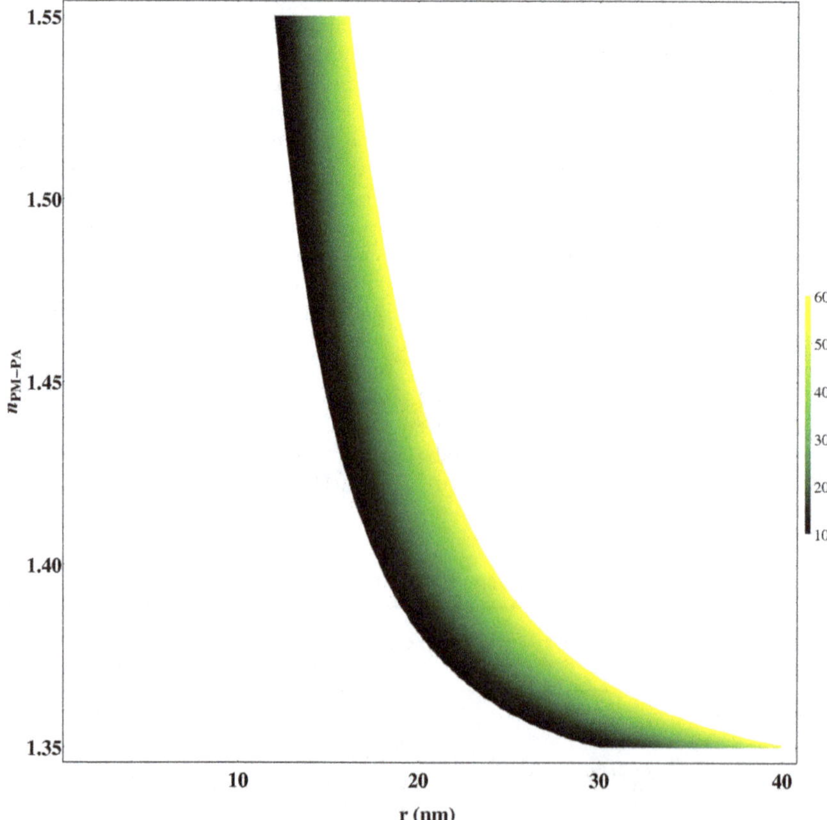

Figure 8. Variation of $\kappa \times 10^{50}$ ($m^6/molecule$) as a function of the PM-PA refractive index and radius obtained by using equation 8 with $n_{water} = 1.334$ [62].

the other hand, a larger radius only can be obtained when the refractive index of PM-PA approximates to the water.

Conclusions

In this study, a model of the formation of thick Purple Membrane-polyacrylamide films has been proposed. During polymerization process of polyacrylamide, polymeric chains grow around purple membrane forming an intermediate PM-doped nanogel. By means of UV-Visible spectroscopy the temporal evolution of the polymerization process of acrylamide for different pH solutions has been experimentally studied. Thus, in order to validate this model, a theoretical treatment has been developed based on semiclassical quantum mechanical techniques for the calculation of absorption spectra. Furthermore, this theoretical analysis takes into account the scattering of the PM and the absorption of bR ground state. However, to describe the observed asymmetrical absorption spectra, Fano line shape has been applied observing a good correspondence between theory and experiment. As a result, the scattering of the PM center and an estimation of its radius has been obtained, verifying that the polymeric chains grow around purple membrane forming an intermediate PM-center. This theoretical analysis of UV-Visible spectra and the possible synthesis route of bR-doped nanogels could have potential applications in biomedicine and photonic technologies. In this sense, challenges exits for preparation of advanced nanogels with novel responsive mechanisms to interact with biological microenvironments. Due to the biocompatibility of the polymers used, the ability to encapsulate and protect purple membrane make these systems attractive for delivery and to use it for example as biosensors [57,58]. Finally, bacteriorhodopsin-based materials have a broad potential application in photonics [59]. Therefore, we believe that the presented insights, particularly the possible synthesis route of bR-doped nanogels and the theoretical analysis brings the possibility to fabricate nanodevices that could give solution to applications such as organic solar cells, nano optical-switches or optical memories [60,61].

Author Contributions

Conceived and designed the experiments: MG SB PA LC. Performed the experiments: MG SB PA LC. Analyzed the data: MG SB PA LC. Contributed reagents/materials/analysis tools: MG SB PA LC. Wrote the paper: MG SB PA LC.

References

1. Birge R, Gillespie N, Izaguirre E, Kusnetzow A, Lawrence A, et al. (1999) Biomolecular electronics: Protein-based associative processors and volumetric memories. J Phys Chem B 103: 10746–10766.
2. Hampp N (2000) Bacteriorhodopsin as a photochromic retinal for optical memories. Chem Rev 100: 1755–1776.
3. Brauchle C, Hampp N, Oesterhelt D (1991) Optical applications of bacteriorhodopsin and its mutated variants. Advanced Materials 3: 420–428.
4. Wise KJ, Gillespie NB, Stuart JA, Krebs MP, Birge RR (2002) Optimization of bacteriorhodopsin for bioelectronic devices. Trends Biotechnol 20: 387–394.
5. Hampp N, Popp A, Bruchle C, Oesterhelt D (1992) Diffraction efficiency of bacteriorhodopsin films for holography containing bacteriorhodopsin wildtype BR_{WT} and its variants BR_{D85E} and BR_{D96N}. J Phys Chem: 4679–4685.
6. Downie JD (1994) Real-time holographic image correction using bacteriorhodopsin. Appl Opt 33: 4353–4357.
7. Yao BL, Wang YL, Lei M, Menke N, Chen GF, et al. (2005) Polarization patterns hide and display using photoinduced anisotropy of photochromic fulgide. Opt Express 13: 20–25.
8. Singh CP, Roy S (2003) All-optical switching in bacteriorhodopsin based on m state dynamics and its application to photonic logic gates. Opt Commun 218: 55–66.
9. Miyasaka T, Koyama K (1993) Image sensing and processing by a bacteriorhodopsin based artificial photoreceptor. Appl Opt 32: 6371–6379.
10. Wu PF, Rao DVGLN (2005) Controllable snail-paced light in biological bacteriorhodopsin thin film. Phys Rev Lett 95.
11. Stoeckenius W, Lozier RH, Bogomolni RA (1979) Bacteriorhodopsin and the purple membrane of halobacteria. Biochim Biophys Acta 505: 215–278.
12. Haupts U, Tittor J, Oesterhelt D (1999) Closing in on bacteriorhodopsin: Progress in understanding the molecule. Annu Rev Biophys Biomol Struct 28: 367–399.
13. Hampp N, Braüchle C, Oesterhelt D (1990) Bacteriorhodopsin wildtype and variant aspartate-96 → aspargine as reversible holographic media. Biophys J 58: 83–93.
14. Birge RR, Einterz CM, Knapp HM, Murray LP (1988) The nature of the primary photochemical events in rhodopsin and isorhodopsin. Biophys J 53: 367–385.
15. Lanyi JK (2000) Molecular mechanism of ion transport in bacteriorhodopsin: Insights from crystallographic, spectroscopic, kinetic, and mutational studies. J Phys Chem B 104: 11441–11448.
16. Becher B, Tokunaga F, Ebrey TG (1978) Ultraviolet and visible absorption-spectra of purple membrane-protein and photocycle intermediates. Biochemistry (Mosc) 17: 2293–2300.
17. Balashov SP, Govindjee R, Ebrey TG (1991) Red shift of the purple membrane absorption-band and the deprotonation of tyrosine residues at high ph - origin of the parallel photocycles of trans-bacteriorhodopsin. Biophys J 60: 475–490.
18. Druzhko AB (2009) Optical characteristics of polymer films based on bacteriorhodopsin for irreversible recording of optical information. Photochem Photobiol 85: 614–616.
19. Fraser RDB, Suzuki E (1969) Resolution of overlapping bands - functions for simulating band shapes. Anal Chem 41: 37–39.
20. Birge RR, Cooper TM, Lawrence AF, Masthay MB, Vasilakis C, et al. (1989) A spectroscopic, photocalorimetric, and theoretical investigation of the quantum efficiency of the primary event in bacteriorhodopsin. J Am Chem Soc 111: 4063–4074.
21. Stavenga DG, Smits RP, Hoenders BJ (1993) Simple exponential functions describing the absorbency bands of visual pigment spectra. Vision Res 33: 1011–1017.
22. van Stokkum IHM, Larsen DS, van Grondelle R (2004) Global and target analysis of time-resolved spectra (vol 1658, pg 82, 2004). Biochimica Et Biophysica Acta-bioenergetics 1658: 262–262.
23. Mercer IP, Gould IR, Klug DR (1999) A quantum mechanical/molecular mechanical approach to relaxation dynamics: Calculation of the optical properties of solvated bacteriochlorophyll-a. J Phys Chem B 103: 7720–7727.
24. Neugebauer J (2009) Subsystem-based theoretical spectroscopy of biomolecules and biomolecular assemblies. ChemPhysChem 10: 3148–3173.
25. Olbrich C, Strumpfer J, Schulten K, Kleinekathofer U (2011) Theory and simulation of the environmental effects on fmo electronic transitions. Journal of Physical Chemistry Letters 2: 1771–1776.
26. Fano U (1961) Effects of configuration interaction on intensities and phase shifts. Phys Rev 124: 1866–1878.
27. Miroshnichenko AE, Flach S, Kivshar YS (2010) Fano resonances in nanoscale structures. Reviews of Modern Physics 82: 2257–2298.
28. Zaitsev SY, Lukashev EP, Solovyeva DO, Chistyakov AA, Oleinikov VA (2014) Controlled influence of quantum dots on purple membranes at interfaces. Colloids and Surfaces B-biointerfaces 117: 248–251.
29. Renugopalakrishnan V, Barbiellini B, King C, Molinari M, Mochalov K, et al. (2014) Engineering a robust photovoltaic device with quantum dots and bacteriorhodopsin. Journal of Physical Chemistry C 118: 16710–16717.
30. Adamov GE, Levchenko KS, Kurbangaleev VR, Shmelin PS, Grebennikov EP (2013) Functional hybrid nanostructures for nanophotonics: Synthesis, properties, and application. Russian Journal of General Chemistry 83: 2195–2202.
31. Hegemann P, Nagel G (2013) From channelrhodopsins to optogenetics. Embo Molecular Medicine 5: 173–176.
32. Kos A, Loohuis NFO, Glennon JC, Celikel T, Martens GJM, et al. (2013) Recent developments in optical neuromodulation technologies. Molecular Neurobiology 47: 172–185.
33. Chuong AS, Miri ML, Busskamp V, Matthews GAC, Acker LC, et al. (2014) Noninvasive optical inhibition with a red-shifted microbial rhodopsin. Nature Neuroscience 17: 1123–1129.
34. Fano U, Cooper JW (1968) Spectral distribution of atomic oscillator strengths. Reviews of Modern Physics 40: 441–507.
35. Sorbello RS (1985) Vibrational-spectra of coupled adsorbed molecules. Physical Review B 32: 6294–6301.
36. Riffe DM (2011) Classical fano oscillator. Physical Review B 84: 064308.
37. Chae DH, Utikal T, Weisenburger S, Giessen H, von Klitzing K, et al. (2011) Excitonic fano resonance in free-standing graphene. Nano Lett 11: 1379–1382.
38. Hopkins B, Poddubny AN, Miroshnichenko AE, Kivshar YS (2013) Revisiting the physics of fano resonances for nanoparticle oligomers. Physical Review A 88: 053819.
39. Gupta R, Xiong Q, Adu CK, Kim UJ, Eklund PC (2003) Laser-induced fano resonance scattering in silicon nanowires. Nano Lett 3: 627–631.
40. Kumar R (2013) Asymmetry to symmetry transition of fano line-shape: analytical description. Indian J Phys 87: 49–52.

41. Fujimoto K, Hayashi S, Hasegawa J, Nakatsuji H (2007) Theoretical studies on the color-tuning mechanism in retinal proteins. J Chem Theory Comput 3: 605–618.
42. Mennucci B, Curutchet C (2011) The role of the environment in electronic energy transfer: a molecular modeling perspective. Phys Chem Chem Phys 13: 11538–11550.
43. Karjalainen EL, Barth A (2012) Vibrational coupling between helices influences the amide i infrared absorption of proteins: Application to bacteriorhodopsin and rhodopsin. J Phys Chem B 116: 4448–4456.
44. Wu JW (1991) Birefringent and electro-optic effects in poled polymer films: steady-state and transient. J Opt Soc Am B: 142–152.
45. Pescitelli G, Woody RW (2012) The exciton origin of the visible circular dichroism spectrum of bacteriorhodopsin. J Phys Chem B 116: 6751–6763.
46. May V, K O (2000) Charge and energy transfer dynamics in molecular system. Wiley-VCH.
47. Huang JY, Chen Z, Lewis A (1989) Second-harmonic generation in purple membrane-poly(vinyl alcohol) films: probing the dipolar characteristics of the bacteriorhodopsin chromophore in bR_{570} and M_{412}. J Phys Chem 93: 3314–3320.
48. Stuart JA, Marcy DL, Wise KJ, Birge RR (2002) Volumetric optical memory based on bacterirhodopsin. Synthetic Metals 127: 3–15.
49. Acebal P, Carretero L, Blaya S, Murciano A, Fimia A (2007) Theoretical approach to photoinduced inhomogeneous anisotropy in bacteriorhodopsin films. Phys Rev E 76: 016608.
50. Bohren CF, Huffman DR (1983) Absorption and Scattering of Light by Small Particles. John Wiley & Sons Inc.
51. Torquato s (1995) Nearest-neighbor statistics for packings of hard-spheres and disks. Physical Review E 51: 3170–3182.
52. Oesterhelt D, Stoeckenous W (1971) Rhodopsin-like protein from purple membrane of halobacterium-halobium. Nature-new Biology 233: 149–152.
53. Kates M, Kushwaha SC, Sprott GD (1982) Lipids of purple membrane from extreme halophiles and of methanogenic bacteria. Methods Enzymol 88: 98–111.
54. Decker C (1992) Kinetic-study of light-induced polymerization by real-time uv and ir spectroscopy. Journal of Polymer Science Part A-polymer Chemistry 30: 913–928.
55. Zhivkov AM (2002) ph-dependence of electric light scattering by water suspension of purple membranes. Colloids and Surfaces A-physicochemical and Engineering Aspects 209: 319–325.
56. Byron ML, Variano EA (2013) Refractive-index-matched hydrogel materials for measuring flow-structure interactions. Experiments In Fluids 54: 1456.
57. Maya S, Sarmento B, Nair A, Rejinold NS, Nair SV, et al. (2013) Smart stimuli sensitive nanogels in cancer drug delivery and imaging: A review. Current Pharmaceutical Design 19: 7203–7218.
58. Liu GY, An ZS (2014) Frontiers in the design and synthesis of advanced nanogels for nanomedicine. Polymer Chemistry 5: 1559–1565.
59. Adamov GE, Devyatkov AG, Gnatyuk LN, Goldobin IS, Grebennikov EP (2008) Bacteriorhodopsin - perspective biomaterial for molecular nanophotonics. Journal of Photochemistry and Photobiology A-chemistry 196: 254–261.
60. Peppas NA, Hilt JZ, Khademhosseini A, Langer R (2006) Hydrogels in biology and medicine: From molecular principles to bionanotechnology. Advanced Materials 18: 1345–1360.
61. Patil AV, Premaraban T, Berthoumieu O, Watts A, Davis JJ (2012) Engineered bacteriorhodopsin: A molecular scale potential switch. Chemistry-a European Journal 18: 5632–5636.
62. Centeno M (1941) The refractive index of liquid water in the near infra-red spectrum. J Opt Soc Am 31: 244–247.

18

TCR Triggering by pMHC Ligands Tethered on Surfaces via Poly(Ethylene Glycol) Depends on Polymer Length

Zhengyu Ma[1]*, David N. LeBard[2], Sharon M. Loverde[3], Kim A. Sharp[4], Michael L. Klein[5], Dennis E. Discher[6], Terri H. Finkel[7,8]

1 Department of Biomedical Research, Nemours/A.I. duPont Hospital for Children, Wilmington, Delaware, United States of America, 2 Department of Chemistry, Yeshiva University, New York, New York, United States of America, 3 Department of Chemistry, College of Staten Island, City University of New York, Staten Island, New York, United States of America, 4 Department of Biochemistry and Biophysics, University of Pennsylvania, Philadelphia, Pennsylvania, United States of America, 5 Institute for Computational Molecular Science and Department of Chemistry, Temple University, Philadelphia, Pennsylvania, United States of America, 6 Department of Chemical and Biomolecular Engineering, University of Pennsylvania, Philadelphia, Pennsylvania, United States of America, 7 Department of Pediatrics, Nemours Children's Hospital, Orlando, Florida, United States of America, 8 Department of Biomedical Sciences, University of Central Florida College of Medicine, Orlando, Florida, United States of America

Abstract

Antigen recognition by T cells relies on the interaction between T cell receptor (TCR) and peptide-major histocompatibility complex (pMHC) at the interface between the T cell and the antigen presenting cell (APC). The pMHC-TCR interaction is two-dimensional (2D), in that both the ligand and receptor are membrane-anchored and their movement is limited to 2D diffusion. The 2D nature of the interaction is critical for the ability of pMHC ligands to trigger TCR. The exact properties of the 2D pMHC-TCR interaction that enable TCR triggering, however, are not fully understood. Here, we altered the 2D pMHC-TCR interaction by tethering pMHC ligands to a rigid plastic surface with flexible poly(ethylene glycol) (PEG) polymers of different lengths, thereby gradually increasing the ligands' range of motion in the third dimension. We found that pMHC ligands tethered by PEG linkers with long contour length were capable of activating T cells. Shorter PEG linkers, however, triggered TCR more efficiently. Molecular dynamics simulation suggested that shorter PEGs exhibit faster TCR binding on-rates and off-rates. Our findings indicate that TCR signaling can be triggered by surface-tethered pMHC ligands within a defined 3D range of motion, and that fast binding rates lead to higher TCR triggering efficiency. These observations are consistent with a model of TCR triggering that incorporates the dynamic interaction between T cell and antigen-presenting cell.

Editor: Marek Cebecauer, J. Heyrovsky Institute of Physical Chemistry, Czech Republic

Funding: This work was supported by NIH 1R21 AI087516, NIH 1R21 AI078387, NIH P20GM103464, the University of Pennsylvania Center for AIDS Research pilot grant (2-P30-AI-045008-11), the Children's Hospital of Philadelphia (CHOP) and the CHOP Research Institute, and the Nemours Foundation and Nemours Children's Hospital. The funders had no role in study design, data collection and analysis, decision to publish, or preparation of the manuscript.

Competing Interests: The authors have declared that no competing interests exist.

* Email: zma@nemours.org

These authors contributed equally to this work.

Introduction

T cells recognize antigens through the binding between T cell receptors (TCRs) and peptide-major histocompatibility complexes (pMHCs) at the interface between T cell and antigen presenting cells (APCs). pMHC-TCR binding triggers TCR signaling that activates T cells. T cell activation initiates T cell-mediated adaptive immune responses, which are responsible for pathogen clearance or autoimmune disease, depending on the source of peptide antigen. Despite its critical importance, it remains unclear how specific pMHC-TCR binding initiates, or triggers, a signal from the TCR in the first place. The mechanism of TCR signal initiation, also called "the TCR triggering puzzle", cannot be explained by classical models such as receptor conformational change or crosslinking [1,2].

A key feature of TCR triggering is the two dimensional (2D) nature of pMHC-TCR interaction. pMHC and TCR are anchored on plasma membranes and their movement is limited to 2D diffusion. The binding between pMHC and TCR, therefore, can only occur when the two plasma membranes are brought together through cell-cell contact and are closely aligned by adhesion molecules. The membrane-membrane contact, however, is not static. The T cell-APC interaction is dynamic and their relative motion inevitably applies mechanical stress to the interacting membranes and pMHC-TCR binding. Several models of TCR triggering have been proposed by taking into consideration certain features of the complex 2D pMHC-TCR interaction. The kinetic segregation model of TCR triggering, for example, proposes that the closely aligned membranes create steric barriers that segregate surface molecules based on their size [3]. The exclusion of large molecules such as tyrosine phosphatase CD45 from the vicinity of bound pMHC and TCR, which are both relatively small, initiates TCR signaling by creating tyrosine kinase-rich zones around the TCR. The receptor deformation model, on the other hand, postulates that the binding between membrane-anchored pMHC and TCR transfers mechanical forces associated with cell locomotion to the TCR/CD3 complex.

The mechanical forces deform the TCR/CD3 into a conformation or configuration that favors signal initiation [4,5].

The binding properties that determine the efficiency of TCR triggering should also be considered in a 2D context. The kinetics of 3D binding is largely determined by the overall binding activation energy and bond formation detail at the binding interface. 2D binding kinetics, on the other hand, is additionally influenced by factors such as ligand and receptor size, lateral diffusion rate, pre-aligned binding interface [6,7], and mechanical stress associated with membrane dynamics [8]. Studies on the relationship between the 2D kinetics of pMHC-TCR binding (on-rate, off-rate and affinity) and its signaling potential have started to emerge recently [9,10,11]. The results, however, have been highly inconsistent.

To delineate how the 2D nature of pMHC-TCR interaction contributes to TCR triggering, here we altered the 2D pMHC-TCR interaction by tethering pMHC on surfaces with flexible poly(ethylene glycol) (PEG) polymer linkers of varying lengths, and compared their effects on T cell activation. With increase in polymer length, tethered pMHC ligands have an increased range of motion in the third dimension. Thus, the pMHC-TCR interaction becomes more 3D-like. We found that pMHC ligands tethered with PEG polymers of up to 380 nm were capable of triggering TCR. The efficiency of triggering, however, gradually decreased with increase in linker length. Molecular dynamics simulation suggested that pMHC tethered with longer polymers binds its receptors with slower on-rates and off-rates. These observations are consistent with the receptor deformation model of TCR triggering.

Results

Tethering pMHC Ligands to a Surface with PEG Polymer Linkers

To tether pMHC ligands to a surface using PEG linkers, we first conjugated pMHC with a PEG polymer, then tethered the pMHC-PEG conjugates onto a plastic surface through biotin-streptavidin interactions (Fig. 1). To this end, mouse MHC class II molecule IEk with covalently linked moth cytochrome c (MCC) peptide (aa88-103) was engineered to have a free cysteine at the C-terminal end of the IEk β chain. The protein was expressed in a baculovirus insect expression system in secreted form and purified with affinity chromatography (Fig. S1A) [12,13]. The purified protein eluted as 50 kDa monomers in gel filtration chromatography (Fig. S1B). The protein was then conjugated with hetero-bifunctional polymer linker Maleimide-PEG-Biotin (Mal-PEG-Bio) through the reaction between the maleimide group of the PEG linker and the sulfhydryl group (−SH) of the protein C-terminal cysteine. Nine PEG linkers of different lengths with molecular weights ranging from 88 to 60000 Da were used (Table 1). In gel filtration, these polymers were eluted in the expected volumes and order (Fig. 2A). After the reaction, the mixture containing IEkMCC-PEG conjugates, unreacted IEkMCC, and unreacted PEG polymers was subjected to chromatography for separation (Fig. 2B). The three products of the reaction with mid-length polymers (PEG 3500, PEG 5000, and PEG 7500) were separated by a single round of gel filtration chromatography and the conjugate peaks were collected. For reactions with large polymers (PEG 15000, PEG 30000, and PEG 60000), the conjugates and the polymers could not be separated by gel filtration. The polymers were therefore first eliminated with an IEk-specific antibody 14-4-4s affinity column. The remaining IEkMCC-PEG conjugates and free IEkMCC were then separated using gel filtration. For reactions with small polymers (PEG 88, PEG 484 and PEG 2000), gel filtration could not separate IEkMCC and conjugates, but polymers could be eliminated. In this case, the peaks containing both IEkMCC and conjugates were collected. The concentrations of those conjugates were calculated based on measured biotin concentration and the knowledge that each IEkMCC molecule can have at most one biotin. Any free IEkMCC in the mixture has no affect on the subsequent tethering step since it cannot bind streptavidin.

As shown with the gel filtration analyses (Fig. 2A), the hydrodynamic sizes of the polymers are much larger than those of globular proteins of similar molecular weights, indicating a relatively extended conformation of the PEG polymers in aqueous solution. For example, PEG 7500 was eluted with a similar volume to a protein with molecular weight of 44000 Da. Conjugation of PEG polymers significantly increased the hydrodynamic size of IEkMCC protein (Fig. 2B). Addition of a PEG 7500 polymer to IEkMCC almost doubled its apparent molecular weight, leading to baseline separation of the conjugates and unreacted IEkMCC. Consistent with their monomeric nature, the conjugates did not activate T cells when used in solution even at high concentrations (data not shown).

FRET Characterization of Surface-tethered pMHCs

IEkMCC-PEG conjugates, each with a biotin at the free end of the PEG polymer, were tethered on plastic surfaces covalently coated with streptavidin (Fig. 1). To analyze the average distance of IEkMCC protein to the surface, the efficiency of fluorescence resonance energy transfer (FRET) between DyLight 549-labeled IEkMCC and DyLight 649-labeled streptavidin was measured (Fig. S2). As shown in Fig. 3A, high FRET efficiency was observed between IEkMCC and streptavidin linked with the shortest linker PEG 88. FRET efficiency gradually decreased with increasing length of the PEG linker. FRET efficiency was no longer measurable when linkers longer than PEG 7500 were used.

The acceptor photobleaching-based FRET assay used here compares the fluorescence intensity of FRET donor (DyLight 549) before and after the photobleaching of the FRET acceptor (DyLight 649). The photobleaching step takes many minutes. Therefore, the FRET efficiency indicates an average distance between IEkMCC and streptavidin, rather than an instantaneous distance. The FRET results are consistent with the average shape of the PEG polymer as a sphere of Flory radius in an aqueous medium. Indeed, after normalization against the FRET efficiency of PEG 88, the measured FRET efficiencies of PEG polymers matched closely with the ones calculated based on the Flory radii of the polymers (Fig. 3B). It should be noted, however, that IEkMCC proteins are not fixed at a particular distance from the surface for each surface-tethered pMHC. Behaving similarly to an ideal chain [14], a PEG polymer is highly dynamic in solution. The IEkMCC protein can therefore be positioned at any distance within the contour length (fully extended length) of the polymer at any given moment. Taken together, the FRET data indicate that IEkMCC ligands tethered on the surface via PEG polymers behaved as predicted based on known PEG behavior.

T Cell Activation by pMHC Ligands Tethered to a Surface

To determine how T cell activation is affected by ligand-surface tethering by PEG polymers, IEkMCC ligands tethered to a surface with nine different PEG polymers were used to stimulate IEkMCC-specific 5C.C7 T cells. The surface densities of ligands tethered with different polymers were comparable when detected using the IEk-specific antibody 14-4-4s (Fig. S3). T cells were stimulated in ligand-coated wells for 6 hours, and IL2 production was measured by intracellular staining and flow cytometry. Dose-

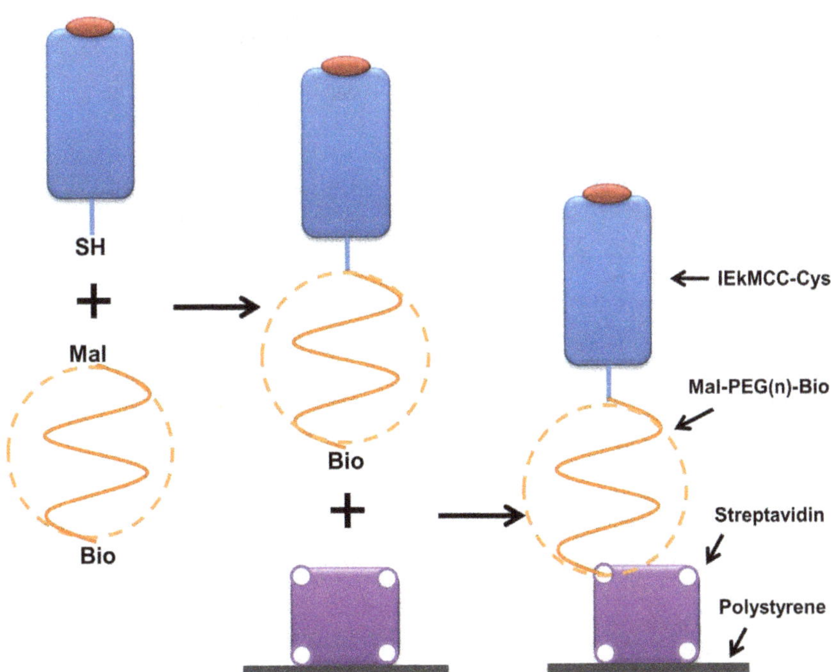

Figure 1. Schematic illustration of IEkMCC ligands tethered onto a plastic surface with PEG polymer linkers. IEkMCC proteins with free c-terminal cysteines were first conjugated with heterobifunctional PEG linkers Mal-PEG-Bio through interactions between the sulfhydryl group and the maleimide group. Conjugates with biotin at the free ends of the polymer were then tethered to a plastic surface coated with streptavidin.

dependent responses were observed for all linkers, as well as an inverse correlation between the percentage of IL2 producing cells and linker length above a critical size (Fig. 4A). IEkMCC tethered with PEG 88, PEG 484, PEG 2000, and PEG 3500appeared to stimulate IL2 production with similar and high efficiency. Linkers larger than PEG 3500 showed a gradually decreasing ability to stimulate IL2 production, although IEkMCC tethered with the longest linker, PEG 60000 (contour length = 380 nm), was still capable of inducing low levels of IL2 production at high coating concentrations (>100 pM).

IEkMCC with the shortest linker, PEG 88, had the same efficiency as IEkMCC without a PEG linker but biotinylated at the lysine residue of a c-terminal AviTag sequence (IEkMCC-bio) [15] (Fig. S4A). This was expected, since PEG 88 has a contour length of only 0.6 nm, or the length of about 5 peptide bonds (0.13 nm). Consistent with the dispensable role of costimulatory molecules in activating pre-activated T cells [16,17], addition of costimulatory molecule B7.1 to the surface did not enhance the ability of IEkMCC-PEG 88 to induce cytokine production (Fig. S4A). Addition of free polymers did not affect T cell activation by pMHC-bio (Fig. S4B). The maleimide group at one end of the free polymer was hydrolyzed to abolish its reactivity. The biotin group on the other end allows binding to streptavidin on the plate. T cells adhered well to plate surfaces coated with all IEkMCC-PEGs regardless of PEG length (Fig. S5). T cell adhesion may have been facilitated by binding to RYD sequences in streptavidin, which mimic the integrin-binding RGD sequence [18].

Table 1. PEG linkers and properties.

Linkers	MW (Dalton)	Number of PEO units	Contour length (nm)[1]	Flory radius (nm)[2]	FRET efficiency (%)
PEG 88	88	2	0.6	0.5	58
PEG 484	484	11	3.1	1.2	53
PEG 2000	2000	45	12.7	2.8	34
PEG 3500	3500	80	22.3	3.9	20
PEG 5000	5000	114	31.8	4.8	14
PEG 7500	7500	170	47.7	6.1	2
PEG 15000	15000	341	95.5	9.3	-
PEG 30000	30000	682	190.9	14.0	-
PEG 60000	60000	1364	381.8	21.3	-

[1]PEG contour length is calculated based on the PEO unit length of 0.28 nm in water [32].
[2]The Flory radius (R_F) of the PEG polymer of N subunits and unit length a was calculated using $R_F = a{\cdot}N^{3/5}$, where a is 0.28 nm.

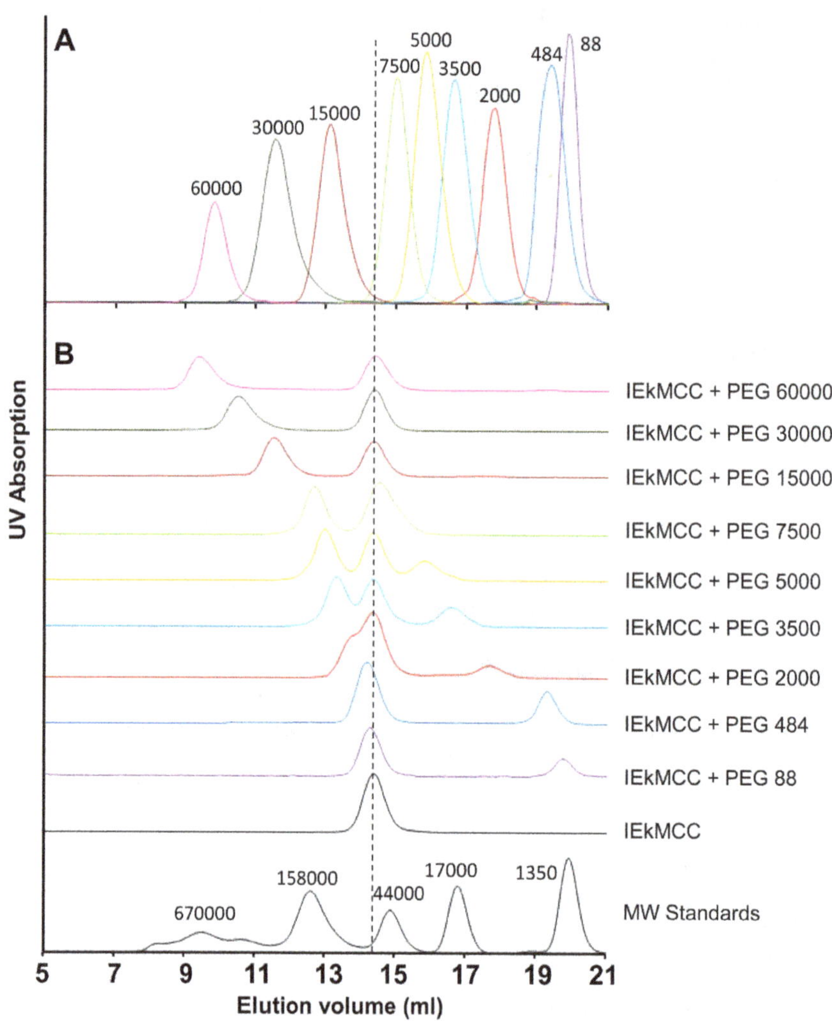

Figure 2. Characterization and separation of PEG polymer linkers and IEkMCC-PEG conjugates. (A) Compiled elution curves of nine PEG polymer linkers from a Superdex 200 10/300 GL gel filtration column. The polymers were detected through the weak UV absorption of the biotin group using a 245 nm UV detector. (B) Separation of the IEkMCC and PEG polymer reaction products. The reaction products were loaded on a Superdex 200 10/300 GL gel filtration column to separate IEkMCC-PEG conjugates, unreacted IEkMCC, and unreacted PEG polymers. The reaction products of PEG 15000, PEG 30000 and PEG 60000 were first purified with an IEk-binding affinity column to eliminate unreacted PEG polymers. The dotted vertical line indicates the elution volume of IEkMCC protein. The late elution peaks of unreacted polymers can be seen for PEGs ranging from PEG 88 to PEG 5000. In reaction with PEG 7500, unreacted IEkMCC and unreacted polymer formed a single peak that was eluted at a position between unconjugated IEkMCC and pure PEG 7500.

T cell responses to different concentrations of ligands at a given time point may not accurately reflect the relationship between T cell response and PEG linker length. It is possible, for example, that similar responses to the shorter polymers are due to saturation of the response at the time of stimulation termination. In other words, these polymers may have stimulated T cells to similar levels of response at a particular end point, but with a different rate. To explore this, T cells were stimulated with a fixed concentration of pMHC ligand and IL2 production assayed at different time points. As shown in Fig. 4B and 4C, the rate of T cells committing to IL2 production depended on the PEG linker length in a manner similar to that observed for dose response. Taken together, these results demonstrate that T cell responses inversely correlate with PEG polymer length, with the exception of PEGs smaller than PEG 5000.

Kinetics of TCR Interaction with pMHC Tethered on a Surface by Molecular Dynamics Simulation

To explore how PEG length influences the kinetics of pMHC-TCR binding, we carried out coarse-grained Molecular Dynamics (CG-MD) simulations to examine the binding *in silico*. As shown in Fig. S6, the interaction between PEG-tethered pMHC ligand and TCR on a cell surface was represented by the interaction between the free end of a fully flexible PEG polymer (with another end fixed in position) and a patch of DMPC lipid bilayer [19,20]. A weak binding potential mimicking the affinity of pMHC-TCR binding (~10 μM) was added between the free end of the PEG polymer and the lipid bilayer. Binding was defined as an event in which the mobile tail bead of the polymer moves within a 1 nm cutoff distance from any portion of the lipid bilayer. The PEG-bilayer system was simulated as a function of PEG length to examine how polymer length (chain entropy) affects rate of binding. PEG 4000, PEG 10000 and PEG 20000 were chosen

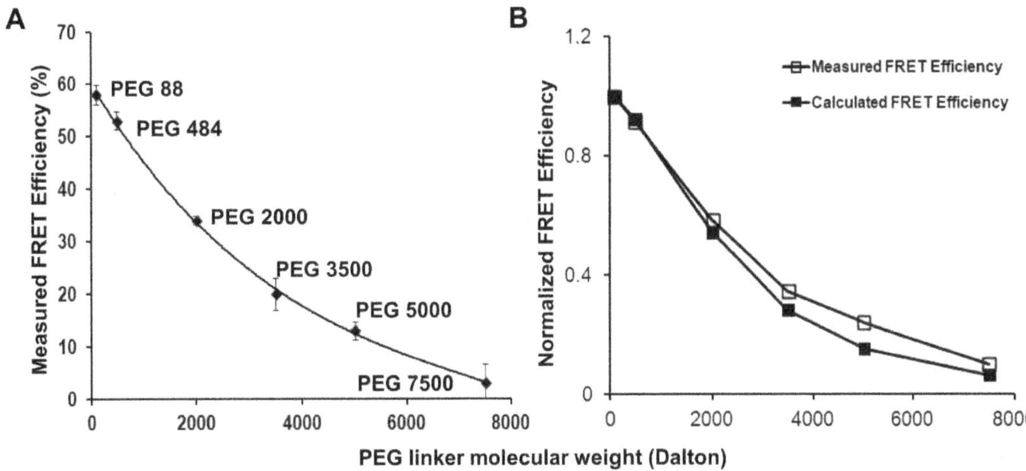

Figure 3. FRET between streptavidin on plastic plates and IEkMCC tethered with PEG polymers. (A) Measured FRET efficiencies of IEkMCC tethered with six different PEG polymers. The intensity of DyLight 549 was captured before and after DyLight 649 was photobleached. The measured FRET efficiency (E_m) was calculated using the intensity of DyLight 549 before (Ib) and after (Ia) DyLight 649 photobleaching ($E_m = \frac{Ia - Ib}{Ia}$). The averaged values of two measurements were plotted with standard deviations. (B) After normalization, the measured FRET efficiencies match those calculated based on the Flory radius (R_F) of the PEG polymers. The R_F of the PEG polymer of N subunits and unit length a was calculated using $R_F = a \cdot N^{3/5}$, where a is 0.28 nm [32]. Theoretical FRET efficiency (E_t) was calculated using the equation $E_t = \frac{1}{1+(r/R_0)^6}$, where the Förster distance (R_0) of the DyLight 549-DyLight 649 donor-acceptor pair is 5 nm and the distance between the pMHC ligand and streptavidin r is R_F of the PEG polymer plus the pMHC radius of 2 nm. The FRET efficiencies were normalized by dividing the FRET efficiencies by the FRET efficiency of PEG 88.

mainly because of the clear PEG length effect on T cell activation in this MW range. Also, MD simulations of longer PEG lengths are unrealistic due to their long relaxation times (≥200 ns). Consideration of protein geometry was omitted to make the simulation feasible with available computational resources. While this simplification hinders derivation of absolute values for binding kinetics, the relative impact of polymer length on binding kinetics may be assessed. Details of the multiple-replica CG-MD [21,22] simulations accelerated by graphics processing units (GPU's) [23] are fully described in Text S1.

To assess the relative impact of polymer length on binding kinetics, we scaled the binding rates of PEG 10000 and PEG 20000 against the rates of PEG 4000. As shown in Figure 5, both the on-rate and off-rate of the binding decreased with linker length. The decrease of the on-rate is possibly due to steric effects of long polymers limiting access to binding sites, as shown in Text S1. The decrease of the off-rate with polymer length may be explained by the lower entropic force applied to the binding by longer polymers, since the entropic force of the polymer is inversely proportional to the length of the polymer chain. The decrease of overall binding rate with polymer length can also be justified in a conceptual context by considering the longer relaxation time of the longer polymers (Text S1). We also estimated the effective force applied to the pMHC-TCR binding from the CG-MD simulations using the potential of mean force acting at the most probable binding distance. Around 40 pN of force was applied by PEG 4000 and the force decreased slightly for longer polymers. It has been shown that 40 pN is large enough to initiate protein conformational changes [24,25]. Together, our CG-MD results suggest that the ability of tethered pMHC to activate T cells correlates with the rates of TCR binding. In addition, pMHC-TCR may sustain a significant amount of mechanical force derived from polymer entropic energy.

Discussion

To investigate how 2D pMHC-TCR interaction contributes to TCR triggering, we gradually increased the range of motion of pMHC ligands in the third dimension using flexible PEG polymers. Our approach differs from previous studies where pMHC or scFv of anti-CD3 antibody were fused on top of CD4 and other transmembrane proteins [26,27]. The TCR binding sites of these molecules are elevated from the cell surface at a fixed distance by the added protein domains. Their movement in the third dimension, however, still fully depends on the movement of the cell membrane. Therefore, these molecules still interact with TCR in a strictly 2D fashion. In contrast, pMHC ligands in our system are tethered to a surface through flexible PEG linkers. Although these polymers take on a spherical equilibrium configuration in an aqueous environment, their chain units move much more freely than the amino acid residues in a folded protein molecule. For instance, PEG 7500 and pMHC were eluted at similar volumes from gel filtration columns, indicating their similar hydrodynamic radii. When attached to a surface on one end, however, the free end of PEG 7500 polymer could be at any position within the contour length of the polymer (~48 nm) at a given moment, whereas the distal end of pMHC would stay 5 nm above the surface. On the other hand, it should be noted that the "free" end of the polymer does not diffuse freely like an untethered molecule. Its positional dynamics is determined by thermal motion, which will be affected by the chain entropy of the polymer.

The interaction between TCR and pMHC tethered to a surface with PEG polymers therefore must have characteristics of both 2D and 3D interactions. Ideally, their binding kinetics should be determined experimentally. This, however, is technically highly challenging and is beyond the scope of this paper. Unlike 3D interaction kinetics that can be routinely characterized with the mature surface plasmon resonance technology using commercially

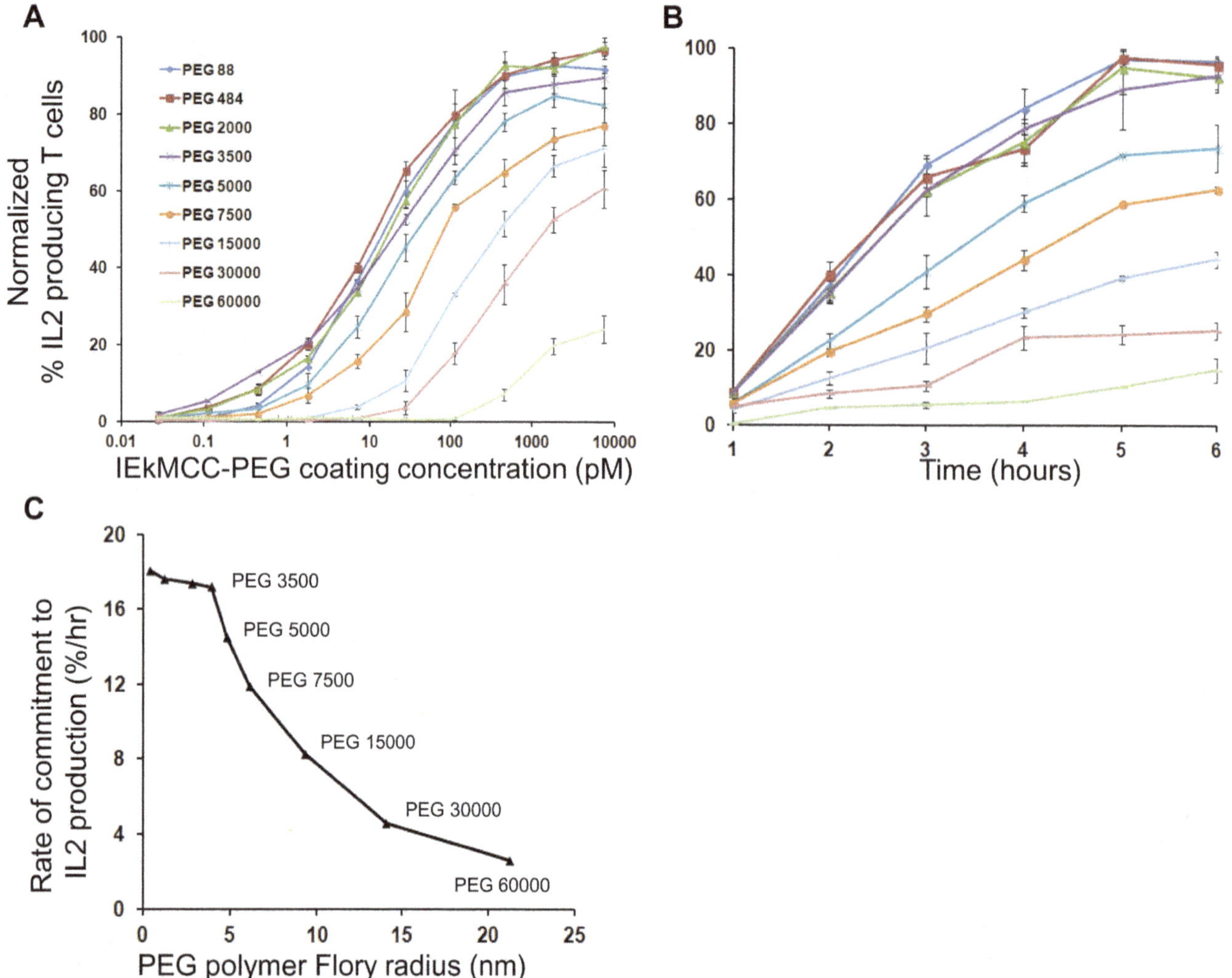

Figure 4. T cell activation by IEkMCC tethered with PEG polymers of different lengths. (A) T cell IL2 production in response to IEkMCC-PEG ligands of varying coating densities after 6 hours of stimulation. Data are representative of three independent experiments. The percent of T cells producing IL2 was determined by intracellular staining and flow cytometry. Three experiments using T cells from three different mice were performed (see Fig. S8 for flow cytometry plots). The percent of T cells producing IL2 was normalized to the highest value in each experiment. The data points are averages of the normalized values with standard errors of the means. (B) The rate of T cell response to IEkMCC ligands tethered with PEG polymers of different lengths. T cell IL2 production in response to stimulation on 96 well plates coated with 110 pM IEkMC-PEG ligands. T cells were harvested every hour for 6 hours and levels of IL2 expression were assayed by flow cytometry. Three experiments using T cells from three different mice were performed (see Fig. S9 for flow cytometry plots). The percent of T cells producing IL2 was normalized to the highest value in each experiment. The data points are averages of the normalized values with standard errors of the means. (C) The rates of T cell IL2 responses to IEkMCC ligands tethered with PEG polymers were extracted from the slope of linear fitting curves in Fig. 4B and plotted against the Flory radius of the polymers. The linear regressions and equations for deriving the rates are shown in Fig. S7.

available systems, such as Biacore, 2D kinetics characterization is still a developing field of research. Different methodologies often give dramatically different results for similar ligand-receptor interactions. For example, 2D pMHC-TCR interactions have been characterized using laminar flow chamber [9], fluorescence microscopy imaging [10], and micropipette and biomembrane force probes [11]. The 2D off-rates obtained varied from being similar to the 3D off-rate [9], to about ten-fold higher than the 3D off-rate [10], to a thousand fold higher than the 3D off-rate [11]. To the best of our knowledge, no attempt has been made to date to experimentally measure protein binding behavior in a system that tethers ligands on a 2D surface using flexible polymer linkers to better understand the parameters governing 2D interactions in T cell activation.

CG-MD simulation offered an alternative approach for assessing the influence of PEG polymers on pMHC-TCR binding at the T cell-surface interface. CG-MD simulations have proven useful for understanding complex molecular level phenomena including pMHC-TCR interactions [28,29,30]. The combination of state-of-the-art GPU computing and a CG-MD approach in our study enabled us to simulate nearly the entire range of PEG lengths to characterize the kinetics of pMHC-TCR binding. Our simulation setup is based on a robust model for the interaction between a free end of PEG polymer and the lipid bilayer, with a binding potential added to match the affinity of pMHC-TCR interaction. It should be noted, however, that certain assumptions and simplifications were required to make the approach computationally feasible. First, the physical dimension and geometry of the pMHC and TCR proteins were not considered. Second, a

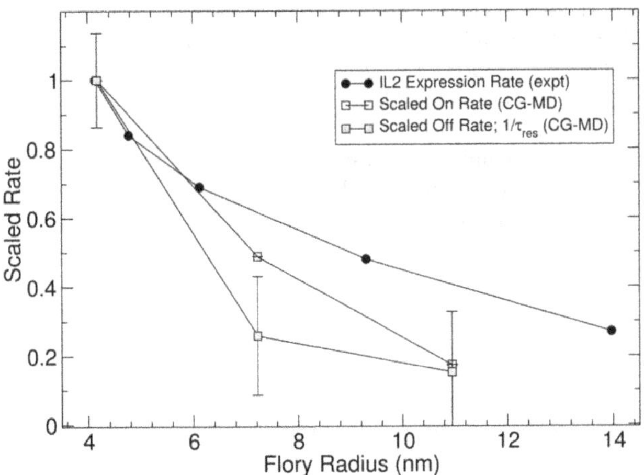

Figure 5. The impact of polymer length on pMHC-TCR binding kinetics based on CG-MD simulation. The on-rates and off-rates derived from the simulation for the PEG 4000, PEG 10000 and PEG 20000 were scaled against the data for the PEG 4000 and plotted as a function of the polymer Flory radius. To display the relationship between binding rates and TCR triggering efficiency, the experimentally determined IL2 expression rates for PEG 3500, PEG 5000, PEG 7500, PEG 150000 and PEG 30000 shown in Fig. 4A were scaled against PEG 3500 and plotted as a function of the polymer Flory radius.

binding event is arbitrarily defined as when the binding partners are within 1 nm of distance. And third, the distance between the two surfaces were fixed. With these simplifications, it is difficult to reliably derive realistic absolute values for kinetics parameters. The qualitative influence of PEG length on binding rates, however, may be reasonably assessed. The simulation showed that the on-rate and off-rate of pMHC-TCR binding decreases with the length of the linker. Also, the PEG polymer entropy applies a significant amount of force to the pMHC-TCR binding to better understand the parameters governing 2D interactions in T cell activation.

Our T cell activation experiments show that T cell responses to PEG-tethered ligands in terms of IL2 production displayed two distinct phases (Fig. 4C). pMHCs tethered by the four smallest PEGs, PEG 88, 484, 2000, and 3500, stimulated T cells similarly and with high efficiency. At this small MW range, the effect of PEG polymer dynamics is relatively small, and the kinetics of pMHC-TCR interaction is dominated by factors associated with intrinsic cell membrane dynamics. The relatively fixed "upright" orientation of pMHC ligands tethered with these short linkers may also play a dominant role in the kinetics of pMHC-TCR binding. In addition, we cannot exclude the possibility of TCR crosslinking for these short polymers, especially at high ligand density. Starting with PEG 5000, T cell responses gradually deceased with increasing PEG linker length. T cells responded to the largest PEG 60000 polymer, although only at high coating concentrations. In addition to IL2 production, we attempted to assay tyrosine phosphorylation using anti-phosphotyrosine antibodies and flow cytometry. The differences between ligands with different polymer lengths, however, could not be clearly resolved due to weak staining.

The results of our study are not consistent with the receptor crosslinking model or the kinetic segregation model. pMHCs tethered with PEG 2000 (contour length = 12.7 nm) or longer were able to trigger TCR. It is unlikely that TCR was triggered through TCR crosslinking by dimeric pMHCs. It has been shown that for activation through pMHC dimerized with linkers, the contour length of the linkers must be less than 10 nm [31]. Moreover, T cells were activated by low densities of pMHCs on a surface that would have only a small probability of forming dimers (Text S2). Our results do not support or disprove the co-receptor crosslinking model. Longer polymers might reduce pMHC potency by interfering with binding between pMHC and co-receptor. We think it is more likely, however, that the polymers interfere with the pMHC-TCR-CD4 three-party binding as a whole, given the closeness of the binding sites. The kinetic segregation model of TCR triggering relies on pMHC-TCR interaction at a tight space between two closely aligned membranes for the exclusion of large tyrosine phosphatase molecules such as CD45. In our study, pMHC ligands tethered through flexible PEG polymers such as PEG 5000 and longer should be able to interact with TCRs on relatively distant T cell plasma membranes. The gap between two membranes in this case should be able to accommodate CD45 and not cause its segregation with TCRs. Moreover, our CG-MD simulation results suggest that ligands with longer polymers bind TCRs more stably but trigger TCR less efficiently. This contradicts the kinetic segregation model, where TCR triggering depends on stable pMHC-TCR interaction [3].

Our results are consistent with the receptor deformation model of TCR triggering. In this model, pMHC-TCR bindings are pulled apart, or ruptured, by mechanical forces associated with the dynamic T cell-APC interaction. Before each rupture, signaling is triggered via the TCR through receptor deformation [5]. The overall triggering efficiency is determined by the frequency of rupture events, which is in turn determined by the on-rate and off-rate of binding. Higher rupture frequency facilitated by faster on-rate and off-rate translates to better integration of signals from multiple rupture events, thus, higher TCR triggering efficiency. While our work does not provide direct evidence for TCR triggering by receptor deformation, our simulation suggests that TCRs are subjected to mechanical forces derived from PEG polymer entropy. In addition, the experimental and simulation data show that pMHC ligands tethered with shorter polymers have higher on-rates and off-rates and trigger TCR more efficiently (Fig. 5). The correlation between binding rates and TCR triggering efficiency in our study is supported by a recent study that experimentally measured the 2D on-rates and off-rates of different pMHC-TCR binding pairs using micropipette and biomembrane force probes [11]. Furthermore, this study showed that 2D off-rates of pMHC-TCR binding were 30 to 8300-fold faster than the 3D off-rate [11], strongly suggesting that pMHC-TCR binding is under mechanical stress at the T cell-APC interface.

In summary, using pMHC ligands tethered to a surface with flexible PEG polymers, our study sheds light on the mechanism of the TCR triggering from a unique perspective. This approach of altering the 2D pMHC-TCR interaction should continue to offer valuable insights, in conjunction with methods to experimentally determine the kinetics of TCR binding to surface-tethered pMHC ligands.

Materials and Methods

Mice, cells and reagents

B10.BR H-2k mice and 5C.C7 TCR transgenic mice were purchased from Jackson Laboratories and Taconic, respectively. To generate T cell blasts, splenocytes from 5C.C7 TCR transgenic mice and irradiated splenocytes from B10.BR mice were mixed in complete Click's medium (Irvine Scientific, Santa Ana, CA) supplemented with 50 U/ml IL2 and 50 μM moth cytochrome c peptide. T cell blasts were used on day 7 to day 10 post stimulation. All animal experiments were approved by the Institutional Animal Care and Use Committee (IACUC) of Nemours/A.I. duPont Hospital for Children. All efforts were made to minimize the number of animals used and their suffering. Heterobiofunctional PEG 88 and PEG 484 linkers with a maleimide group at one end and a biotin group at the other end were from Pierce (Rockford, IL). All other heterobifunctional linkers with longer PEG linkers were purchased from Jenkem Technology USA (Allen, TX). EDC (1-Ethyl-3-[3-dimethylaminopropyl] carbodiimide Hydrochloride), Sulfo-NHS, TCEP (Tris(2-Carboxyethyl) phosphine Hydrochloride), streptavidin-DyLight 649, DyLight 549, and PBS (0.1 M phosphate and 0.15 M NaCl) were from Pierce (Rockford, IL). Brefeldin A, streptavidin, BSA (bovine serum albumin), MES (2-(*N*-morpholino)ethanesulfonic acid), and APTES (3-Aminopropyltriethoxysilane) were from Sigma-Aldrich (St. Louis, MO). DPBS (Dulbecco's PBS) was from Invitrogen (Carlsbad, CA). 14-4-4s mAb was generated from a hybridoma kindly provided by John Kappler (National Jewish Health).

Protein Preparation

Plasmid construct for expressing the extracellular domain of IEk with covalently linked MCC peptide was a gift from John Kappler (National Jewish Health). For conjugation with polymers, the construct was modified by PCR to add a cysteine at the C-terminus of the β chain. Baculoviruses were generated with the construct and BaculoGold Linearized baculovirus DNA (BD Biosciences, San Diego, CA) in Sf9 insect cells (Invitrogen, Calsbad, CA). High titer stocks of cloned recombinant baculovirus were used to infect Hi5 insect cells (Invitrogen, Calsbad, CA) cultured in spinner flasks. IEkMCC protein was purified from the supernatant of infected Hi5 cell cultures using an affinity chromatography column conjugated with 14-4-4s antibody. The protein was further purified using a Superdex 200 10/300 GL gel filtration column (GE Healthcare, Piscataway, NJ) before conjugating with PEG linkers. The design and production of IEkMCC with a C-terminal AviTag was described previously [15]. The protein was biotinylated at the lysine residue of the AviTag using BirA enzyme as described previously [15].

Conjugate Formation and Purification

IEkMCC-Cys protein was first treated with 0.1 mM TCEP for 1 hour at room temperature to reduce the sulfhydryl group of the c-terminal free cysteine. The protein was then reacted with PEG polymer at 1:20 molar ratio in PBS buffer pH 7.4 for 4 hours. For reactions with PEG 88, PEG 484 and PEG 2000, unreacted PEG linkers were separated from the conjugates and unreacted IEkMCC by gel filtration using a Superdex 200 10/300 GL gel filtration column, and the mixture of conjugates and unreacted IEkMCC protein was collected. The molar ratio of biotin to IEkMCC of the mixture was determined using a HABA (4'-hydroxyazobenzene-2-carboxylic acid)-based assay (Pierce, Rockford, IL). For reactions with PEG 3500, PEG 5000, and PEG 7500, IEkMCC-PEG conjugates were separated from unreacted IEkMCC and PEG linkers in one step using a Superdex 200 10/300 GL column. For reactions with PEG 15000, PEG 30000, and PEG 60000, unreacted PEG linkers were first eliminated using an affinity column conjugated with IEk specific antibody 14-4-4s. IEkMCC-PEG conjugates were then separated from unreacted IEkMCC by gel filtration using Superdex 200 10/300 GL.

Covalent Coating of Streptavidin to Polystyrene Surfaces

96-well strip well tissue culture treated polystyrene plates (Corning, Lowell, MA) were treated with 5 mM EDC and 5 mM Sulfo-NHS in MES buffer (0.1 M MES, 0.15 M NaCl, pH 6.0) for 40 minutes to add NHS ester to the carboxyl groups on the polystyrene surface. After washing, 100 μg/ml streptavidin in PBS pH 7.4 was added and incubated overnight at room temperature. The plate was then blocked with DPBS with 1% BSA. The amount of bound streptavidin was determined as 40 ng per well using an assay based on the quenching of biotin-FITC (Invitrogen, Calsbad, CA) fluorescence by streptavidin binding.

FRET Imaging and Analyses

For FRET imaging, DyLight 649-labeled streptavidin was coated on biotinylated plates (Pierce, Rockford, IL). IEkMCC-PEG(n)-Bio conjugates were labeled with DyLight 549 and coated on the plates via biotin-streptavidin binding. The bottom of the wells was imaged under a 60× water immersion objective on a Zeiss Axioplan 2 upright microscope. Using a 300 W xenon light source, we were able to achieve 95% photobleaching of DyLight 649 in about 5 minutes, and no photoconversion from DyLight 649 to DyLight 549 was observed (Fig. S2C). The intensity of DyLight 549 was recorded before and after DyLight 649 was photobleached. All images were analyzed with SlideBook software (Intelligent Imaging Innovations, Denver, CO).

T Cell Stimulation and Flow Cytometry

IEkMCC-PEG conjugates in DPBS with 1 mg/ml BSA were incubated in streptavidin-coated wells overnight at 4°C. Plates were washed, and 2.5×10^5 T cells in complete Click's medium were added. For IL2 production, 20 μg/ml brefeldin A was added to the medium. After stimulation, T cells were harvested, fixed with 3% formaldehyde in DPBS, permeabilized with DPBS buffer containing 1% BSA and 0.1% saponin, and stained with APC-labeled anti-mouse IL2 antibody JES6-5H4 (Biolegend, San Diego, CA).

Coarse-grained Molecular Dynamics Simulation

MD simulations were performed on the sticky membrane PEG systems, in which polymer length was varied from 4000 to 20000 Da. Multiple replica random-walker MD simulations were employed and run in the NPT ensemble using a GPU-accelerated CG model at a temperature of 300 K and pressure of 1 atm. Individual simulations were run for 500 ns each, and a total of 20 replicas were used per system, producing 10 microseconds of aggregate trajectory data per polymer. Details of these simulations and the resulting theoretical calculations can be found in Text S1.

Supporting Information

Figure S1 Characterization of IEkMCC proteins. (A) Purified IEkMCC protein with a free c-terminal cysteine was analyzed by SDS-PAGE and Coomassie Blue staining. The denatured protein migrated as two distinct bands with molecular weights consistent with the α and β chains. (B) Purified protein was analyzed by gel filtration chromatography using a Superdex 200 10/300 GL column. The protein was eluted as a single peak with a molecular weight of ~55 KDa.

Figure S2 Measurement of FRET between surface-tethered IEkMCC ligands and streptavidin on a plastic surface. (A) Schematic of FRET setup. The IEkMCC protein was labeled with DyLIght 549 (red star). The streptavidin on a plastic surface was labeled with DyLight 649. (B) Acceptor photobleaching FRET for linkers PEG 88 and PEG 7500. For the short PEG 88 linker, a significant increase in DyLight 549 intensity was seen after

DyLight 649 was photobleached. For the long PEG 7500 linker, only a slight increase in DyLight 549 intensity was observed after DyLight 649 photobleaching. (C) Photobleaching of DyLight 649 does not lead to photoconversion to DyLight 549. Streptavidin labeled with DyLight 649 (Cy5-like dye) was continuously imaged at both DyLight 549 and DyLight 649 channels with 1 s exposure time and 0 second intervals.

Figure S3 Similar densities of IEk-MCC tethered on plastic through PEG linkers of different lengths. IEkMCC-PEG conjugates at the indicated concentrations were incubated overnight at 4°C on streptavidin-coated 96-well plates. After washing, IEkMCC was detected using the IEk-specific antibody 14-4-4s and goat anti-mouse HRP.

Figure S4 T cell activation by IEkMCC-PEG 88 is independent of PEG tether, constimulatory molecules, or free PEG linkers. Streptavidin plates were coated with IEkMCC-PEG 88 or IEkMCC-bio (IEkMCC without a PEG linker but biotinylated at the lysine residue of a c-terminal AviTag sequence) at indicated concentrations. To test the effect of costimulatory molecules, plates coated with IEkMCC-PEG 88 were washed and further coated with B7.1-Fc fusion protein (1 nM; R&D Systems, biotinylated at 2.3 biotins per molecule). T cell IL2 production was measured by intracellular staining and flow cytometry after 6 hrs of stimulation. (B) IEkMCC-bio without PEG linker (7.2 nM) was anchored on streptavidin coated plates by incubating overnight at 4°C. Polymers (200 pM) were then added for 1 hr at room temperature. The PEG polymers were pre-incubated in PBS overnight at room temperature to hydrolyze the maleimide group. T cells were added to the washed plates and incubated for 6 hrs. Percent of IL2 producing cells was determined using intracellular cytokine staining and flow cytometry. The data points are averages of the values from replicate samples with standard deviations.

Figure S5 T cell adhesion to plates with tethered ligands is independent of linker length. T cells were added to plates coated with IEkMCC-PEG88, IEkMCC-PEG 5000, or IEkMCC-PEG 60000 at 7.2 nM. Cells were imaged 1 hr later using Evos microscope (Life Technologies) with a 20× objective.

Figure S6 Snapshot of the molecular dynamics simulation. The PEG-bilayer system was simulated as a function of PEG length (from 4000 to 20000 Dalton as shown in red) with a distance of 9 nm between the fixed end of the polymer and the center of the bilayer to examine how polymer length (chain entropy) affects both the rate and affinity of binding.

Figure S7 Linear regression trend lines and equations are displayed for Fig. 4B to show the rate of T cell commitment to IL2 production.

Figure S8 Flow cytometry plots for the three experiments done for Fig. 4A. At least 5000 live cells were collected for each sample. Note that in Experiment #1, IEkMCC-PEG 88 was not assayed.

Figure S9 Flow cytometry plots for the three experiments done for Fig. 4B. At least 5000 live cells were collected for each sample. Note that only three time points were assayed in Experiment #1.

Text S1 CG-MD Simulation.

Text S2 Calculation of the number of pMHC-PEG-bio bound to each streptavidin on a plastic surface.

Acknowledgments

We thank D. Shivers, Huihui Han and X. Zhang and for technical assistance and J. Kappler of National Jewish Health for DNA constructs encoding IEkMCC.

Author Contributions

Conceived and designed the experiments: ZM. Performed the experiments: ZM DNL SML. Analyzed the data: ZM DNL SML KAP MLK DED THF. Wrote the paper: ZM DNL SML KAP THF.

References

1. van der Merwe PA (2001) The TCR triggering puzzle. Immunity 14: 665–668.
2. van der Merwe PA, Dushek O (2011) Mechanisms for T cell receptor triggering. Nat Rev Immunol 11: 47–55.
3. Davis SJ, van der Merwe PA (2006) The kinetic-segregation model: TCR triggering and beyond. Nat Immunol 7: 803–809.
4. Ma Z, Discher DE, Finkel TH (2012) Mechanical Force in T Cell Receptor Signal Initiation. Front Immunol 3: 217.
5. Ma Z, Janmey PA, Finkel TH (2008) The receptor deformation model of TCR triggering. Faseb J 22: 1002–1008.
6. Dustin ML, Bromley SK, Davis MM, Zhu C (2001) Identification of self through two-dimensional chemistry and synapses. Annu Rev Cell Dev Biol 17: 133–157.
7. Shaw AS, Dustin ML (1997) Making the T cell receptor go the distance: a topological view of T cell activation. Immunity 6: 361–369.
8. Mempel TR, Henrickson SE, Von Andrian UH (2004) T-cell priming by dendritic cells in lymph nodes occurs in three distinct phases. Nature 427: 154–159.
9. Robert P, Aleksic M, Dushek O, Cerundolo V, Bongrand P, et al. (2012) Kinetics and mechanics of two-dimensional interactions between T cell receptors and different activating ligands. Biophys J 102: 248–257.
10. Huppa JB, Axmann M, Mortelmaier MA, Lillemeier BF, Newell EW, et al. (2010) TCR-peptide-MHC interactions in situ show accelerated kinetics and increased affinity. Nature 463: 963–967.
11. Huang J, Zarnitsyna VI, Liu BY, Edwards LJ, Jiang N, et al. (2010) The kinetics of two-dimensional TCR and pMHC interactions determine T-cell responsiveness. Nature 464: 932–U156.
12. Crawford F, Kozono H, White J, Marrack P, Kappler J (1998) Detection of antigen-specific T cells with multivalent soluble class II MHC covalent peptide complexes. Immunity 8: 675–682.
13. Kozono H, White J, Clements J, Marrack P, Kappler J (1994) Production of soluble MHC class II proteins with covalently bound single peptides. Nature 369: 151–154.
14. Rubenstein M, Colby RH (2003) Polymer Physics: Oxford University Press.
15. Ma Z, Sharp KA, Janmey PA, Finkel TH (2008) Surface-anchored monomeric agonist pMHCs alone trigger TCR with high sensitivity. PLoS Biol 6: e43.
16. London CA, Lodge MP, Abbas AK (2000) Functional responses and costimulator dependence of memory CD4+ T cells. J Immunol 164: 265–272.
17. Pardigon N, Bercovici N, Calbo S, Santos-Lima EC, Liblau R, et al. (1998) Role of co-stimulation in CD8+ T cell activation. Int Immunol 10: 619–630.
18. Alon R, Bayer EA, Wilchek M (1990) Streptavidin Contains an Ryd Sequence Which Mimics the Rgd Receptor Domain of Fibronectin. Biochemical and Biophysical Research Communications 170: 1236–1241.
19. Shinoda W, DeVane R, Klein ML (2007) Multi-property fitting and parameterization of a coarse grained model for aqueous surfactants. Molecular Simulation 33: 27–36.
20. Shinoda W, DeVane R, Klein ML (2010) Zwitterionic lipid assemblies: molecular dynamics studies of monolayers, bilayers, and vesicles using a new coarse grain force field. J Phys Chem B 114: 6836–6849.
21. LeBard DN, Levine BG, Mertmann P, Barr SA, Jusufi A, et al. (2012) Self-assembly of coarse-grained ionic surfactants accelerated by graphics processing units. Soft Matter 8: 2385–2397.
22. Levine BG, LeBard DN, DeVane R, Shinoda W, Kohlmeyer A, et al. (2011) Micellization Studied by GPU-Accelerated Coarse-Grained Molecular Dynamics. Journal of Chemical Theory and Computation 7: 4135–4145.
23. Anderson JA, Lorenz CD, Travesset A (2008) General purpose molecular dynamics simulations fully implemented on graphics processing units. J Comput Phys 227: 5342–5359.

24. Johnson CP, Tang HY, Carag C, Speicher DW, Discher DE (2007) Forced unfolding of proteins within cells. Science 317: 663–666.
25. del Rio A, Perez-Jimenez R, Liu R, Roca-Cusachs P, Fernandez JM, et al. (2009) Stretching single talin rod molecules activates vinculin binding. Science 323: 638–641.
26. Choudhuri K, van der Merwe PA (2007) Molecular mechanisms involved in T cell receptor triggering. Semin Immunol 19: 255–261.
27. Li YC, Chen BM, Wu PC, Cheng TL, Kao LS, et al. (2010) Cutting Edge: mechanical forces acting on T cells immobilized via the TCR complex can trigger TCR signaling. J Immunol 184: 5959–5963.
28. Cuendet MA, Michielin O (2008) Protein-protein interaction investigated by steered molecular dynamics: The TCR-pMHC complex. Biophysical Journal 95: 3575–3590.
29. Cuendet MA, Zoete V, Michielin O (2011) How T cell receptors interact with peptide-MHCs: A multiple steered molecular dynamics study. Proteins-Structure Function and Bioinformatics 79: 3007–3024.
30. Knapp B, Omasits U, Schreiner W, Epstein MM (2010) A Comparative Approach Linking Molecular Dynamics of Altered Peptide Ligands and MHC with In Vivo Immune Responses. PLoS One 5.
31. Cochran JR, Cameron TO, Stern LJ (2000) The relationship of MHC-peptide binding and T cell activation probed using chemically defined MHC class II oligomers. Immunity 12: 241–250.
32. Oesterhelt F, Rief M, Gaub HE (1999) Single molecule force spectroscopy by AFM indicates helical structure of poly(ethylene-glycol) in water. New J Phys 1: 6.1–6.11.

19

Conformational Antibody Binding to a Native, Cell-Free Expressed GPCR in Block Copolymer Membranes

Hans-Peter M. de Hoog[1], Esther M. Lin JieRong[1], Sourabh Banerjee[1], Fabien M. Décaillot[2], Madhavan Nallani[1,3]*

1 ACM Biolabs Pte Ltd, Research Techno Plaza XF-6, Singapore, Singapore, **2** Singapore Immunology Network, Agency for Science, Technology and Research (A*STAR), Singapore, Singapore, **3** Centre for Biomimetic Sensor Science, School of Materials Science and Engineering, Nanyang Technological University, Singapore, Singapore

Abstract

G-protein coupled receptors (GPCRs) play a key role in physiological processes and are attractive drug targets. Their biophysical characterization is, however, highly challenging because of their innate instability outside a stabilizing membrane and the difficulty of finding a suitable expression system. We here show the cell-free expression of a GPCR, CXCR4, and its direct embedding in diblock copolymer membranes. The polymer-stabilized CXCR4 is readily immobilized onto biosensor chips for label-free binding analysis. Kinetic characterization using a conformationally sensitive antibody shows the receptor to exist in the correctly folded conformation, showing binding behaviour that is commensurate with heterologously expressed CXCR4.

Editor: Sadashiva Karnik, Cleveland Clinic Lerner Research Institute, United States of America

Funding: The authors are grateful to Exploit Technologies, Agency for Science, Technology and Research (A*STAR), Singapore, for financial support. This project was partly funded by a SPRING Singapore PoV grant. ACM Biolabs provided support in the form of salaries for authors HH, ELJ and SB but did not have any additional role in the study design, data collection and analysis, decision to publish, or preparation of the manuscript.

Competing Interests: HH, ELJ, and SB are employees of ACM Biolabs, which offers membrane protein products and services to its customers. There are no further patents, product and services to declare.

* Email: mnallani@ntu.edu.sg

Introduction

G protein-coupled receptors (GPCRs) are cell-surface receptors that mediate the communication of the cell with its environment and, as such, form important targets for therapeutic intervention. GPCRs are notoriously hard to obtain in a format amenable to biophysical studies, which depending on the characterization method, requires moderately to highly pure receptor preparations. The limited success of obtaining sufficiently pure receptor preparations results from low levels of expression of the native proteins and their low stability in lipid membranes. To boost expression levels, researchers have resorted to the use of engineered cell lines (e.g., HEK, CHO, Sf9 cells) as well as engineered proteins, which may involve mutagenesis, deletion of destabilizing sequence elements, and production of GPCR chimeras. [1] In a few instances, this has resulted in comparatively stable receptors that can be expressed at high concentrations and are even amenable to crystallization although the effects of such types of protein engineering on the native function of these intricate proteins remains a topic of debate. [2,3] More importantly, these methods are highly labor and time intensive and there is currently no method to quickly produce stable receptor preparations in amounts suitable for thorough physical characterization.

A second, comparatively unexplored, approach would be to engineer the membrane surrounding the protein, in order to exceed the limited physical stability afforded to it by a lipid bilayer.[4–7] Our research, along with others, has shown that membrane proteins can insert into the fully synthetic membrane of block copolymer vesicles or polymersomes,[8–10] whose bilayer architecture is akin to that of the plasma membrane but with unprecedented physical stability. [11] Extension of this concept to include cell-free expression of the protein allows direct insertion of the native, full-length membrane proteins into the polymer membrane, as we have shown qualitatively in the past for the D2 receptor. [12,13].

We here show evidence that, at least for the GPCR currently under study, this 'artificial cell membrane' (ACMs), when subjected to cell-free synthesis, presents properly folded, native protein and exhibit sufficient stability to allow label-free biosensor analysis. As the GPCR of choice we employ CXCR4, a relatively well-characterized chemokine receptor and a key target for immunological intervention as well as a co-receptor for entry of HIV into T cells. [14] The native CXCR4 is expressed and inserted into polybutadiene-*b*-poly(ethylene oxide) polymersomes, and immobilized on a surface-plasmon resonance (SPR) biosensor chip surface for analysis of the binding characteristics of antibodies specific (Ab) to CXCR4.[15–17] Screening of the CXCR4-ACMS against a conformationally sensitive Ab, as well as its natural ligand SDF-1 shows that such in-vitro expressed receptors exist in the correctly folded conformation. From a fundamental point of view the presented approach highlights the role of the polymer membrane in extending stability to CXCR4, and suggests that membrane engineering could form a viable alternative to protein engineering for exploring the biophysics of this class of receptors.

Materials and Methods

For a detailed description of the materials used, preparation of the polymersomes, cloning and in-vitro expression of CXCR4, as well as radioligand binding assays and melting curves, see the supporting information.

Biosensor analysis

All experiments were performed on a Biacore T200 (GE healthcare).

To capture the streptavidin on the Biacore AU chip (GE healthcare, UK), a thoroughly cleaned Au chip was first functionalized with 11-mercaptoundecanoic acid (0.1 M in ethanol). The surface was then activated with a mixture of 0.2 M 1-ethyl-3-(3-dimethylaminopropyl) carbodiimide (EDC) and 0.05 M *N*-hydroxysuccinimide (NHS) at a flow rate of 10 μL min^{-1} for 7 min. Streptavidin (0.1 mg mL^{-1} in PBS) was subsequently flowed across the surface to an immobilization level of 2000 RU. Ethanolamine (GE Healthcare; 1M, pH = 8.5; 10 μL min^{-1} for 7 min) was used to block any unreacted activated esters. CXCR4 ACMs (0.01 mg mL^{-1} in running buffer) were then immobilized at 5 μL min^{-1} to the desired immobilizations level. BSA (5 mg mL^{-1}) was added to the running buffer to reduce nonspecific binding.

CXCR4 VLPs (Integral Molecular, PA) were captured by direct amine-coupling to the carboxylic acid-functional surface. The VLPs (1:100 dilution of 400 units in PBS) were then flowed across the surface, resulting in the immobilization of the VLPs. Immobilization level was 5000 RU. Active ester groups were blocked with ethanolamine (1 M, pH = 8.5).

Binding analysis

For analysis, the indicated concentrations of CXCR4 conformational antibody (CD184, BD Pharmigen, NJ) in running buffer were flowed across the surface for 100 s at 70 μL min^{-1} in sequence of increasing concentration with a single final dissociation phase of 10 min. As negative controls, blank polymersomes were used that had been subjected to the in-vitro synthesis procedure in the absence of c-DNA (null particles in the case of VLPs). Data presented is double-referenced against running buffer and reference surface. Binding analysis was performed using the supplied software (Bia-evaluation for T-200).

Results and Discussion

We initially set out to express CXCR4 according to our previously published procedures, i.e., expression of the protein using a coupled transcription-translation wheat germ extract (WGE; Figure 1A). [12] For immobilization on gold chips, it could be argued that the commonly employed amine-coupling chemistry may lead to the immobilization of some unfolded receptors in the form of aggregates, whether or not stabilized by components of the wheat germ extract. In order to rule out the occurrence of these possibilities, we modified our experimental procedures. First, we adapted the purification protocol of ACMs to exclude any unincorporated/free CXCR4 from solution. Using this method, a clear CXCR4 band in the western blot was seen only when ACMs were present in the in-vitro synthesis (IVS) reaction (Figure 1A). Second, we adapted the surface immobilization method in SPR to show that conformational binding *solely* originated from CXCR4 receptor inserted into the polymersome membrane. Henceforth, we coupled streptavidin to the gold chip by amine coupling, and captured the CXCR4-ACMs by interacting with a small fraction (1%) of biotinylated lipids (1,2-distearoyl-*sn*-glycero-3-phosphoethanolamine-*N*-[biotinyl(polyethylene glycol)-2000 (DSPE-PEG-biotin) that was mixed in with the polymersome membrane (Figure 1B). As a result, CXCR4-ACMs were stably immobilized on the biacore chip, presenting only receptors integrated in the polymer membrane.

Following this approach we immobilized the C4-ACMs at immobilization levels of ca. 5000 RU and evaluated the binding of the monoclonal antibody (mAb) 12G5 to the receptor by running a concentration series of increasing concentration over the receptor surface (Figure 2). Here it should be noted that the ACMs display diameters from 150 to 200 nm such that the greater proportion of the membranes is within the evanescent regime. [12,18] The mAb 12G5, directed against CXCR4, recognizes a conformation-dependent epitope involving the second and third extracellular domains (ECL1 and ECL2) of CXCR4, as well as the N-terminal domain. [19] Therefore, binding of 12G5 to CXCR4-ACMs would indicate the presence of the correctly folded receptor, oriented with the extracellular domain facing the outside solution. For comparison, we employed commercially available virus-like particles (VLPs; particles that are derived from cell membranes and carry enriched receptor) presenting CXCR4. CXCR4 proteoliposomes (supplier) (structurally similar to ACMs but having a lipid bilayer membrane) gave a relatively small signal during initial testing in our hands (data not shown) so that we decided to pursue our studies using CXCR4 VLPs as a comparison. Both VLP and proteoliposome preparations have been shown to bind ligands, with CXCR4 VLPs having been successfully used in biosensor analysis. [20,21,22].

Using CXCR4-ACMs, we observed a concentration-dependent increase in response, which fitted well to a 1:1 binding interaction and exhibiting clear association and dissociation phases between injections. We did not observe more complex kinetics resulting from bivalent binding. The shape of the sensorgrams, being linear rather than exponential especially at higher mAb concentrations, indicated the occurrence of mass transfer such that, at the current immobilization levels, receptor concentration was probably too high. Nevertheless, these experiments indicate that the mAbs bound readily to CXCR4-ACMs, signifying two important facts: (1) in-vitro CXCR4 receptor inserted into polymersome membranes retain their native conformation of the extracellular side, and (2) the receptor is present in appreciable concentrations with the extracellular side facing the bulk solution.

To further assess the functionality of the receptor incorporated in ACMs, we performed additional experiments to study binding of radiolabeled (I-125) SDF-1α (stromal cell-derived factor-1 alpha, the natural ligand of CXCR4) to CXCR4 in ACMs and native membrane preparations (Figure 2B; See supporting information for details). Dissociation constants values for SDF-1α were found to be 8.4 nM and 1.4 nM for CXCR4-ACMs and CXCR4 membrane preparations, respectively (Figure S1 and S2). Combined with the data of binding of conformationally-sensitive Ab, these results show that CXCR4 embedded in polymersome membranes binds ligands and Abs with affinities comparable with native protein in natural lipid membranes. Apart from the difference in properties between the polymer membrane and the native membrane, the somewhat lower affinity of SDF-1α for ACMs may result from the fact that in this synthetic system G proteins are absent. These proteins are known to affect the receptor's conformation and may therefore alter ligand-binding affinity. [23].

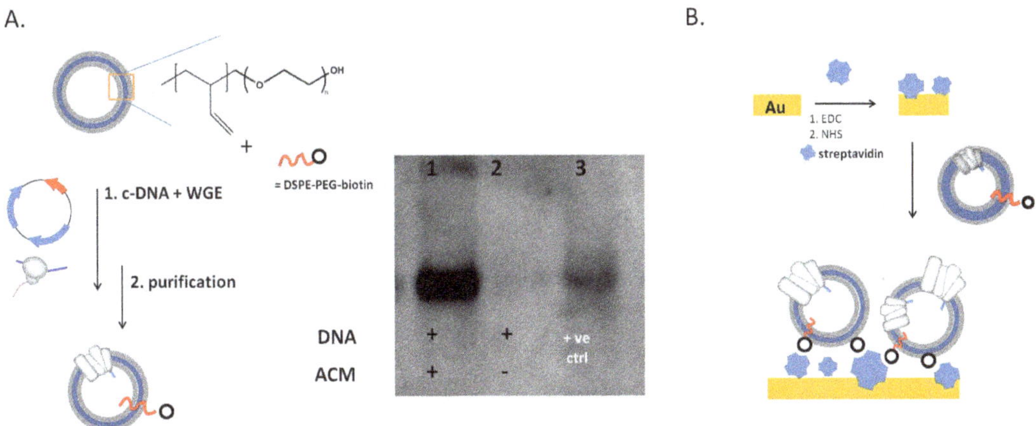

Figure 1. In-vitro synthesis and direct insertion of CXCR4 into polymersomes. A: PB-PEO polymersomes and CXCR4 c-DNA were subjected to in-vitro synthesis using a wheat germ coupled translation-transcription extract (WGE) and purified by a filtration step. After filtration, insertion of CXCR4 in the polymersome membrane was verified by Western blot. CXCR4 c-DNA in absence of polymersomes, which went through the same process, did not show the presence of CXCR4. The positive control is a commercially available CXCR4 cell membrane preparation. B: After purification, ACMs were immobilized onto a biosensor gold chip by first coupling streptavidin using standard EDC/NHS coupling, and then capturing the CXCR4 ACMs by the interaction with streptavidin of biotinylated lipid mixed into the polymersome membrane.

For CXCR4-VLPs, we immobilized the VLPs by standard amine-coupling, as reported previously. [20] Even though the approach may not necessarily be optimal for analyzing VLP-ligand interactions by biosensor analysis, this still allowed us to

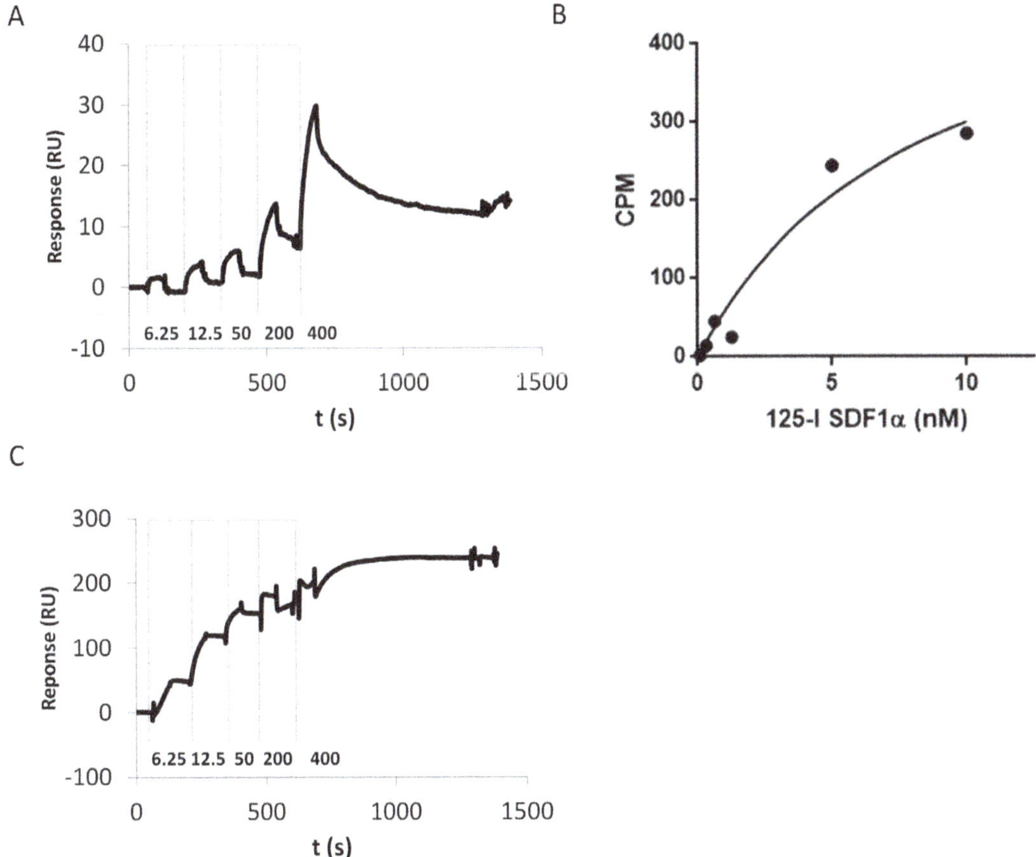

Figure 2. Kinetic screening of 12G5 mAb binding to CXCR4-ACMs immobilized onto biosensor chips. A: Ab was injected at increasing concentrations (6.25–400 nM) over 100 s, followed by a buffer wash (without regeneration) between injections (immobilization level: ca. 5000 RU; biotin/streptavidin immobilization). B. Saturation binding of 125-I SDF1α to CXCR4-ACMs. A dissociation constant of 8.4 nM was determined. C. The same series of measurements as shown in Fig. 2 A, conducted using immobilized VLPS (immobilization level: 5000 RU).

assess the performance of our approach versus other receptor preparations. 12G5 binding to VLPs gave good binding responses initially, with clear association curves. Dissociation was, however, subdued pointing to accumulation of material on the chip surface and hampering further screening. This phenomenon is in line with the previously reported observation that virus-receptor interactions are multivalent and generally irreversible. [18] Therefore, these preparations seem more amenable to a sandwich set-up where the receptor-presenting particles are captured by an Ab-covered surface, after which the ligand of interest is injected. After binding, the surface is regenerated to expose the first Ab and the cycle is repeated. Such an approach prevents reproducibility issues caused by rapid deactivation of the receptor preparation when reconstituted via traditional approaches (detergent, lipid). Indeed, a similar approach has been shown to be successful for SPR-based binding analysis of crude CXCR4 cell membrane preparations. [14,15].

Having demonstrated the presence of properly folded receptor in ACMs, the receptor preparation was subjected to repeated injection cycles of CXCR4 mAb. To prevent mass-transfer from occurring, CXCR4-ACMs were immobilized at lower immobilization levels (~1500 RU on the sensor chip), and the immobilized receptor surface was subjected to multiple injections of 12G5 (100 nM), over a total of 25 injections lasting a total of 4 hours. At these immobilization levels, although binding levels were much reduced, a good quality of binding curves was obtained as exemplified by association phases that showed clear exponential binding behaviour and almost full dissociation (Figure 3). Moreover, the curves fitted well to a 1:1 binding model. Interestingly, repeated cycles of conformational antibody injection demonstrated an unprecedented conformational stability of the receptor preparations under the analysis conditions: over the time span of 24 cycles (ca 4 h), only a minimal decrease (7%) in the activity of the receptor surface was observed. Given the stability of polymersomes preparation in general, [11,24] the cause for the eventual, albeit minor, degradation of the signal is most likely protein-related and may either be attributed to loss of ACMs from the surface or unfolding of the receptor. Since there was no correlation of the decreasing binding signal to the drop in baseline (which showed an initial drop, after which it stabilized (data not shown)), and assuming that the decrease in baseline points to loss of material from the sensor surface, the decrease in activity of the receptor comes from degradation of the receptor itself, most likely unfolding.

To demonstrate how the stabilized native CXCR4 could be employed, a small screen was conducted with, apart from 12G5, two additional monoclonal antibodies against CXCR4, i.e. clone 7L25 and C064025, a monoclonal IgG2B antibody. The three mAbs were screened versus the same CXCR4-ACM preparation, separately injecting each antibody in a series of increasing concentration (for kinetic analysis) (Fig. 4). The 12G5 antibody was screened last in the sequence to compare the binding parameters obtained with that obtained in Figure 3. Testament to the conformational stability of the receptor preparation, all binding curves exhibited clear association and dissociation phases, irrespective of the receptor density or cycle number of the injection, once again confirming the presence of a stable, properly folded receptor in the polymer membrane. Generally, the fits to the curves at 400 RU were more accurate than that at 1500 RU, which we attribute to the fact that, at lower receptor densities, the fit of concentration with binding level all saturated. From the fits at 400 RU, the detailed binding parameters were extracted (k_{on}, k_{off}, and K_D), as summarized in Table 1. Comparison of the binding parameters for the 12G5 mAb yielded values of 69 (+/−) nM and

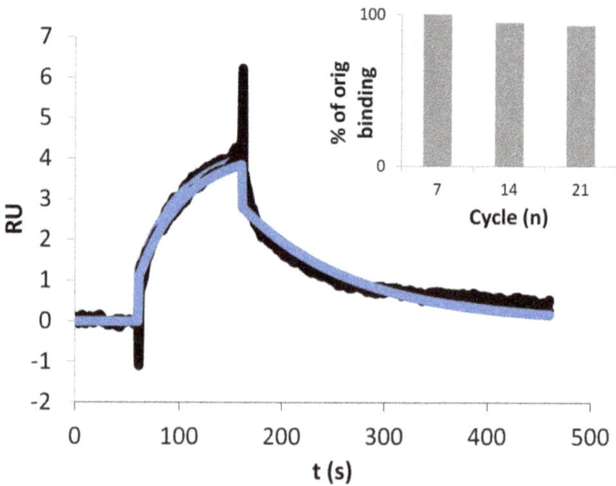

Figure 3. Sensorgrams of mAb 12G5 binding (100 nM) to CXCR4-ACMs immobilized at low RU (ca. 1400) via biotin/spteptavidin immobilization of the embedding polymersome matrix. The analyte was injected in triplicate, at cycle 7, 14, and 21. Intermediate cycles involved blank injections. The blue line shows the fit to the curve assuming 1:1 binding kinetics. The inset shows the relative decrease in binding activity of the surface as measured by the binding level 4 s before the end of the injection.

52 nM for the 100 nM screen and kinetic titration, respectively, indicating that the developed protocol was reproducible. Here some caution is warranted because the values were obtained at different immobilization levels (1500 vs 400 RU), and the fit to the sensorgram at 1500 RU involved only a single (albeit triplicate) concentration. The values for the other two antibodies were somewhat higher with roughly similar k_{on} and k_{off} values. The results indicate that 12G5 is somewhat stronger than binding of 7L25 and C064025, although the variation in either of the binding parameters (k_{on}, k_{off}, and K_D) was not that large. This appears to be in line with earlier observations on screening of CXCR4-specific antibodies where it was observed that most antibodies exhibited fairly similar affinities (in this instance, EC 50 values). [25].

The biosensor data demonstrate that CXCR4 embedded via cell-free expression in polymeric membranes possess an on-chip stability that is at par, or even enhanced, as compared to receptors expressed by cell-based expression and subjected to detergent-assisted lipid reconstitution. [16] Importantly, the binding data observed are commensurate with that observed for cell-based CXCR4 expression (as observed for binding of SDF-1α, suggesting that merely the presence of an amphiphilic membrane is able to provide the stable integration of (complex) membrane proteins. Concerning the physical stability of the preparation (as judged from its amenability for extended biosensor interrogation) we expect the polymer-based bilayer membrane to provide support for the CXCR4 by two independent mechanisms that result from the enhanced physical stability of bilayer polymer membranes. First, the polymer membrane can withstand the demanding physical restraints (high concentration of solutes, osmolarity) of direct incubation with the cell-free extracts and remains intact during purification. Second, the bilayer polymer membrane is sufficiently stable to allow its intact surface immobilization, providing a stable matrix for multiple-hour interrogation of the binding characteristics of the membrane protein by surface-plasmon resonance. The approach therefore allows to circumvent

Table 1. Kinetic parameters extracted from fitting the binding curves of concentration series of three different mAbs against a single preparation of immobilized CXCR4 ACMs.

	k_{on} ($\times 10^3$ Ms^{-1})	k_{off} ($\times 10^{-3}$ s^{-1})	K_D (nM)
12G5[a]	120	6.3	52.0
C064025	91.1	25.5	280
7L25	78.7	25.3	322

[a]KD for 12G5 binding to CXCR4-ACMS shown in Figure 2 was 60.2±17 nM.

the deleterious effect that detergent-solubilization may have on receptor structure through direct embedding in the more native environment of an amphiphilic membrane. Evidently, one topic of future investigation should be how the approach may benefit the biophysical characterization of other members of the GPCR family, especially when it comes to poorly characterized (i.e., fragile) receptors such as olfactory receptors, which at the same time suffer from low expression levels, where we argue that the outcome would mainly be dependent on the properties of the receptor itself, as both the IVS and biosensor characterization protocol is expected to be readily applicable to the GPCR family, since it is essentially only the DNA sequence that changes. Indeed, it is encouraging that the expression protocol we employed is essentially the same as what we reported for the (unrelated) dopamine receptor D2. [12].

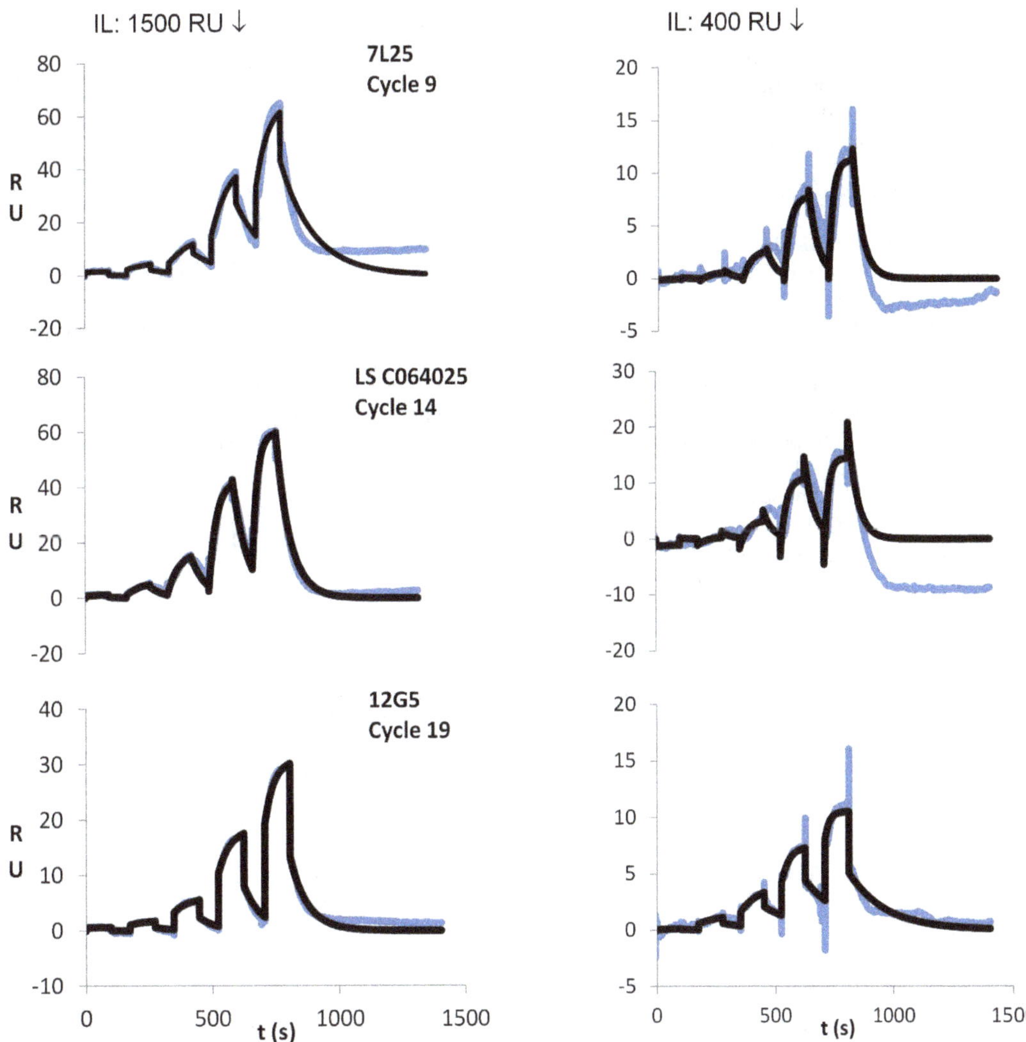

Figure 4. SPR sensorgrams obtained by kinetic screening of three different types of conformationally sensitive ABs to a single CXCR4 ACM preparation immobilized on a sensor chip. Conditions were the same as in Figure 2, except for the immobilization level, which is as indicated in the figure.

To gauge, nevertheless, the stabilization of CXCR4 embedded in the polymer bilayer as compared to detergent-solubilized receptor, we applied the developed protocol to measurement of the melting temperature of the receptor by label-free biosensing and tested both CXCR4-ACMs and detergent (P20) solubilized receptors, expressed by IVS (See SI for a detailed protocol). Although this detergent may not be optimal for receptor-stabilization, it was compatible with the IVS extract, whereas milder detergents such as CHAPS were not. As expected, binding levels observed for the detergent-solubilized receptor were low, but did give rise to a measurable signal, which after normalization could be compared to the curve obtained for ACMs (Figure S3). For detergent-solubilized CXCR4 and CXCR4-ACMs, melting temperatures were obtained of 29.7 and 38.8°C, respectively, demonstrating that the polymer bilayer shifts the melting temperature of the protein fold by almost 10°C. Hence, the data supports our hypothesis that direct expression into native-like bilayer membranes stabilizes the receptor when compared to a process whereby the heterologously expressed receptor has to go through a detergent-solubilization step.

In summary, the results show that native CXCR4 receptor produced by in-vitro expression and inserted in polymer membranes can serve as an alternative to currently available methods of cell-based production, receptor engineering and complex reconstitution procedures, with the main advantage being that the receptor can be produced in sufficient quantity in a manner of hours. We are currently exploring the generality of the developed approach to include other members of the GPCR family.

Supporting Information

Figure S1 Displacement of radio-labeled SDF1-a using CXCR4 cell membrane preparations. Radioligand binding experiments were conducted at GVK Biosciences, Hyderabad, India. CxCR4 ligand binding studies were performed in 96-well plate in a total volume of 200 μL consisting of 50 μL of 125I-SDF, 50 μL of assay buffer (50 mM HEPES pH 7.4, 5 mM MgCl2, 1 mM CaCl2, 0.2% BSA) and 100 μL of membrane preparation diluted in appropriate buffer. Non-specific binding was determined in the presence of 10 μM of unlabeled SDF or untransfected control membrane. The plate was incubated at room temperature for 2 h. Reactions were terminated by flash filtration and inverse transfer to 96-well glass fibre filter plates. The plate was then dried for 30 minutes at 60°C and sealed at the bottom with an adhesive sheet. Subsequently, 50 μL of scintillation fluid was added to each well, the plates sealed on top and the radioactivity counted in a 96-well plate counter (Top count NXT, Perkin Elmer). The assay was first validated using commercially available CXCR4 membrane preparations. A fixed concentration of radiolabeled 125I-SDF (0.5 nM) was used to determine the IC50 by incubating the CXCR4 membrane with different concentration of unlabeled SDF. Linear regression was performed using Graphpad Prism. The IC50 value was 1.4 nM, correlating with the value stated by the supplier (0.9 nM).

Figure S2 Saturation binding of 125-I SDF1α to CXCR4-ACMs. For CXCR4 ACMs, first the optimal concentration was determined with respect to specific binding. At 0.625 μg/well, the TB/NSB ratio measured 3.8 and the percentage specific binding was 74% (data not shown). The CXCR4 ACMs were then subjected to a saturation assay (constant receptor concentration, varying ligand concentration), with and without 0.5 μM unlabeled SDF. The dissociation rate constant, K_d, and the maximal number of receptor binding sites, B_{max}, was calculated using GraphPad Prism.

Figure S3 Melting curves of CXCR4 ACMs and CXCR4 solubilized in detergent. CXCR4 ACMs were freshly prepared as described above and kept at 4°C. For IVS of CXCR4 in P20 solution, the same protocol was followed but instead of ACMs, detergent was added to a concentration of 1%. Although the detergent selected may not be optimal for receptor-stabilization, it was compatible with the IVS extract, whereas milder detergents such as CHAPS were not. Expression was verified by Western blot after removal of the insoluble fraction by centrifugation. To preserve activity, the resulting detergent solubilized receptor was used on the same day without further purification. For determination of the melting temperature, samples were incubated for 15 minutes in a PCR thermal cycler at temperature increments of 10°C, from 10–80°C. After incubation the samples were cooled to 4°C and directly analyzed for their binding activity. For binding analysis, 12G5 mAB to CXCR4 was immobilized to the custom-made chip essentially as detailed for immobilization of streptavidin, upon which the CXCR4 preparations (ACM or detergent) were injected. Temperature data was analyzed using GraphPad Prism. Data is represented in percentage of initial response to correct for the differences in activity of CXR4 ACMs and detergent-solubilized receptor.

Text S1 Materials and methods used in the experiments.

Author Contributions

Conceived and designed the experiments: MN. Performed the experiments: ELJ HH. Analyzed the data: ELJ HH SB MN. Contributed reagents/materials/analysis tools: FD. Contributed to the writing of the manuscript: HH ELJ SB FD MN.

References

1. Robertson N, Jazayeri A, Errey J, Baig A, Hurrell E, et al. (2010) The properties of thermostabilised G protein-coupled receptors (StaRs) and their use in drug discovery. Neuropharmacology 60: 36–44.
2. Wu B, Chien EYT, Mol CD, Fenalti G, Liu W, Katritch V, et al. (2010) Structures of the CXCR4 chemokine GPCR with small-molecule and cyclic peptide antagonists. Science 330: 1066–1071.
3. Discher DE, Eisenberg A (2002) Polymer vesicles. Science 297: 967–973.
4. Discher BM, Won YY, Ege DS, Lee JCM, Bates FS, et al. (1999) Polymersomes: Tough vesicles made from diblock copolymers. Science 284: 1143–1146.
5. Oates J, Watts A (2011) Uncovering the intimate relationship between lipids, cholesterol and GPCR activation. Curr Opin Struct Biol 21: 802–807.
6. Phillips R, Ursell T, Wiggins P, Sens P (2009) Emerging roles for lipids in shaping membrane-protein function. Nature 459: 379–385.
7. Kumar M, Grzelakowski M, Zilles J, Clark M, Meier W (2007) Highly permeable polymeric membranes based on the incorporation of the functional water channel protein Aquaporin Z. PNAS 104: 20719–20724.
8. Nallani M, Benito S, Onaca O, Graff A, Lindemann M, et al (2006) A nanocompartment system (Synthosome) designed for biotechnological applications. J Biotechnol 123: 50–59.
9. Choi HJ, Montemagno CD (2005) Artificial organelle: ATP synthesis from cellular mimetic polymersomes. Nano Lett 5: 2538–2542.
10. Kita-Tokarczyk K, Grumelard J, Haefele T, Meier W (2005) Block copolymer vesicles - Using concepts from polymer chemistry to mimic biomembranes. Polymer 46: 3540–3563.
11. Nallani M, Andreasson-Ochsner M, Tan CWD, Sinner EK, Wisantoso Y, et al. (2011) Proteopolymersomes: In vitro production of a membrane protein in polymersome membranes. Biointerphases 6: 153–157.

12. May S, Andreasson-Ochsner M, Fu Z, Low YX, Tan D, et al. (2013) In vitro expressed GPCR inserted in polymersome membranes for ligand-binding studies. Angew. Chem. – Int. Ed. 52: 749–753.
13. Allen SJ, Crown SE, Handel TM (2007) Chemokine: Receptor structure, interactions, and ntagonism Annu. Rev. Immunol. 25: 787–820.
14. Karlsson OP, Löfås S (2002) Flow-mediated on-surface reconstitution of G-protein coupled receptors for applications in surface plasmon resonance biosensors. Anal. Biochem. 300: 132–138.
15. Stenlund P, Babcock GJ, Sodroski J, Myszka DG (2003) Capture and reconstitution of G protein-coupled receptors on a biosensor surface. Anal. Biochem. 316: 243–250.
16. Navratilova I, Sodroski J, Myszka DG (2005) Solubilization, stabilization, and purification of chemokine receptors using biosensor technology. Anal. Biochem. 339: 271–281.
17. Knoll W (1998) Interfaces and thin films as seen by bound electromagnetic waves. Ann. Rev. Phys. Chem. 49: 569–638.
18. Strizki JM, Turner JD, Collman RG, Hoxie J, González-Scarano F (1987) A monoclonal antibody (12G5) directed against CXCR-4 inhibits infection with the dual-tropic human immunodeficiency virus type 1 isolate HIV-1(89.6) but not the T-tropic isolate HIV-1(HxB). J. Virol. 71: 5678–5683.
19. Carnec X, Quan L, Olson WC, Hazan U, Dragic T (2005) Anti-CXCR4 monoclonal antibodies recognizing overlapping epitopes differ significantly in their ability to inhibit entry of human immunodeficiency virus type 1. J. Virol. 79: 1930–1933.
20. Willis S, Davidoff C, Schilling J, Wanless A, Doranz BJ, et al (2008) Virus-like particles as quantitative probes of membrane protein interactions. Biochemistry 47: 6988–6990.
21. Hoffman TL, Canziani G, Jia L, Rucker J, Doms RW (2000) A biosensor assay for studying ligand-membrane receptor interactions: Binding of antibodies and HIV-1 Env chemokine receptors. PNAS 97: 11215–11220.
22. Whorton MR, Bokoch MP, Rasmussen SGF, Huang B, Zare RN, et al. (2007) A monomeric G protein-coupled receptor isolated in a high-density lipoprotein particle efficiently activates its G protein. PNAS 104: 7682–7687.
23. Baribaud F, Edwards TG, Sharron M, Brelot A, Heveker N, et al. (2007) Antigenically distinct conformations of CXCR4. J. Virol. 75: 8957–8967.
24. Yaşayan G, Redhead M, Magnusson JP, Spain SG, Allen S, et al. (2012) Well-defined polymeric vesicles with high stability and modulation of cell uptake by a simple coating protocol. Polym. Chem. 9: 2596–2604.
25. Hanson MA, Cherezov V, Griffith MT, Roth CB, Jaakola VP, et al. (2008) A specific cholesterol binding site is established by the 2.8 A structure of the human beta2-adrenergic receptor. Structure 16: 897–905.

20

Effects of Surfactants on the Improvement of Sludge Dewaterability Using Cationic Flocculants

Yongjun Sun[1,2*], Huaili Zheng[1,2*], Jun Zhai[1,2], Houkai Teng[3], Chun Zhao[1,2], Chuanliang Zhao[1,2], Yong Liao[1,2]

1 Key laboratory of the Three Gorges Reservoir Region's Eco-Environment, State Ministry of Education, Chongqing University, Chongqing, China, **2** National Centre for International Research of Low-carbon and Green Buildings, Chongqing University, Chongqing, China, **3** CNOOC Tianjin Chemical Research and Design Institute, Tianjin, China

Abstract

The effects of the cationic surfactant (cationic cetyl trimethyl ammonium bromide, CTAB) on the improvement of the sludge dewaterability using the cationic flocculant (cationic polyacrylamide, CPAM) were analyzed. Residual turbidity of supernatant, dry solid (DS) content, extracellular polymeric substances (EPS), specific resistance to filtration (SRF), zeta potential, floc size, and settling rate were investigated, respectively. The result showed that the CTAB positively affected the sludge conditioning and dewatering. Compared to not using surfactant, the DS and the settling rate increased by 8%–21.2% and 9.2%–15.1%, respectively, at 40 mg·L^{-1} CPAM, 10×10^{-3} mg·L^{-1} CTAB, and pH 3. The residual turbidities of the supernatant and SRF were reduced by 14.6%–31.1% and 6.9%–7.8% compared with turbidities and SRF without surfactant. Furthermore, the release of sludge EPS, the increases in size of the sludge flocs, and the sludge settling rate were found to be the main reasons for the CTAB improvement of sludge dewatering performance.

Editor: Vishal Shah, Dowling College, United States of America

Funding: This research was supported by the National Natural Science Foundation of China (Project No. 21177164), Major projects on control and rectification of water body pollution (Project No. 2013ZX07312-001-03-03), and the 111 Project (Project No. B13041). The funders had no role in study design, data collection and analysis, decision to publish, or preparation of the manuscript.

Competing Interests: The authors have declared that no competing interests exist.

* Email: sunyongjun008@163.com (YS); zhl@cqu.edu.cn (HZ)

These authors contributed equally to this work.

Introduction

About 3.102×10^9 tons of municipal wastewater are generated in China in 2007, 49.1% of which need to be treated by some biological processes. Large amount of sewage sludge have been produced in sewage treatment facilities [1]. Raw wastewater sludge contains huge amount of water along with organic solids, which causes problems in transportation, treatment, and disposal [2]. One important stage of sludge treatment prior to disposal is the reduction of sludge volume by sludge dewatering; this process reduces transportation and handling costs [3]. Nevertheless, sludge dewatering remains expensive [4].

Chemicals, such as alum, polymeric ferric sulfate, polyacrylamide, and chitosan, are typically incorporated in sludge to improve the dewaterability [5–7]. Flocculants, given at the fixed dosages, are commonly used in conditioning the physical and chemical properties of sludge to improve sludge the dewatering performance, while the polyacrylamide added to sludge is a widely used pretreatment procedure in the wastewater treatment plants (WWTP). The previous reports indicate that the surfactants and polyacrylamide could be used as the dewatering reagents to substantially decrease the moisture content in filter cakes [8–9]. Sludge conditioned by the surfactants has gained considerable attention because of their excellent performance in improving sludge dewatering. The surfactants alter the microorganism cell structure by allowing the cell materials to leave the attached surface and simultaneously dissolve this surface in aqueous solutions to improve sludge dewaterability [10]. Huang et al. [11] found that the presence of the surfactants in alum sludge systems was able to improve the sludge quality by reducing SRF and bound water content, as well as increasing settling rate, dewatering rate, and the solid content of the sludge cakes as long as the polyacrylamides were not used. Chu et al. [12] concluded that cationic (CTAB) and anionic surfactants sodium dodecyl sulfate (SDS) enhance the filtration efficiency; however, the former and the latter respectively increase and decrease the consolidation rate, respectively. CTAB and SDS mainly interact with the aggregate surface and interior, respectively. Besra et al. [13–14] reported that the mechanism of the reduction of surface tension was responsible for the enhancement of the dewatering kaolin suspensions. The mechanism of the absorption of the complex surfactants in the presence of flocculants exhibits an important function in enhancing the dewatering of kaolin suspensions. In some cases, the reduction in surface tension and the increase in hydrophobicity or a combination of both are able to improve surfactant dewatering [15]. Interfacial tension can be reduced in the sludge system by adding CTAB, a good dewatering performance is obtained [13].

It was also reported that the surfactants can alter the settleability of the activated sludge by changing the release of extracellular

Table 1. The effect of different dewaterability parameters.

Dewaterability parameters		Effects
Abbreviation	**Full name**	
DS	Dry solid content	The dry solid content of filter cake after dewatering, indicating the dewatering degree
Turbidity	Residual turbidity of supernate	Representing the quality of flocculation process
EPS	Extracellular polymeric substances	Affecting the distribution of moisture in the sludge particles. Polysaccharide, protein are the main components of EPS
SRF	Specific resistance to filtration	Its value is an important parameter to measure difficulty lever of filtration.
Zeta potential		Charge character of sludge particle surface
Floc size distribution		Representing sizes of sludge flocs, larger floc size indicates the better dewatering performance.
Settling behavior		Representing sedimentation properties of activated sludge

polymer substances (EPS). The EPS often comprises polysaccharides, proteins, and DNA, which entrap water and cause high viscosity. EPS is related to the settlement of activated sludge, in which the sludge aggregates are difficult to be packed because of its large size, which would result in interstitially bound water contained in the activated sludge and poor settleability. [16] The evidence mentioned above suggested the need to evaluate the effects of the surfactants on the improvement of the sludge dewaterability using the cationic flocculants. In addition, the effects of the surfactants on improving the actual dewaterability of activated sludge through CPAM are rarely studied. The effects of the surfactant during flocculation on EPS distribution have not been reported.

In this study, the effects of surfactants on improvement of sludge dewaterability by cationic flocculants were investigated. Dry solid (DS) content and the residual turbidity of supernatant were used to evaluate the sludge dewaterability. Specific resistance of filtration (SRF), extracellular polymeric substance (EPS) content of sludge supernatant, floc size, Zeta potential, and settling rate were also measured to explain their changes observed during sludge dewatering. A comparison between sludge dewatering behaviors resulting from flocculation with or without CTAB was made. The causes for the dewaterability improvement after adding CTAB were also elucidated.

Materials and Methods

Test materials

Raw municipal sludge samples were collected from the sludge thickener of Dadukou Drainage Co., Ltd. (Chongqing, China). The sludge thickener was just used to store the activated sludge without adding any chemical reagent. The samples were then transported to the laboratory within 30 min after sampling. The samples were subsequently stored at 4°C in the refrigerator; all experiments were completed in 48 h. The initial characteristics of the sludge were as follows: pH 6.80±0.12, Dry solid content (DS)

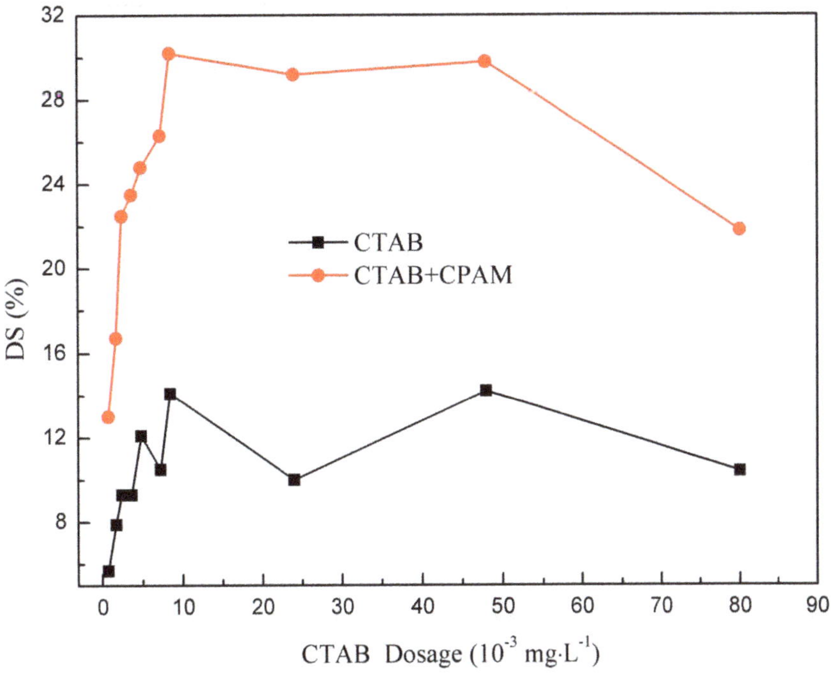

Figure 1. Effect of surfactant dosage on DS.

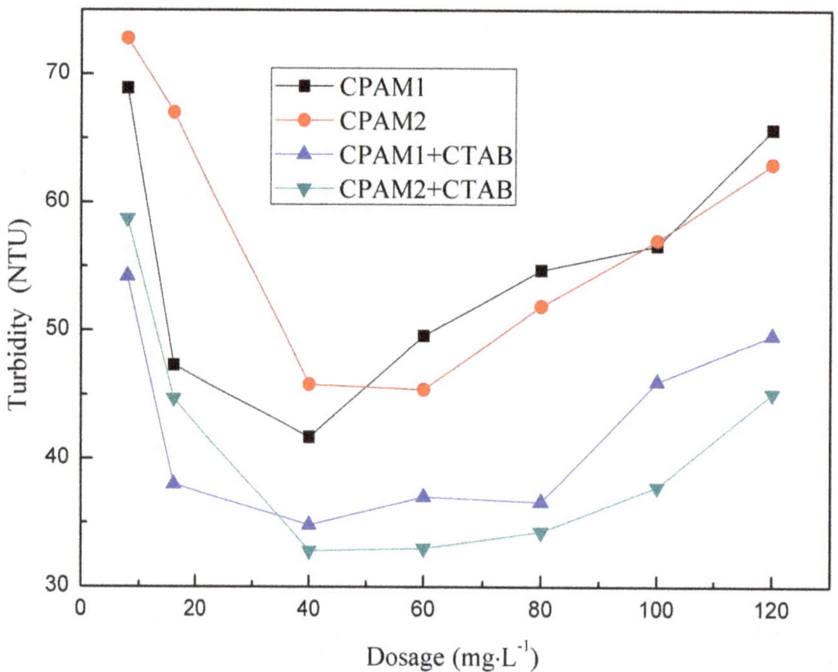

Figure 2. Effect of flocculant dosage on supernatant turbidity.

1.80%±0.36%, total suspended solid (TSS) 55%±9%, mass density (TCOD and SCOD) 0.885 mL·g^{-1}. CTAB (molecular formula, $C_{16}H_{33}N(CH_3)_3{}^{+}Br^{-}$; molecular weight, 364.4; ionic type, cationic; aggregation number, 91) was applied to improve dewaterability. CPAM was synthesized in laboratory; the detailed synthesis and characterization of CPAM have been previously reported [5,17]. The flocculants used in the dewatering tests were CPAM1 (2.0 dL·g^{-1}intrinsic viscosity and 30% cationic degree) and CPAM2 (2.0 dL·g^{-1} intrinsic viscosity and 40% cationic degree).

Flocculation and dewatering experiments

A program-controlled Jar-test apparatus (ZR4-6 Jar Tester, Zhongrun Water Industry Technology Development Co., Ltd., China) was used for sludge dewatering experiments. The

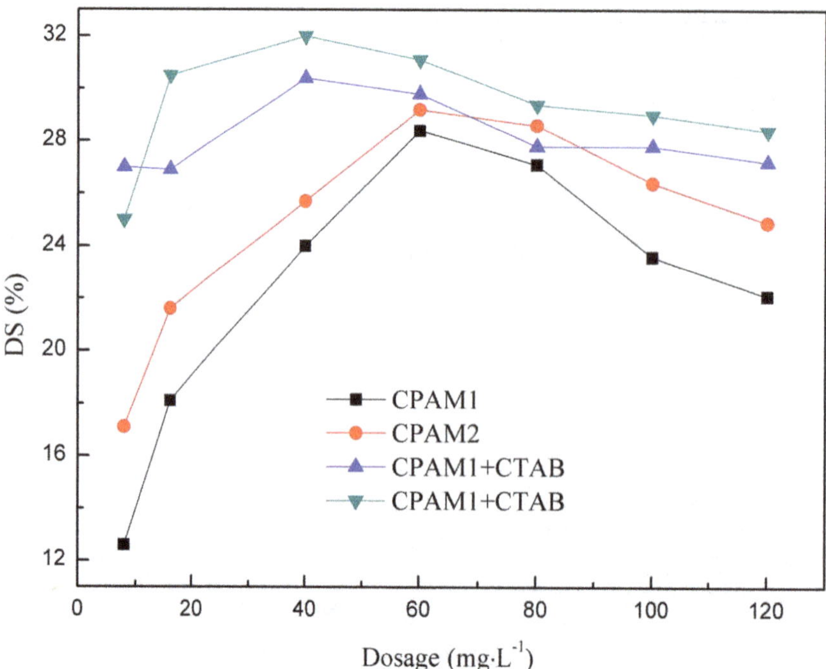

Figure 3. Effect of flocculant dosage on DS content.

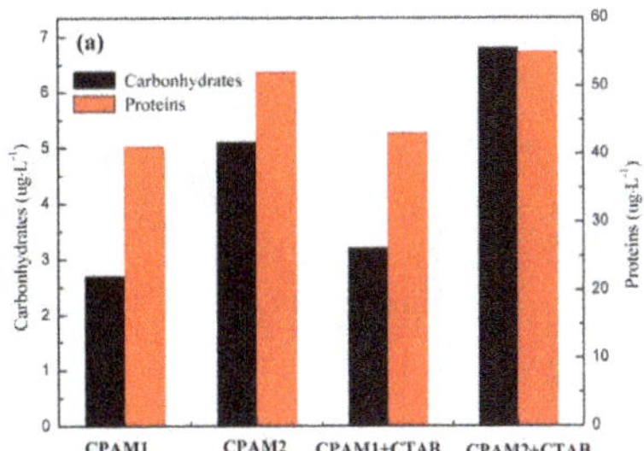

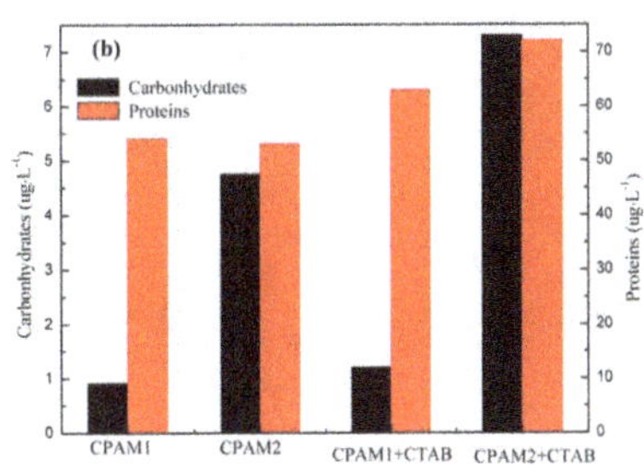

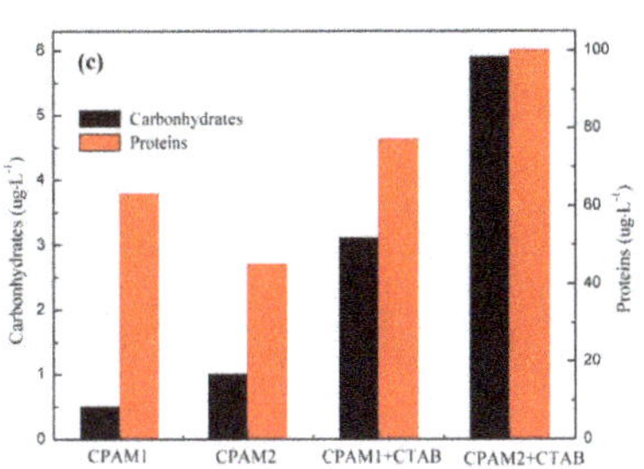

Figure 4. Effect of flocculant dosage (a) 20, (b) 40, and (c) 80 mg·L^{-1} on EPS.

dewaterabilities of the sludge flocculated with CPAM alone and with CPAM and surfactant were assessed in terms of residual turbidity of the supernatant, dry solid content, specific resistance to filtration (SRF), settling rate, zeta potential, and floc size. Dewatering tests for waste sludge were carried out at room temperature. Each result was an average of three repeated tests under similar experimental conditions. All the standard deviations were controlled in less than 5%. The pH of the sludge system was adjusted by adding NaOH (1.0 mol·L^{-1}) or HCl (1.0 mol·L^{-1}). Waste sludge (volume, 500 mL) was transferred into beakers before the required amount of flocculants (CPAM) or the mixture with surfactants (CTAB) was added. The sludge solution was rapidly mixed at 100 rpm for 30 s, followed by slow stirring at 30 rpm for 2 min and sedimentation for 10 min.

Analytical methods

Proteins and polysaccharides were selected using the method of Raynaud et al. [18] to characterize the composition of EPS. The proteins were determined according to the method of Lowry et al. [19] with BSA as the standard. The polysaccharides were assayed by phenol–sulfuric acid method of Dubois et al. [20] with glucose as the standard. All assays were conducted in threefolds. If the proteins and polysaccharides appeared in the filtrate following the addition of surfactants/flocculants, it was assumed that the EPS was from the activated sludge [21–22].

Vacuum filtration was selected to measure the dewaterability of the flocculated sludge. The supernatant of the sludge after flocculation was extracted for the turbidity measurements using a turbidimeter (HACH 2100Q, USA). After rapid agitation, Zeta potential of the supernatant was measured with Zetasizer Nano ZS90 (Malvern, U.K.). Meanwhile, the size of the sludge floc was measured simultaneously with Winner 2000 laser particle size analyzer (Jinan Micro-Nano Technology Co., Ltd., China). The settling rate of the sludge was expressed in terms of the height of the sludge–water interface as a function of time. The sludge sample (volume, 500 mL) was then transferred to a graduated cylinder and allowed to settle after four times of inversion. The height of the liquid slurry interface was noted at a regular time interval without disturbance. The settling rate was the average settling rate calculated in the first 2.5 min of settling time, and the standard deviation was less than 5%.

The flocculated sludge was poured into a Buchner funnel for filtration under a vacuum pressure of 0.06 MPa for 15 min or until the vacuum cannot be maintained in <15 min. The filterability of the sludge was measured by the specific resistance of the sludge. DS was determined by the following equation given as

$$\mathrm{DS} = \frac{W_2}{W_1} \times 100\% \tag{1}$$

where W_1 is the weight of wet filter cake after filtration, and W_2 is the weight of filter cake after drying at 105°C for 24 h.

SRF was calculated as following [21]:

$$SRF = \frac{2bPA^2}{\mu c} \tag{2}$$

where P is the pressure of filtration (N/m^2), A is the filtration area (m^2), m is the viscosity of the filtrate (N s/m^2), c is the weight of solids per unit volume of filtrate (kg/m^3, $c = 1/C_i/((100C_i)–C_f)/(100C_f))$, C_i is the initial moisture content (%), C_f is the final moisture content (%), b is the slope determined from the $t/V_f(y)$–$V_f(x)$ plot, V_f is the volume of filtrate (m^3), and t is the filtration time (s). The effect of different dewaterability parameters was shown in Table 1.

Results and Discussion

Effect of surfactant dosage on dewatering performance

Figure 1 show the effect of addition of the surfactant on sludge dewatering performance. In addition, no apparent sludge–liquid separation surface is observed, and the supernatant turbidity cannot be measured, as the surfactant merely promoted destabilization and failed to bridge the sludge particles [22].

Mixing the surfactants and flocculants and adding these materials to the sludge sample reduced the volume of dewatered sludge and enhances the DS content. The surfactant significantly improved the sludge dewatering performance in the case of CPAM1 (Figure 1). However, considering keeping the surfactant costs as low as possible at a highest possible DS content, the optimal CPAM1 dosage was 40 mg·L^{-1}, while the surfactant dosage was selected at 10×10^{-3} mg·L^{-1} (2 wt% CPAM1), and the CTAB content was kept constant at 2 wt% CPAM1 in the subsequent tests.

Effect of flocculant dosage on dewatering performance

Figures 2 and 3 show the effect of flocculant dosage on supernatant turbidity and DS content in the presence and absence of CTAB, respectively. Initially, increasing the flocculant dosage quickly decreases the supernatant turbidity and then sharply increases it. However, DS content was markedly increased and then slightly decreased. The flocculation performance significantly improved upon addition of CTAB during this process; incorporation of CTAB decreased the turbidity of the supernatant and increased the DS content. Compared to not using surfactant, DS increased by 8%–21.2%. An explanation for the obtained results could be as follows: At low dosages, CTAB failed to fully neutralize the negative charge on surfaces of sludge particles; CPAM was unsuccessful in linking or bridging the particles together. The formed sludge flocs are too small, loose, and fragile [23].

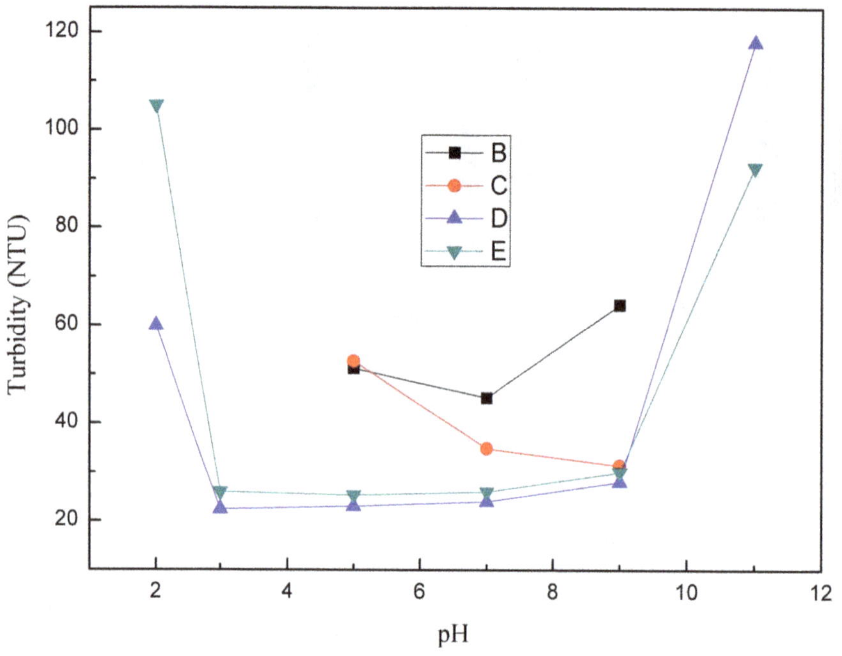

Figure 5. Effect of pH on supernatant turbidity.

Flocculant overdose gradually increased the positive charge of the colloidal system; electrostatic repulsion causes detachment of some segments to other sludge particles. At last, the optimal CPAM dosage was fixed at 40 mg•L^{-1}, and the CTAB dosage was 2wt% CPAM. This result was similar to the previous study about enhancing the papermaking sludge dewatering by surfactant [24]. Combined with the role of CTAB and CPAM, the sludge system flocculated by CTAB and CPAM into the slurry system, a good dewatering result could be obtained and the sludge dewatering properties and settlement could be significantly improved.

Figure 4 showed that the carbohydrate and protein concentrations in the supernatant gradually increased in the presence of CTAB, but the increase degree of the concentration of carbohydrates and proteins was small. The concentrations of the carbohydrates and proteins in the supernatant increased to 18.6%~520% and 7%~122% after addition of CTAB,

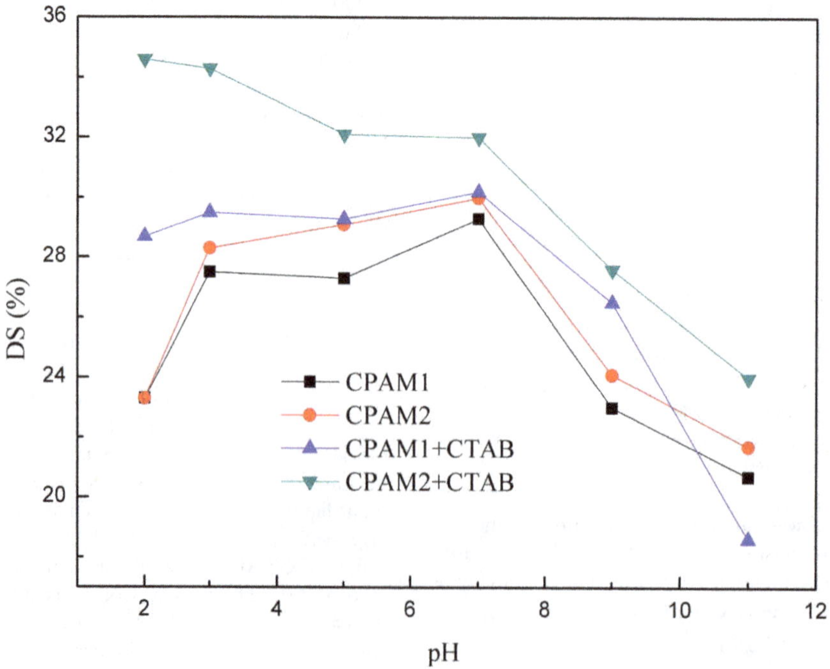

Figure 6. Effect of pH on DS content.

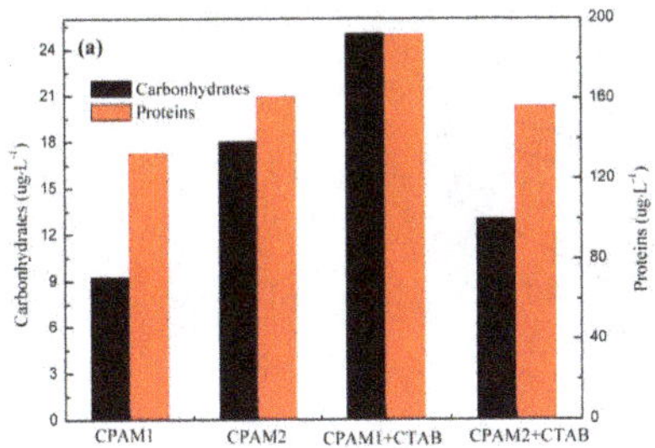

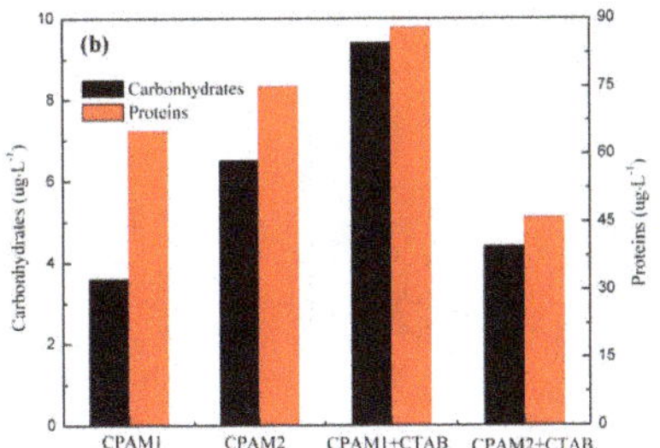

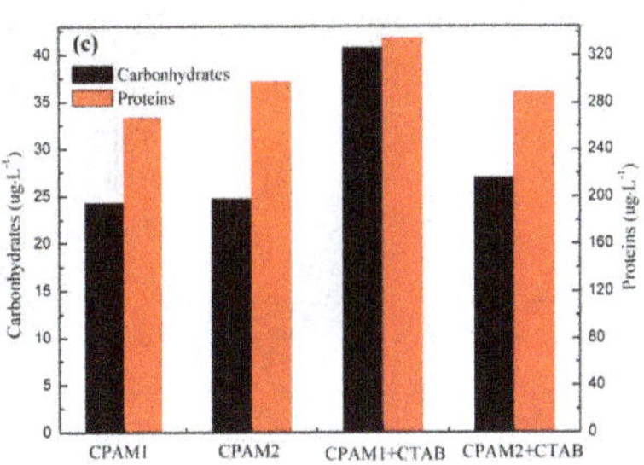

Figure 7. Effects of pH value on EPS at (a) pH = 3, (b) 7, and (c) 11.

respectively; a high amount of EPS was dissolved when the sludge was flocculated with a high cationic degree. In the presence of CTAB, the carbohydrate and protein concentrations in the supernatant by CPAM2 at 80 mg/L are 5.9 and 100 μg•L^{-1}, respectively.

An explanation could be that the suffusion resulted in the absorption of the external surface of sludge flocs by water molecules, leading to an increase of the amount cell adhesion water, and the formation of thick water film on the protein layer surface [25]. Therefore, the presence of EPS in the sludge flocs increased the difficulty of the sludge dewatering. One possible explanation for the enhanced surfactant-mediated sludge dewatering is the solubilization of EPS in the supernatant; EPS leaves the activated sludge surface and dissolves into water. Reduction of EPS results in a more compact activated sludge under the same mechanical force or at the same dewatering time [26].

Effect of pH on dewatering performance

Figures. 5–7 illustrate that the changes in pH significantly affect the flocculation performance and sludge dewatering ability. Increasing the pH initially decreased the turbidity of the supernatant and then markedly increased it. Meanwhile, when the pH value was in the range of 2 to 5 and 9 to 11, the residual turbidity of supernate flocculated by CPAM1 and CPAM2 was too high to be determined. However, the same change in pH slightly increased the DS content and decreased thereafter. The optimal DS contents were obtained at pH 6–8 for CPAM+CTAB. Figure 6 shows that the DS content is higher in the presence of CTAB compared with that in the absence of CTAB under an acidic conditions. Strong acid and alkaline conditions deteriorated the formation of the flocs and the performance of the CPAM-treated sludge dewatering. Small flocs were formed and there was little supernatant volume obtained; a blurry sludge–water separation was observed. H^+ enhanced the positive charges of the sludge surface at low pH, resulting in an increased electrostatic repulsion and colloidal stability [16]. Increasing the specific resistance of sludge to filtration would yield a low DS content. The negative charges on the surfaces of the sludge particles increased at higher pH; these charges could not be fully neutralized by CPAM at a fixed dosage, thereby increasing the repulsive force between the sludge particles and their specific resistances to filtration. Addition of CTAB decreased some negative charges on the sludge surface and reduced the repulsion between sludge particles. The optimum pH for sludge dewatering ranged between 5 and 8 (Figure 6).

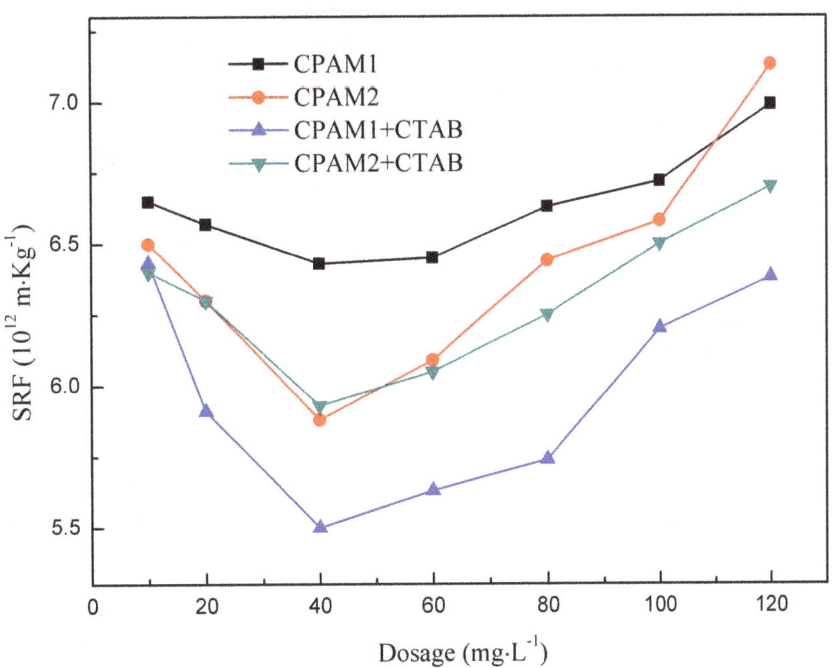

Figure 8. Effect of flocculant dosage on SRF.

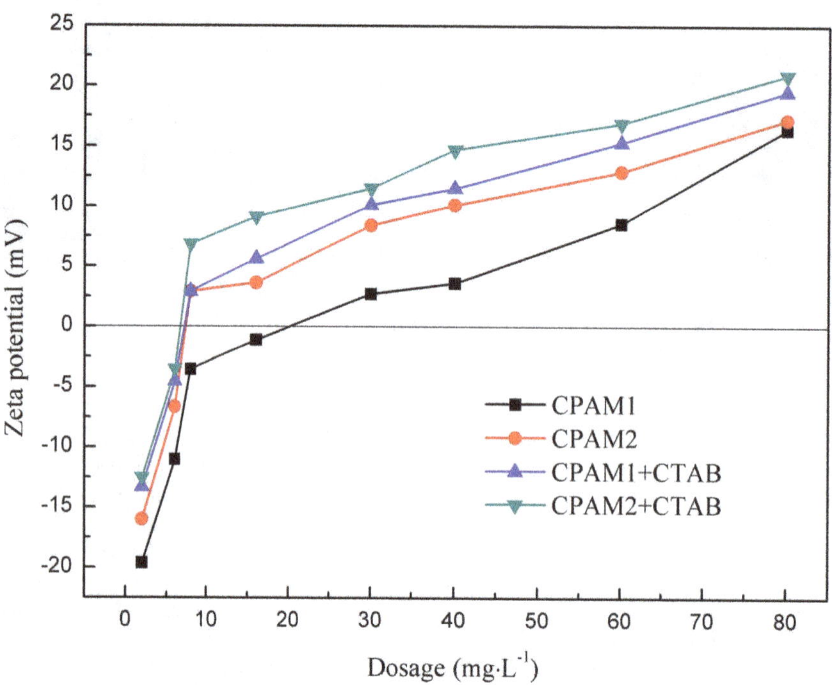

Figure 9. Effect of flocculant dosage on Zeta potential.

Figure 7 shows the changes in EPS concentration in the supernatant after surfactant addition under different pH. High EPS concentrations were released from the sludge particles under acidic and the alkaline conditions, especially for CPAM1, which increased the viscosity of sludge and deteriorated the dewaterability [27]. For example, the concentration of the carbohydrates flocculated by CPAM2 at pH 3, 7, and 11 were 18.0 ug•L^{-1}, 6.5 ug•L^{-1}, and 24.7 ug•L^{-1}, respectively. While the concentration of the proteins flocculated by CPAM2 at pH 3, 7, and 11 were 161 ug•L^{-1}, 75 ug•L^{-1}, and 298 ug•L^{-1}, respectively. The EPS in increasing the concentrations followed a decreasing order: basic> acidic>neutral, because the surfactant was connected between water molecules and the EPS on surfaces of sludge flocs. EPS was released from these surfaces upon stirring. The surfactants also increased the solubility of EPS in water phase, which resulted in an easy dissolution and release of EPS from the surfaces of sludge flocs. The permeability of microorganisms in the sludge was changed by adjusting acid or alkali; the intracellular proteins were denatured and the adhesion force between microorganisms in the sludge particles and the EPS was reduced. The release of the EPS reduced a large number of hydrophilic groups on the surfaces of sludge flocs; the content of the bound water in the sludge likewise decreased, which enhanced the performance of the sludge dewatering. The results indicated that the enhanced dewatering performance was attributed to the reduction of EPS in the sludge upon addition of the surfactants.

The improvement of the dewatering upon basic/acidic and surfactant treatments was attributed to the activated sludge surface left by a portion of EPS; the sludge aggregates are thus easily packed and the water content of dewatered sludge is reduced [16].

Effect of flocculant dosage on SRF

Figure 8 shows that the addition of CTAB initially reduced the specific resistance of the sludge. The respective SRFs of CPAM1 and CPAM2 were 6.43×10^{12} and 5.88×10^{12} m•kg^{-1} at 40 mg/L dosage. Addition of CTAB reduced the SRFs of CPAM1 and CPAM2 to 5.93×10^{12} and 5.50×10^{12} m•kg^{-1}, respectively. Compared to not using surfactant, The SRFs were reduced by 6.9%–7.8%. SRF is mainly affected by the characteristics of sludge flocs. Reduction in SRF was attributed to the formation of large but strong flocs with narrow size distribution similar to the result of Besra et al. [28] Sludge particles that were too small caused the filtration pores clogged and increase in SRF. The effect of flocculation increased the size of the sludge flocs and formed the rugged flocs construction by adding CPAM; the porous structure of sludge flocs was maintained during filtration and dehydration, thereby reducing the specific resistance of sludge.

The correlation between EPS concentration and SRF was the same as in the previous works on sludge co-conditioning by CTAB and CPAM (Figures 4 and 8) [29]. In addition, the dewaterability of sludge treated with both CPAM and CTAB supplementation was better than that of sludge treated with CTAB or CPAM alone [30], which was attributed to the increase in concentrations of free and unbound EPS, because surfactants could significantly release these substances. EPS and internal bound water were released by adding CTAB, and improved the performance of sludge dewatering.

Zeta potential

The effect of flocculant dosage on Zeta potential is shown in Figure 9; the charge neutralization capacity of the flocculants administered with or without CTAB decreased in the following order: CPAM2+CTAB> CPAM1+CTAB> CPAM2> CPAM1. A low CPAM dosage is necessary to attain zero charge upon addition of CTAB during flocculation; a portion of the negative surface charge on the sludge surface was already neutralized by the cationic CTAB. The Zeta potential of CPAM+CTAB was slightly shifted to a value lower than that of CPAM.

With the increase in flocculant dosage, the Zeta potential significantly increased to an isoelectric point where the potential

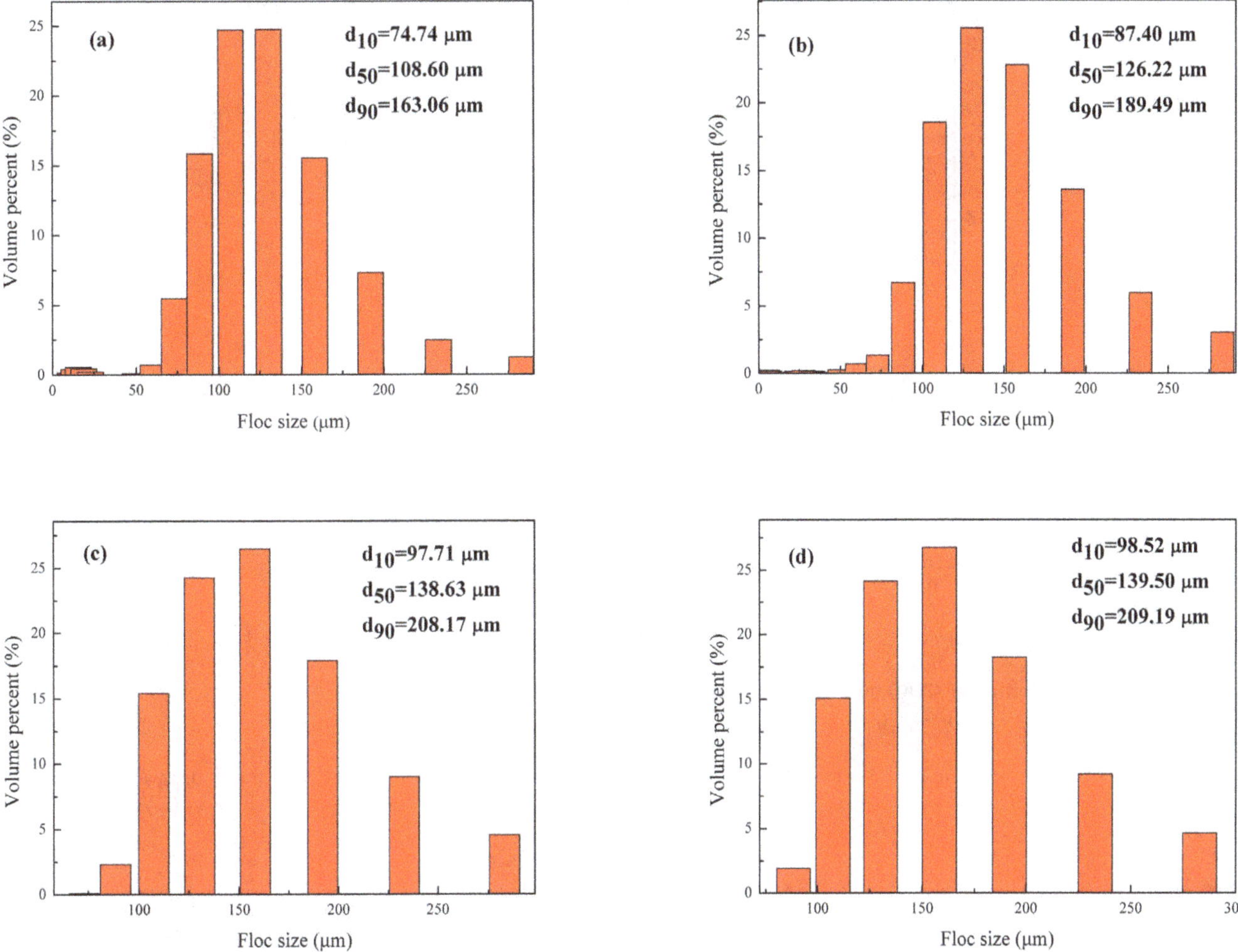

Figure 10. Floc size distribution after flocculation for (a) CPAM1, (b) CPAM2, (c) CPAM1+CTAB, and (d) CPAM2+CTAB.

was close to zero (0 mV). The dosage of the isoelectric point was less than 20 mg/L, while the optimal dosage for sludge dewatering was 40 mg/L. Pefferkorn et al. [31] reported that if charge neutralization was the only flocculation path, the optimal efficiency was achieved when the Zeta potential was close to zero. The potentials treated by CPAM+CTAB at optimal dosages were much higher than the isoelectric point, indicating that charge neutralization insignificantly contributed to sludge dewatering. In contrast to CPAM alone, the Zeta potential significantly increased when CPAM was with CTAB. The addition of CTAB increased the charge neutralization capacity of CPAM, because the former was cationic and neutralized the negative charge on surfaces of sludge flocs. The result implied that the sludge was flocculated by absorption bridging, charge neutralization, and the mechanism of the electrostatic path [32–33].

Floc size

Figure 10 shows the floc size distributions at the end of flocculation, which were expressed as respective particle sizes corresponding to 10%, 50%, and 90% of the size histogram; 10% of the flocs had sizes in the range [0 μm–d_{10}], 50% of the flocs had sizes in the range [0 μm–d_{50}], and 90% of the flocs had sizes in the range [0 μm–d_{90}]. Larger flocs provide were easier for dewatering than smaller flocs; smaller flocs clog the cake during filtration, thereby reducing dewaterability [17]. The respective values of d_{50} for CPAM1, CPAM2, CPAM1+CTAB, and CPAM2+CTAB with the same dosage at pH 7were 108.6, 126.22, 138.63, and 139.50. Comparison of CPAM1 with CPAM2 implied that a high cationic degree yielded a large particle diameter. Comparison between CPAM1 (CPAM2) and CPAM1+CTAB (CPAM2+CTAB) suggested that the addition of CTAB resulted in large floc sizes at the same conditions with a strong capability for charge neutralization. Addition of CTAB increased the EPS concentration in the supernate (Figures 4 and 7). When comparing Figures 4 and 9, high EPS concentrations in the supernate increased the sizes of sludge floc sizes; EPS significantly affected the size and stability of latter sludge flocs. The dewaterabilities of sludge by CPAM and CTAB showed that the cationic surfactants were adsorbed on the sludge surface by electrostatic and Van der Waals forces; the changes in the characteristics of the sludge flocs were also observed, especially in EPS distribution and floc size [34].

Effect of addition of CTAB on settling behavior

Figure 11 shows the effect of addition of CTAB on settling behaviors. The settling rate was the average settling rate calculated in the first 2.5 min of settling time. The settling rates for the sludge conditioned with CPAM1, CPAM2, CPAM1+CTAB, and CPAM2+CTAB were 2.79, 3.04, 3.21, and 3.32 cm•min^{-1}, respectively. The comparison of CPAM1 and CPAM2 indicated that the flocculants with high cationic degrees yielded rapid

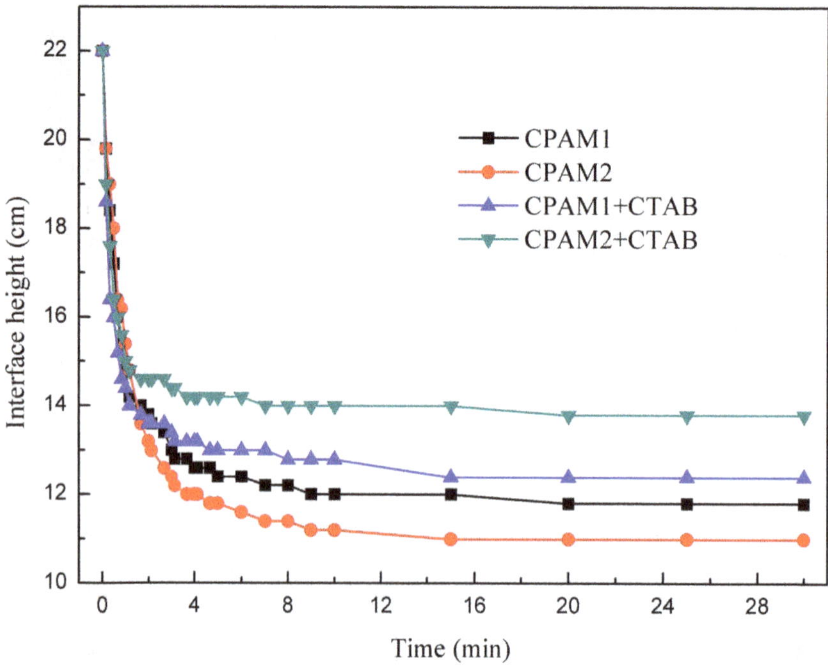

Figure 11. Effect of addition of CTAB on settling behavior.

settling rates. Compared to not using surfactant, the settling rate was increased by 9.2%–15.1%. The comparison made between CPAM1 (CPAM2) and CPAM1+CTAB (CPAM1+CTAB) revealed that the addition of CTAB hastened the settling rate at the same flocculant dosage; however, the ultimate sediment height flocculated with CTAB was larger than that flocculated without CTAB, reflecting the sludge flocculated without CTAB has better sedimentation properties.

Floc size and EPS mainly affected sludge settling; large floc size of the settlement at the initial stage yielded a rapid settling. The sludge combined with a high EPS concentration worsened the properties of sludge settling. EPS in the supernatant enhanced the flocculation and increased the floc size (Figure 10). The long-chain structure of the cationic surfactant destabilized the colloidal particles by neutralizing the negative charge on the surfaces of sludge flocs [35]. The ultimate height of the sludge-water interface reflected the overall settlement performance. In fact, the relationship between the settling rate of sludge in the first 2.5 min and the ultimate height of sludge-water interface was not very significant. A possible explanation for this opposite phenomenon was that the flocs produced by charge neutralization were always denser and smaller than that generated by the adsorption-bridging effect [36]. According to the discussion about ζ-potential and floc size, the results further demonstrated that the main mechanism flocculated by CPAM1, CPAM2, CPAM1+CTAB, and CPAM2+CTAB was adsorption-bridging effect. However, the charge neutralization ability was strengthened by addition of the CTAB, resulting in generated more denser flocs by CTAB+CPAM in initial stage with rapid settling rate. The settling rate flocculated by CPAM only was lower than that obtained by CPAM+CTAB in the first 2.5 min of settling time. With the increase of the settling time, the adsorption-bridging effect was stronger than the charge neutralization, leading to large and loose flocs. Meanwhile Wang et al. concluded that the settleability of the sludge would be greatly weakened when more EPS (Figure 4) were released in sludge matrix [30]. As a consequence, the ultimate height of the sludge-water interface by CTAB+CPAM was higher than the height by CPAM flocculation.

Conclusions

The effects of the surfactants on improving the sludge dewaterability by the cationic flocculants were investigated. The experimental results indicated that the addition of flocculants and surfactants in sludge flocculation could improve the dewatering performance in terms of DS content, supernatant turbidity, SRF, floc size, and EPS concentration in the supernatant. DS content dewatered by CPAM without CTAB ranged from 27.3%–28.3%; however, the DS content was increased to 29.5%–34.3% by adding CTAB at pH 3. Sludge sedimentation rate and floc size were significantly accelerated and increased through surfactant application. The improvement of the performance of the sludge dewatering resulting from reduction of EPS and the increase of the floc size in sludge flocculation process. The settling rate flocculated by the CPAM only was lower than that obtained by CPAM+CTAB. However, an opposite phenomenon was observed, the ultimate height of sludge-water interface flocculated by CPAM only was higher than that obtained by CPAM+CTAB. Meanwhile, the sludge flocs were compact, and a large amount of water could be removed during filtration and dewatering. Incorporation of the surfactants had a positive effect on the sludge dewatering.

Acknowledgments

We acknowledge Guocheng Zhu for experiment design, and Zuping Xiong for his help on experiment operation; and anonymous reviewers for helpful comments on our manuscript.

Author Contributions

Conceived and designed the experiments: YS HZ. Performed the experiments: YS Chuanliang Zhao YL. Analyzed the data: YS JZ HT. Contributed reagents/materials/analysis tools: HZ JZ. Wrote the paper: YS HZ. Equipment maintenance: JZ HT Chun Zhao YL.

References

1. Wang W, Luo Y, Qiao W (2010) Possible solutions for sludge dewatering in China. Front. Environ Sci Eng China. 4: 102–107.
2. More TT, Yan S, Tyagi RD, Surampalli RY (2010) Potential use of filamentous fungi for wastewater sludge treatment. Bioresource Technol 101: 7691–7700.
3. Qi Y, Thapa KB, Hoadley AFA (2011) Application of filtration aids for improving sludge dewatering properties –A review. Chem Eng J 171: 373–384.
4. Pei HY, Hu WR, Liu QH (2010) Effect of protease and cellulase on the characteristic of activated sludge. J Hazard Mater 178: 397–403.
5. Zheng HL, Sun YJ, Zhu CJ, Guo JS, Zhao C, et al. (2013) UV-initiated polymerization of hydrophobically associating cationic flocculants: Synthesis, characterization, and dewatering properties. Chem Eng J 234: 318–326.
6. Zhu JR, Zheng HL, Zhang Z, Jiang ZZ, Liu LW, et al. (2012) Synthesis and characterization of a dewatering reagent: cationic polyacrylamide (P(AM–DMC–DAC)) for activated sludge dewatering treatment. Desalin Water Treat 51: 2791–2801.
7. Ma JY, Zheng HL, Tan MZ, Liu LW, Chen W, et al. (2013) Synthesis, Characterization, and Flocculation Performance of Anionic Polyacrylamide P (AM-AA-AMPS). J Appl Polym Sci 129: 1984–1991.
8. Yu GH, He PJ, Shao LM, He PP (2008) Stratification structure of sludge flocs with implications to dewaterability. Environ Sci Technol 42: 7944–7949.
9. Neyens E, Baeyens J (2003) A review of thermal sludge pre-treatment processes to improve dewaterability. J Hazard Mater B98: 51–67
10. Yuan H, Zhu N, Song F (2011) Dewaterability characteristics of sludge conditioned with surfactants pretreatment by electrolysis. Bioresource Technol 102: 2308–2315.
11. Huang C, Pan JR, Fu C, Wu C (2002) Effects of Surfactant Addition on Dewatering of Alum Sludges. J Environ Eng 128: 1121–1127.
12. Chu CP, Lee DJ, Huang C (1998) The Role of Ionic Surfactants in Compression Dewatering of Alum Sludge. J Colloid Interf Sci 206: 181–188.
13. Besra L, Singh BP, Reddy PSR, Sengupta D K (1998) Influence of surfactants on filter cake parameters during vacuum filtration of flocculated iron ore sludge. Powder Technol 96: 240–247.
14. Besra L, Sengupta DK, Roy SK, Ay P (2002) Polymer adsorption: its correlation with flocculation and dewatering of kaolin suspension in the presence and absence of surfactants. Int J Miner Process 66: 183–202.
15. Besra L, Sengupta DK, Roy SK, Ay P (2002) Studies on f locculation and dewatering of kaolin suspensions by anionic polyacrylamide f locculant in the presence of some surfactants. Int J Miner Process 66: 1–28
16. Chen YG, Yang HZ, Gu GW (2001) Effect of acid and surfactant treatment on activated sludge dewatering and settling. Water Res 35: 2615–2620.
17. Zheng HL, Sun YJ, Guo JS, Li FT, Fan W, et al. (2014) Characterization and evaluation of dewatering properties of PADB, a highly efficient cationic flocculant. Ind Eng Chem Res 53: 2572–2582.
18. Raynaud M, Vaxelaire J, Olivier J, Fauvel ED, Baudez JC (2012) Compression dewatering of municipal activated sludge: Effects of salt and pH. Water Res 46: 4448–4456.
19. Lowry OH, Rosebrough NJ, Tarr AL, Randall RJ (1951) Protein measurement with the Folin phenol reagent. J Biol Chem 193: 265–275.
20. Dubois M, Gilles KA, Hamilton JK, Rebers PA, Smith F (1956) Colorimetric method for determination of sugars and related substances. Anal Chem 28: 350–356.
21. Zhang ZQ, Xia SQ, Zhang J (2010) Enhanced dewatering of waste sludge with microbial flocculant TJ-F1 as a novel conditioner. Water Res 44: 3087–3092.
22. Chen YG, Chen YS, Gu GW (2004) Influence of pretreating activated sludge with acid and surfactant prior to conventional conditioning on filtration dewatering. Chem Eng J 99: 137–143.
23. Zheng HL, Zhu GC, Jiang SJ, Tshukudu T, Xiang XY, et al. (2011) Investigations of coagulation–flocculation process by performance optimization, model prediction and fractal structure of flocs. Desalination 269: 148–156.
24. Li QL, Han Q, Yan KS (2011) Application of surfactant in the sludge dewatering. Pap. Pap. Making 30: 59–62. (in chinese)
25. Ji ZY, Chen GL, Chen YG (2010) Effects of waste activated sludge and surfactant addition on primary sludge hydrolysis and short-chain fatty acids accumulation, Bioresource Technol 101: 3457–3462
26. Jiang S, Chen YG, Zhou Q (2007) Effect of sodium dodecyl sulfate on waste activated sludge hydrolysis and acidification. Chem Eng J 132: 311–317.
27. Wile BM, Jin B, Lant P (2003) The influence of key chemical constituents in activated sludge on surface and flocculating properties. Water Res 37: 2127–2139.
28. Besra L, Sengupta DK, Roy SK, Ay P (2003) Influence of surfactants on flocculation and dewatering of kaolin suspensions by cationic polyacrylamide (PAM-C) flocculant. Sep Purif Technol 30: 251–264.
29. Mikkelsen LH, Keiding K (2002) Physico-chemical characteristics of full scale sewage sludges with implications to dewatering. Water Res 36: 2451–2462.
30. Wang LF, Wang LL, Li WW, He DQ, Jiang H, et al. (2014) Surfactant-mediated settleability and dewaterability of activated sludge. Chem Eng Sci 116: 228–234.
31. Pefferkorn E (2006) Clay and oxide destabilization induced by mixed alum/macromolecular flocculation aids. Adv Colloid Interfac Sci 120: 33–45.
32. Zhu GC, Zheng HL, Chen WY, Fan W, Zhang P, et al. (2012) Preparation of a composite coagulant: Polymeric aluminum ferric sulfate (PAFS) for wastewater treatment. Desalination 285: 315–323.
33. Zhu GC, Zheng HL, Zhang Z, Tshukudu T, Zhang P, et al. (2011) Characterization and coagulation-flocculation behavior of polymeric aluminum ferric sulfate (PAFS). Chem Eng J 178: 50–59.
34. Liao BQ, Allen DG, Droppo IG, Leppard GG, Liss SN (2001) Surface properties of sludge and their role in bioflocculation and settleability. Water Res 35: 339–350.
35. Chen H, Zheng X, Chen YG, Li M, Liu K, et al. (2014) Influence of Copper Nanoparticles on the Physical-Chemical Properties of Activated Sludge. PLoS One 9(3): e92871. doi:10.1371/journal.pone.0092871
36. Yang Z, Yuan B, Huang X, Zhou JY, Cai J, et al. (2012) Evaluation of the flocculation performance of carboxymethyl chitosan-graft-polyacrylamide, a novel amphoteric chemically bonded composite flocculant. Water Res 46: 107–114.

L-Histidine Inhibits Biofilm Formation and *FLO11*-Associated Phenotypes in *Saccharomyces cerevisiae* Flor Yeasts

Marc Bou Zeidan[1], Giacomo Zara[1], Carlo Viti[2], Francesca Decorosi[2], Ilaria Mannazzu[1], Marilena Budroni[1], Luciana Giovannetti[2], Severino Zara[1]*

1 Dipartimento di Agraria, University of Sassari, Sassari, Italy, 2 Dipartimento di Scienze delle Produzioni Agroalimentari e dell'Ambiente, University of Florence, Firenze, Italy

Abstract

Flor yeasts of *Saccharomyces cerevisiae* have an innate diversity of *FLO11* which codes for a highly hydrophobic and anionic cell-wall glycoprotein with a fundamental role in biofilm formation. In this study, 380 nitrogen compounds were administered to three *S. cerevisiae* flor strains handling *FLO11* alleles with different expression levels. *S. cerevisiae* strain S288c was used as the reference strain as it cannot produce FLO11p. The flor strains generally metabolized amino acids and dipeptides as the sole nitrogen source, although with some exceptions regarding L-histidine and histidine containing dipeptides. L-histidine completely inhibited growth and its effect on viability was inversely related to *FLO11* expression. Accordingly, L-histidine did not affect the viability of the Δ*flo11* and S288c strains. Also, L-histidine dramatically decreased air–liquid biofilm formation and adhesion to polystyrene of the flor yeasts with no effect on the transcription level of the *FLO11* gene. Moreover, L-histidine modified the chitin and glycans content on the cell-wall of flor yeasts. These findings reveal a novel biological activity of L-histidine in controlling the multicellular behavior of yeasts.

Editor: Alvaro Galli, CNR, Italy

Funding: This work was funded by the Sardinian Region (Italy) ("Legge Regionale 7 Agosto 2007, n. 7: Promozione della ricerca scientifica e dell'innovazione tecnologica in Sardegna"). The funders had no role in study design, data collection and analysis, decision to publish, or preparation of the manuscript.

Competing Interests: The authors have declared that no competing interests exist.

* Email: szara@uniss.it

Introduction

Nitrogen starvation triggers cell adhesion and multicellular growth in different yeast species [1–2]. In *Saccharomyces cerevisiae* flor strains, nitrogen limitation induces activation of the *FLO11* gene and formation of the air–liquid biofilm, or flor velum [3]. The General Amino Acids Control (GAAC) pathway and/or the plasma membrane localized Ssy1-Ptr3-Ssy5 (SPS) sensor, responsible for nitrogen sensing, are also involved in the regulation of *FLO11* gene expression [4–7]. The *FLO11* gene codes for an extensively O-mannosylated cell-wall protein that triggers cell–cell and cell–surface adhesion and air–liquid biofilm formation in flor yeast strains [8–9]. The phosphorylation of the mannosyl side chains on the outer surface of yeast creates abundant negatively charged groups and provides yeast with an anionic surface charge at pH≥3 [10–11]. Therefore, nonspecific interactions, such as hydrophobic and electrostatic interactions, are also involved in cellular adhesion and binding [12–14]. Indeed, *flo11* mutants show a drastic decrease in cell-wall O-mannosylation sites, loss of adhesion and biofilm formation ability, and loss of affinity for hydrophobic solvents [8,15–20]. These phenotypes also appear to be greatly influenced by gene length and expression levels of the *FLO11* gene [21]. Along with the *FLO11* gene response to adverse environments, cell-wall components such as chitin, β-glucan, and mannosyl residues are also involved in the process of adaptation to environmental stress, which is orchestrated mainly by the cell-wall integrity pathway [22–23].

In the present study, the effects of 380 nitrogen sources were evaluated for flor strains with different *FLO11* alleles, using Phenotype Microarray (PM) technique. The data show high variability in nitrogen metabolism among the tested strains. The flor strains metabolized a wide range of nitrogen sources, but remarkably, did not metabolize dipeptides containing L-histidine. Interestingly, subsequent biofilm formation and adhesion to polystyrene analysis explored a novel role of L-histidine in reducing dramatically these *FLO11*-related phenotypes.

Materials and Methods

Yeast strains

The yeast strains used in this study are reported in Table 1. The A9, M23 and V80 strains are flor strains that were isolated from different Vernaccia wineries in Sardinia, and that differ in their *FLO11* gene lengths (5, 3.1 and 6 kb) and expression levels (19.04, 7.2, 0.25 AU) [21]. Strain 3238-32 is a haploid derivative of A9, and strain 3238-32Δ*flo11* is a derivative of 3238-32 that was obtained by Zara *et al.* (2005) [16] and that lacks functional

Table 1. *Saccharomyces cerevisiae* strains used in this study.

Strain	Genetic background	Reference
A9	Wild flor strain of *S. cerevisiae* isolated from Arvisonadu wine	Zara et al., 2009
M23	Wild flor strain of *S. cerevisiae* isolated from Malvasia wine	Zara et al., 2009
V80	Wild flor strain of *S. cerevisiae* isolated from Vernaccia wine	Zara et al., 2009
3238-32	*MATα leu2-Δ1 lys2-801 ura3-52*	Zara et al., 2005
3238-32Δ*flo11*	*MATα leu2-Δ1 lys2-801 flo11Δ::URA3 ura3-52*	Zara et al., 2005
S288c	*MATα SUC2 gal2 mal mel flo1 flo8-1 hap1 ho bio1 bio6*	Mortimer & Johnston, 1986

Flo11p. The S288c strain has a mutation in the *FLO8* gene that disables *FLO11* expression [24].

Media and culture preparation

The media used in this study were YPD medium (1% yeast extract, 2% peptone, 2% glucose), 20% YPD medium (0.2% yeast extract, 0.4% peptone, 0.4% glucose), Biolog specific IFY-0 with the appropriate additives (1× IFY-0 culture medium, 20 mM D-glucose, 5 mM KH_2PO_4, 2 mM $NaSO_4$ and 1× DyeD Biolog), buffered SC minimal medium (0.17% yeast nitrogen base [YNB] without ammonium sulfate and amino acids, 0.5% ammonium sulfate, 20 mM glucose, and aliquots of 0.1 M $C_6H_8O_7.H_2O$ and 0.2 M Na_2HPO_4 stock solutions that were added to buffer the medium at pH 3, 4, 5 and 6, based on instructions from the Sigma-Aldrich buffer reference center) and flor medium (0.17% YNB without ammonium sulfate and amino acids, 0.5% ammonium sulfate, 4% ethanol) [16]. The supplemented amino acids were added at standard concentrations, as required. Unless otherwise stated, the cell cultures were prepared by overnight incubation in 5 mL YPD at 30°C and with 200 rpm agitation, and then aliquots of the cultures were inoculated in fresh YPD medium for 4 h under the same conditions, to reach the exponential phase (optical density at 600 nm [OD_{600}], 0.4 to 0.5). The cell cultures were than washed twice with sterile water, the OD_{600} was measured, and the appropriate cell concentrations were inoculated into the different media.

Phenotype microarray

The phenotype microarray (PM) was carried out on microtiter plates (PM3B, PM6, PM7 and PM8) purchased from Biolog, Omnilog (Hayward, CA, USA), which allowed the screening of 380 different nitrogen sources, including single amino acids, di/tripeptides, purines, pyrimidines and monoamines [25]. The PM technology measures active metabolism by recording the irreversible reduction of tetrazolium violet to formazan, as indirect evidence of respiratory activity. The strains were grown on YPD agar plates overnight at 30°C and resuspended in 15 mL nutrient supplement solution (9.12 mM L-leucine, 5.76 mM L-lysine, 2.59 mM uracil) using a sterile cotton swab, and the cell density was adjusted to 62% transmittance on a Biolog turbidimeter, as equivalent to OD_{600} 0.22 ($2–3\times10^6$ cells/mL). The final inoculating fluids were prepared by diluting the cell suspension 48-fold (62% transmittance in nutrient supplement solution) in IFY-0 apposite culture medium. Then 100 uL of the final inoculating fluids were seeded into the Biolog PM3B, PM6, PM7 and PM8 plates. Next, the PM plates were sealed with Breath-easy gas permeable membrane (Sigma-Aldrich, Milan, Italy), and incubated statically at 30°C in an Omnilog Reader for 96 h. Each experiment was performed in duplicate. The quantitative color changes were recorded automatically every 15 min using a CCD camera, to generate a growth curve for each well. The metabolism of the control wells was considered as the zero point for the other wells. The kinetic responses of the strains in each well were analyzed using the Omnilog-PM software (Biolog, Inc., Hayward, CA, USA).

For the analysis with the nitrogen metabolic assays, two kinetic parameters were used: S, the slope of the kinetic curve; and ΔH, the difference between the maximum and the minimum heights of the kinetic curve. Both of these parameters were combined to calculate the nitrogen activity index, I_N, defined as in Equation (1):

$$I_N = \left(\frac{\Delta H}{\Delta H_{N\max}}\right) \times \left(\frac{S}{S_{N\max}}\right) \quad (1)$$

where $\Delta H_{N\max}$ and $S_{N\max}$ are the highest ΔH and the highest slope, respectively, recorded in the nitrogen panels (PM3B, PM6, PM7 and PM8). The I_N ranged between 0 (no metabolic activity) and 1 (maximum metabolic activity), and was used as a parameter for the cluster analysis of the metabolic profiles of the strains grown on the nitrogen sources. As the negative controls (wells A01 in PM3, PM6, PM7, PM8, without a nitrogen source) showed high background; the nitrogen sources were considered to be used when I_N was >0.33. Cluster analysis was performed using the Bionumeric software (Applied Maths, Inc, Austin, TX, USA), using Pearson's coefficient and the Unweighted Pair Group Method with Arithmetic Mean (UPGMA). The cophenetic correlation coefficient was computed to evaluate the quality of the cluster analysis.

Antimicrobial activity of L-histidine and L-histidine-containing dipeptides

Dose response assays were carried out in 96-well microtiter plates. Aliquots of 135 μL of the cell suspensions containing 10^4 cells/mL in 20% YPD were mixed in the microtiter plate wells with 15 μL 10× concentrated L-histidine from serial two-fold dilutions. Distilled water was used instead of L-histidine in the control wells. All of the samples were prepared in triplicate. The same test was repeated with the dipeptides histidine–methionine (HM), histidine–valine (HV) and histidine–serine (HS) at ≥95% purity (GenScript, NJ, USA), chosen for being representative of the L-histidine containing dipeptides tested by the PM analysis, and for their different physico-chemical features. The microtiter plates were incubated statically at 30°C for 48 h. Growth was measured automatically every 30 min at OD_{600} using a SPECTROstar nano microplate spectrophotometer (BMG Labtech, Ortenberg, Germany). The average of specific growth rates and the lag time of the curves obtained were analyzed using the DMFit software [26].

Biofilm formation, adhesion ability, and cell viability in the presence of L-histidine and L-histidine–contained dipeptides

Biofilm formation was analyzed in 24-well microtiter plates in the presence of L-histidine, HM, HV and HS, as follows. Cell suspensions containing 5×10^6 cells/mL were prepared in flor medium, and aliquots of 1350 μL were mixed in 24-well microtiter plates with 150 μL 10× concentrated L-histidine or dipeptides stock, to a final concentration of 10 mM; distilled water was added to the control wells. The plates were prepared in duplicate and were incubated statically at 30°C for 5 days. The biofilm weights were measured and calculated as described by Zara et al. (2010) [27], and the cell viability was determined by serial dilution spot tests on YPD agar plates.

The yeast adherence to polystyrene was evaluated as described by Reynolds and Fink (2001) [15], with some modifications. Briefly, cell cultures were prepared as for the biofilm formation test, and 90 μL cell suspensions containing 5×10^6 cells/mL in flor medium were placed into the 96-well polystyrene microtiter plates with 10 μL 10× concentrated L-histidine, HM, HV and HS solutions, to a final concentration of 10 mM. The cell suspensions were incubated statically at 30°C for 48 h. An equal volume of 1% (w/v) crystal violet was added to each well. After 30 min, the wells were washed with sterile water, and the adherence of cells was quantified by solubilizing the retained crystal violet in 100 mL 10% (w/v) SDS and an equal volume of sterile water. After 30 min, 50 μL of these solutions were transferred to fresh 96-well polystyrene microtiter plates, and then A_{570} and A_{590} were measured spectrophotometrically.

Quantitative real-time PCR

The yeast strain A9 was grown overnight and refreshed as described above. Aliquots of 2.7 mL flor medium containing 5×10^6 cells/mL were mixed with 300 μL of sterile water (Ctrl) or with a 10× L-histidine (final concentration 10 mM), and further incubated for 48 hours at 30°C without agitation. Three independent biological replicates were conducted for each sample. Cells were collected by centrifugation and kept at −80°C until processed for RNA isolation. Total RNA was extracted using the Aurum Total RNA Mini Kit (Bio-Rad, Milan, Italy). Two micrograms of total RNA were retro transcribed with iscript cDNA synthesis kit (Invitrogen Life Technologies, Milan, Italy). Quantitative real time PCR (qPCR) was performed using a CFX Connect Real-Time PCR System (Bio-Rad, Milan, Italy), according to manufacturer's protocols using the Syber GreenER qPCR SuperMix for iCycler (Invitrogen Life Technologies, Milan, Italy), with the following thermal profile: activation step (95°C for 10 min); amplification step (40 cycles of 95°C for 10 s, 56°C for 10 s, 72°C for 10 s); melting curve program (95°C for 10 s, 60°C for 15 s, 95°C with a heating rate of 0.1°C/s); and cooling step (40°C for 30 s). Primers for the target gene *FLO11*, as well as *ALG9*, *TAF10* and *UBC6* as independent reference genes [28–29], were designed to an equal annealing temperature of 56°C (Table S1). The quantification cycle point (Cq) for each transcript was obtained using the Bio-Rad CFX Manager software (Bio-Rad, Milan, Italy). Three technical repeats of each one of the three biological replicates were conducted. Normalization of the expression levels among different samples was carried out by considering the geometric mean of the expression levels of the three reference genes *ALG9*, *TAF10* and *UBC6*. *FLO11* relative expression levels were determined using the formula proposed by Pfaffl et al. (2001) [30].

Flow cytometry analysis of mannose residues

Flow cytometry techniques were used to quantify the mannose residues of the cells in the presence of L-histidine. Cell suspensions of 5×10^6 cells/mL in flor medium were incubated for 3 h without or with 10 mM L-histidine, washed, and resuspended in phosphate-buffered saline, pH 7.2 (1.18 g/L of Na_2HPO_4-$2H_2O$, 0.22 g/L NaH_2PO_4, 8.5 g/L NaCl). Then 10 μL concanavalin A lectin labeled with fluorescein isothiocyanate (ConA-FITC; FITC contents 3.6 mol/mol lectin; Sigma-Aldrich Milan, Italy; stock solution, 1 mg conjugate/mL) was added, and the cells were incubated for 20 min at room temperature, in the dark. After this incubation, the samples were immediately analyzed, using a BD FACSCalibur flow cytometer (BD Biosciences, San Jose, USA). The acquisition protocol of 20,000 cells/sample was defined at FL1-h after measuring the background fluorescence and the maximum fluorescence of each strain, to standardize the fluorescence activity between them. The data were analyzed using the Expo32 software included with the cytometer.

Fluorescence microscopy

Fluorescence microscopy was used to quantify the chitin residues of all of the tested yeast strains in the presence of L-histidine. One mL flor medium containing 5×10^6 cells/mL was incubated without or with 10 mM L-histidine for 3 h at 30°C. In separate experiments, strains 3238-32 and 3238-32Δ*flo11* were incubated in SC minimal medium buffered to pH 3 and 6 with 1 mM tetramethylrhodamine-labeled histidine–histidine dipeptide (TMR-HH; ≥95% purity; GenScript, NJ, USA) for 3 h at 30°C. After the incubations, aliquots of 25 μM calcofluor white (CFW) were added for 5 min. The cells were washed and examined using a YM10 monochrome fluorescence CCD camera (BX61 motorized system microscope, Olympus, Tokyo, Japan) with excitation/emission wavelengths of 395/440 nm for CFW detection, and 550/573 nm for TMR detection. Differential interference contrast and fluorescence images were captured under the 100× objective using the imaging software Cell* for life science microscopy (Olympus, Tokyo, Japan). The captured photographs were merged using MacBiophotonics MBF ImageJ software.

Cell surface charge variation of 3238-32 and the flo11 mutant in minimal medium at different pHs

The growth and the cell surface net charge of strains 3238-32 and its isogenic 3238Δ*flo11* in minimal medium plus 5 mM L-histidine were measured at different pHs. The cells (10^4 cells/mL) were incubated in a series of buffered SC minimal media plus 5 mM L-histidine. The cells were grown in 96 wells microtiter plates, statically at 30°C for 48 h. Their growth was monitored by measuring the OD_{600} in a SPECTROstar nano microplate spectrophotometer (BMG Labtech, Ortenberg, Germany). Replicates of each experiment were used to measure the cell surface net charge Z-potential using a Zetasizer Nano (Zetasizer Ver. 6.20 Malvern Instruments, Malvern, UK), after 48 h of incubation. All of the measurements represent means and standard deviations of three replicates.

Results

Flor strains differ significantly in catabolism of nitrogen sources

According to the PM technique, the electron flow that results from the catabolism of nutritional substrates induces a shift in the tetrazolium dye to a purple color. When catabolism occurs at a subnormal rate, this results in a decrease in the electron flow and decreased intensity of the purple color [31]. On this basis, PM plates were used to test the ability of the strains to catabolize 380 different nitrogen sources, photographed every 15 min, to generate a growth curve for each well that primarily reflected dye reduction [25]. After 96 h of static incubation, PM analysis showed that the four strains differed greatly in the use of the nitrogen sources. In particular, while A9 and M23 catabolized 128 and 121 nitrogen sources, respectively, this number dramatically decreased for V80 and S288c, which used just 14 and 40 nitrogen sources, respectively (Fig. S1). Differences were observed also for the catabolism rate of nitrogen sources among strains. On the basis of cluster analysis of PM results, the four strains were ascribed to two groups, one consisting of A9 and M23, and the other one containing V80 and S288c (Fig. 1A). A9 and M23, but not V80 and S288c, grew slightly on different nucleotides, such as cytosine and adenine. They metabolized single L-amino acids, such as L-arginine, L-glutamine, L-phenylalanine, L-serine and L-tryptophan and showed high metabolic rates when fed with dipeptides containing alanine, valine, serine and threonine on their N-terminus. In parallel, all of the strains showed clear inability to metabolize dipeptides containing proline, asparagine, cysteine and lysine at their N-terminus. Notably, none of the strains grew in the L-histidine wells (Fig. S1). At the same time, A9 and M23 clearly did not grow in the presence of dipeptides containing L-histidine at their C- and/or N-terminus. On the contrary, strains V80 and S288c showed high and specific metabolic rates toward these dipeptides (Fig. 1B).

L-histidine affects growth of *S. cerevisiae* flor yeasts

To further evaluate the inhibitory effects of L-histidine on *S. cerevisiae*, dye-independent growth measurements were carried out in 20% YPD medium added with up to 80 mM L-histidine. The OD_{600} of cell suspensions was measured after 48 h treatment with increasing concentrations of this amino acid. The L-histidine minimal inhibitory concentrations (MICs) ranged from 20 mM to 25 mM, and the half maximal inhibitory concentrations (IC_{50}) were from 10 mM to 15 mM (Fig. 2). The diploid A9, M23 and V80 strains were slightly more resistant to higher L-histidine concentrations, with respect to the S288c haploid strain. Moreover, the four strains differed markedly in their tolerance to L-histidine. In the presence of 2.5 mM and 5.0 mM L-histidine, all of them increased the duration of the lag phase (Table 2). However, at these concentrations of L-histidine, strains V80 and S288c showed greater tolerance with respect to strains A9 and M23. Accordingly, the specific growth rate (μ) was not affected or was increased in strains V80 and S288c, while it was dramatically decreased in strains A9 and M23 (Table 2).

Previous studies have shown that V80 is characterized by low expression levels of Flo11p [21], while S288C does not express the *FLO11* gene due to a mutation in *FLO8* [24]. On the contrary, A9 and M23 showed high expression levels of the *FLO11* gene [21]. Thus, to evaluate possible correlations between tolerance to L-histidine and the expression levels of *FLO11*, the effects of L-histidine were evaluated also for the 3238-32 and 3238-32Δ*flo11* strains. 3238-32 is a haploid derivative of A9 that is characterized by high expression levels of *FLO11*, while 3238-32Δ*flo11* lacks Flo11p [3,16]. Interestingly, these two strains showed dramatically different specific growth rate inhibition in the presence of L-histidine, and while 3238-32 behaved as A9 and M23, the behavior of 3238-32Δ*flo11* was comparable to that of V80 and S288C (Table 2).

L-histidine–containing dipeptides also had inhibitory effects on all of the tested strains, although at higher concentrations with respect to L-histidine (data not shown).

L-histidine affects FLO11-associated phenotypes

To further investigate possible interactions between L-histidine and Flo11p, the effects of L-histidine were tested for biofilm forming ability and adherence to polystyrene for all of the strains in the flor medium. After 5 days of incubation in the presence of

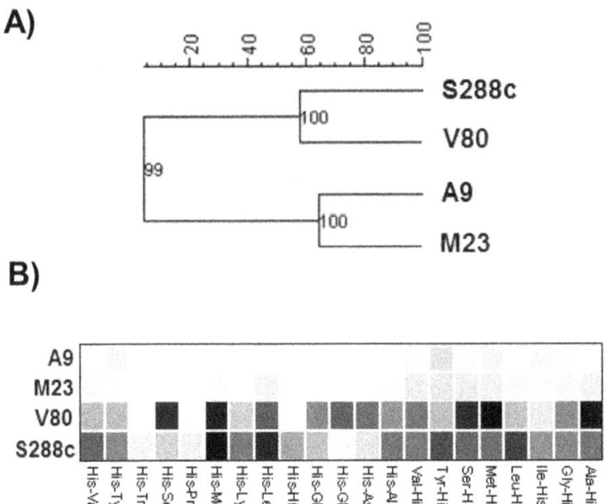

Figure 1. High throughput and cluster analysis of nitrogen metabolism of different *S. cerevisiae* strains. The nitrogen uptake of the A9, M23, V80 and S288c strains was measured using the phenotype microarray technique. (**A**) Cluster analysis (Pearson coefficient, UPGMA) for similarity regrouping of tested strains on all nitrogen sources. I_N was used as a parameter. Values at the nodes represent cophenic correlation coefficients. (**B**) Each square represents the growth of one strain in the PM wells supplied with the indicated L-histidine containing dipeptide, as a nitrogen source. The extent of growth was generated from the tetrazolium dye reduction during 96 h and represented by the intensity of coloration; white squares mean no growth and dark black squares mean abundant growth. Dipeptides are grouped respect to the N-terminus amino acid.

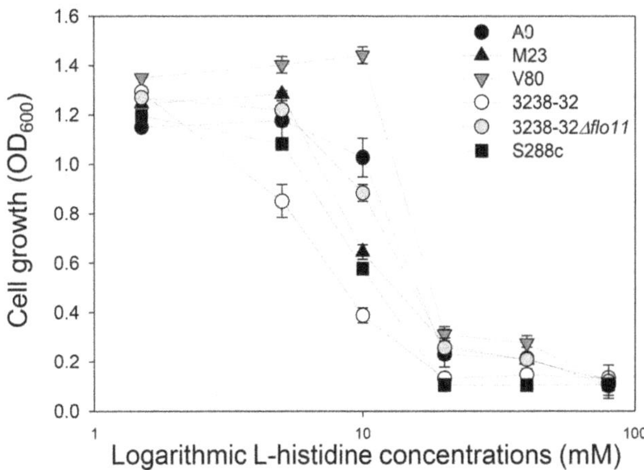

Figure 2. L-histidine affects the growth of different *S. cerevisiae* strains in YPD rich medium. Tested strains (10^4 cells/mL) were incubated statically in 20% YPD for 48 h at 30°C, without (Ctrl) or with a serial dilution of L-histidine with concentrations range from 1 to 80 mM,. Dose-response curves show mean OD_{600} ± SD after 48 h of inoculation. *S. cerevisiae* flor strains are A9 (black circles), M23 (black triangles), V80 (grey down-pointing triangles), 3238-38 (white circles), 3238-32Δ*flo11* (grey circles) and S288c (black squares).

Table 2. Strains growth rate inhibition and Lag phase delay in the presence of different L-histidine concentrations.

Strain	L-histidine [mM] 2.5 µ inhibition (%)	2.5 Lag phase delay (h)	5 µ inhibition (%)	5 Lag phase delay (h)	10 µ inhibition (%)	10 Lag phase delay (h)	20 µ inhibition (%)	20 Lag phase delay (h)
A9	6.41	0.97	9.22 (a)	1.4	26.54 (b)	2.41	69.84	16.36
M23	8.4 (a)	1.65	25.09 (b)	1.68	30.92	5.58	63.27	8.37
V80	(−1.5) (c)	1.38	(−0.96) (c)	2.84	15.86	2.84	72.14 (d)	14.68
3238-32	(−2.01)	6.68	37.53	6.68	73.87 (d)	18.29	100 (e)	-
3238-32Δ*flo11*	(−15.31)	3.79	(−8.5)	12.17	70.75	17.94	100 (e)	-
S288c	(−25.89)	0.66	(−5.31)	2.86	(−3.79)	9.47	98.35 (e)	-

The average of Log-phase specific growth rate was calculated by the DMFit software [26], and specific growth rate inhibition (%) and Lag-phase delay (h) of L-histidine treated cells were calculated in respect to control cells. Negative values in parentheses represent results with no growth rate inhibition. Values with the same letter are not statistically different (Multiple comparison analysis; 95% confidence). Minus symbol (-) represent a complete Lag-phase delay.

10 mM L-histidine, strains A9, M23, V80, and 3238-32 showed dramatic reductions in air-liquid biofilm formation (Fig. 3A and 3B). This phenomenon was accompanied by minor reductions in the cell viabilities of these strains (Fig. 3C). S288c and 3238-32Δ*flo11* did not form an air–liquid biofilm, due to the lack of Flo11p, and their viability was not significantly affected by 10 mM L-histidine, as shown by the CFU recovery (Fig. 3C). Also, the addition of 10 mM dipeptides resulted in variations in biofilm formation after 5 days. Strains A9, M23 and 3238-32 did not form biofilms in the presence of all of the three dipeptides, and showed only a small reduction in CFU (Fig. 3), which was similar to that observed in the presence of L-histidine. On the contrary, V80 increased biofilm weight and viability in the presence of the dipeptides HM, HV and HS (Fig. 3).

Adhesion to polystyrene was evaluated after 48 h incubation in flor medium without or with 10 mM L-histidine or 10 mM dipeptides. S288c and 3238-32Δ*flo11* showed very low adhesion after 48 h, as expected for strains lacking Flo11p. However, strains A9, M23, V80, and 3238-32, which were highly adhesive in the absence of L-histidine, showed drastic reductions in their adhesion to polystyrene in the presence of L-histidine and the three dipeptides HV, HM and HS (Fig. 4).

Despite the above noted effect of L-histidine in inhibiting the *FLO11* associated phenotypes, the transcription analysis of *FLO11* in strain A9 (used as representative flor yeast strain) revealed that the addition of 10 mM L-histidine did not significantly ($P = 0.763$) affect *FLO11* transcription levels in flor medium.

L-histidine induces modifications to the cell wall

To further investigate the inhibitory effects of L-histidine on biofilm formation, the fluorescence of concanavalin A–FITC-treated cells was analyzed using flow cytometry. This approach detects the levels of cell-wall protein mannosylation, which is a crucial factor in the biofilm formation process [32]. All of the strains showed enhancement in concanavalin A binding upon treatment with 10 mM L-histidine. The fluorescence intensity emitted by cells of A9, M23, V80, 3238-32, 3238-32Δ*flo11* and S288c varied (measured as arbitrary fluorescent units; afu), as shown in Figure 5. As the variation in cell fluorescence intensity directly reflects the variations in the contents of cell-wall glycans, which are mainly mannose residues, an enhancement of fluorescence in L-histidine–treated cells indicates an increase in cell-wall protein mannosylation.

To determine whether these changes in mannosylation were accompanied by general cell-wall modifications induced by L-histidine, variations in the chitin content were also analyzed, according to Watanabe *et al.*, (2005) [33]. Fluorescence microscopy of CFW-stained cells showed remarkable differences in staining intensity among the strains that depended on the presence of L-histidine. In the absence of L-histidine, A9, M23, V80, 3238-32, and S288c showed low chitin content (Fig. 6). However, strains A9, M23 and 3238-32 increased chitin content upon L-histidine treatment, while V80, 3238-32Δ*flo11* and S288c did not show any significant variations. At the same time, CFW staining of the 3238-32Δ*flo11* strain was comparable both in the absence and presence of L-histidine. This might be due to constitutive over-production of chitin in cell wall related mutants, which will affect cell-wall integrity [34–35]

FLO11 - L-histidine interaction model

To determine whether the effect of L-histidine is *FLO11*-dependent and/or mediated by physico-chemical interactions, strains 3238-32 and 3238-32Δ*flo11* were grown in SC medium

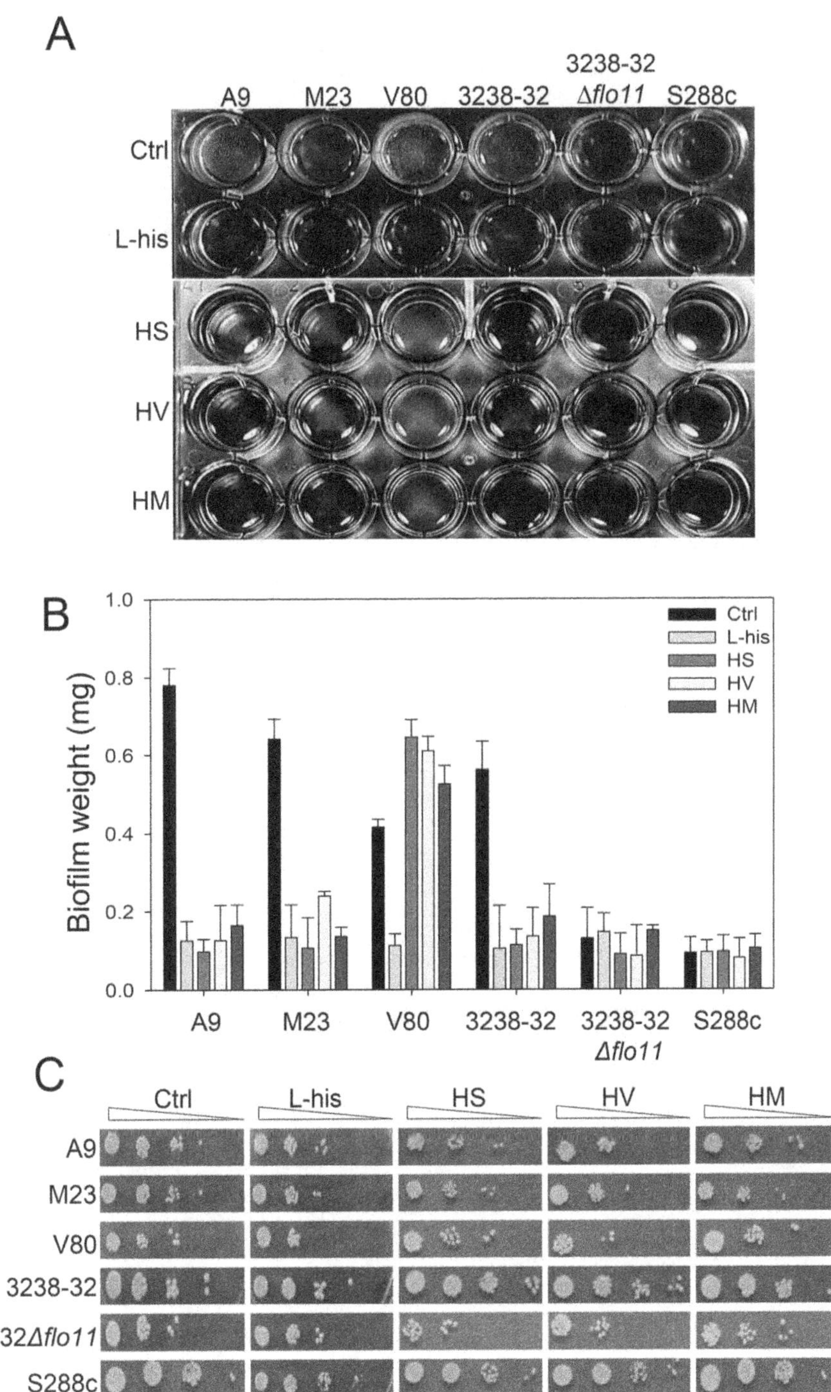

Figure 3. Biofilm formation of flor strains is inhibited by L-histidine. (**A**) Biofilm formation at the air-liquid interface in 24-well microtiter plates for strains A9, M23, V80, 3238-32, 3238-32Δ*flo11* and S288c after 5 days of static incubation in 1.5 mL flor medium at 30°C in the absence (Ctrl) and presence of 10 mM of L-histidine (L-his) and L-histidine–containing dipeptides. The biofilm is visualized as opaque floating material at the top of each well. (**B**) Dry weight determinations of the biofilms formed by the strains in (A) without (Ctrl) and with treatment with 10 mM L-histidine and the L-histidine–containing dipeptides (as indicated). Data are means +SD of three replicate treatments. (**C**) CFU recovery after plating on YPD agar using serial dilutions of a duplicate of all the strains/L-histidine and strains/dipeptides combinations.

buffered at pHs from 3.0 to 6.0. The 3238-32 growth performance changed depending on pH, and the cell density increased significantly with an increase in pH, reaching a maximum at pH 6.0. On the contrary, growth of 3238-32Δ*flo11* remained stable and independent of pH. At pH 6.0, the 3238-32 and 3238-32Δ*flo11* strains showed comparable growth (Fig. 7).

The cell surface net charge of 3238-32 also varied significantly, whereby at pH 3.0, it was slightly positive (+0.1±0.062 mV), while it decreased at increasing pH, to reach −8.5±0.087 mV at pH 6.0. In contrast, the cell surface net charge of 3238-32Δ*flo11* was stable, varying from −3.04±0.142 mV at pH 3.0 to −4.22±0.081 mV at pH 6.0. Interestingly, the increase in the

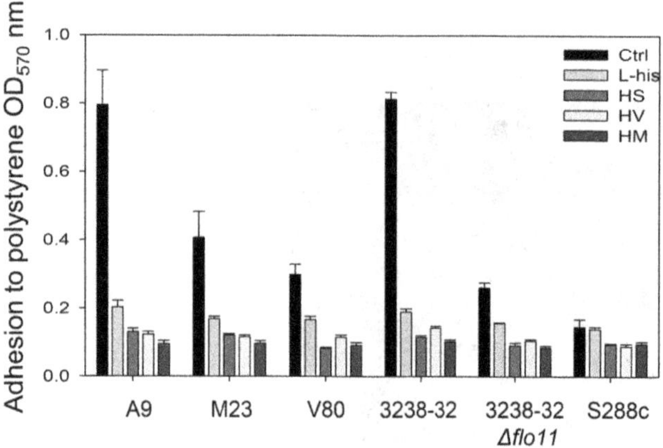

Figure 4. Loss of adhesion in presence of L-histidine and other dipeptides. Adhesion is expressed as OD_{570} and was measured using crystal violet dye after 48 h incubation of 5×10^6 cell/mL of the *S. cerevisiae* strains in flor medium without (Ctrl) and with 10 mM L-histidine or the L-histidine–containing peptides.

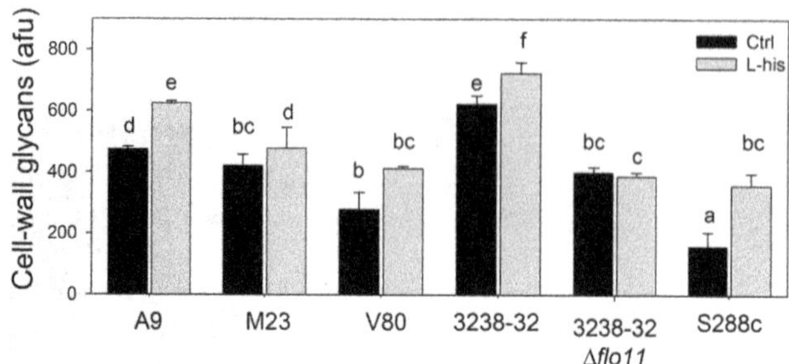

Figure 5. Modulation of cell-wall glycans of *S. cerevisiae* strains in the absence and presence of L-histidine. Cell-wall glycan levels (as arbitrary fluorescent units) without (Ctrl) and with 10 mM L-histidine treatment for A9, M23, V80, 3238-32, 3238-32Δ*flo11* and S288c strains (5×10^6 cell/mL) in flor medium after 2 h. Data are means +SD from three replicate samples, of the fluorescence intensity of ConA-FITC bound to cell-wall glycans of 20.000 cells/sample. Multiple comparison analysis was conducted. Bars with the same letters are no statistically different (95% confidence).

anionic charge correlated with the increase in growth with the 3238-32 strain (Fig. 7).

Accordingly, TMR-HH stained cells of strains 3238-32 and 3238-32Δ*flo11* varied with pH. At pH 3.0, strains 3238-32 and 3238-32Δ*flo11* had low and comparable fluorescence, while at pH 6.0, the fluorescence intensity was notably enhanced in 3238-32, but not in 3238-32Δ*flo11* (Fig. S2).

Discussion

Flor strains of *S. cerevisiae* yeast have the unique ability to form biofilms at the air–liquid interface of wine at the end of fermentation, when the medium is depleted of nutrients and further growth becomes dependent on oxygen. This multicellular growth is directly correlated with a series of rearrangements to the cell wall, in terms of the hydrophobicity and adhesion [21]. Indeed, *S. cerevisiae* can use either anaerobic or aerobic modes of substrate metabolism, which can induce specific changes to the cell at the level of the cell-wall organization, nutrient consumption, and cellular interactions with the surrounding environment [36–37].

Nitrogen is a fundamental nutrient in living cells, and its metabolism is involved in major developmental decisions in *S. cerevisiae* [38]. In nitrogen-starvation environment, some signaling pathways (i.e. TORC1, SPS-sensor and GAAC) that are largely related to nitrogen and amino-acid sensing and regulation have been shown to be involved in *FLO11* gene expression and multicellular growth in *S. cerevisiae* [39].

A previous study reported that clinical and vineyard isolates of *S. cerevisiae* can grow on a wide range of nitrogen sources, with respect to laboratory strains [40]. Accordingly, PM analysis in the present study showed similar behavior of the A9 and M23 flor strains, which metabolized more nitrogen sources with respect to the S288c laboratory strain. This reflects the high adaptation ability of these strains. However, this was not the case for the V80 strain, which showed a different behavior, similar to the laboratory strain S288c. Indeed, while V80 and S288c clearly metabolized L-histidine-containing dipeptides, A9 and M23 were definitely unable to grow on these dipeptides.

Dose-response analysis in nutrient-rich (dye-independent) medium showed that L-histidine not only does not support cellular growth as a nitrogen source, but its presence (concentrations ≥ 10 mM) reduces the growth rate, delays the lag-phase, and finally inhibits the growth of the tested strains. These effects were also observed in strains treated with higher concentrations of L-histidine–containing dipeptides. Other authors reported that L-carnosine, a L-histidine–containing dipeptide with potential antineoplastic effects [41], is able to slow cell growth rates and increase death of yeast cells in fermentative metabolism [37]. Interestingly, according to Letzien et al., (2014) [41], L-histidine mimicks the effect of L-carnosine although showing a stronger effect, similar to that observed in the present work.

In nutrient-depleted media, *S. cerevisiae* can trigger a series of stress-signaling pathways and responses, which include modulation of the cell wall, expression of the *FLO11* gene, and formation of biofilms [15,18,27]. This phenomenon was also observed in this study in the control wells of the biofilm-forming strains A9, M23, V80 and 3238-32, but not in the wells that contained L-histidine. In fact, the presence of 10 mM L-histidine was sufficient to completely inhibit biofilm formation and adhesion to polystyrene for all of the tested strains, and these major inhibitory effects were accompanied by minor reductions in cell viability. These inhibitory effects did not correlate with the transcription level of *FLO11*, which remained stable in the absence or the presence of L-histidine. The stability of *FLO11* expression levels evidences that L-histidine cannot be used as a nitrogen source, because if so, it would have been sensed by the GAAC pathway and/or the SPS sensor, leading to a repression of *FLO11* [4–7].

As stated before, cellular adhesion and binding are likely to be influenced by nonspecific interactions, such hydrophobic and electrostatic interactions [12–14]. Among the 20 naturally-occurring amino acids, L-histidine is a cationic amino acid with a unique imidazole ring as a side chain. These particular physico-chemical features make it a good candidate for nonspecific interactions. These would mainly be stacking and hydrogen-bond interactions, which would provide L-histidine with high affinity for cationic metals, aromatic amino acids, and many other compounds [42–43]. These features of L-histidine might induce the loss of cell adhesion and biofilm formation of the flor strains, by providing nonspecific physical interactions with the embedded cell-wall components in general, and with the highly O-mannosylated cell-wall mannoprotein Flo11p in particular. This leads to the failure of air–liquid biofilm formation and cell adhesion.

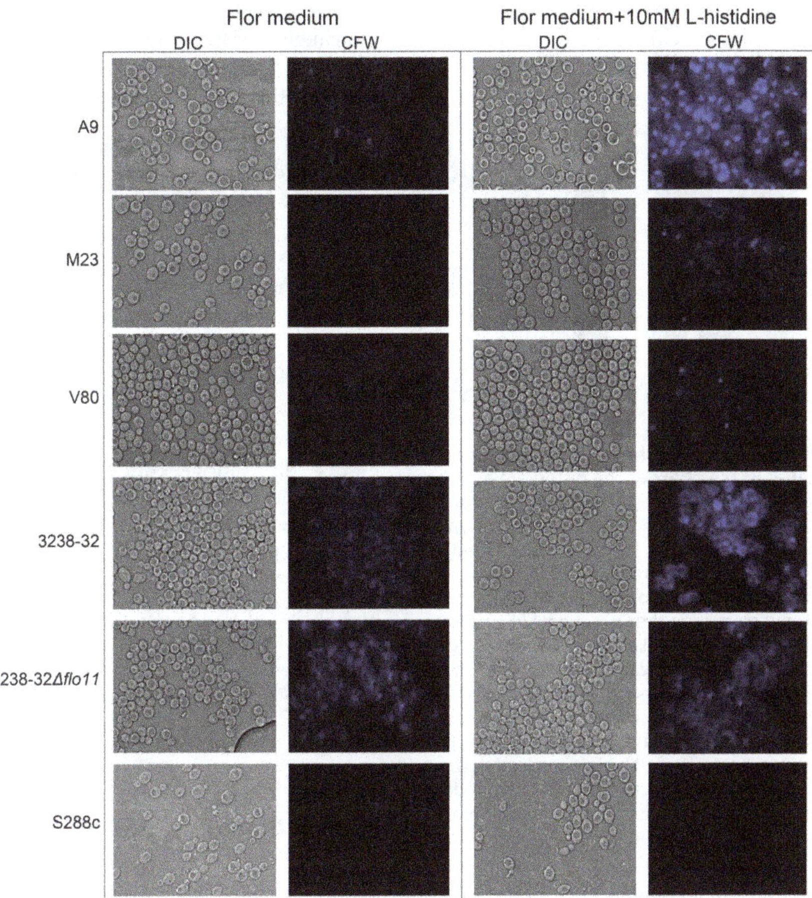

Figure 6. CFW staining of *S. cerevisiae* strains in the absence and presence of L-histidine. The A9, M23, V80, 3238-32, 3238-32Δ*flo11* and S288c strains (5×10^6 cell/mL) were incubated for 2 h in flor medium without or with 10 mM L-histidine. After incubation, the samples were stained with 25 μM CFW for 5 min and observed by fluorescence microscopy. Bright-field differential interference contrast (DIC) and CFW images of the same field are shown. All of the images were captured under the same acquisition parameters and therefore reflect actual differences in CFW staining.

Cell-wall glycans and chitin are mainly responsible for cell permeability, and they are related to the cell-wall integrity pathways for responses to adverse conditions [10,44]. The enhancement of fluorescence intensity of these cell-wall compounds in L-histidine–treated cells reveals the antimicrobial effects of this amino acid and reduces the permeability of the cell, which favors nonspecific interactions with cell-wall mannoproteins.

The proposed interaction model of 3238-32 and its isogenic 3238-32Δ*flo11* with L-histidine, showed that the effect of this amino acid is *FLO11*-dependent, and related to pH and cell-surface charge. In more detail, at pH 3.0, the repulsive interactions between the high cationic charge of L-histidine and the slightly positive cell-surface charge of the 3238-32 strain resulted in a reduction of 3238-32 growth. In parallel, at pH 6.0, the attractive interactions between the low cationic charged and neutralized imidazole ring of L-histidine (at the side-chain isoelectric point) and the high anionic cell surface charge of the 3238-32 strain led to the decreasing of the antimicrobial effects of this amino acid and thus an increasing of 3238-32 growth. These results highlight the role of *FLO11*, as 3238-32Δ*flo11* did not change its cell-surface charge and its interactions with L-histidine, and its growth. Similar behavior was seen by microscopic observations for these strains and TMR-HH: at pH 3.0, the low fluorescence intensity emitted from 3238-32 cells reflects the low adsorption of this dipeptide, which then increases at pH 6.0. Here again, 3238-32Δ*flo11* showed stable fluorescence intensity, and thus a stable interaction with the dipeptide.

The molecular mechanisms of this novel role of L-histidine are still unknown. Many studies have shown similar modes of action of several small cationic peptide sequences, with antimicrobial effects toward different fungi species. This is seen for human histatins and histidine-rich glycoproteins, which are directly involved in the host response to invasive growth of *Candida albicans*, with their binding to the cell-wall glycoprotein Msb2p [45]. A similar anti-adhesive behavior was reported for filastatin against some *Candida spp.* [46]. In contrast, hydrophobic interactions with the high cationic and hydrophobic hexapeptide PAF26 served as a bridge between some *S. cerevisiae* flor strains, to enhance biofilm formation [47]. To our knowledge, no previous studies have reported this mode of action of L-histidine. Interestingly, a recent study described a novel role of some D-amino acids in the triggering of bacterial biofilm disassembly. These D-amino acids did not affect the growth rate of bacterial cultures, and their mode of action is associated to their incorporation into the peptide side chains of the cell-wall peptidoglycan [48].

In conclusion, the main result in the present study relate to biofilm formation and adhesion ability. These findings reveal a novel biological activity of L-histidine that might be of high biotechnological interest. These data also suggest that glycosylated mucin-like proteins at the fungal cell wall, such as Flo11p, might

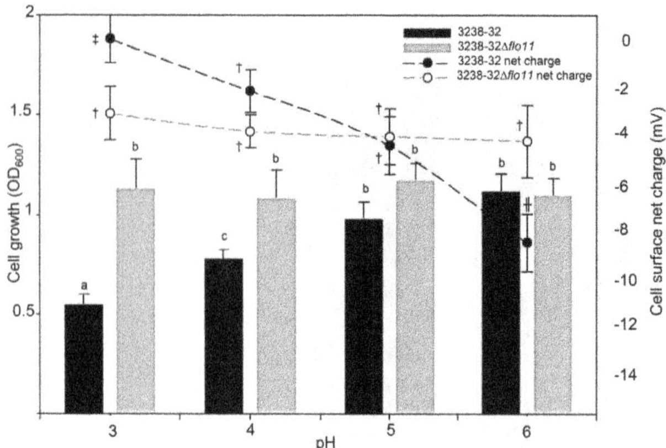

Figure 7. L-histidine effect on cell growth is *FLO11*-dependent, and related to pH and cell-surface charge. The histogram and the line graph illustrate the cell growth and the cell surface net charge, respectively, of the 3238-32 and 3238Δ*flo11* strains. Cells (10^4 cells/mL) were incubated in SC minimal medium. The media were buffered at pH 3, 4, 5 and 6 using 0.1 M citric acid monohydrate and 0.2 M sodium phosphate. Cells were grown in 96-well microplates, statically at 30°C for 48 h. Growth was monitored measuring the OD_{600} in a SPECTROstar nano-microplate spectrophotometer (BMG Labtech, Germany). Cell surface net charge (Z-potential) was measured under the same conditions using the Zetasizer Ver. 6.20 (Malvern Instruments Ltd) after 48 h of treatment. Data are means ±SD of three replicates. A one-way analysis of varience (ANOVA) was performed and followed by Tukey honestly significant difference test ($P<0.05$). The analyses were performed independently for growth and cell surface charge data and values with the same letter/symbol are not statistically different.

be interacting partners for this unique amino acid. Future work will aim to explore the significance of these interactions in relation to the antimicrobial mechanisms of L-histidine with other non-flor yeast and filamentous fungi, and to determine the importance of protein glycosylation in this mechanism.

Supporting Information

Figure S1 High throughput analysis of nitrogen metabolism of different *S. cerevisiae* strains. The nitrogen uptake of the A9, M23, V80 and S288c strains was measured using the phenotype microarray technique. Growth on nitrogen sources groups is showed and each square represents the growth of one strain in the PM wells supplied with a nitrogen source. The extent of growth was generated from the tetrazolium dye reduction during 96 h and represented by the intensity of coloration; white squares mean no growth and dark black squares mean abundant growth.

Figure S2 Fluorescence microscopy of *S. cerevisiae* strains 3238-32 and 3238-32Δ*flo11* exposed to TMR-HH. Cells (5×10^6 cells/ml) were incubated in minimal medium with 1 mM of TMR-HH at 30°C for 2 h and subsequently with 25 μM CFW at 20°C for 5 min. Representative DIC bright-field as well as CFW, TMR, and CFW/TMR-overlay fluorescence micrographs of the same field are shown, for the different strains, as indicated.

Table S1 Oligonucleotide primers used in this study.

Author Contributions

Conceived and designed the experiments: SZ MBZ GZ. Performed the experiments: MBZ FD. Analyzed the data: MBZ SZ GZ FD CV. Contributed reagents/materials/analysis tools: IM MB LG. Wrote the paper: SZ MBZ GZ FD CV IM MB LG.

References

1. Verstrepen KJ, Klis FM (2006) Flocculation, adhesion and biofilm formation in yeasts. Mol Microbiol 60: 5–15.
2. Granek JA, Magwene PM (2010) Environmental and genetic determinants of colony morphology in yeast. PLoS Genet 6: e1000823.
3. Zara G, Budroni M, Mannazzu I, Zara S (2011) Air-liquid biofilm formation is dependent on ammonium depletion in a *Saccharomyces cerevisiae* flor strain. Yeast 28: 809–814.
4. Braus GH (2003) Amino Acid Starvation and Gcn4p Regulate Adhesive Growth and *FLO11* Gene Expression in Saccharomyces cerevisiae. MOL BIOL CELL 14: 4272–4284.
5. Ljungdahl PO (2009) Amino-acid-induced signalling via the SPS-sensing pathway in yeast. Biochem Soc Trans 37: 242–247.
6. Bruckner S, Mosch HU (2012) Choosing the right lifestyle: adhesion and development in *Saccharomyces cerevisiae*. FEMS Microbiol Rev 36: 25–58.
7. Torbensen R, Moller HD, Gresham D, Alizadeh S, Ochmann D, et al. (2012) Amino acid transporter genes are essential for *FLO11*-dependent and *FLO11*-independent biofilm formation and invasive growth in *Saccharomyces cerevisiae*. PLoS One 7: e41272.
8. Fidalgo M, Barrales RR, Ibeas JI, Jimenez J (2006) Adaptive evolution by mutations in the *FLO11* gene. Proc Natl Acad Sci U S A 103: 11228–11233.
9. Alexandre H (2013) Flor yeasts of *Saccharomyces cerevisiae*—Their ecology, genetics and metabolism. International Journal of Food Microbiology 167: 269–275.
10. Lipke PN, Ovalle R (1998) Cell wall architecture in yeast: new structure and new challenges. J Bacteriol 180: 3735–3740.
11. Klis FM, de Jong M, Brul S, de Groot PW (2007) Extraction of cell surface-associated proteins from living yeast cells. Yeast 24: 253–258.
12. Caridi A (2006) Enological functions of parietal yeast mannoproteins. Antonie Van Leeuwenhoek 89: 417–422.
13. Holle AV, Machado MD, Soares EV (2012) Flocculation in ale brewing strains of *Saccharomyces cerevisiae*: re-evaluation of the role of cell surface charge and hydrophobicity. Appl Microbiol Biotechnol 93: 1221–1229.
14. Kregiel D, Berlowska J, Szubzda B (2012) Novel permittivity test for determination of yeast surface charge and flocculation abilities. J Ind Microbiol Biotechnol 39: 1881–1886.
15. Reynolds TB, Fink GR (2001) Bakers' yeast, a model for fungal biofilm formation. Science 291: 878–881.
16. Zara S, Bakalinsky AT, Zara G, Pirino G, Demontis MA, et al. (2005) *FLO11*-Based Model for Air-Liquid Interfacial Biofilm Formation by *Saccharomyces cerevisiae*. Applied and Environmental Microbiology 71: 2934–2939.
17. Dranginis AM, Rauceo JM, Coronado JE, Lipke PN (2007) A biochemical guide to yeast adhesins: glycoproteins for social and antisocial occasions. Microbiol Mol Biol Rev 71: 282–294.
18. Barrales RR, Jimenez J, Ibeas JI (2008) Identification of Novel Activation Mechanisms for *FLO11* Regulation in *Saccharomyces cerevisiae*. Genetics 178: 145–156.
19. Fidalgo M, Barrales RR, Jimenez J (2008) Coding repeat instability in the *FLO11* gene of *Saccharomyces* yeasts. Yeast 25: 879–889.
20. Goossens KV, Willaert RG (2012) The N-terminal domain of the Flo11 protein from *Saccharomyces cerevisiae* is an adhesin without mannose-binding activity. FEMS Yeast Res 12: 78–87.
21. Zara G, Zara S, Pinna C, Marceddu S, Budroni M (2009) *FLO11* gene length and transcriptional level affect biofilm-forming ability of wild flor strains of *Saccharomyces cerevisiae*. Microbiology 155: 3838–3846.
22. Cid VJ, Duran A, del Rey F, Snyder MP, Nombela C, et al. (1995) Molecular basis of cell integrity and morphogenesis in *Saccharomyces cerevisiae*. Microbiol Rev 59: 345–386.
23. Levin DE (2005) Cell wall integrity signaling in *Saccharomyces cerevisiae*. Microbiol Mol Biol Rev 69: 262–291.
24. Mortimer RK, Johnston JR (1986) Genealogy of principal strains of the yeast genetic stock center. Genetics 113: 35–43.
25. Bochner BR (2009) Global phenotypic characterization of bacteria. FEMS Microbiology Reviews 33: 191–205.
26. Baranyi J, Roberts TA (1994) A dynamic approach to predicting bacterial growth in food. Int J Food Microbiol 23: 277–294.
27. Zara S, Gross MK, Zara G, Budroni M, Bakalinsky AT (2010) Ethanol-Independent Biofilm Formation by a Flor Wine Yeast Strain of Saccharomyces cerevisiae. Applied and Environmental Microbiology 76: 4089–4091.

28. Vandesompele J, De Preter K, Pattyn F, Poppe B, Van Roy N, et al. (2002) Accurate normalization of real-time quantitative RT-PCR data by geometric averaging of multiple internal control genes. Genome Biol 3: RESEARCH0034.
29. Teste MA, Duquenne M, Francois JM, Parrou JL (2009) Validation of reference genes for quantitative expression analysis by real-time RT-PCR in *Saccharomyces cerevisiae*. BMC Mol Biol 10: 99.
30. Pfaffl MW, Horgan GW, Dempfle L (2002) Relative expression software tool (REST) for group-wise comparison and statistical analysis of relative expression results in real-time PCR. Nucleic Acids Res 30: e36.
31. Bochner BR, Gadzinski P, Panomitros E (2001) Phenotype microarrays for high-throughput phenotypic testing and assay of gene function. Genome Res 11: 1246–1255.
32. Martinez-Esparza M, Sarazin A, Poulain D, Jouault T (2009) A method for examining glycans surface expression of yeasts by flow cytometry. Methods Mol Biol 470: 85–94.
33. Watanabe H, Azuma M, Igarashi K, Ooshima H (2005) Analysis of chitin at the hyphal tip of *Candida albicans* using calcofluor white. Biosci Biotechnol Biochem 69: 1798–1801.
34. Popolo L, Gilardelli D, Bonfante P, Vai M (1997) Increase in chitin as an essential response to defects in assembly of cell wall polymers in the ggp1delta mutant of *Saccharomyces cerevisiae*. J Bacteriol 179: 463–469.
35. Garcia-Rodriguez LJ, Trilla JA, Castro C, Valdivieso MH, Duran A, et al. (2000) Characterization of the chitin biosynthesis process as a compensatory mechanism in the fks1 mutant of *Saccharomyces cerevisiae*. FEBS Lett 478: 84–88.
36. Aguilar-Uscanga B, Francois JM (2003) A study of the yeast cell wall composition and structure in response to growth conditions and mode of cultivation. Lett Appl Microbiol 37: 268–274.
37. Cartwright SP, Bill RM, Hipkiss AR (2012) L-carnosine affects the growth of *Saccharomyces cerevisiae* in a metabolism-dependent manner. PLoS One 7: e45006.
38. Forsberg H, Ljungdahl PO (2001) Sensors of extracellular nutrients in *Saccharomyces cerevisiae*. Curr Genet 40: 91–109.
39. Ljungdahl PO, Daignan-Fornier B (2012) Regulation of Amino Acid, Nucleotide, and Phosphate Metabolism in *Saccharomyces cerevisiae*. Genetics 190: 885–929.
40. Homann OR, Cai H, Becker JM, Lindquist SL (2005) Harnessing natural diversity to probe metabolic pathways. PLoS Genet 1: e80.
41. Letzien U, Oppermann H, Meixensberger J, Gaunitz F (2014) The antineoplastic effect of carnosine is accompanied by induction of PDK4 and can be mimicked by L-histidine. Amino Acids 46: 1009–1019.
42. Shimba N, Serber Z, Ledwidge R, Miller SM, Craik CS, et al. (2003) Quantitative identification of the protonation state of histidines in vitro and in vivo. Biochemistry 42: 9227–9234.
43. Liao SM, Du QS, Meng JZ, Pang ZW, Huang RB (2013) The multiple roles of histidine in protein interactions. Chem Cent J 7: 44.
44. Latge JP (2007) The cell wall: a carbohydrate armour for the fungal cell. Mol Microbiol 66: 279–290.
45. Szafranski-Schneider E, Swidergall M, Cottier F, Tielker D, Roman E, et al. (2012) Msb2 shedding protects *Candida albicans* against antimicrobial peptides. PLoS Pathog 8: e1002501
46. Fazly A, Jain C, Dehner AC, Issi L, Lilly EA, et al. (2013) Chemical screening identifies filastatin, a small molecule inhibitor of *Candida albicans* adhesion, morphogenesis, and pathogenesis. Proc Natl Acad Sci U S A 110: 13594–13599.
47. Bou Zeidan M, Carmona L, Zara S, Marcos JF (2013) *FLO11* Gene Is Involved in the Interaction of Flor Strains of *Saccharomyces cerevisiae* with a Biofilm-Promoting Synthetic Hexapeptide. Appl Environ Microbiol 79: 6023–6032.
48. Kolodkin-Gal I, Romero D, Cao S, Clardy J, Kolter R, et al. (2010) D-amino acids trigger biofilm disassembly. Science 328: 627–629.

MacSyFinder: A Program to Mine Genomes for Molecular Systems with an Application to CRISPR-Cas Systems

Sophie S. Abby[1,2]*, Bertrand Néron[3], Hervé Ménager[3], Marie Touchon[1,2], Eduardo P. C. Rocha[1,2]

1 Microbial Evolutionary Genomics, Institut Pasteur, Paris, France, **2** UMR3525, CNRS, Paris, France, **3** Centre d'Informatique pour la Biologie, Institut Pasteur, Paris, France

Abstract

Motivation: Biologists often wish to use their knowledge on a few experimental models of a given molecular system to identify homologs in genomic data. We developed a generic tool for this purpose.

Results: **Mac**romolecular **Sy**stem **Finder** (MacSyFinder) provides a flexible framework to model the properties of molecular systems (cellular machinery or pathway) including their components, evolutionary associations with other systems and genetic architecture. Modelled features also include functional analogs, and the multiple uses of a same component by different systems. Models are used to search for molecular systems in complete genomes or in unstructured data like metagenomes. The components of the systems are searched by sequence similarity using Hidden Markov model (HMM) protein profiles. The assignment of hits to a given system is decided based on compliance with the content and organization of the system model. A graphical interface, MacSyView, facilitates the analysis of the results by showing overviews of component content and genomic context. To exemplify the use of MacSyFinder we built models to detect and class CRISPR-Cas systems following a previously established classification. We show that MacSyFinder allows to easily define an accurate "Cas-finder" using publicly available protein profiles.

Availability and Implementation: MacSyFinder is a standalone application implemented in Python. It requires Python 2.7, Hmmer and makeblastdb (version 2.2.28 or higher). It is freely available with its source code under a GPLv3 license at https://github.com/gem-pasteur/macsyfinder. It is compatible with all platforms supporting Python and Hmmer/makeblastdb. The "Cas-finder" (models and HMM profiles) is distributed as a compressed tarball archive as Supporting Information.

Editor: Néstor V. Torres, Universidad de La Laguna, Spain

Funding: This project was funded by the Institut Pasteur (http://www.pasteur.fr/en), the French "Centre National de la Recherche Scientifique" (http://www.cnrs.fr/index.php) and the European Research Council (http://erc.europa.eu/) (grant EVOMOBILOME, number 281605). The funders had no role in study design, data collection and analysis, decision to publish, or preparation of the manuscript.

Competing Interests: The authors have declared that no competing interests exist.

* Email: sophie.abby@pasteur.fr

Introduction

Macromolecular systems are involved in key aspects of cell biology [1,2]. They can be constituted of nanomachines, like the ribosome or the flagellum, or molecular pathways, like the ones allowing the degradation of foreign genetic elements by CRISPR-Cas systems. The identification and classification of macromolecular systems is important to characterize biological traits, and is routinely done in many laboratories. However, it is difficult to do on a systematic basis by a number of reasons. Firstly, systems are made of many different components with different levels of dispensability, some being essential and others accessory. For example, homologous recombination in bacteria involves some key essential components (like RecA), and several associated alternative pathways (like RecBCD and RecFOR) [3]. Secondly, key components may have homologs in other systems, complicating their unambiguous assignment to a given system. This is for instance the case of the non-flagellar type III secretion system for which eight of the nine core genes have homologs in the bacterial flagellum [4]. Thirdly, the components of the systems evolve at very diverse rates, complicating the identification of homology by sequence similarity. For example, many proteins involved in reproduction are highly conserved, whereas others endure selection for fast evolution [5]. These difficulties can be partly circumvented by searching for the whole set of components of the system because the integration of all the information leads to more accurate inference. This is especially relevant if the genes encoding these components are organized in highly conserved ways. In Prokaryotes, organelles and viruses, macromolecular systems are often encoded in one or a few conserved neighbouring operons ensuring tight regulation and correct assembly/functioning. This facilitates the assignment of certain components to a system [6–9].

We have developed a program named **Mac**romolecular **Sy**stem **Finder** (MacSyFinder) to detect molecular systems in genome data from user-defined biological models. The components of the systems are searched using protein profiles encoded as hidden Markov models (HMM), such as those available in databases like PFAM, TIGRFAM or PRODOM [10–12]. Protein profiles provide a compressed way to represent a database of homologous sequences, giving increased sensitivity and specificity [13]. MacSyFinder identifies the presence of a given system according to the specifications of the input model, which includes customizable information on the type and number of components, on their genetic organization, and other relevant discriminating traits. We implemented MacSyFinder as a generic portable tool that can be installed in-house for large genomic or metagenomic projects. The companion program, MacSyView, allows the visualization of the results of MacSyFinder. To show a typical situation where MacSyFinder can be useful, we built a set of models to identify Cas proteins. Clustered regularly interspaced short palindromic repeats (CRISPR) arrays and their associated Cas (CRISPR-associated) proteins form the CRISPR-Cas system. CRISPR-Cas are sophisticated adaptive immune systems that rely on small RNAs for sequence-specific targeting of foreign nucleic acids such as viruses and plasmids [14]. Cas proteins have been intensively studied in the recent years for their role in the interaction between Prokaryotes and their mobile genetic elements and for their biotechnological interest [15,16]. Tools are available to detect and analyse CRISPR arrays [17–19], however, no program is available to detect and class *cas* operons themselves. This example shows that using information from the literature and available protein profiles, one can easily build an accurate and efficient "Cas-finder" with MacSyFinder.

MacSyFinder's Rationale

Definition of the models

MacSyFinder models, written using an XML grammar, describe the components and genetic organization of a given macromolecular system (see the documentation in File S1 for a full description of the grammar). Each model is defined in a dedicated file named after the type of system (*e.g.,* CAS-TypeI.xml'), which contains system-wise and component-wise features (Figure 1). MacSyFinder considers three classes of components: *mandatory*, *accessory*, and *forbidden*. Components that are ubiquitous and identifiable in all systems are defined as *mandatory*. Other components of the system are defined as *accessory*. These *accessory* components can be essential for the assembly/functioning of the system, while not being identifiable by sequence similarity because of rapid evolution or because they are non-homologous among variants of the system. Discrimination between partly homologous systems is easier when some specific components are defined as *forbidden* in the models of the systems lacking them (Fig. 1).

Systems that respect a pre-defined minimal quorum of components are identified as complete. The quorum is either the number of *mandatory* components and/or the sum of *mandatory* and *accessory* components (see the documentation on attributes *min_mandatory_genes_required* and *min_genes_required* in File S1). Components defined as functionally *exchangeable* are only counted once in the quorum. These components can be part of systems defined in other models using the *system_ref* keyword. Genes encoding components that participate in multiple systems of the same type, such as proteins interacting with different instances of a system, are labelled *multi_system*.

The genetic architecture of the components is defined using several attributes. Two components are co-localized when their genes are closer than a given number of genes (system-wise parameter *inter_gene_max_space*, Fig. 1). A component defined with the *loner* attribute does not need to be co-localized with other components to be part of a system. One can also specify component-specific values of *inter_gene_max_space*. The system-wise parameter *multi_loci* allows MacSyFinder to detect systems encoded by several distant clusters of genes.

Implementation, system requirements and availability

MacSyFinder was coded in Python, and details on its object-oriented implementation are available in Text S1 and in File S1. MacSyFinder requires Python version 2.7, the formatdb or makeblastdb tools (version 2.2.28 or better for the latter) [20,21] and the program Hmmer [13,22]. MacSyFinder is freely available. Its source code is distributed under a GPLv3 license at https://github.com/gem-pasteur/macsyfinder and updated versions will be accessible there. MacSyFinder is compatible with all platforms supporting Python, Hmmer, and makeblastdb. The MacSyFinder release used in this paper is provided in Data S1. MacSyView's source code is freely available at https://github.com/gem-pasteur/macsyview but it is also distributed in the MacSyFinder's package. MacSyView was coded in Javascript and uses third-party libraries that are included in the package, and accredited in the COPYRIGHT file (See Text S1). It was tested on Chromium and Firefox for Linux, and on Chrome, Firefox and Safari for Mac OS X. A documentation file including installation and users' instructions, details on modelling procedures and examples for MacSyView and MacSyFinder is available in File S1.

Input and output

The MacSyFinder program (Data S1) receives as input a list of systems defined in XML files (see above), protein profiles, command-line parameters and a file with protein sequences in fasta format (see the documentation in File S1). The parameters can be specified in the command-line or in a configuration file. System and component parameters specified in the command-line override model specifications in the XML files.

MacSyFinder manages three different types of protein datasets. The **unordered** dataset lacks information on gene order and genome origin. This mode is useful to study large sequence databanks or metagenomic data. Naturally, in these datasets the notions of co-localization and quorum are not relevant. The **unordered replicon** dataset includes protein sequences from one single genome. This is useful to analyse unassembled genomes with large numbers of contigs. In this case the notion of quorum is relevant (albeit with certain limitations), but co-localization is not. The **ordered replicon** dataset includes proteins from one single replicon that are ordered according to the position of the corresponding genes in the genome. This is the most powerful mode and can be used to analyse complete or nearly complete genomes. Another related mode (*gembase*) requires a specific input file format and allows the analysis of multiple ordered replicons in a single step (see File S1).

The output of MacSyFinder includes log files, intermediate results, the number of detected systems, and the information on each detected component from each instance of the system. This information is made available in the form of text tables and JSON files. We have built MacSyView, a standalone web-browser application that uses output JSON files to visualise the systems and their genomic context. MacSyView generates exportable SVG files containing views of the detected systems (Fig. 2).

Functioning

The user runs MacSyFinder from the command-line on a protein sequence dataset for a number of systems of interest. The non-redundant list of components to search is extracted from the XML files. The presence of a given component is determined by similarity search with HMM protein profiles using the program Hmmer [13]. The hits are filtered according to user-defined

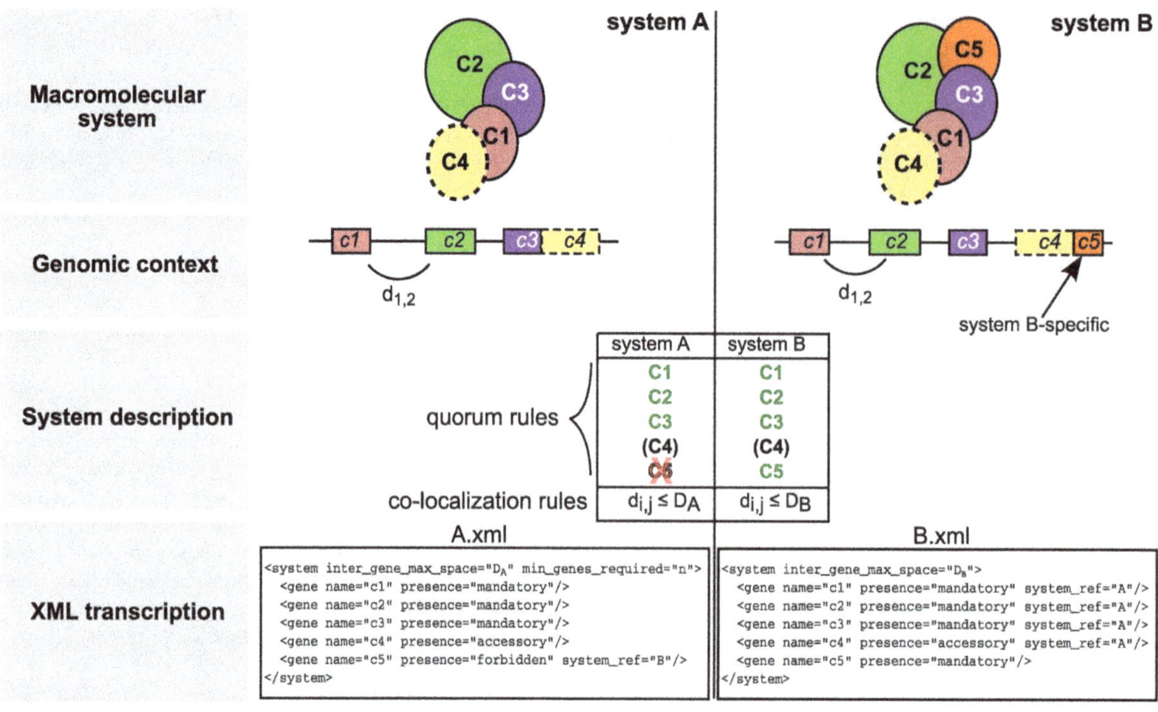

Figure 1. Modelling systems with MacSyFinder. The components of a system assemble into macromolecular systems or correspond to a biological pathway. They are typically encoded in genomes in one or a few different loci ("Genomic context"). We illustrate how systems can be modelled and distinguished with two imaginary systems "A" and "B" that have four homologous components (C1–C4, similar colours for the two systems). The system "B" has one component that is not found in "A"(C5). The parameter *inter_gene_max_space* (D) defines the maximal number of genes between two consecutive components ($d_{i,j}$). The two systems are defined by a set of *mandatory* (green), *accessory* (black) and *forbidden* (red) components. The quorum rules allow relaxing the definition of the system without altering the list of its components (*min_genes_required* and *min_mandatory_genes_required* parameters in XML files). If they are not specified, a default value is computed from the number of components described in the XML files. The bottom part of the figure shows the description of the systems in the XML grammar (see the documentation in File S1). Components listed here refer to protein profiles (Fig. 3). When a component is found in several systems, it is defined only once, and can be reused in another system with the *system_ref* keyword. Much more complex features can be defined, including exchangeable genes, distant genes and component-specific parameters (File S1).

i-evalue (for statistical significance) and to the minimal coverage of the profile in the alignment (to control for the minimal size of the profile that must be matched to obtain biologically relevant hits). The components defined in the models are searched in parallel for rapidity (Fig. 3A). If multiple profiles match the same protein, MacSyFinder selects the hit with the highest score. The subsequent steps depend on whether the input dataset is an ordered replicon, an unordered genome or an unordered genomic dataset (*e.g.*, a metagenome).

If the dataset is an **ordered replicon**, the hits are clustered according to the genetic organization specified in the model. Clusters including the components of a single type of system are used to fill inventories of "compatible" systems (Fig. 3B). If multiple systems are compatible with the set of components in the clusters, then the different candidate systems are examined. The order of exam is given by decreasing number of components shared between the cluster and the compatible systems. The cluster will be assigned to the first system in the list that fits its content. A system is regarded as complete if the quorum is respected. When a complete instance of the system has components from a single locus, further new occurrences of the same components in the cluster are used to produce a novel instance. When a single cluster is not enough to make a complete instance and the *multi_loci* parameter is turned on, the hits are stored to fill up an instance of the system encoded by multiple distant loci. Clusters with components from multiple systems are split in sub-clusters containing components from a single system. These sub-clusters are then re-analysed in terms of their components (Fig. 3B). MacSyFinder can only resolve these complex cases if the components of each system are contiguous, instead of scattered on the cluster.

Unordered sequence datasets cannot be analysed with the co-localization criteria. Therefore, hits from the similarity searches are directly used to fill inventories of each system. Systems are complete if the required quorum is respected. The presence of *forbidden* components is ignored in this mode, even if such occurrences are stored to inform the user. A single system instance will be filled per system and dataset, independently of the number of component occurrences found. This is because components cannot be individually assigned to particular instances in the absence of the genomic context. Nevertheless, the analysis of the number of identified components can be used to estimate the number of instances in the dataset.

Application

Data

The complete genomes of bacteria (2484) and archaea (159) were downloaded from NCBI RefSeq (ftp://ftp.ncbi.nih.gov/genomes/, November 2013). Profiles for the Cas protein families were obtained from the TIGRFAM database, version 13.0 (http://www.jcvi.org/cgi-bin/tigrfams/index.cgi, August 15

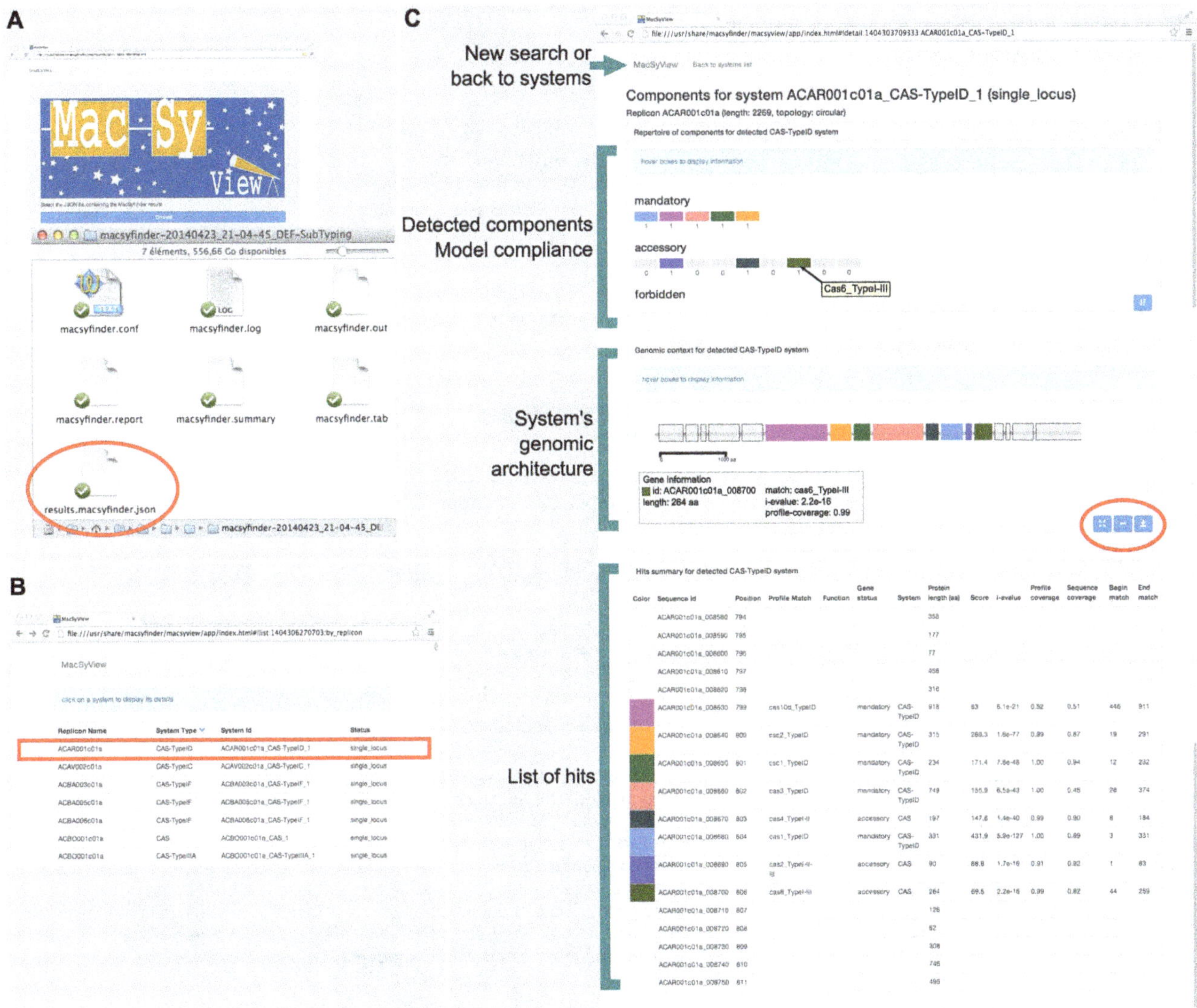

Figure 2. Snapshot of MacSyFinder's results as viewed with MacSyView. A. The MacSyView web-browser based application allows the visualization of MacSyFinder's output file "results.macsyfinder.json". B. MacSyView displays the list of systems available in the results file. The user picks a system to visualize by clicking on it in the list. C. The page displaying the system is made of a header, and three panels. The header allows to select another input file, or to go back to the list of systems. It displays information on the system that is being visualized. The first panel shows how the detected system fits the model compliance in terms of its components. Boxes represent the number of each *mandatory*, *accessory*, and *forbidden* components. A tooltip gives the name of the component when the mouse hovers a box. Component boxes can be sorted by decreasing number of components. The second panel shows the genetic context of the system (as transcribed from the input fasta file), with components drawn to scale. When the mouse hovers a box, a tooltip displays information on the corresponding component, including scores of the Hmmer hit. This view can be exported as a SVG file for drawing purposes (tools circled in red). The third panel gives detailed information on the components of the system.

2012) [11,23]. Among the 89 profiles available, 53 were constructed by Haft and colleagues [24] and correspond to 45 Cas protein families that were used to propose a CRISPR-Cas systems classification [25]. Subsequently, these authors constructed 36 additional protein profiles more specific to given subtypes that are available in TIGRFAM [23]. We renamed these profiles to make them more informative for the user (see Table S1 for correspondence with established classification).

Developing a new system's model

The goal of MacSyFinder is to query genomic data using biologically meaningful models of a system. The first step of the model building procedure is therefore to use the available knowledge to identify the system components, their frequency and their genetic architecture. The second step is to obtain protein profiles for the components either by building them specifically for this purpose or by retrieving them from public databases. Protein profiles can be built easily from multiple sequence alignments of homologous proteins using Hmmer [13]. The third step is to write the model in the simple MacSyFinder's XML grammar (see above and Fig. 1 for an example). The final step is to include information about homologous systems in the model. The use of system-specific profiles and *forbidden* attributes facilitates the discrimination between systems (Fig. 1). Our experience is that complex models should be built by iterating several times on these steps from simpler models. Indeed, the fine-tuning of the quorum definitions and genetic architectures can vastly increase the quality of identification of a system. Often, one is confronted with systems

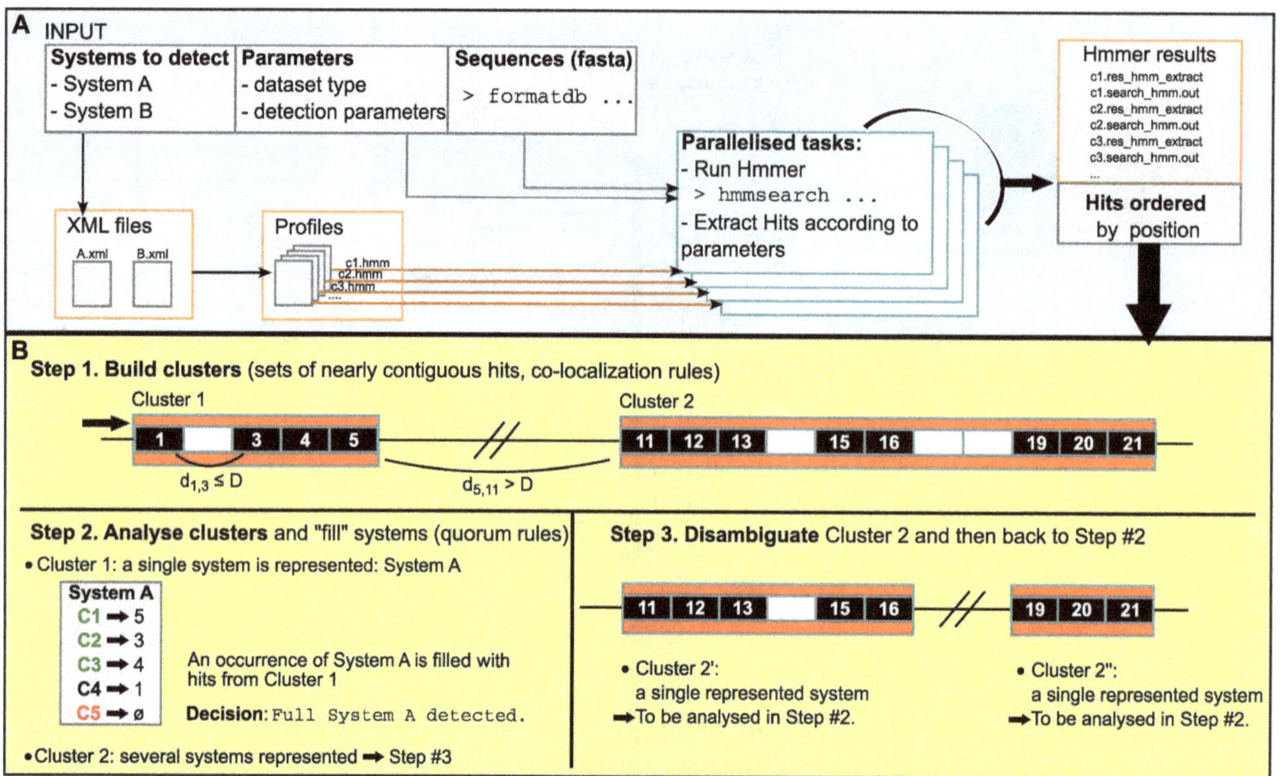

Figure 3. Functioning of MacSyFinder. A. The user launches MacSyFinder to detect macromolecular systems A and B (example of Fig. 1). System-specific parameters are read from the corresponding XML definition files. This includes the list of the components of the systems and the corresponding HMM profiles. Other detection parameters are picked by order of priority: on the command-line, in the configuration file, and in the XML files. Sequences are indexed with the "formatdb" or "makeblastdb" tools for similarity search with the Hmmer program. MacSyFinder runs (optionally in parallel) the Hmmer searches on a non-redundant list of components' profiles. If the sequence dataset is "unordered" MacSyFinder only outputs the hits and the components detected for each type of system. B. Step #1: the co-localization criterion can be used in the ordered datasets. It involves clustering the hits separated by less than D protein-coding genes. The components described as "loner" in the XML definition files can be at any distance from other components. Step #2: the components of each cluster are used to fill the occurrences of the systems. Depending on the quorum, a cluster can describe a "full" system, or a "scattered" system. Step #3: clusters with components belonging to more than one system are split in unique systems and then re-directed separately to step #2.

for which very few instances have been experimentally studied. In this case, iteration of the modelling steps provides both more reliable models and a better knowledge of the systems diversity. To exemplify the use of MacSyFinder we built models to identify Cas proteins and classify CRISPR-Cas systems. This is a very typical example of systems that are intensively studied, for which there are many protein profiles in the databases, but no software dedicated to their detection.

Detection and classification of CRISPR-Cas systems

The known *cas* operons have from 3 to 13 genes encoding very diverse proteins, among which several nucleases and helicases with DNA and/or RNA binding domains [24,25]. A unified classification of CRISPR-Cas systems has been recently established based on the presence or absence of peculiar Cas protein families, and on the genetic architecture of the *cas* operon [25]. Three major types and several subtypes of CRISPR-Cas systems have been described. *cas1* and *cas2* universally occur across types and subtypes, whereas *cas3*/*cas7*, *cas9*, and *cas10* have been defined as the signature genes for type I, type II, and type III, respectively (Fig. 4). Protein profiles matching most of these Cas protein families are publicly available in the TIGRFAM database [11,24]. We used this information to exemplify how MacSyFinder can be used to identify and classify these systems.

General model and choice of parameters. In a first round of analysis, we defined a **general simple model** to identify all possible clusters of Cas proteins in 2643 prokaryotic genomes. In this general definition, all the CRISPR-Cas-HMM profiles available in TIGRFAM database were used whatever their type or subtype specificity (Table S1). At this stage, we used relatively relaxed criteria: all the components were defined as *accessory* and all clusters with at least 3 different components (*min_genes_required* = 3) distant from at most 5 genes (*inter_gene_max_space* "D" = 5) were retained. With this procedure, we identified 1628 clusters of Cas proteins and could annotate 10663 Cas proteins (*i.e.*, with significant matches to protein profiles). The total number of genes in the detected clusters ranged from 3 to 36 with an average of 7.7±3.5 genes (Fig. S1A). In these clusters, most of the genes (86%) encode known Cas proteins (*i.e.*, described in the general definition) and 56% of clusters have components strictly contiguous (Fig. S1B). While these preliminary results suggest that most clusters are Cas systems, a small fraction of them (7%) is larger than the larger described systems (>13 genes, Fig. S1A), suggesting that the above-mentioned parameters might be too permissive (Fig. S1C). These large clusters might correspond to contiguous or intertwined systems (*i.e.*, chimeric variants). To test this hypothesis, we explored the effect of changing D on the identification of clusters (*i.e.*, D = 4, 5, and 6, see Table S2). A

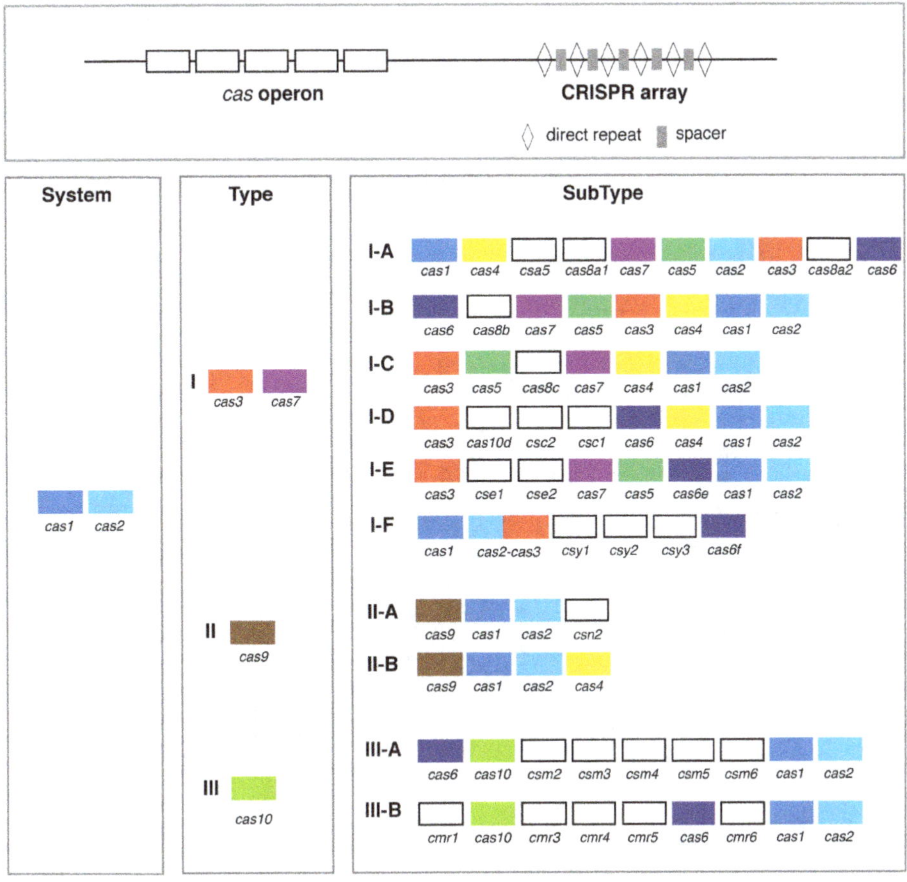

Figure 4. Simplified operon organization of the three major types and ten subtypes of CRISPR-Cas systems. Each *cas* gene family is indicated with a distinct colour, those specific to a subtype are in white. Only the main *cas* gene families are represented.

more stringent co-localization criterion (D = 4), resulted in a decrease of the overall number of Cas proteins assigned to systems, the subdivision of several previously detected clusters, and the persistence of large clusters (Table S2). A less stringent criterion (D = 6) led to the fusion of several clusters with a small gain of Cas proteins assigned to systems (Table S2). We therefore set the final co-localization criterion to 5, and the minimal number of genes to 3. Doubts about multiple closely co-occurring systems can often be removed using more specific "typing" and "subtyping" models, because in this case contiguous systems of different types will be set apart (see below). While the general definition of the system is very simple, it fetches systems with Cas1 or a Cas2 protein in respectively 88% and 73% of the clusters, even if weak constraints were imposed on their presence (*accessory* proteins in the general definition of the system). We identified Cas clusters in 78% of archaeal genomes and 39% of bacterial genomes. This is very similar to previous observations and therefore suggests that even the general model accurately identifies Cas systems [26] (Fig. S1D, and see the paragraph on the validation of our models).

Typing and Subtyping CRISPR-Cas systems. To exemplify the ability of MacSyFinder to characterise sub-systems we built models for each type and subtype of Cas systems from the pre-existing classification [25] (Fig. 4). We first tested the specificity of the 89 available protein profiles for a given type and subtype by analysing the co-occurrence of pairs of Cas proteins in clusters detected with the general model (Fig. 5 and Text S1). Then we designed the corresponding models accordingly. In the final models (Fig. S2 and Data S2), all profiles specific to a system were defined as *mandatory* (signature gene) while all the others were defined as *accessory*. Because some systems have very similar content and organization (*e.g.*, Type II-A and II-B), profiles distinguishing them are *accessory* or *mandatory* in a system, and *forbidden* in the other (see Fig. 1 and Fig. S2 for examples). Although the types and subtypes have different numbers of genes, we set the *min_genes_required* parameter to 3, and the *inter_gene_max_space* parameter to 5 for all models to make the detection as large as possible and comparable with that resulting from the general model. We defined 5"typing" models and 15"subtyping" models for *cas* loci detection and classification (see Fig. S2 and Data S2).

Using the **subtyping models** we classed the previously detected Cas clusters, but were also able to split different contiguous systems (Fig. 5). Thus, among the 1628 Cas clusters, 95% correspond to a single system, 3% to contiguous distinct systems (including the type III-B well known to be associated with other systems-type [26]), the remaining 2% correspond to chimeric variants. Most of the Cas clusters could readily be assigned to proposed types (97%) and subtypes (94%) with our models. The remaining corresponded to *cas* locus with no gene signature, or to chimeric variants (Table S3).

Validation of detected systems. We made two analyses to obtain a more precise assessment of the accuracy of the method. Firstly, we quantified how often Cas systems detected with the General definition (command-lines available in Text S1)

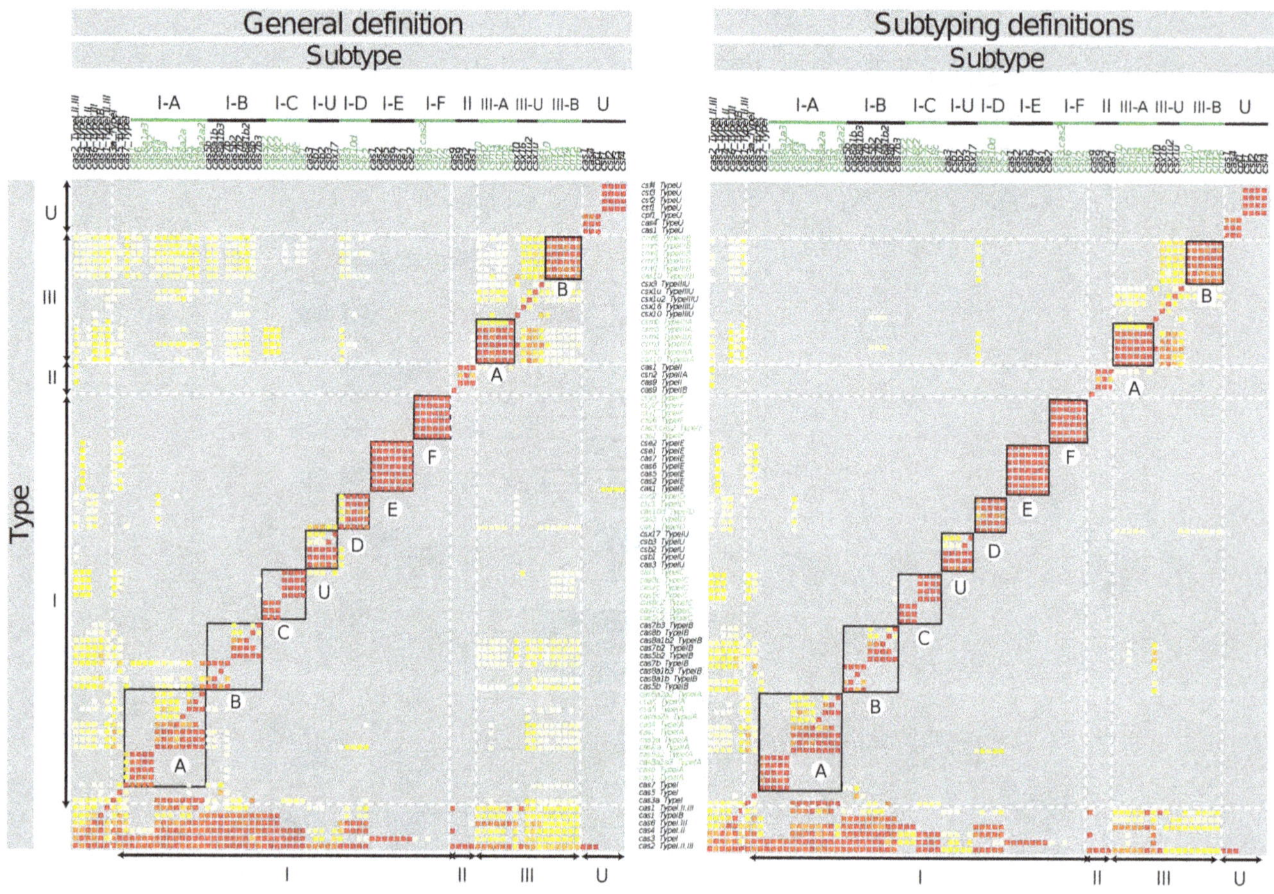

Figure 5. Frequency of co-occurrence between Cas proteins present in clusters detected with the general model (left) and the subtyping models (right). Each matrix was normalized by the maximum of each column. The higher the frequency is, the warmer the colour is: the red diagonal corresponds to a 100% co-occurrence. Only frequencies above 1% were represented, others are in grey.

co-occurred with CRISPR arrays as is the case in all fully functional described systems. We searched for CRISPR-arrays with [17] as described in [27] and found that 88% of the detected Cas systems are close (<1kb, same result for <5kb) to a CRISPR-array and that 98% are present in a replicon containing at least one CRISPR-array. The absence of CRISPR in so few Cas-containing genomes suggests the method has a low rate of false positives. Secondly, we took from the literature the list of CRISPR-Cas with experimentally characterized *in vivo* effects [28]. In this list we could detect 100% of the 25 known Cas systems of genomes included in our dataset (Table S4) with our "general", "typing" and "subtyping" models (see command-lines in Text S1). Furthermore, we could assign the correct subtype to 23 of them, and we propose a subtype for the system of *Mycoplasma gallisepticum*. This suggests a low rate of false negatives. Altogether these results suggest the method is very accurate and that most clusters correspond to CRISPR-Cas systems. Type I systems are more abundant in both bacteria (in ~31% of the bacterial genomes) and archaea (~71%), Type II are only found in bacteria, while Type III are more prevalent in archaea (~38%) (Table 1 and Fig. S3). These results are consistent with previous analyses [25]. Subtypes I-C, I-E and I-F are more commonly found in bacteria, while subtypes I-A, I-B and I-D are frequent in archaea, as previously noted [26]. Overall, these results suggest that our models are able to accurately identify and type Cas systems. Profiles and models for the "Cas-Finder" are provided in Data S2. Users can easily add or remove components and change the genetic organization specifications.

Discussion

The use of MacSyFinder will often involve preliminary steps to model the biological systems of interest. This allows the researcher to produce structured knowledge and is particularly useful when these systems have distinguishable traits, such as a specific genetic architecture. Often there are few studies suggesting the parameters to use in the models. Under these circumstances, one should start with very simple models, *e.g.*, noting all components as *accessory* and using low quorums. The analysis of the results of these preliminary models often provides important clues on how to produce more complex and accurate models. For example, by relaxing the criteria of the requirements to identify type III secretion systems (T3SS) we were recently able to identify a new homologous system in *Myxobacteria* [4]. Modelling itself can thus lead to new biological findings.

MacSyFinder ignores phylogenetic information when putting together components of systems scattered in a replicon or in unordered datasets. In contrast, the preliminary distinction between homologous proteins can often be done using MacSy-Finder without the need for lengthy phylogenetic analyses. This works in two steps. First, one must produce a multiple alignment gathering the different families of homologous proteins. This alignment must be divided into sub-alignments according to the

Table 1. Taxonomic distribution of CRISPR-Cas types and sub-types in prokaryotes expressed in number and percentage of the genomes harboring the systems.

Typing	Type I							Type II			Type III			Type U
Bacteria	769 (31%)							177 (7%)			222 (9%)			7
Archaea	113 (71%)							0			60 (38%)			0
Subtyping	**I-A**	**I-B**	**I-C**	**I-D**	**I-E**	**I-F**	**I-U**	**II-A**	**II-B**	**II-U**	**III-A**	**III-B**	**III-U**	**U**
Bacteria	16 (<1%)	173 (7%)	196 (8%)	31 (1%)	282 (11%)	111 (5%)	47 (2%)	81 (3%)	5 (<1%)	93 (4%)	123 (5%)	112 (5%)	3 (<1%)	7 (<1%)
Archaea	55 (35%)	48 (30%)	2 (<1%)	16 (10%)	5 (3%)	0	2 (1%)	0	0	0	28 (18%)	40 (25%)	0	0

different systems, leading to the production of different profiles for the different sub-families of homologs. Finally, and as a rule, for a given protein, the best-scoring profile corresponds to the relevant homologous family (see Fig. 5, and Fig. 2 in [4]).

It is difficult to estimate *a priori* how accurate MacSyFinder will be for any given biological system because this will depend on several system-specific variables. First, it will depend on the number of components of the system, their frequency in the system and their degree of sequence conservation. Systems with many highly conserved and frequent components will be much easier to identify than systems with many infrequent and fast-evolving components. Second, it will depend on the existence of other systems sharing homologous components. Systems including many components with homologs in other systems will be harder to identify. We have shown MacSyFinder can type CRISPR-Cas systems, even if they share homologs. Hence, even in these difficult situations MacSyFinder provides accurate models. The situation is necessarily more complicated when identifying systems with many homologs encoded by genes scattered in the genomes. In this case, phylogenetic methods may help in the reconstruction of the different systems.

Considering MacSyFinder's running time, the limiting step is usually the identification of hits by Hmmer, which is currently very efficient [13]. To speed up this step, MacSyFinder is able to compute and analyse Hmmer hits in parallel. MacSyFinder and its companion MacSyView are easy to install standalone tools. This is an advantage when it is necessary to keep the data private or when projects are so large that network transfer time is prohibitive. MacSyFinder was built to be simple to use. It is thus ideal for biologists without extensive knowledge of programming or scripting wishing to unravel the diversity of certain systems or to annotate genetic data. Often, bioinformaticians produce methods to identify machineries and would like to easily package them for reproducibility and distribution among biologists. This can be easily done with MacSyFinder *via* the distribution of XML files and the relevant protein profiles. The "Cas-finder" we present here is a particularly relevant case. At the time we started the project, there was public information available on the protein profiles and on the genetic organization of the systems. We only had to define the models and use them in such a way that we could identify the systems and class them. The result is a highly accurate application to identify *cas* operons that can be easily distributed (Data S2).

Supporting Information

Figure S1 Genomic architecture and taxonomic distribution of detected *cas* genes clusters (general model). A. Distribution of the number of different genes in detected clusters. B. Distribution of the maximal distance between two components observed in each detected cluster. C. Boxplot of the number of different genes in each cluster vs. the maximal distance between two components observed in each cluster. D. Proportion of bacterial and archaeal genomes without (cluster−) and with at least one cluster of *cas* genes (cluster+).

Figure S2 Schematic and simplified representation of the subtype models. Each box corresponds to a *cas* gene family and the name of the corresponding HMM protein profiles are listed below. Some *cas* gene families have multiple HMM profiles available in the TIGRFAM database. Each *cas* gene family has its boxes filled (subtype non-specific) or surrounded (subtype-specific) by a distinct colour. Only the main *cas* gene families are represented. For full subtype models, see the XML files in Data S2.

Figure S3 Taxonomic distribution of the three CRISPR-Cas systems types. For each clade, the number of representative genomes is given, along with bar plots showing the percentage of these genomes containing the three types of CRISPR-Cas systems.

Table S1 List of HMM profiles used for the "Cas-Finder".

Table S2 Impact of the co-localization parameter on the detection.

Table S3 Detection results.

Table S4 Validation of the CRISPR-Cas systems detection on systems with *in vivo* effects listed in the review by Bondy-Denomy et al. 2014 [28].

Text S1 Supporting text (PDF file).

File S1 MacSyFinder's documentation file (PDF file).

Data S1 The MacSyFinder/MacSyView package (compressed tarball archive).

Data S2 The Cas-Finder: models and profiles (compressed tarball archive).

Acknowledgments

Julien Guglielmini and Jean Cury for fruitful discussions.

Author Contributions

Conceived and designed the experiments: SSA EPCR MT. Performed the experiments: SSA MT. Analyzed the data: MT. Contributed to the writing of the manuscript: SSA EPCR MT. Designed the MacSyFinder software: SSA BN. Designed the MacSyView application: SSA BN HM.

References

1. Alberts B (1998) The cell as a collection of protein machines: preparing the next generation of molecular biologists. Cell 92: 291–294.
2. Pereira-Leal JB, Levy ED, Teichmann SA (2006) The origins and evolution of functional modules: lessons from protein complexes. Philos Trans R Soc Lond B Biol Sci 361: 507–517.
3. Michel B, Grompone G, Florès MJ, Bidnenko V (2004) Multiple pathways process stalled replication forks. Proc Natl Acad Sci U S A 101: 12783–12788.
4. Abby SS, Rocha EP (2012) The non-flagellar type III secretion system evolved from the bacterial flagellum and diversified into host-cell adapted systems. PLoS Genet 8: e1002983.
5. Galagan JE, Nusbaum C, Roy A, Endrizzi MG, Macdonald P, et al. (2002) The genome of M. acetivorans reveals extensive metabolic and physiological diversity. Genome Res 12: 532–542.
6. Huynen M, Snel B, Lathe W 3rd, Bork P (2000) Predicting protein function by genomic context: quantitative evaluation and qualitative inferences. Genome Res 10: 1204–1210.
7. Overbeek R, Fonstein M, D'Souza M, Pusch GD, Maltsev N (1999) The use of gene clusters to infer functional coupling. Proc Natl Acad Sci USA 96: 2896–2901.
8. Lathe WC, Snel B, Bork P (2000) Gene context conservation of a higher order than operons. Trends Biochem Sci 25: 474–479.
9. Zaslaver A, Mayo A, Ronen M, Alon U (2006) Optimal gene partition into operons correlates with gene functional order. Phys Biol 3: 183–189.
10. Finn RD, Tate J, Mistry J, Coggill PC, Sammut SJ, et al. (2008) The Pfam protein families database. Nucleic Acids Res 36: D281–288.
11. Haft DH, Loftus BJ, Richardson DL, Yang F, Eisen JA, et al. (2001) TIGRFAMs: a protein family resource for the functional identification of proteins. Nucleic Acids Res 29: 41–43.
12. Servant F, Bru C, Carrere S, Courcelle E, Gouzy J, et al. (2002) ProDom: automated clustering of homologous domains. Brief Bioinform 3: 246–251.
13. Eddy SR (2011) Accelerated Profile HMM Searches. PLoS Comput Biol 7: e1002195.
14. Barrangou R, Marraffini LA (2014) CRISPR-Cas systems: Prokaryotes upgrade to adaptive immunity. Mol Cell 54: 234–244.
15. Hsu PD, Lander ES, Zhang F (2014) Development and Applications of CRISPR-Cas9 for Genome Engineering. Cell 157: 1262–1278.
16. Barrangou R (2014) RNA events. Cas9 targeting and the CRISPR revolution. Science 344: 707–708.
17. Bland C, Ramsey TL, Sabree F, Lowe M, Brown K, et al. (2007) CRISPR recognition tool (CRT): a tool for automatic detection of clustered regularly interspaced palindromic repeats. BMC Bioinformatics 8: 209.
18. Grissa I, Vergnaud G, Pourcel C (2007) CRISPRFinder: a web tool to identify clustered regularly interspaced short palindromic repeats. Nucleic Acids Res 35: W52–57.
19. Edgar RC (2007) PILER-CR: fast and accurate identification of CRISPR repeats. BMC Bioinformatics 8: 18.
20. NCBI {BLAST} executables download website (includes makeblastdb). Available: ftp://ftp.ncbi.nih.gov/blast/executables/LATEST/. Accessed 2014 Sep 28.
21. Camacho C, Coulouris G, Avagyan V, Ma N, Papadopoulos J, et al. (2009) BLAST+: architecture and applications. BMC Bioinformatics 10: 421.
22. Eddy SR (1998) Profile hidden Markov models. Bioinformatics 14: 755–763.
23. Haft DH, Selengut JD, Richter RA, Harkins D, Basu MK, et al. (2013) TIGRFAMs and Genome Properties in 2013. Nucleic Acids Res 41: D387–395.
24. Haft DH, Selengut J, Mongodin EF, Nelson KE (2005) A guild of 45 CRISPR-associated (Cas) protein families and multiple CRISPR/Cas subtypes exist in prokaryotic genomes. PLoS Comput Biol 1: e60.
25. Makarova KS, Haft DH, Barrangou R, Brouns SJ, Charpentier E, et al. (2011) Evolution and classification of the CRISPR-Cas systems. Nat Rev Microbiol 9: 467–477.
26. Staals RHJ, Brouns SJJ (2013) Distribution and Mechanism of the Type I CRISPR-Cas Systems. In: Barrangou R, Oost Jvd, editors. CRISPR-Cas Systems - RNA-mediated Adaptive Immunity in Bacteria and Archaea. Berlin Heidelberg: Springer Berlin Heidelberg.
27. Touchon M, Rocha EP (2010) The small, slow and specialized CRISPR and anti-CRISPR of Escherichia and Salmonella. PLoS ONE 5: e11126.
28. Bondy-Denomy J, Davidson AR (2014) To acquire or resist: the complex biological effects of CRISPR-Cas systems. Trends Microbiol 22: 218–225.

Rapidly-Deposited Polydopamine Coating via High Temperature and Vigorous Stirring: Formation, Characterization and Biofunctional Evaluation

Ping Zhou[1,2], Yi Deng[1,2], Beier Lyu[3], Ranran Zhang[4], Hai Zhang[4,5], Hongwei Ma[3], Yalin Lyu[4]*, Shicheng Wei[1,2]*

1 Department of Oral and Maxillofacial Surgery, School and Hospital of Stomatology, Peking University, Beijing, China, **2** Center for Biomedical Materials and Tissue Engineering, Academy for Advanced Interdisciplinary Studies, Peking University, Beijing, China, **3** Suzhou Institute of Nano-Tech and Nano-Bionics, Chinese Academy of Sciences, Suzhou, China, **4** Department of Stomatology, Beijing Anzhen Hospital, Capital Medical University, Beijing, China, **5** Department of Restorative Dentistry, School of Dentistry, University of Washington, Washington, United States of America

Abstract

Polydopamine (PDA) coating provides a promising approach for immobilization of biomolecules onto almost all kinds of solid substrates. However, the deposition kinetics of PDA coating as a function of temperature and reaction method is not well elucidated. Since dopamine self-polymerization usually takes a long time, therefore, rapid-formation of PDA film becomes imperative for surface modification of biomaterials and medical devices. In the present study, a practical method for preparation of rapidly-deposited PDA coating was developed using a uniquely designed device, and the kinetics of dopamine self-polymerization was investigated by QCM sensor system. It was found that high temperature and vigorous stirring could dramatically speed up the formation of PDA film on QCM chip surface. Surface characterization, BSA binding study, cell viability assay and antibacterial test demonstrates that the polydopamine coating after polymerization for 30 min by our approach exhibits similar properties to those of 24 h counterpart. The method has a great potential for rapid-deposition of polydopamine films to modify biomaterial surfaces.

Editor: Jie Zheng, University of Akron, United States of America

Funding: This work was supported by the National Natural Science Foundation (81371697), www.nsfc.gov.cn, to SCW; Beijing Natural Science Foundation (7132124), www.bjnsf.org, to SCW; Peking University's 985 Grant, www.pku.edu.cn, to SCW. The funders had no role in study design, data collection and analysis, decision to publish, or preparation of the manuscript.

Competing Interests: The authors have declared that no competing interests exist.

* Email: sc-wei@pku.edu.cn (SCW); lvyalin@vip.sina.com (YPI)

Introduction

Inspired by the composition of adhesive proteins in mussels, Messersmith et al [1] (2007) employed the innate self-polymerization attribute of dopamine to form thin and surface-adherent polydopamine (PDA) film, which later on has been widely applied as a surface modification agent. Polydopamine possesses covalent and non-covalent bonding capabilities for a broad range of organic, inorganic and metallic substrates [2], harboring potential applications in the challenging field of antibacterial [3], antifungal [4], antifouling [5,6], drug delivery vehicle [7], biosensor [8], cell culture [9], tissue engineering [10–12] and so on. However, to author's knowledge, dopamine solution is usually polymerized for at least several hours prior to surface modification/deposition. It is too long from practical point of view and it is crucial to develop a rapid PDA film formation method.

Although PDA coating has been used for numerous years and tremendous effort has been invested in understanding its structure, it has yet to be unambiguously determined. It is known that the process of dopamine-polymerization first involves oxidation of a catechol to a benzoquinone [13] and PDA film is a complex network with free catechol groups available for further chemical surface modification [14]. However, recent research suggests that PDA is a supra-molecular aggregate of monomers rather than a covalent polymer [15].

Despite the wide application and numerous structure analyses for polydopamine coating, much less effort has been focused on factors that would affect the deposition kinetics of polydopamine film, which has become an obstacle for further optimization. Dopamine self-polymerizes to polydopamine usually under slightly alkaline condition, a pH typical of marine environments (2 $g\cdot L^{-1}$ dopamine in 10 mM Tris-HCl, pH = 8.5). The thickness of PDA film is growing as a function of time in 24 h detected by spectroscopic ellipsometry [16]. Zhao, C et al [17] found that addition of oxidizing agents increased the rate of PDA formation under basic conditions. Moreover, the effect of pH and concentration on the deposition kinetics of dopamine solution has also been studied by spectroscopic ellipsometry, AFM [18] and surface plasmon resonance (SPR) [19]. A constant increase in the maximal film mass, thickness and roughness was observed as an augment in the dopamine concentration from 0.1 $g\cdot L^{-1}$ to 10 $g\cdot L^{-1}$, and a pH value of 8.5 was demonstrated to be the best pH for Tris-HCl buffer solution [18,19]. Furthermore, Xu,Y et al [20] observed that the self-polymerization speed of dopamine grew as temperature increased from 20°C to 60°C. Besides these influence factors, the reaction method could also play a key role in the deposition kinetics of PDA. There is, however, barely any effort devoted in the subject area, which results in inconsistent reaction temperature and time reported in a variety of research articles [21–26]. Currently, there are three major dopamine polymerization processes: static, shake, and stir. With regard to the static method, substrates are submerged in dopamine solution for 16 h [10], 18 h [3], overnight [21,22] or 24 h [12] at room temperature. In shake method, dopamine solutions are shaken to achieve uniform PDA film on the substrate at quite different reaction temperature and time (such as 1 h at ambient temperature [23], 24 h at ambient temperature [24] and 8 h at 30°C [25]). Stir method is popularly used recently, which protects substrates from the particle deposition in the solution [16]. Unfortunately, like static and shake method, the stirring time for dopamine-polymerization is undefined and varies from several hours [26] to 24 h [18]. Undefined conditions in all three methods generate PDA films of various thickness, mass and quality, and make it impossible to compare studies between different labs.

The main aim of this study is to investigate the deposition kinetics of PDA film as a function of temperature and stirring, and to develop a method for rapidly-deposited PDA coating under standard environment (2 $g{\cdot}L^{-1}$ dopamine in 10 mM Tris-HCl, pH = 8.5).In this investigation, the impact of temperature and reaction method on the deposition kinetics of PDA coating was evaluated by quartz crystal microbalance (QCM) sensor system, which is commonly used to monitor changes in mass at the quartz crystal sensing surface and capable of real-time *in situ* detection in solution [27]. It could be a powerful analytical tool to study the deposition kinetics and properties of PDA coating. In addition, a unique device was designed for preparation of rapidly-deposited PDA film, in which QCM chip substrates were vertically placed in dopamine solution under vigorous stirring (300 $r{\cdot}min^{-1}$) at 60°C and QCM sensor system was used to monitor the mass change of PDA coating on chip surface. PDA films were characterized using a variety of analytical tools and bio-performance of these films was also evaluated. Results show that using high temperature and vigorous stirring could dramatically speed up dopamine-polymerization and thirty minutes reaction time was found appropriate for surface modification. Our method has demonstrated great potential of being a rapid and practical approach for surface modification of biomaterials and medical devices using self-polymerized PDA coating.

Materials and Methods

Materials

Tris (hydroxymethyl) aminomethane (Tris) was purchased from Beijing Chemical Reagents Company (Beijing, China). Dopamine hydrochloride was obtained from Sigma (Missouri, USA). Bovine serum albumin (BSA) fraction V was purchased from Aladdin (Shanghai, China). Commercial pure titanium (Ti), grade 2, was provided from Northwest Institute for Non-ferrous Metal Research (Xi'an, China). Pure polyetheretherketone (PEEK) was purchased from Hua-jun Special Engineering Plastic Products co., LTD (Changzhou, China). Cefotaxime sodium (CS, $C_{16}H_{16}N_5O_7S_2Na$) was provided by Amresco Inc. (Ohio, USA).All reagents were of analytical grade. All aqueous solutions were prepared with distilled water.

Quartz crystal microbalance and chips

The QCM and chips were purchased from Dongwei Biological Technology Co., LTD (Hangzhou, China).The chips were AT-cut planar silicon oxide-coated QCM crystals (14 mm in diameter) with a 5 MHz (4.95 MHz±50 KHz) nominal resonance frequency. Prior to use, all QCM chips were cleaned in an UV/ozone Tip-Cleaner (BioForce Nanosciences, Ames, IA,USA) for 30 min and thoroughly rinsed with distilled water and ethanol for three times, then dried in a stream of nitrogen. Reaction solution was pumped through sensor with QCM chip by a peristaltic pump (BT100K, Baoding, China). The system and reaction solution were conditioned to a required temperature prior to experiment.

The QCM is based on the piezoelectric influence, where a deposited mass is registered as changes in frequency of an oscillating QCM crystal. The adsorbed mass can then be calculated by the Sauerbrey equation [28]:

$$\Delta m = -C\Delta f/n$$

where, Δm equals change in mass per unit surface area, C is the instrument sensitivity constant (17.7 $ng{\cdot}cm^{-2}\ Hz^{-1}$), Δf is the frequency change of the specific harmonic and n denotes the number of overtone. The fundamental frequency of our study was simultaneously acquired at of 15 MHz (n = 3).

The influence of temperature and reaction method on the deposition kinetics and characteristics of polydopamine films

The effect of temperature on the polymerization kinetics of PDA coating was detected *in-situ* by QCM. The clean QCM chip was placed onto the QCM sensor system. QCM sensor system was conditioned at 25°C, 37°C and 60°C respectively prior to experiment. After the baseline became stable, 500 μL 2 $g{\cdot}L^{-1}$ dopamine solution (10 mM Tris-HCl buffer, pH = 8.5) was placed on the sensor surface to study the influence of temperature on dopamine-polymerization *in situ* and in real-time. Throughout this investigation, all solutions were preheated in a Thermostatic Water Bath (HWSX-650/T, Jinhua, China) at the same temperature to the QCM sensor system.

To investigate the influence of reaction method on dopamine-polymerization, the mass change of chips decorated by PDA films formed by conditions of static, shaking and stirring were measured. For the static method, 3 ml 2 $g{\cdot}L^{-1}$ dopamine solution (10 mM Tris-HCl buffer, pH = 8.5) was added into each well of 12-well plate (Corning, New York, USA) with cleaned chips and reacted at 60°C for 1 h. In the shaking method, cleaned chips were placed in a 12-well plate. 3 ml standard dopamine solution was injected into each well for 1 h at 60°C with shaking (300 $r{\cdot}min^{-1}$). In the stirring method, on the other hand, cleaned chips were vertically placed into 200 ml standard dopamine solution at 60°C with stirring (300 $r{\cdot}min^{-1}$) for 1 h. 300 $r{\cdot}min^{-1}$ was chosen as the maximal speed that would not affect the vertically placed QCM chips in dopamine solution to investigate the effect of stirring to dopamine-polymerization. The plate and buffer were preheated prior to experiment. After rinsed with distilled water and ethanol, each chip was dried in a stream of nitrogen and the mass change was investigated by QCM sensor system. Each group had three chips and their average values were calculated.

Preparation of rapidly deposited polydopamine coating

Figure 1 illustrates the formation of polydopamine coating on QCM chips surface by normal method and rapidly-deposited method. In the rapidly-deposited method, 200 ml 10 mM Tris-

HCl solution (pH = 8.5) was preheated in a constant temperature magnetic stirrer (SHJ-11, Jinhua, China) at 60°C, and the clean QCM chips were vertically placed into the solution in a uniquely designed device, which was made up of an retort stand, a wire loop and ten small clips. The chips substrates were fixed by the clips with the same interval and their surfaces were in a circle. For PDA coating, 0.4 g dopamine was dissolved and polymerized with vigorous stirring (300 r·min^{-1}) for 5 min, 10 min, 20 min,30 min, 1 h, 2 h, 4 h and 8 h respectively. PDA formed by our method named stir-PDA (sPDA). The mass changes of QCM chips decorated by PDA formed at 37°C with gently shaking at the speed of 70 r·min^{-1} for 24 h was studied as the control, named normal method (nPDA). Briefly, the cleaned chips were placed in 12-well plates (Corning, New York, USA). 3 ml 2 g·L^{-1} dopamine solution (10 mM Tris-HCl buffer, pH = 8.5) was injected into each well for 24 h at 37°C with shaking (70 r·min^{-1}). The plates and buffer were preheated prior to experiment. After rinsed with distilled water and ethanol, each chip was dried in a stream of nitrogen and the mass change was monitored by QCM sensor. To totally clear away the physically adsorbed PDA particles that not possess covalent bonding capability on substrate surface, PDA coated chips were cleaned in the distilled water by ultrasonic (KQ-5200DE, Kunshan, China) for different periods (1 min, 2 min, 5 min and 10 min) and the mass changes were investigated by QCM.

Moreover, the film thickness of sPDA coated chips with various polymerization times and 24 h-nPDA coated chips that had ultrasonically cleaned for 10 min was further investigated by spectroscopic ellipsometry (MD2000D, J.A.Woollam, Nebraska, USA).

Characterization

Particles of sPDA and nPDA that acquired from previously reacted dopamine solution were used to evaluate the chemical properties. Briefly, dopamine solutions polymerized by our newly developed method and normal method were filtered by 0.22 um filter respectively, and the obtained PDA particles were dried in a vacuum oven (SDZF-6050, Shanghai, China) at 37°C for 1 h. Fourier-transform infrared spectrometry (FTIR) (Magna-IR 750, Nicolet, USA) was used to identify the functional groups of sPDA and nPDA in the form of pellets (KBr pellet).The spectra were recorded from 4000 cm^{-1} to 400 cm^{-1}. The chemical constituents of sPDA and nPDA particles were analyzed by X-ray photoelectronic spectroscopy (XPS) (Kratos, Manchester, UK).The surface hydrophilicity and surface energy (SE) of QCM chips modified by sPDA or nPDA that hat had ultrasonically cleaned for 10 min were characterized by water contact angle measurements (Dataphysics Instrument, Stuttgart, Germany). Three measurements were performed for each chip at ambient temperature based on the sessile drop method, and the mean value was taken as the reported result. The SE value was determined according to OWRK method using distilled water. The surface topographies of sPDA and nPDA films on the surface of QCM chips that had ultrasonically cleaned for 10 min were analyzed by an atomic force microscope (AFM) (Buker, Massachusetts, USA). AFM images were obtained in contact mode. The scan range was 1.0 um × 1.0 um, and scan rate was 1 Hz. Ra was used to evaluate the surface roughness on the basis of scan area.Before AFM measurement, the chips were rinsed with distilled water and ethanol, and allowed to air dry.

BSA immobilization study

BSA was selected as the characteristic protein to investigate the adsorbing properties of sPDA film with various polymerization times (5 min, 10 min, 20 min, 30 min, 1 h, 2 h, 4 h and 8 h), and to compare with that of 24 h nPDA film. Briefly, sPDA and nPDA film decorated QCM chips that had ultrasonically cleaned for 10 min were, respectively, put onto the flow chamber, which was then continuously flushed with 0.01 M PBS buffer (pH = 7.4). After the baseline became stable, the solution was replaced by fresh 5 mg/ml BSA solution in PBS buffer. The reaction was conducted at 25°C by temperature monitor and all used solutions were preheated.

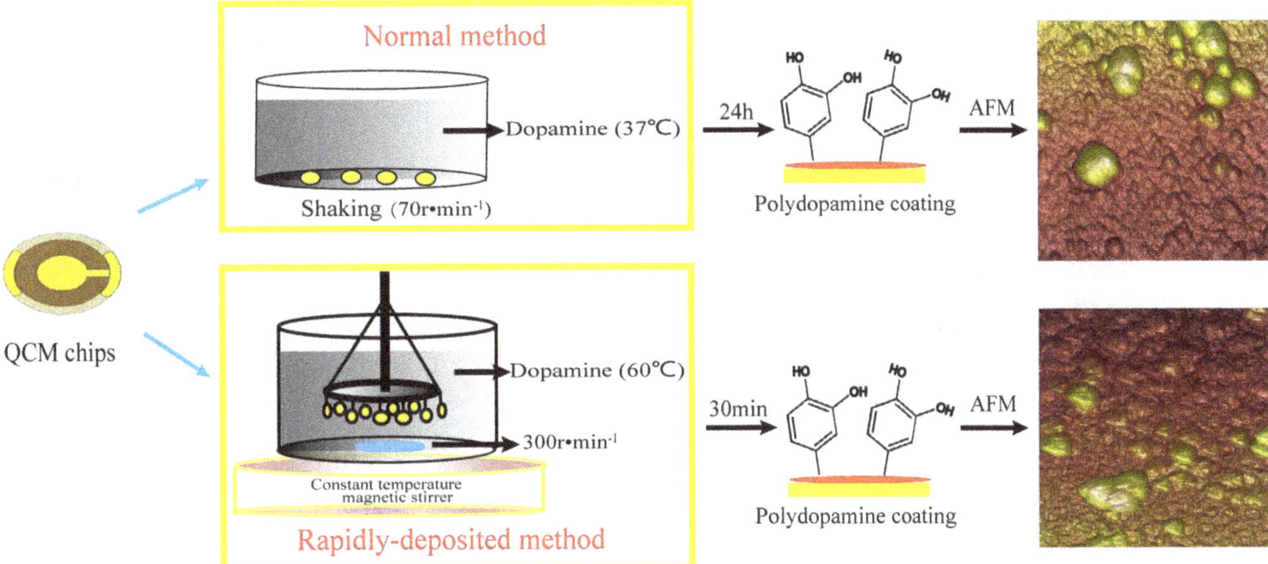

Figure 1. Scheme of polydopamine coating process on QCM chips surfaces by normal method and rapidly-deposited method. Substrates are vertically placed in standard dopamine solution at the temperature of 60°C with stirring (300 r·min^{-1}) using our uniquely designed device, which can significantly improve the deposition of polydopamine on QCM chips.

Cytotoxicity assay

The pristine QCM chips, 30 min-sPDA film coated chips and 24 h-nPDA film coated chips were sterilized under UV irradiation for 30 min, placed in a 24-well plate. The influence of 30 min-sPDA coating and 24 h-nPDA coating on cell viability was evaluated by CCK-8 assay (CCK-8, Dojindo, Kumamoto, Japan), using MG-63 osteoblast cells obtained from the American Type Culture Collection. Cells were grown in Dulbecco's modified Eagle medium (DMEM) (Hyclone, UT,USA), supplemented with 10% fetal bovine serum (Hyclone, UT,USA), 100 U·mL^{-1} penicillin (Amresco, Cleveland, USA), and 0.1 mg·mL^{-1} streptomycin (Amresco, Cleveland, USA) under standard cell culture conditions (37°C, 100% humidity, 95% air and 5% CO_2).

After cell counting, MG-63 osteoblast cell was seeded into 24-well culture plates with sterilized PDA film modified QCM chips at a density of 2×10^4 cells/well. The medium was replaced with fresh one every 24 h.

The control experiments were carried out using the complete growth culture medium without substrates (non-toxic control). After incubating for 24 h and 72 h, respectively, 500 μL medium containing 50 μL of CCK-8 reagent was added into each well and incubation went on for two more hours. Then, 200 μL of supernatant from each well was transferred to a new 96-well cell culture plate. The absorbance value (OD value) was measured at 450 nm with a microplate reader (Model 680, Bio-Rad, Hercules, CA). Three specimens were tested for each incubation period, and each test was performed in triplicate.

Evaluation of antibiotics-binding ability

To study the ability of sPDA film for the immobilization of biomolecules, the inhibition of bacterial adhesion on CS decorated Ti and PEEK surfaces by 30 min-sPDA and 24 h-nPDA was compared. The method for antibacterial activity evaluation is the same as previous reported [29]. Briefly, prior to surface modification, Ti and PEEK were cut into discs of 15 mm in diameter and 2 mm in thickness. These disc samples were polished with a series of SiC abrasive papers (400, 1000, 1500 and 2000 grit). They were then ultrasonically cleaned in acetone, anhydrous ethanol and distilled water, respectively, for 10 min and dried in a stream of nitrogen. 30 min-sPDA coated Ti and PEEK, 24 h-nPDA coated Ti and PEEK were prepared as described previously. 3 ml fresh 50 mg·mL^{-1} CS in distilled water was injected into each well of 12-well plate with modified Ti and PEEK at 37°C with shaking (70 r·min^{-1}) for 24 h. They were then rinsed with distilled water and dried in a stream of nitrogen. 30 min-sPDA-CS decorated Ti, 24 h-nPDA-CS decorated Ti, 30 min-sPDA-CS decorated PEEK and 24 h-nPDA-CS decorated PEEK were sterilized under UV irradiation for 30 min on each side. *Escherichia coli* (*E. Coli* 1.1369, obtained from China General Microbiological Culture Collection Center, China) and *Streptococcus mutans* (*S. Mutans* UA159, obtained from American type Culture Collection, USA) were selected as the experimental strains. *E.Coli* and *S.Mutans* were cultured in Luria-Bertani (LB) medium and Brian Heart Infusion (BHI) medium respectively. Bacterial adhesion on the surfaces of pristine Ti, pristine PEEK, 30 min-sPDA-CS modified Ti, 30 min-sPDA-CS modified PEEK, 24 h-nPDA-CS modified Ti and 24 h-nPDA-CS modified PEEK for various period (4 h, 24 h and 72 h) was assessed *via* Microbial Viability Assay Kit-WST (Dojindo, Kumamoto, Japan) and a microplate reader (Elx808, Bio-tek, Vermont, USA).Moreover, the activity of bacterial in the medium was also measured to evaluate the release action of CS on those samples. Minimum bactericidal concentration of CS was added into the mixture of bacterial solution and culture medium as the control.

Statistical analysis

The results were expressed as mean ± standard deviations (SD) derived from experiment and assessed statistically using Student's t-test. Statistical analysis was carried out with software SPSS13.0 for Windows. Statistical significance was accepted at $p<0.05$.

Results and Discussion

Dopamine-polymerization as a function of temperature and reaction method

The influence of temperature on the deposition kinetics of PDA coating was *in-situ* and real-time monitored for the first time by QCM (Figure 2A). The Δm signals of PDA films on the sensor surface markedly augment with an increase in the polymerization temperature between 25°C and 60°C, indicating temperature has significant impact on dopamine-polymerization and subsequent deposition. According to the thermodynamic principle, higher temperature would promote the oxidation of dopamine as well as the deposition process of PDA onto substrate surface. This result is consistent with investigation of dopamine polymerization under various temperatures measured by spectroscopic ellipsometry [20].

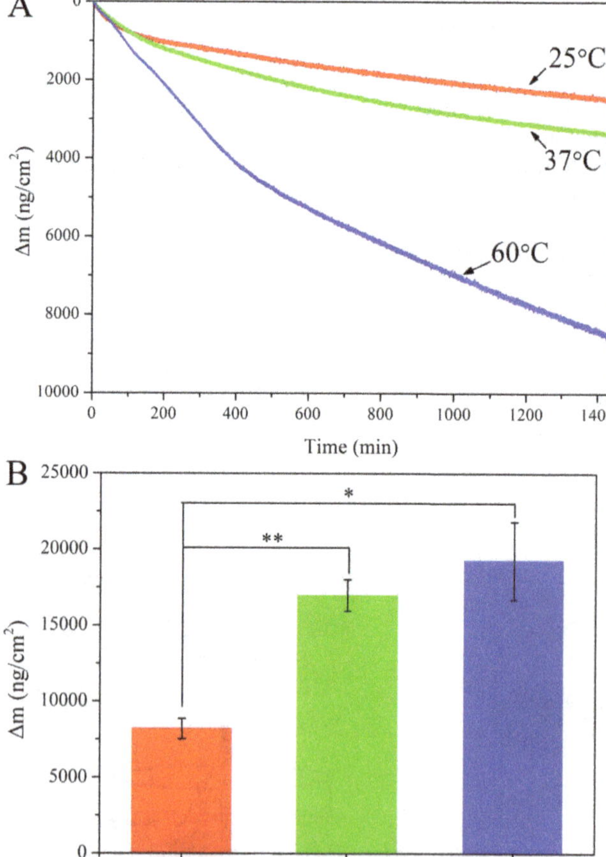

Figure 2. The influence of temperature and reaction method on the deposition kinetics of polydopamine films. (A): Representative mass change vs. time (min) curves as a function of temperature for polymerization of dopamine solution. (B): The mass change of QCM chips coated by PDA formed by methods of static, shaking (300 r·min^{-1}) and stirring (300 r·min^{-1}) for 1 h respectively. *Represents $p<0.05$ and ** Represents $p<0.01$ compared with chips coated by PDA formed at static group, n = 3.

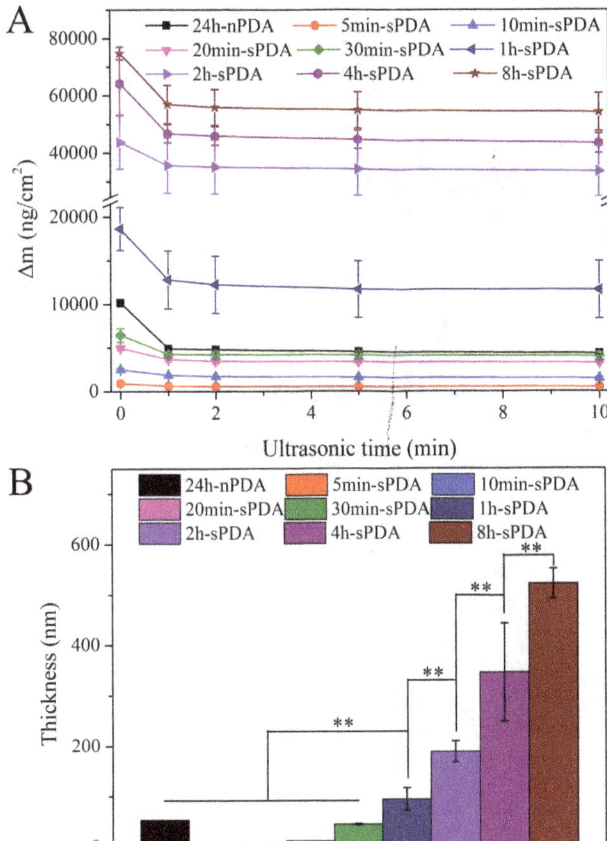

Figure 3. The mass and thickness measurement of sPDA films and nPDA films. (A): The mass change of QCM chips decorated by sPDA films of varying polymerization times (5 min, 10 min, 20 min, 30 min, 1 h, 2 h, 4 h and 8 h) as a function of ultrasonic cleaning time, and compared with that of 24 h-nPDA film coated chips. (B): The film thickness of QCM chips coated by sPDA of different polymerization times (5 min, 10 min, 20 min, 30 min, 1 h, 2 h, 4 h and 8 h) and 24 h-nPDA. Those chips had ultrasonically cleaned in distilled water for 10 min before measurement. ** Represents $p<0.01$, $n=3$.

The mass change of QCM chips decorated by PDA films formed from methods of static, shaking and stirring was measured to evaluate the effect of reaction method on the polymerization behavior of PDA coating (Figure 2B). The mass change of chips decorated by PDA film polymerized for 1 h at 60°C with stirring method (19275 ± 2575 ng·cm^{-2}) is a little higher than that of PDA film from shaking method (16981 ± 1037 ng·cm^{-2}) ($p>0.05$), and both are significantly higher than that of PDA film formed at static condition (8202 ± 665 ng·cm^{-2}) ($p<0.05$), indicating vigorous shaking and stirring could dramatically increase the formation speed of PDA film on the sensor surface. That is presumably due to more fresh oxygen in air dissolved in solution with shaking or stirring, which contributes to the self-oxidation of dopamine. However, some disadvantages, such as solution will be splashed out and the chips may turned over to the back side, were observed for dopamine-polymerization with vigorous shaking. This may be the reason why most reported papers we know had used shaking method at low speed for the polymerization of dopamine [23,25,29]. Therefore, vigorous stirring possess good potential to be the reaction method to facilitate the polymerization of dopamine solution.

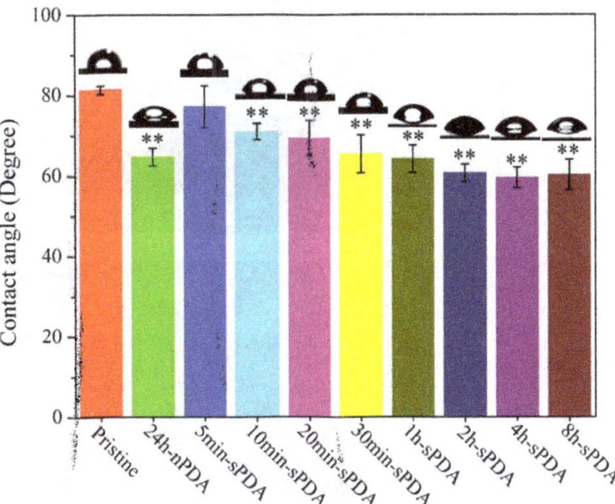

Figure 4. Water contact angles of QCM chips modified by sPDA films and nPDA films. The Contact angles of pristine QCM chips and QCM chips modified by coating of 24 h-nPDA, 5 min-sPDA, 10 min-sPDA, 20 min-sPDA 30 min-sPDA, 1 h-sPDA, 2 h-sPDA, 4 h-sPDA and 8 h-sPDA. All samples were ultrasonically cleaned for 10 min and dried. ** Represents $p<0.01$ compared with the pristine chips group, $n=3$.

Mass and thickness of polydopamine film

For surface modification, dopamine solution is generally polymerized for several hours in the presence of substrates, which is too long for any practical applications. Knowing that both high temperature and vigorous stirring could promote the polymerization of dopamine, a unique device was designed, which uses a constant temperature magnetic stirrer for the formation of rapidly-deposited PDA coating. The mass change of chips decorated by sPDA films with varying polymerization times (5 min, 10 min, 20 min,30 min,1 h, 2 h, 4 h and 8 h) was measured by QCM. To better evaluate PDA coating formed by our method, dopamine solution polymerization under normal condition was conducted as control. As mentioned in the introduction, the dopamine-polymerization process is quite different in a variety of research articles [21–26]. In our study, condition of 37°C with gently shaking at the speed of 70 r·min^{-1} for 24 h was used as the normal methodfor the reasons as follows: 1). 37°C is the most used reaction temperature; 2). Gentle shaking is popularly applied for the polymerization of dopamine solution [23,25,29]; 3). 24 h is the longest reported reaction time reported for PDA coating at which the mass and film thickness will get equilibrium [1].

It has been recognized that excess physically adsorbed PDA particles on the surface of substrate do not have the covalent bonding capability. It is crucial to eliminate the potentially detrimental effect from physically adsorbed PDA particles on the cell functions. PDA films on the substrates are usually washed with distilled water for three times [30,31]. Compared with conventional water rinsing procedure, ultrasonic cleaning may be more efficient and stable. Unfortunately, various ultrasonic times, such as 5 min [32],10 min thrice [12,33] and 15 min [34], were reported to clean out physically adsorbed PDA particles on substrate surface, and the cleaning efficiency had not been clearly demonstrated. In the present study, we assessed the mass changes of PDA film decorated chips undergone ultrasonic cleaning for 1 min, 2 min, 5 min and 10 min, respectively. A great amount of physically adsorbed PDA was obliterated after ultrasonic cleaning for 5 min, and mass change of PDA-coated chips reached

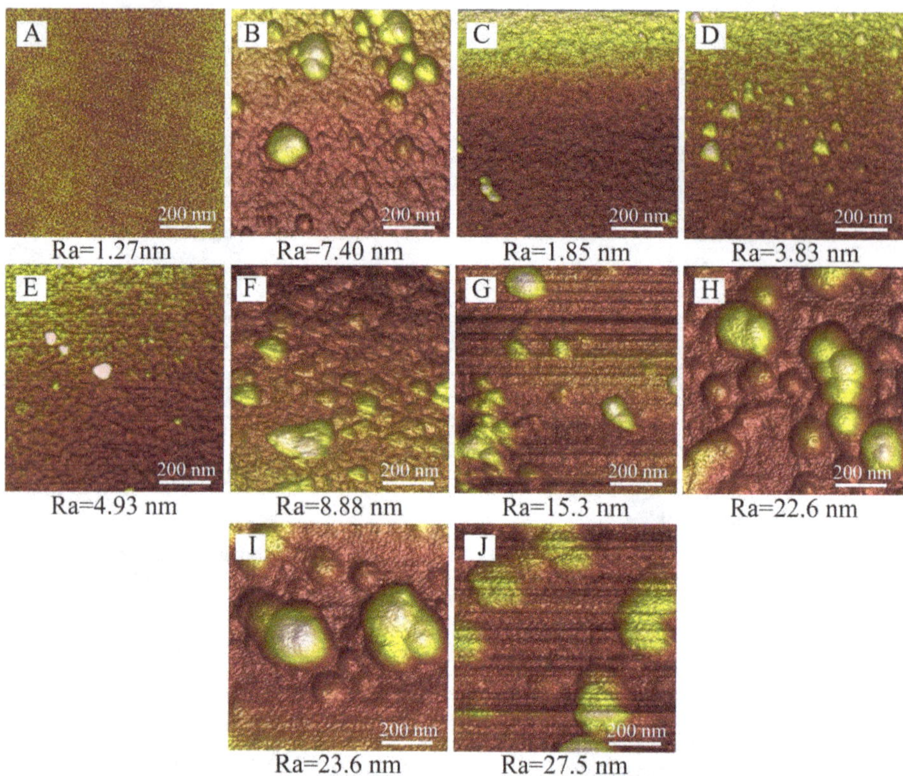

Figure 5. The AFM surface topographies of QCM chips modified by sPDA films and nPDA films. (A): Pristine chip. (B) 24 h-nPDA decorated chip. (C): 5 min-sPDA decorated chip. (D): 10 min-sPDA decorated chip. (E): 20 min-sPDA decorated chip. (F): 30 min-sPDA decorated chip. (G): 1 h-sPDA decorated chip. (H): 2 h-sPDA decorated chip. (I): 4 h-sPDA decorated chip. (J): 8 h-sPDA decorated chip. All those samples had ultrasonically cleaned in distilled water for 10 min. Scale bar: 200 nm.

equilibrium with the extension of time (Figure 3A), suggesting that 5 min treatment is enough to get rid of excessive PDA particles on the sensor surface. This is in consistent with the ultrasonic cleaning time on the surface of silicon wafers reported by Junfei Ou et al [32].

After ultrasonically cleaned, it was found that the mass change of chips modified by PDA coating attained by our method was dramatically increased with the augment of the polymerization time, and reached a plateau after reacted for 8 h (Figure S1 in File S1). After polymerization for 1 h, the mass change of chips decorated by sPDA films is much bigger than that of 24 h-nPDA ($p<0.01$). Interestingly, the mass change of 30 min-sPDA coated chips (6448 ± 788 ng·cm^{-2}) was much less than that of 24 h-nPDA (10186 ± 453 ng·cm^{-2}) prior to cleaning (Figure 3A). However, after ultrasonically cleaned, the mass change of chips modified by 30 min-sPDA films (3992 ± 454 ng·cm^{-2}) is in close vicinity to that of 24 h-nPDA (4324 ± 178 ng·cm^{-2}) (Figure 3A). These results indicate fewer physically absorbed sPDA particles existed on the chip surface than that of nPDA, since vigorous stirring could protect substrates from the particle aggregation in dopamine solution. Besides, there is no PDA on the surface of substrate at the first 30 min by common method [35], But evident mass change was observed when dopamine-polymerization went on for 5 min by our method.

The film thickness of chips coated by sPDA films with different polymerization times (5 min, 10 min, 20 min,30 min,1 h, 2 h, 4 h and 8 h) and 24 h-nPDA that had been ultrasonically cleaned for 10 min was further investigated by spectroscopic ellipsometry. In accordance with the mass change, the film thickness of sPDA on the sensor surfaces increases with the increase in polymerization time and get equilibrium after reacted for 8 h (Figure S1 in File S1). As shown in Figure 3, part B, the thickness of 30 min-sPDA film (44.95 ± 1.15 nm) on the surface of chips is slightly under that of 24 h-nPDA (53.55 ± 0.25 nm) ($p>0.05$), which is close to 50 nm that reported by Messersmith et al [1]. However, the thickness of sPDA films after polymerized for 1 h are much higher than that of 24 h-nPDA ($p<0.01$) (Figure 3B). As shown in Figure S3 in File S1, a good correlation was observed for the curve of sPDA mass versus time and that thickness of versus time, indicating the average molecular weight and density of PDA exhibit no significant difference with the augment of polymerization time, since PDA particles is presumed to be hierarchical aggregation comprised with plate-like aggregates at a size of 1–2 nm through π–π stacking [36].

Those mass and film thickness measurements of PDA coating prove that our method using thermal-dynamic treatment of high temperature and vigorous stirring could not only dramatically promote the polymerization speed of dopamine solution, but also deposit much more PDA onto the substrate surface. Moreover, the mass and film thickness of PDA coating after polymerization for 30 min by our approach was similar to those of 24 h counterpart, possess good prospects to be the condition for rapid-deposition of PDA films to modify biomaterial surfaces.

Characterization of PDA films

The chemical attributes of sPDA and nPDA were probed using Fourier-transform infrared spectrometry (FTIR), X-ray photoelectronic spectroscopy (XPS) and water contact angle measurement. FTIR spectra (KBr) of sPDA and nPDA powders are similar

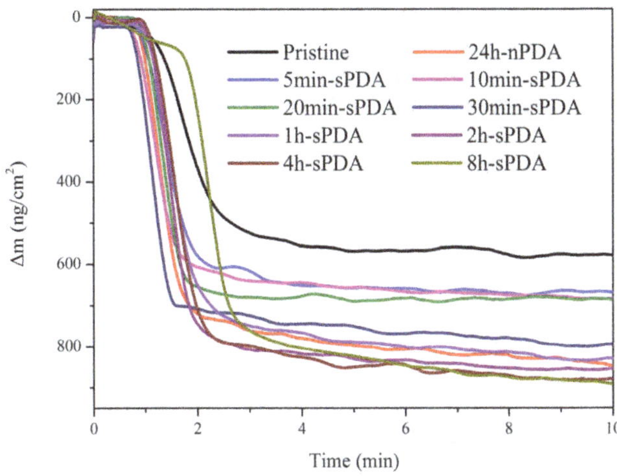

Figure 6. BSA binding evaluation of sPDA films and nPDA films. Representative mass change vs. time (min) curves of pristine chip, 24 h-nPDA coated chip, 5 min-sPDA coated chip, 10 min-sPDA coated chip, 20 min-sPDA coated chip, 30 min-sPDA coated chip, 1 h-sPDA coated chip, 2 h-sPDA coated chip, 4 h-sPDA coated chip and 8 h-sPDA coated chip reaction with 5 mg·ml^{-1} BSA solution.

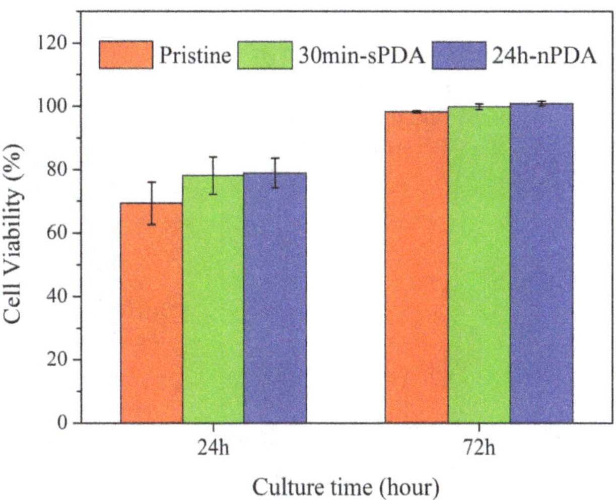

Figure 7. Cytotoxicity assay of QCM chips modified by 30 min-sPDA films and 24 h-nPDA films. Viability of MG63 osteoblasts incubated with pristine QCM chips, 30 min-sPDA decorated chips and 24 h-nPDA decorated chips for 24 h and 72 h. Viability is expressed as a percentage relative to the result obtained with the non-toxic control (MG63 osteoblasts incubated without substrates). n = 3.

(Figure S2 in File S1). According to Daniel R. Dreyer et al's previous study [15], the peaks at 1510 cm^{-1} and 1600 cm^{-1} are consistent with the indole or indoline structures. The large peak at 3376 cm^{-1} is in accordance with the presence of hydroxyl structures as well as water. No evident difference in chemical element content of C, N and O was observed between sPDA powder and nPDA powder in X-ray photo-electronic spectroscopy (XPS) (Figure S3A in File S1).Moreover, the high-resolution C 1 s spectra of sPDA powder and nPDA powder were deconvoluted into five different curves respectively and they are similar(Figure S3B in File S1). The binding energies centered at 284.60 eV, 285.59 eV, 286.01 eV,287.71 eV and 291.43 eV were assigned to the carbon skeleton (-C-C-/-C-H-), amino group (-C-N-), hydroxyl group (-C-OH), carbonyl group (-C = O and -C(O)O-) respectively [29,37].These results indicate that rapidly-deposited PDA film formed by the treatment of high temperature and vigorous stirring, compared with nPDA, exhibits no chemical alteration. Figure 4 shows the water contact angles of the pristine QCM chip, sPDA film modified chips with different polymerization times and 24 h-nPDA film modified chips that hat had ultrasonically cleaned for 10 min. The pristine chip is hydrophobic, with a contact angle of 81.30±1.02°. After coated by 5 min-sPDA film, the contact angle of chips presents no evident change. However, significant decrease was observed for the contact angle of chips (70.95±2.03°) when the dopamine-polymerization time increased to 10 min. It remained in the close vicinity to those of 30 min-sPDA (65.34±4.96°), 1 h-sPDA film (64.18±3.40°)and 24 h-nPDA film coated chips (64.67±2.19°). The contact angles reduced to 60.66±2.18° when polymerization time increased to 2 h and reached an equilibrium plateau with the extension of time. These results indicate sPDA coating possesses good hydrophilicity.

The surface topographies of sPDA films of varying polymerization times (5 min, 10 min, 20 min,30 min,1 h, 2 h, 4 h and 8 h) and 24 h-nPDA films on chips that had ultrasonically cleaned for 10 min were analyzed by an atomic force microscope (AFM) (Figure 5). The Ra value of chip decorated by sPDA coating augments with the increase of polymerization time. For sPDA coating at 5 min of dopamine polymerization time by our method, partial aggregates were formed on the surface of QCM chips (Figure 5C). When the polymerization time increased to 10 min and 20 min, more PDA aggregates appeared on chip surface, but there was no significant change in particle size and Ra value, which is consistent with the thickness measurement (Figure 5D–E). The surface topography of 30 min-sPDA film modified chip (Ra = 8.88 nm) (Figure 5F) is similar to 24 h-nPDA film coated chip (Ra = 7.40 nm) (Figure 5B), with a significant higher roughness than that of 20 min-sPDA film (Ra = 4.93 nm). Much higher surface roughness is observed for chip coated by sPDA after polymerized for 1 h and the Ra value reaches up to 27.5 nm for 8 h-sPDA (Figure 5G–J). These results are in accordance with the mass change and film thickness measurements, and further confirms that high temperature and vigorous stirring could dramatically improve the polymerization speed of dopamine solution.

BSA binding ability evaluation

BSA was selected to be the characteristic protein in real-time evaluation of the protein adhesion properties on the sPDA film coated chips surface by QCM, and in comparison with that of 24 h-nPDA film coated chips (Figure 6). In accordance with the mass measurement, the Δm signals of sPDA films reaction with 5 mg·ml^{-1} BSA solution augment with an increase in polymerization time between 5 min and 8 h, and present higher mass change than that of the pristine chips. The curves of mass change versus time for chips coated by 5 min-sPDA, 10 min-sPDA, 20 min-sPDA reacted with BSA are similar. When the dopamine polymerization time of our method increases to 30 min, much more BSA is bonded onto the chip surface, and the Δm reaches up to 800 ng·cm^{-2}, which is slightly less than that of 24 h-nPDA decorated chip (840 ng·cm^{-2}).The curve of mass change versus time for chip coated by 1 h-sPDA reacted with BSA is almost overlap with that of 24 h-nPDA. The mass changes of chips coated by 2 h-sPDA (859 ng·cm^{-2}), 4 h-sPDA (876 ng·cm^{-2}) and 8 h-sPDA (891 ng·cm^{-2}) reacted with BSA are similar. The small difference among sPDA coating at a time scale of twenty minutes in film thickness and surface topography contributes to the slow growth of Δm signals reacted with BSA solution. In the same way,

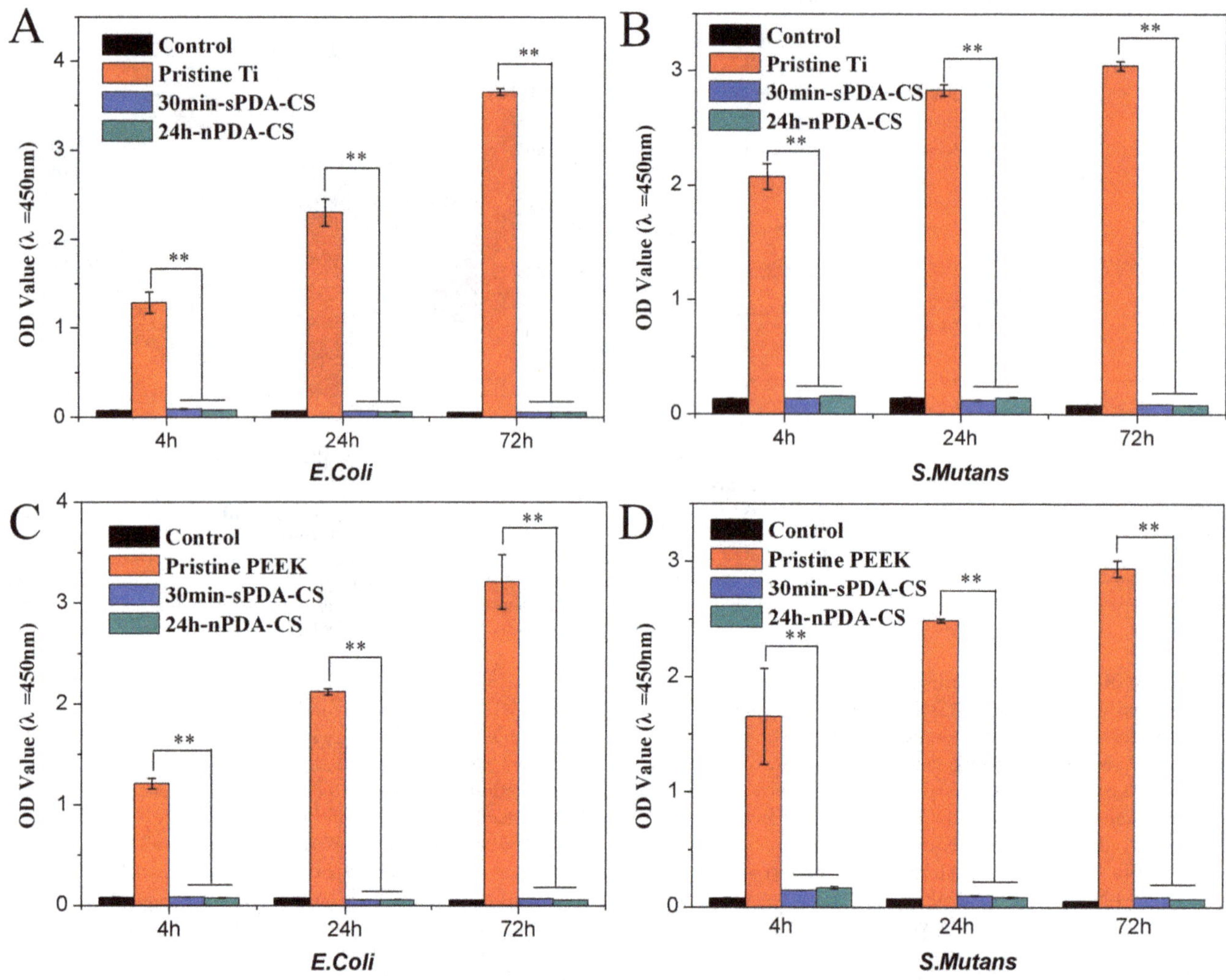

Figure 8. Bacterial adhesion on the surface of 30 min-sPDA-CS and 24 h-nPDA-CS decorated Ti and PEEK. Number of living *E.Coli* and *S.Mutans* adhered on decorated Ti and PEEK surfaces after exposed to bacterial suspension for 4 h, 24 h and 72 h. ** represents $p<0.01$ compared with the pristine group, $n=3$.

the film thickness and surface topography is significantly changed after polymerized for 30 min and a much higher Δm signal was detected. Even though evident augment was observed for the mass changes and film thickness of chips coated by sPDA with the polymerization time increased from 1 h to 8 h, their BSA absorbing ability possess no big difference. The reason is that the surface of sPDA decorated chips has been covered with PDA film after polymerized for 30 min, and there is no significant change in surface coverage with the extension of reaction time.

Cytotoxicity assay

Previous results indicate coating of sPDA and nPDA exhibit same chemical property. The mass change, film thickness and BSA binding ability of 30 min-sPDA coating is in close vicinity to that of 24 h-nPDA. Moreover, 30 min-sPDA coating and 24 h-nPDA possess similar surface morphology and hydrophilicity. Therefore, 30 min was chosen as the standard polymerization time for our rapidly-deposited PDA coating in follow-up cytotoxicity study and surface functional modification.

The effect of 30 min-sPDA film and 24 h-nPDA film on cell viability was evaluated by CCK-8 assay with MG-63 osteoblasts (Figure 7). After incubated for 24 h, the cell viability of 30 min-sPDA coated chips is slightly higher than that of pristine chips, and similar to that of 24 h-nPDA coated chips. The cell viability of all samples is significantly lower than that of the culture plate ($p<0.01$). However, MG63 osteoblasts on all samples exhibit almost the same cell viability when incubated for 72 h. Our study suggest that both sPDA film and nPDA film possess good cytocompatibility, which is consistent with previous report [25].

Antibiotics evaluation

Antibiotic property is one of the most important properties for biomaterials, especially for implant applications. In order to explore the practical use of our newly-developed method, inhibition of bacterial adhesion on the surface of 30 min-sPDA-CS and 24 h-nPDA-CS decorated Ti and PEEK for up to 72 h was investigated and the results are shown in Figure 8. For the amount of adhered viable *E.Coli* and *S.Mutans*, no statistically significant difference was found between 30 min-sPDA-CS modified Ti and 24 h-nPDA-CS modified Ti for 4 h, 24 h and 72 h ($p>0.05$). Moreover, their bacteria adhesion amounts are dramatically lower than that of the pristine Ti ($p<0.01$) (Figure 8A–B). Inhibition of bacterial adhesion on the surface of 30 min-sPDA-CS decorated PEEK and 24 h-nPDA-CS decorated

PEEK, on the other hand, is the same to that of Ti (Figure 8C–D). Furthermore, both for Ti and PEEK, the number of *E.Coli* and *S.Mutans* in the medium with 30 min-sPDA-CS modified substrates is similar to that of 24 h-nPDA-CS modified substrates for up to 72 h ($p>0.05$), which is significantly lower than that of the pristine ($p<0.01$) (Figure S4 in File S1). These results demonstrate superior bioactivity of PDA films formed by our method, which has great potential to be universal approach for surface modification.

Conclusions

In this study, a new method for the formation of rapidly-deposited PDA film was developed, in which substrates are vertically placed in standard dopamine solution at the temperature of 60°C with vigorous stirring (300 $r \cdot min^{-1}$) using an uniquely designed device. The influence of temperature and stirring on the deposition kinetics of PDA film was investigated by the quartz crystal microbalance (QCM). Our results indicate that high temperature and vigorous stirring both can dramatically speed up the rate of dopamine-polymerization. Physical treatment of high temperature and vigorous stirring would result in no chemical changes for polydopamine. The mass, thickness and BSA binding ability of PDA film formed by our approach for 30 min are comparable to those of shaking counterpart for 24 h. Moreover, cell viability assay and antibacterial test demonstrate that our rapidly-deposited PDA coating possesses equally excellent surface modification properties compared with PDA formed by normal method. Furthermore, substrates placed vertically in our approach could not only reduce the deposited PDA aggregates from solution, but also be decorated with PDA film on all surfaces. Our method for rapid deposition of polydopamine film may be used as a potential approach for the surface modification of biomaterials and medical devices.

Supporting Information

File S1 Contains the following files: Figure S1, Mass change versus time and film thickness versus time for sPDA decorated chips. The black curve shows the mass changes versus time of QCM chips coated by sPDA with various polymerization time (5 min, 10 min, 20 min, 30 min, 1 h, 2 h, 4 h and 8 h) that had ultrasonically cleaned for 10 min. The red curve shows the film thickness versus time of the same samples. $n = 3$. Figure S2, FTIR spectra analysis. The FTIR spectra (KBr) of nPDA powder and sPDA powder. Figure S3, XPS survey scan spectra of nPDA powder and sPDA powder. (A): XPS wide spectra of nPDA powder and sPDA powder, insert table shows the contents of C, N and O. (B): High-resolution spectrum of carbon peaks (C 1 s) for nPDA powder and sPDA powder respectively. Figure S4, Bacterial adhesion in the medium with 30 min-sPDA-CS and 24 h-nPDA-CS decorated Ti and PEEK. Number of living *E.Coli* and *S.Mutans* in the medium cultured with modified Ti and PEEK after exposed to bacterial suspension for 4 h, 24 h and 72 h. ** represents $p<0.01$ compared with the pristine group, $n = 3$.

Acknowledgments

We thank Mrs. Chao Xu from State Key Laboratory of Advanced Optical Communication Systems & Networks for the help of AFM measurement in this manuscript.

Author Contributions

Conceived and designed the experiments: PZ HZ HWM YPL SCW. Performed the experiments: PZ YD BRL RRZ. Analyzed the data: PZ YD BRL. Contributed reagents/materials/analysis tools: PZ BRL HWM YPL SCW. Wrote the paper: PZ YPL SCW.

References

1. Lee H, Dellatore SM, Miller WM, Messersmith PB (2007) Mussel-Inspired surface chemistry for multifunctional coatings. Science 318: 426–430.
2. Brubaker CE, Messersmith PB (2012) The present and future of biologically inspired adhesive interfaces and materials. Langmuir 28: 2200–2205.
3. Sileika TS, Kim HD, Maniak P, Messersmith PB (2011) Antibacterial performance of polydopamine-modified polymer surfaces containing passive and active components. ACS Appl Mater Interfaces 3: 4602–4610.
4. Paulo CSO, Vidal M, Ferreira LS (2010) Antifungal nanoparticles and surfaces. Biomacromolecules 11: 2810–2817.
5. Ding X, Yang C, Lim TP, Hsu LY, Engler AC, et al. (2012) Antibacterial and antifouling catheter coatings using surface grafted PEG-b-cationic polycarbonate diblock copolymers. Biomaterials 33: 6593–6603.
6. Cho JH, Shanmuganathan K, Ellison CJ (2013) Bioinspired catecholic copolymers for antifouling surface coatings. ACS Appl Mater Interfaces 5: 3794–3802.
7. Cui J, Yan Y, Such GK, Liang K, Ochs CJ, et al. (2012) Immobilization and intracellular delivery of an anticancer drug using mussel-inspired polydopamine capsules. Biomacromolecules 13: 2225–2228.
8. Peng HP, Liang RP, Zhang L, Qiu JD (2013) Facile preparation of novel core-shell enzyme-Au-polydopamine-Fe_3O_4 magnetic bionanoparticles for glucose sensor. Biosens Bioelectron 42: 293–299.
9. Rim NG, Kim SJ, Shin YM, Jun I, Lim DW, et al. (2012) Mussel-inspired surface modification of poly (L-lactide) electrospun fibers for modulation of osteogenic differentiation of human mesenchymal stem cells. Colloids Surf B - Biointerfaces 91: 189–197.
10. Ku SH, Park CB (2010) Human endothelial cell growth on mussel-inspired nanofiber scaffold for vascular tissue engineering. Biomaterials 31: 9431–9437.
11. Tsai WB, Chen WT, Chien HW, Kuo WH, Wang MJ (2011) Poly(dopamine) coating of scaffolds for articular cartilage tissue engineering. Acta Biomater 7: 4187–4194.
12. Zhou YZ, Cao Y, Liu W, Chu CH, Li QL (2012) Polydopamine-induced tooth remineralization. ACS Appl Mater Interfaces 4: 6901–6910.
13. Lynge ME, van der Westen R, Postma A, Stadler B (2011) Polydopamine–a nature-inspired polymer coating for biomedical science. Nanoscale 3: 4916–4928.
14. Herlinger E, Jameson RF, Linert W (1995) Spontaneous autoxidation of dopamine. Journal of the Chemical Society, Perkin Transactions 2: 259–263.
15. Dreyer DR, Miller DJ, Freeman BD, Paul DR, Bielawski CW(2012) Elucidating the structure of poly(dopamine). Langmuir 28: 6428–6435.
16. Kang K, Choi IS, Nam Y (2011) A biofunctionalization scheme for neural interfaces using polydopamine polymer. Biomaterials 32: 6374–6380.
17. Wei Q, Zhang F, Li J, Li B, Zhao C (2010) Oxidant-induced dopamine polymerization for multifunctional coatings. Polym Chem 1: 1430.
18. Vincent B, Doriane DF, Valérie T, David R (2012) Kinetics of polydopamine film deposition as a function of pH and dopamine concentration: Insights in the polydopamine deposition mechanism. Colloid Interface Sci 386: 366–372.
19. Li H, Cui D, Cai H, Zhang L, Chen X, et al. (2013) Use of surface plasmon resonance to investigate lateral wall deposition kinetics and properties of polydopamine films. Biosens Bioelectron 41: 809–814.
20. Jiang J, Zhu L, Zhu L, Zhu B, Xu Y (2011) Surface characteristics of a self-polymerized dopamine coating deposited on hydrophobic polymer films. Langmuir 27: 14180–14187.
21. Bhang SH, Kwon SH, Lee S, Kim GC, Han AM, et al. (2013) Enhanced neuronal differentiation of pheochromocytoma 12 cells on polydopamine-modified surface. Biochem Biophys Res Commun 430: 294–1300.
22. Wang LQ, Jeong KJ, Chiang HH, Zurakowski D, Behlau I, et al. (2011) Hydroxyapatite for keratoprosthesis biointegration. INVEST OPHTH VIS SCI 52: 7392–7399.
23. Lee YB, Shin YM, Lee JH, Jun I, Kang JK, et al. (2012) Polydopamine-mediated immobilization of multiple bioactive molecules. Biomaterials 33: 8343–8352.
24. Wei Q, Li B, Yi N, Su B, Yin Z, et al. (2011) Improving the blood compatibility of material surfaces via biomolecule-immobilized mussel-inspired coatings. J Biomed Mater Res A 96: 38–45.
25. Zhu LP, Jiang JH, Zhu BK, Xu YY (2011) Immobilization of bovine serum albumin onto porous polyethylene membranes using strongly attached polydopamine as a spacer. Colloids Surf B Biointerfaces 86: 111–118.
26. Zheng L, Liu Q, Xiong L, Li Y, Han K, et al. (2012) Controlled preparation of titania nanofilm by a template of polydopamine film and its reversible wettability. Thin Solid Films 520: 2776–2780.

27. Fatisson J, Azari F, Tufenkji N (2011) Real-time QCM-D monitoring of cellular responses to different cytomorphic agents. Biosens Bioelectron 26: 3207–3212.
28. G S (1959) Verwendung von schwingquarzen zur wagung dunner schichten und zur mikrowagung. Z Angew Phys 155: 206–222.
29. He S, Zhou P, Wang L, Xiong X, Zhang Y, et al. (2014) Antibiotic-decorated titanium with enhanced antibacterial activity through adhesive polydopamine for dental/bone implant. J R Soc Interface 11: 20140169.
30. Lynge ME, Ogaki R, Laursen AO, Lovmand J, Sutherland DS, et al. (2011) Polydopamine/liposome coatings and their interaction with myoblast cells. ACS Appl Mater Interfaces 3: 2142–2147.
31. Shin YM, Lee YB, Shin H (2011) Time-dependent mussel-inspired functionalization of poly (L-lactide-co-varepsilon-caprolactone) substrates for tunable cell behaviors. Colloids Surf B Biointerfaces 87: 79–87.
32. Junfei Ou, Jinqing Wang, Sheng Liu, Zhou J, Yang S (2009) Self-assembly and tribological property of a novel 3-layer organic film on silicon wafer with polydopamine coating as the Interlayer. J Phys Chem C 113: 20429–20434.
33. Luo R, Tang L, Zhong S, Yang Z, Wang J, et al. (2013) In vitro investigation of enhanced hemocompatibility and endothelial cell proliferation associated with quinone-rich polydopamine coating. ACS Appl Mater Interfaces 5: 1704–1714.
34. Pop-Georgievski O, Verreault D, Diesner MO, Proks V, Heissler S, et al. (2012) Nonfouling poly (ethylene oxide) layers end-tethered to polydopamine. Langmuir 28: 14273–14283.
35. Bernsmann F, Ponche A, Ringwald C, Hemmerle J, Raya J, et al. (2009) Characterization of dopamine-melanin growth on silicon oxide. J Phys Chem C 113: 8234–8242.
36. Lynge ME, van der Westen R, Postma A, Städler B (2011) Polydopamine—a nature-inspired polymer coating for biomedical science. Nanoscale 3: 4916–4928.
37. Delpeux S, Beguin F, Benoit R, Erre R, Manolova N, et al. (1998) Fullerene core star-like polymers-1. Preparation from fullerenes and monoazidopolyehers. Eur Polym J 34: 905–915.

24

Biologically Active Polymers from Spontaneous Carotenoid Oxidation: A New Frontier in Carotenoid Activity

James B. Johnston[1], James G. Nickerson[2], Janusz Daroszewski[3], Trevor J. Mogg[3], Graham W. Burton[3]*

1 National Research Council of Canada, Charlottetown, Prince Edward Island, Canada, **2** Avivagen Inc., Charlottetown, Prince Edward Island, Canada, **3** Avivagen Inc., Ottawa, Ontario, Canada

Abstract

In animals carotenoids show biological activity unrelated to vitamin A that has been considered to arise directly from the behavior of the parent compound, particularly as an antioxidant. However, the very property that confers antioxidant activity on some carotenoids in plants also confers susceptibility to oxidative transformation. As an alternative, it has been suggested that carotenoid oxidative breakdown or metabolic products could be the actual agents of activity in animals. However, an important and neglected aspect of the behavior of the highly unsaturated carotenoids is their potential to undergo addition of oxygen to form copolymers. Recently we reported that spontaneous oxidation of ß-carotene transforms it into a product dominated by ß-carotene-oxygen copolymers. We now report that the polymeric product is biologically active. Results suggest an overall ability to prime innate immune function to more rapidly respond to subsequent microbial challenges. An underlying structural resemblance to sporopollenin, found in the outer shell of spores and pollen, may allow the polymer to modulate innate immune responses through interactions with the pattern recognition receptor system. Oxygen copolymer formation appears common to all carotenoids, is anticipated to be widespread, and the products may contribute to the health benefits of carotenoid-rich fruits and vegetables.

Editor: Hiroyoshi Ariga, Hokkaido University, Japan

Funding: Funding support provided by the Atlantic Canada Opportunities Agency Atlantic Canada Innovation Fund, Contract No. 189074. http://www.acoa-apeca.gc.ca/eng/ImLookingFor/ProgramInformation/AtlanticInnovationFund/Pages/AtlanticInnovationFund.aspx to GWB JGN JD TJM and Innovation PEI Discovery and Development Fund (http://www.innovationpei.com/dcfund). The funders had no role in study design, data collection and analysis, decision to publish, or preparation of the manuscript.

Competing Interests: The authors have read the journal's policy and have the following competing interests: JGN, JD, TJM and GWB are employees of Avivagen Inc. JD and GWB own shares in Avivagen Inc. Avivagen Inc. has developed 3 products based on oxidized ß-carotene for use as companion animal supplements: 1. Oximunol Chewables for dogs, marketed by Bayer in the US to veterinarians. 2. Vivamune Health Chews for cats and dogs, marketed online in the US. 3. A feed additive for food animals currently being tested in several countries in Asia to improve immune health of poultry, swine and shrimp under production conditions, potentially providing an alternative to antibiotic use in feeds. JBJ, JGN, JD, and GWB are authors on patents applied for by Avivagen Inc. relating to aspects of oxidized carotenoids: 1. Compositions and methods for promoting weight gain and feed conversion (GWB JD). International publication: WO 2006/034570. Granted in US, Australia, Canada, Indonesia, Mexico, NZ, Russia, Singapore, South Africa and South Korea. Pending in Argentina, Brazil, Chile, Europe. 2. Compositions and methods for enhancing immune response (JBJ, GWB). International publication: WO 2009/052629. Granted in NZ. Pending in US, Australia, Canada, Europe, India, Japan, South Korea. 3. Methods and compositions for improving the health of animals (JD). International publication: WO 2010/124391. Granted in NZ. Pending in US, Australia, Canada, Europe. 4. Methods and compositions for use in aquaculture (JGN). International publication: WO 2011/103464. Pending in US, Canada, Chile, Europe.

* Email: g.burton@avivagen.com

Introduction

Fruit and vegetable intake-based epidemiological studies relating to the incidence of chronic diseases such as cancer and heart disease [1] have created considerable interest in potential non-vitamin A benefits of carotenoids and of ß-carotene in particular. Various possible mechanisms operating at the functional, cellular and molecular levels have been proposed [1,2,3,4]. Of these, a possible antioxidant function [1,5] initially attracted much interest [4]. However, the lack of support or even failure of early human intervention trials [6,7,8] cast serious doubt upon the value of pharmaceutical-level ß-carotene supplementation for ameliorating chronic diseases. Furthermore, the recent observation that antioxidant supplementation enhances cancer progression in mice [9] undermines an antioxidant role.

Actual demonstrated non-vitamin A health benefits of carotenoids in animals point to involvement of immune function [2,10,11,12,13,14]. It has been suggested carotenoids can participate in and modulate processes involving reactive oxygen species [2,3,12]. This behavior would still be consistent with the dual antioxidant/pro-oxidant character of carotenoids inherent in the extensive system of conjugated double bonds [4,5]. In this scenario, carotenoid oxidative breakdown products [15] or their metabolites [16] have been suggested to be the actual bioactive agents. However, progress in this area has been hampered by a lack of identified candidate compounds.

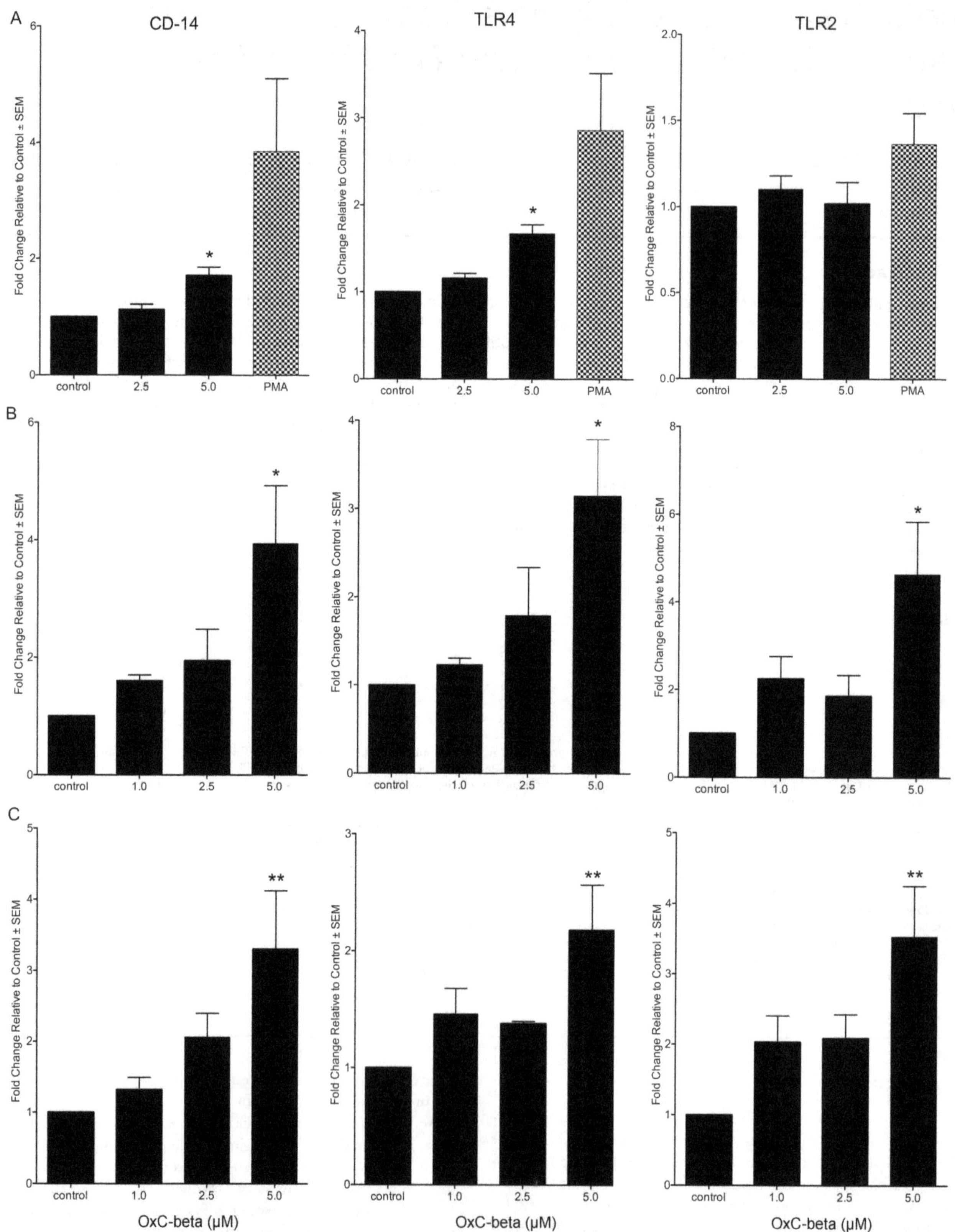

Figure 1. Effect of OxC-beta on CD14, TLR-4, and TLR-2 levels *in vitro*. Human THP-1 monocytes (A), fibroblasts (B), and endothelial cells (C), were treated with the indicated concentrations of OxC-beta or vehicle control (DMSO) for 24 hours. Immune receptor content was measured 24 hours post-treatment by FACS analysis. OxC-beta-induced increase in receptor level was assessed relative to untreated control cells using a one-way analysis of variance with Tukey's post test for multiple comparisons. DMSO had no effect on receptor level (result not shown). Phorbol myristate acetate (PMA) was used as a positive control in experiments with THP-1 cells (hatched bars).

Recently, we reported the discovery that spontaneous oxidation of ß-carotene is dominated not by cleavage reactions but by *addition* of oxygen to form potentially bioactive, oxygen-rich, ß-carotene-oxygen copolymers [17]. We also found that this dominance appears to be common to most, if not all, carotenoid compounds, even early on in the oxidation. Given this finding and the ubiquity of carotenoids and their susceptibility to oxidation during exposure to air, it is anticipated that carotenoid-oxygen co-polymers would occur naturally in a variety of situations. In this regard and as an example, we have noted [17] strong similarities in the elemental compositions and infrared spectra of the products from fully oxidized ß-carotene and lycopene on one hand, and, on the other, sporopollenin, an almost chemically and biochemically intractable polymeric component of the highly robust outer walls (exines) of pollens and spores [18]. We surmised that the carotenoid-oxygen copolymer compounds are similar to an early stage or precursor form of the highly elaborated, naturally occurring sporopollenin exine, sharing a common underlying chemical motif. Indeed Brooks and Shaw first proposed carotenoid-oxygen emulsion copolymerization is a key, early step in the formation of sporopollenin [18,19,20,21].

Many animal species would be expected to be exposed to these compounds in low and varying amounts in foods and the environment. The question arises: is this previously overlooked class of carotenoid-derived oxygen copolymers biologically active?

As a first step in addressing this question we took fully autoxidized ß-carotene (OxC-beta) as a representative cross-section of carotenoid oxidation products [17] to test for evidence of biological activity *in vitro*. OxC-beta, containing more than 30% oxygen by weight, has a defined product composition comprising ß-carotene oxygen copolymers (85% w/w) and minor amounts of short chain, mostly familiar, norisoprenoid cleavage compounds. Vitamin A and higher molecular weight norisoprenoid compounds are absent. Because ß-carotene also is absent, any biological activity must arise directly from non-vitamin A oxidation products and necessarily precludes any involvement of ß-carotene as an antioxidant. In a preliminary evaluation of biological activity using QRT-PCR, we reported OxC-beta increases expression of several genes associated with pathogen recognition and host defense, including toll-like receptors subtype 2 and 4 (TLR-2, TLR-4), CD14, and several genes involved in the TLR signaling pathway [17].

An earlier, preliminary feeding trial with broiler chickens to evaluate the potential of OxC-beta as a non-antibiotic enhancer of food animal productivity gave results that are consistent with an immune priming effect. We therefore carried out studies, reported here, that undertook to (a) directly establish OxC-beta's effect upon measures of innate immune function associated with host detection and response to bacterial pathogens and (b) identify the polymer fraction as the main source of activity within OxC-beta and other oxidized carotenoids.

Results, Discussion and Conclusions

Flow cytometry confirms the earlier PCR assay results [17], showing up-regulation of TLR and CD14 expression. Treatment with OxC-beta increases plasma membrane content of TLRs and CD14 in several cultured cell types, including monocytes, fibroblasts and endothelial cells (Fig. 1). In monocytes OxC-beta (5.0 μM) significantly increases plasma membrane content of both CD14 and TLR-4 (1.7-fold for both), without altering TLR-2 levels (Fig. 1A). The effect on fibroblasts is more pronounced (Fig. 1B). OxC-beta (5.0 μM) induces increases in CD14 and TLR-4 of 3.9 and 3.1-fold, respectively. TLR-2 also increases significantly by 4.6-fold. CD14, TLR-4 and TLR-2 levels also increase significantly in endothelial cells (Fig. 2C): treatment with 5.0 μM OxC-beta produces increases of 3.3, 2.2 and 3.5-fold, respectively. These results indicate that the previously reported PCR gene expression results are reflected at the protein level.

The *in vitro* receptor results are corroborated *in vivo*. Mice fed OxC-beta show increased intestinal immune receptor content. Oral supplementation of OxC-beta at 10 mg/kg body weight daily for 2 or 4 weeks show marked increases in levels of CD14 and TLR-4 in the small intestine (see Fig. 2 for results at 4 weeks). Furthermore, OxC-beta's effect on receptors appears dose dependent: the increase in receptors for 1 mg/kg, though apparent, was less pronounced as qualitatively indicated by the intensity of staining (results not shown).

The ability to modulate host-innate-responsiveness to bacterial infection has been assessed by evaluating OxC-beta's effect upon the early stage cytokine response and phagocytic activity of monocytes under both naïve and simulated challenge situations. Given OxC-beta's ability to increase TLR-4 and CD14 levels and the critical role that these moieties play in the receptor complex that detects bacterial lipopolysaccharide (LPS), we used an LPS challenge to mimic bacterial infection *in vitro*.

OxC-beta treatment of naïve monocytes has no apparent effect on the level of various cytokines, including TNFα, IL-1β, IL-6, IFNγ, MCP-1 (CCL-2), and IL-8 (results not shown). However, when OxC-beta treatment is followed by an LPS-challenge the pretreatment potentiates the LPS-induced increase in the production of several cytokines, including TNFα (25%), IL-6 (48%), and IL-1β (40%), relative to LPS- challenge alone (Table 1). TNFα and IL-1β are pivotal pleiotropic cytokines directing multiple facets of host response to infection whereas IL-6 is an important mediator of the acute phase response and is required for maintaining microbial resistance.

At the level of monocyte function, OxC-beta treatment increases the phagocytic activity of monocytes. The effect is most prominent under LPS-challenge conditions where pretreatment with OxC-beta increases activity to a level comparable to the positive control, phorbol myristate acetate (PMA, Fig. 3). By contrast the effect of OxC-beta on phagocytic activity in naïve monocytes is more modest (result not shown).

Taken together the results on immune receptor levels, cytokine levels, and phagocytic activity provide mechanistic and function-based evidence that OxC-beta modulates innate immunity in a biologically relevant way. Furthermore, the results suggest that the overall impact would be to prime the innate immune system to more rapidly respond to subsequent challenges. The fact that OxC-beta has little to no apparent effect on cytokine levels and phagocytic activity in the absence of a challenge distinguishes it from traditional immune stimulants, which directly trigger an inflammatory immune response.

Given that activation of the retinoic acid receptor (RAR) pathway is known to affect immune function [22], it was important to rule out the possibility that unidentified compounds within OxC-beta possessed RAR agonist activity. Treatment of MCF-7 cells with OxC-beta failed to induce expression of CYP26A, a known RAR responsive gene [23] (Fig. 4). This result is consistent with the reported absence of retinoic acid and closely related compounds in OxC-beta [17].

Next we turned our attention towards identifying the compound(s) present in OxC-beta responsible for the observed immunological activity. A FACS assay based on OxC-beta's ability to up-regulate CD14 in monocytes indicates the polymeric fraction is the source of activity. Treatment with equal amounts of OxC-beta or the isolated polymeric fraction induced significant

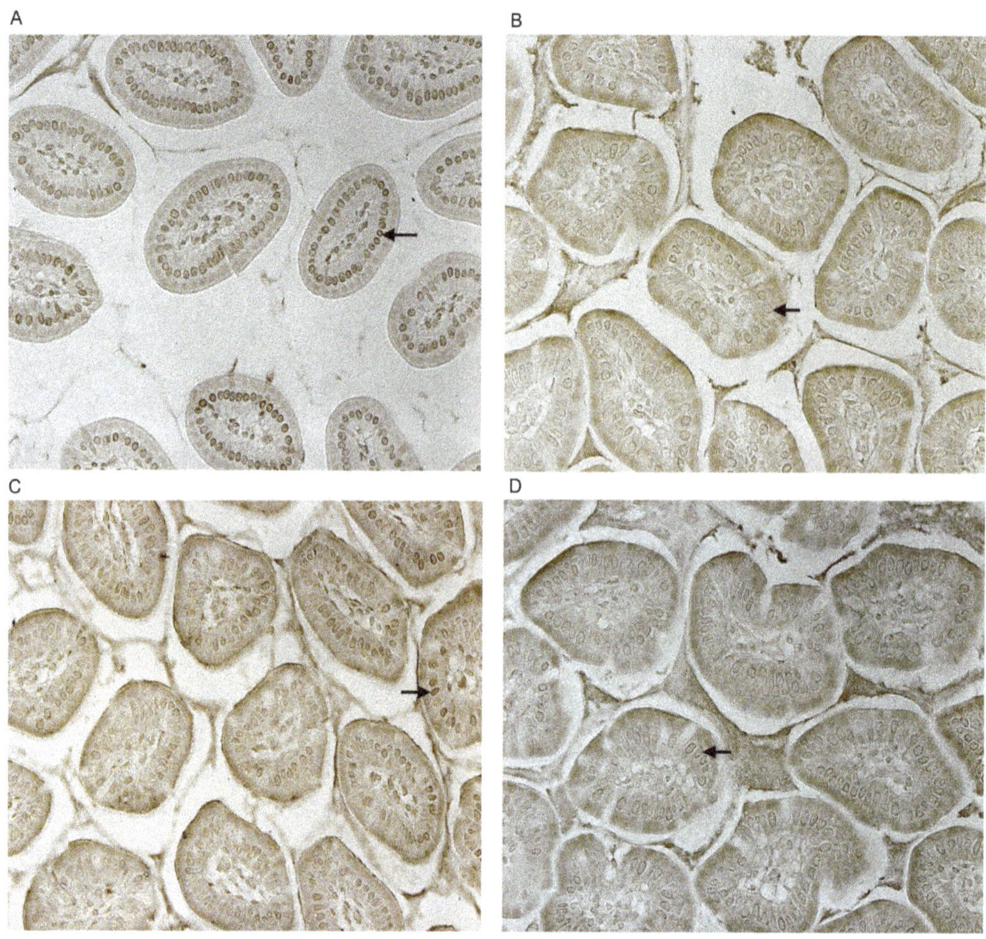

Figure 2. CD14 and TLR-4 staining in gut epithelial cells. Balb/c mice were not supplemented (control) or supplemented daily by oral gavage with OxC-beta (10 mg/kg). After 4 weeks, intestinal tissues were harvested and CD14 and TLR-4 expression was determined by immunocytochemistry. Increased CD14 (A) and TLR-4 (C) expression is readily apparent in epithelial cells in the OxC-beta-supplemented animals compared to the controls receiving vehicle alone (B and D, respectively). Arrows indicate the location of enterocytes within the cross section of microvilli. Magnification 40x.

and comparable increases in the surface content of monocyte CD14 relative to untreated control cells (Fig. 5A). The monomer fraction was much less active relative to either OxC-beta or the polymer, especially considering that OxC-beta and the polymer and monomer fractions were each compared over similar concentration ranges, with the monomer fraction concentration being approximately 7-fold higher relative to its actual level of 15% in OxC-beta. Any activity in the monomer fraction is attributed to the presence of lower molecular weight residual copolymer compounds that were not completely removed in the separation process. The CD14 FACS assay also shows fully oxidized lycopene (OxC-lyc) and OxC-beta have essentially the same up-regulating activity (Fig. 5B). Given OxC-lyc also is predominantly composed of oxygen co-polymers [17], it is reasonable to conclude these too are the source of activity.

The similarity of the activities of OxC-beta and OxC-lyc also makes it unlikely that the minor amounts of norisoprenoid compounds present in each, together with the dissimilarity of their composition, contribute any significant activity. The most abundant norisoprenoid in OxC-beta, geronic acid (2.4%) [17], is inactive (results not shown) and is not present in OxC-lyc. The next most abundant compounds in OxC-beta, ß-ionone-5,6-epoxide (1.2%), dihydroactinidiolide (1.1%) and ß-ionone (0.8%) [17], also are not found in OxC-lyc, as applies for any other norisoprenoid compounds derived from the cyclohexyl rings present in ß-carotene but absent in lycopene. Although several aldehydes have been identified in OxC-beta, they are present at very low levels, for example, 2-methyl-6-oxo-2,4-heptadienal (0.23%) [17]. The numerous other norisoprenoid compounds present in OxC-beta are present at much less than 1% levels, making it unlikely they make any significant contribution to activity, individually or collectively, especially in light of the data for the polymeric fraction presented in Fig. 5A.

The CD14 assay results have two-fold significance. Firstly, they demonstrate for the first time that carotenoid oxygen copolymers, the dominant products of spontaneous carotenoid oxidation, are biologically active. Given that OxC-beta, OxC-lyc and also fully oxidized canthaxanthin all have been found to be predominantly composed of oxygen copolymers [17], it appears reasonable to conclude that through oxidation the highly unsaturated backbone of carotenoids provides a route to a previously unrecognized class of biologically active, polymeric compounds. This very facile natural reaction opens up a new approach to understanding carotenoid non-vitamin A behavior, moving the focus from putative protective antioxidant *processes* to a discrete class of

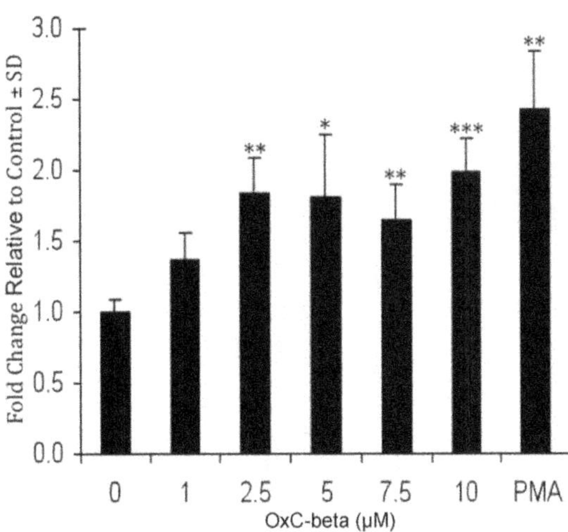

Figure 3. Phagocytosis in OxC-beta treated and LPS-stimulated THP-1 cells. THP-1 monocytes were incubated with the indicated concentration of OxC-beta or DMSO control for 24 hours before being treated with LPS (15 ng/mL). Phagocytosis was evaluated 24 hours after LPS stimulation. Values represent fold changes relative to controls. PMA was used at 25 ng/mL. * $p<0.05$, ** $p<0.02$, *** $p<0.002$, Student's t-test versus controls.

carotenoid oxidation *products* with unanticipated biological activities.

Secondly, the fact that the copolymers are capable of modulating innate immunity, a fundamental physiological system common to all animals, suggests the innate immune system has the ability to recognize structural elements within the polymer fraction and highlights the important role that these compounds potentially play in nature. Of note in this regard is the similar innate immune priming effect observed in an *in vitro* macrophage model [24] for carotenoid-containing spores of the common probiotic bacteria *Bacillus subtilus* [25], although it is not yet known if the activity arises from oxygen-carotenoid copolymers.

The absence of vitamin-A and higher homologs in OxC-beta and its inability to induce expression of an RAR responsive gene, together with the identification of the polymer as the active constituent in both OxC-beta and OxC-lyc, suggest that biological activity arises via a novel pathway. Given the apparent underlying structural similarity of the polymer to sporopollenin [17,18,19] and the likelihood of the latter to be associated with the reproductive structures of varied microbes (e.g., spores and pollen), it seems plausible that the innate immune system may have evolved mechanisms to recognize conserved molecular patterns contained in sporopollenin-like moieties.

The extent of the natural occurrence of oxidized carotenoids containing copolymeric components is not yet known, although sporopollenin, for example, in the form of spores and pollen is ubiquitous and widely distributed in the environment and foods [18,26]. Even in small amounts the cumulative effects of frequent exposure to carotenoid copolymers may be sufficient to partly explain the epidemiological evidence for the benefits of diets rich in carotenoid-containing fruits and vegetables [1]. Also, the likely low level of oxidation products in ß-carotene supplements formulated to protect against oxidation may explain the lack of efficacy of high levels of supplementation in intervention trials [6,7,8]. The fact that intact sporopollenin exine shells have been shown to be readily absorbed across the gut wall and broken down quickly in blood [27] implies that carotenoid-oxygen copolymers would be systemically available for distribution to a wide range of

Table 1. Cytokine expression following OxC-beta treatment and LPS challenge.

Conc.[2]	TNFα		IL-6		IL-1β		IFNγ	
µM	pg/mL[3]	Fold[4]	pg/mL[3]	Fold[4]	pg/mL[3]	Fold[4]	pg/mL[3]	Fold[4]
0.0	1471±171	1.00	37.7±1.9	1.00	1194±46	1.00	20.9±0.6	1.00
0.1	1840±3#	1.25	49.4±3.9*	1.31	1425±117	1.19	21.0±0.1	1.00
0.5	1846±7#	1.25	56.1±2.8*	1.48	1669±78*	1.40	20.5±2.7	0.98
1.0	1832±0.4#	1.24	37.3±2.2	1.00	1065±95	0.89	19.4±3.7	0.93

All cells were challenged with 15 ng/mL LPS.
[2]OxC-beta concentration expressed using the molecular weight of ß-carotene.
[3]Values determined by reference standards included in each assay.
[4]Fold change relative to untreated cells (no OxC-beta).
# $p<0.005$, * $p<0.05$ (Student's *t*-test versus untreated).

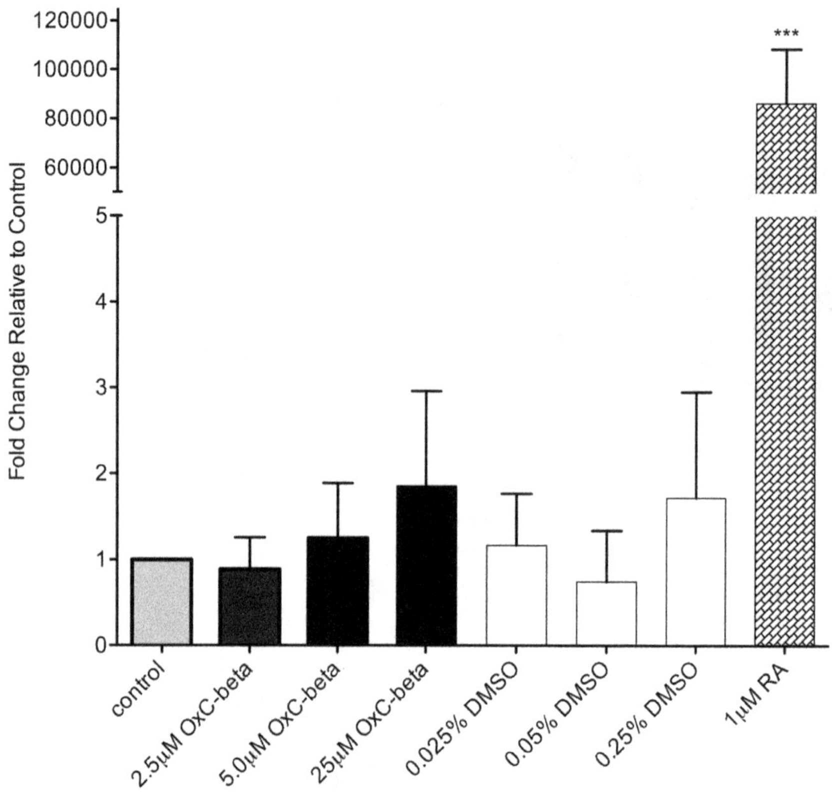

Figure 4. Effect of OxC-beta treatment on CYP26A gene expression in MCF-7 cells. Cells were incubated in the presence of the indicated concentrations of OxC-beta or vehicle control for 24 hours. For vehicle controls the concentrations of DMSO used were equivalent to the concentration of DMSO in the OxC-beta treatment groups, i.e., control groups labeled as 0.025%, 0.05% and 0.25% DMSO had the same DMSO concentration (v/v) as was used in the 2.5 µM, 5.0 µM and 25 µM OxC-beta treatment groups, respectively. Untreated cells were used as negative controls and cells treated with 1.0 µM of *all trans* retinoic acid (RA) served as positive controls. CYP26A gene expression was measured relative to β-actin using quantitative real-time PCR with the two standard curve method. Bars represent the mean ratio of CYP26A expression relative to β-actin from 4 separate experiments. Error bars represent the standard error of the mean. One-way ANOVA with Tukey's test for multiple comparisons indicated no significant difference in relative CYP26A expression between cells treated with OxC-beta, DMSO, or untreated control cells. Treatment with 1.0 µM RA induced a significant increase in CYP26A gene expression compared to all other treatment groups. CYP26A expression ratios are shown calibrated to untreated control cells. *** $p<0.001$.

tissues in the body. Indeed, the ability to convert spores and pollen into empty sporopollenin exine shells that can serve as vehicles for oral delivery of a variety of compounds into the blood is being developed as a novel technology for delivery of drugs and nutrients (http://www.sporomex.co.uk).

Indirect support for the systemic availability of OxC-beta is provided in a preliminary study to be reported elsewhere, with prima facie evidence of OxC-beta activity *in vivo* in exploratory animal feeding trials. Low part-per-million levels of OxC-beta in feed improved measures of production efficiency, i.e., feed conversion and growth, in both swine and broiler chickens. Interestingly, benefits in poultry were more pronounced in a *Clostridium perfringens* challenge model of necrotic enteritis, a finding consistent with the innate immune priming activity of OxC-beta reported above.

Further support for the systemic availability of orally-administered OxC-beta is provided in a study showing enhanced resolution of inflammation in a model of bovine respiratory disease [28]. This study illustrates *in vivo* the ability of OxC-beta to exert a moderating influence upon inflammation in the lung, an immunological activity that was suggested earlier by the *in vitro* results of a PCR gene expression array assay [17]. Both the innate immune priming effect and the anti-inflammatory/pro-resolution effects suggested by the PCR gene expression array results are supported by the results of multiple animal trials that are beyond the scope of this manuscript and will be the subject of future publications.

Materials and Methods

Preparation of OxC-beta and Isolation of Polymer and Monomer Fractions

The preparations of OxC-beta, its polymer and monomer fractions and OxC-lyc have been described [17].

Cell Culture

THP-1, 1079SK-fibroblasts, and MCF-7 cell lines were obtained from American Type Cell Culture (ATCC). Human umbilical vein endothelial cells (HUVEC) were purchased from Life Technologies. All cell lines were cultured in complete media at 37°C with 5% CO_2. For the human THP-1 monocyte line (ATCC TIB-202) complete media was RPMI-1640 media supplemented with 2 mM L-glutamine, 10 mM HEPES, 1.0 mM sodium pyruvate, 10% fetal bovine serum (FBS) and antibiotics. Fibroblasts (ATCC CRL-2097) were cultured in Eagle's Minimum Essential Media (EMEM) supplemented with 0.1 mM non-essential amino acids, 0.1 mM sodium pyruvate, 10% FBS and antibiotics. MCF-7 cells (ATCC HTB-22) were

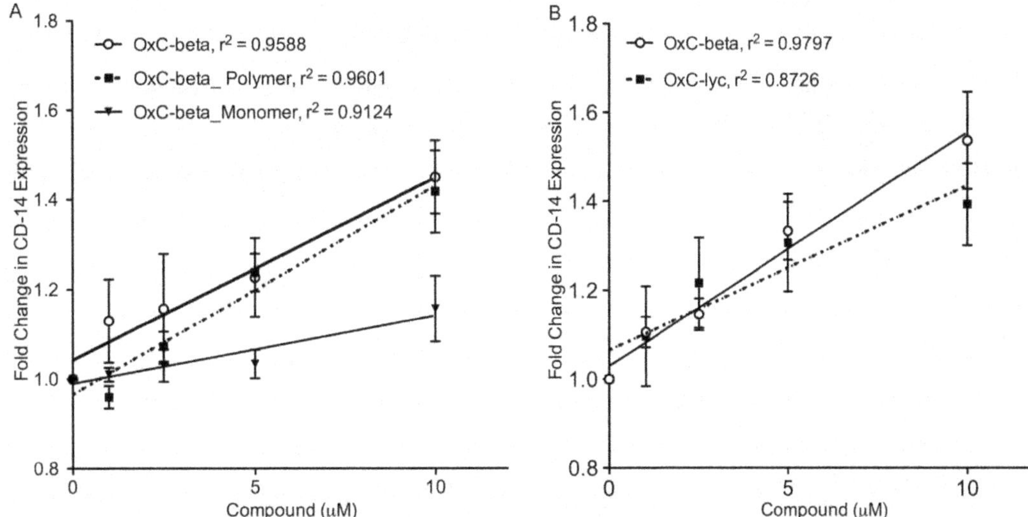

Figure 5. Determination of activities relative to OxC-beta of (A) OxC-beta polymer and monomer fractions, and (B) oxidized lycopene (OxC-lyc), using a CD14 receptor expression assay. THP-1 cells were treated for 24 hours with the indicated concentrations of compounds. CD14 expression was quantified using FACS analysis. The effect of each compound is shown relative to untreated cells. Points represent the mean and standard error from three separate experiments. (A) Correlation analysis indicates a significant dose effect for each compound on CD14 expression with p-values of 0.0036 for OxC-beta, 0.0034 for the polymer, and 0.0113 for the monomer. Comparison of the relative activity of each compound indicates that the monomer is significantly less active than the polymer ($p<0.001$) and OxC-beta ($p<0.01$) while there is no significant difference between the activities of the polymer and OxC-beta. The apparent activity of the monomer may be due to the presence of residual polymers that could not be completely removed from the monomer fraction. (B) OxC-lyc also had a significant dose effect on CD14 surface content ($p = 0.020$) that was not significantly different from the effect of OxC-beta.

cultured in EMEM supplemented with 10% FBS and antibiotics. HUVEC cells (Life Technologies C-003-5C) were cultured in Media 200 supplemented with low serum growth supplement (Life Technologies S00310) and antibiotics.

Test Compound Preparation

10 mM (ß-carotene equivalents) stock solutions of OxC-beta, OxC-beta polymer fraction, OxC-beta monomer fraction, and OxC-lyc were prepared by dissolving 5.37 mg/mL of each compound in DMSO. Stock solutions were stored in 500 μL aliquots at −80°C. Working solutions (200 μM carotene equivalents) were prepared by further dilution of the stock with the appropriate basal media followed by filter sterilization (0.22 μm pore size). Cells were treated with the indicated concentration of each compound by further dilution of the 200 μM working stocks with the appropriate culture media.

Measurement of OxC-beta Effect on Expression of CD14, TLR-4, and TLR-2 *In Vitro*

Cells were seeded in T-25 flasks at a density of 3×10^5 cells/flask and allowed to recover for 24 hours in complete media. Following recovery cells were treated with the indicated concentrations of OxC-beta or corresponding concentration of DMSO (vehicle control) for 24 hours. For experiments with THP-1 cells, treatment with 25 ng/mL phorbol myristate acetate (PMA) was used as a positive control. Following treatment cells were harvested via trypsin treatment (for adherent fibroblast and HUVEC cultures) and centrifugation at 300×g for 5 minutes. Cells were then washed by resuspension in 1 mL of phosphate buffered saline (PBS) and centrifugation at 300×g for 5 minutes. Washed cells were resuspended in 200 μL of staining buffer (Dulbecco's phosphate-buffered saline, pH 7.4, 0.2% bovine serum albumin, 0.09% sodium azide) and the cell density of each suspension was determined by manual counting with a hemocytometer. Suspensions from each treatment group were then transferred, in 100 μL aliquots, into triplicate wells on a 96-well, round bottom microplate (1×10^5 cells/well) and incubated for 1 hour at 4°C in the presence of 10 μL normal mouse serum (eBioscience 24-5544-94). Cells were then spun at 300×g for 5 minutes, washed by resuspension in PBS and centrifugation, and resuspended in 100 μL of staining buffer. Cells were stained for CD14, TLR-4, or TLR-2 by incubation for 1 hour at 4°C with the following phycoerythrin (PE) conjugated antibodies: anti-human CD14 (eBioscience 12–0149), anti-human TLR-4 (eBioscience 12–9917), and anti-human TLR-2 (eBioscience 12–9922). Cells stained with PE conjugated anti-mouse IgG1 (eBioscience 12–4714) for CD14 or anti-mouse IgG2a (eBioscience 12–4724) for TLR-4 and TLR-2 served as isotype controls. After labeling cells were pelleted by centrifugation at 300×g for 5 minutes and washed once in 200 μL of PBS. Washed cells were then resuspended in 100 μL of stain buffer and fluorescence was determined by flow cytometry using a FACS Array Bioanalyzer (BD Biosciences). FACS data were analyzed using FLOWJO software version 8.8.6. Effects on receptor content were expressed as fold-change in the staining intensity of treated versus control cells. Treatment effect on receptor levels was assessed relative to untreated control cells using a one-way analysis of variance with Tukey's post-test for multiple comparisons.

Determination of OxC-beta Effect on Cytokine Profiles in LPS Challenged Monocytes

We hypothesized that OxC-beta may tolerize or prime immune cells to respond to evidence of an invading pathogen based on observations from an earlier animal trial (unpublished). We chose LPS challenge of THP-1 cells as an *in vitro* model to evaluate the potential priming effects of OxC-beta. THP-1 cells were pretreated with the indicated concentrations of OxC-beta (0.1, 0.5, 1.0 μM) for 24 hours, at which point OxC-beta was removed and cells were cultured in the absence of OxC-beta for an additional 5 days. Cells were then challenged with LPS (15 ng/mL) for 24 hours prior to collection of conditioned media for evaluation of cytokine levels. OxC-beta treatment concentrations and the timeline were selected to best model availability of the compound within the host in the earlier animal study.

Analysis of cytokine levels in conditioned media was performed using Endogen Human ELISA kits (Pierce) according to manufacturer's instructions. Conditioned media were prepared by centrifugation to remove cellular debris and used neat, diluted with complete medium, or concentrated using Nanosep 3K centrifugal concentrators (Pall) to ensure that cytokine levels fell within the linear ranges of each assay. Where appropriate, samples were stored at −80°C and thawed by gradual equilibration at room temperature prior to use. Briefly, 50 μL samples were added to each well of a microplate to which antibody specific to the cytokine of interest had been adsorbed and incubated at room temperature for 1–3 hours. Plates were washed three times to remove nonspecifically bound material and incubated for an additional 1–3 hours with biotinylated antibody specific to the cytokine of interest. After washing, plates were incubated for 30 minutes with streptavidin-horse radish peroxidase reagent followed by an additional washing cycle. Washed plates were incubated for 30 minutes with 3,3′,5,5′-tetramethylbenzidine (TMB) substrate, the reaction was stopped and absorbance was measured at 450 nm (550 nm reference). Reference curves were generated for each cytokine using the supplied recombinant standard.

Phagocytosis

THP-1 cells were seeded in 96-well plates (1×10^5 cells/well) in complete media and allowed to recover for 24 hours prior to treatment. Cells were then treated with the indicated concentrations of OxC-beta, the equivalent concentration (v/v) of DMSO (vehicle control), or 25 ng/mL PMA for 24 hours, at which point the compounds were removed and the cells were cultured for an additional 24 hours in complete media containing 15 ng/mL lipopolysaccharide (LPS). Phagocytosis was evaluated using a Vybrant Phagocytosis assay kit (Invitrogen, V6694) based upon the ingestion of fluorescein-labeled *E. coli* (strain K12) bacterial particles. Briefly, treated cells were incubated at 37°C for 5 hours with a 100 μL suspension of fluorescent bioparticles in Hank's buffered salt solution. Following incubation, the suspension was removed and replaced with 100 μL of 2% trypan blue solution for 1 minute. The trypan blue solution was removed and the number of ingested particles determined using a fluorescence microplate reader (480 nm excitation, 520 nm emission). Wells containing only medium (no cells) served as negative reaction controls against which each experimental replicates were equalized. The OxC-beta effect on phagocytosis was assessed relative to the corresponding DMSO control using Student's *t*-test.

Measurement of RAR Agonist Activity

MCF-7 cells were seeded in T-25 flasks at a density of 5×10^5 cells/flask in 5 mL of complete media. Following a 24-hour recovery period cells were exposed to one of 8 treatment conditions for 24 hours. Three OxC-beta concentrations were tested; 2.5 μM (1.34 μg/mL), 5.0 μM (2.67 μg/mL), and 25 μM (13.4 μg/mL) along with the corresponding three DMSO vehicle controls (0.025%, 0.05%, and 0.25% v/v). Cells treated with 1.0 μM of all-trans retinoic acid served as the positive control while untreated cells served as the negative control.

After 24 hours of treatment cells were trypsinized, collected by centrifugation, and processed for total RNA isolation using the standard protocol of the RNeasy kit (Qiagen 74104). The yield and quality of RNA were verified by absorbance at 260 nm and 280 nm and gel electrophoresis. First strand cDNA was synthesized, according to the manufacturer's instructions, using Super Script III (Invitrogen 18080-051) reverse transcriptase with random primers and 1 μg of total RNA as template in a reaction volume of 20 μL. Following completion of the synthesis reaction the cDNA was diluted 1:1 with 20 μL of sterile water and 5 μL of the diluted cDNA was used as template in quantitative real-time PCR (QRT-PCR) assays.

OxC-beta's effect on the expression of CYP26A, a known RAR responsive gene [23] was assessed by QRT-PCR using β-actin as a reference gene. Primer sequences for each gene are as follows:

CYP26A Forward TTCAACCGAACTCCTCTTTGGA.

CYP26A Reverse CTTACTCTTCAGCTCTTCTCG-CACTT.

β-actin Forward TCATGAAGTGTGACGTGGACATC.

β-actin Reverse CAGGAGGAGCAATGATCTTGATCT.

Reactions were carried out in a total volume of 25 μL containing 12.5 μL of 2 x Platinum® SYBR® Green qPCR Super Mix-UDG (Invitrogen 11733-046), 0.5 μL each of forward and reverse primer (10 μM), 2.5 μL of diluted cDNA, and 9 μL sterile water. All reactions were run in triplicate on a Rotor-Gene 6500 HRM instrument (Corbett Life Sciences) running software version 1.7 with the following cycling parameters: 10 minutes at 50°C, 5 minutes at 92°C, 40 cycles of (45 seconds at 95°C, 25 seconds at 62°C, and 30 seconds at 70°C). The presence of a single amplified product in each reaction was verified by melt analysis.

Expression of CYP26A was calculated relative to β-actin using the two standard curve method. The mean ratio of relative expression was calculated based on 3 replicate experiments and treatment effects were tested for statistical significance by one-way ANOVA with Tukey's test for multiple comparisons.

CD14 Receptor Assay

The CD14 receptor assay is based on research demonstrating that OxC-beta treatment of THP-1 cells induced an increase in the surface content of CD14 receptors. The procedure for the receptor assay is described below.

Human THP-1 monocytes were seeded (5×10^5 cells/well) in 3 mL of media into 6-well culture plates and allowed to recover for 24 hours prior to treatment. Following recovery cells were treated for 24 hours with 0, 1.0, 2.5, 5.0, and 10 μM of the test compound before being processed for CD14 membrane content via FACS analysis. Cells were harvested for FACS analysis by centrifugation at 300×g for 5 minutes and then washed once by resuspension in 1 mL of 1 x phosphate buffered saline (PBS) and centrifugation at 300×g for 5 minutes. Washed cells were next resuspended in staining buffer (1 x PBS, 0.1% BSA) and the cell density of each suspension was determined by manual counting with a hemocytometer. Suspensions from each treatment group were then transferred, in 100 μL aliquots, into triplicate wells on a 96-well, round bottom microplate (1×10^5 cells/well). 10 μL of normal mouse serum was then added to each well and cells were incubated for 30 minutes at 4°C. Cells were then stained for CD14 by the addition of 10 μL (0.2 μg) of phycoerythrin (PE) conjugated human anti CD14 antibody (eBioscience 12–0149) and incubation for 1 hour at 4°C. Cells stained with PE conjugated anti-mouse IgG1 (eBioscience 12–4714) served as isotype controls. Following incubation with the antibodies cells were pelleted by centrifugation at 300×g for 5 minutes and washed once in 200 μL of wash buffer. Following washing cells were resuspended in 100 μL of stain buffer and evaluated for CD14 content by FACS analysis using a FACS Array Bioanalyzer (BD Biosciences). FACS data were analyzed using FLOWJO software version 8.8.6. Test compound effects on CD14 levels were expressed as fold-change in the number of CD14 positive cells relative to untreated controls.

Effect of OxC-beta Feeding on Gut Immune Receptor Expression

Animals. Twenty-eight Balb/c mice, 2-weeks of age with body weights ranging from 18–20 g, were purchased from Charles River Laboratories, Canada. Mice were placed in appropriate holding facilities at the Atlantic Veterinary College (Charlottetown, PE, Canada) and were handled according to the institutional guidelines for animal experimentation. The animal use protocol for this study was approved by the University of Prince Edward Island's Animal Care Committee. Mice were housed under constant environmental conditions with 12 hours light/12 hours dark cycles and ad libitum access to food and water. An acclimation period of 1-week was completed prior to beginning the study. On day 0 mice were distributed into 7 cages with 4 mice per cage. Each cage contained 3 treated mice and 1 control mouse. Individual mice within each cage were identified by banding the tail using a felt marker. Treated mice were divided into low dose (1 mg OxC-beta/kg body weight) or high dose (10 mg OxC-beta/kg body weight) groups and each dose group was then further divided into either a 2-week or 4-week treatment period. OxC-beta or vehicle control was administered daily by oral gavage for the duration of the study. Gavage solutions were prepared by diluting stock solutions of OxC-beta in ethanol with saline to give the appropriate dose of OxC-beta in 1% ethanol. In order to account for growth of the mice during the trial mice were weighed every second day and the volume of gavage solution administered was adjusted to maintain a constant dose.

Processing of Tissue Samples

At the conclusion of the 2 and 4-week supplementation periods mice were euthanized and necropsied for isolation of the small intestine. Following removal of the digestive contents the intestinal tissue was opened with a lateral incision and the mucosal surface was gently rinsed with saline. The tissue sample was next separated into 3 sections, ileum, jejunum and duodenum and each section was fixed with 10% buffered formalin for 16 hours. Following fixing, tissue samples were embedded in paraffin blocks and were sectioned for preparation of slides.

Immunohistochemical Staining for CD14 and TLR-4

Tissue sections were stained using primary polyclonal antibodies against murine CD14 (Abcam ab25090) or TLR-4 (Abcam ab53629) and the VectaStain Elite ABC kit (Vector Laboratories PK 6100 and PK 6101) according to the manufacturer's instructions. Bound antibodies were detected by incubation in DAB peroxidase substrate solution (Vector Laboratories SK-4100). Cover slips were applied to slides using permount medium and slides were allowed to dry before analysis using light microscopy. Slides were analyzed by observation under a light microscope at 10x, 20x and 40x magnifications.

Author Contributions

Conceived and designed the experiments: JBJ JGN GWB. Performed the experiments: JGN JD TJM. Analyzed the data: JBJ JGN GWB. Contributed reagents/materials/analysis tools: JD TJM JBJ JGN. Contributed to the writing of the manuscript: GWB JGN JBJ.

References

1. Peto R, Doll R, Buckley JD, Sporn MB (1981) Can dietary beta-carotene materially reduce human cancer rates? Nature 290: 201–208.
2. Chew B, Park J (2009) The Immune System. In: Britton G, Liaaen-Jensen S, Pfander H, editors. Carotenoids. Volume 5: Nutrition and Health. Basel: Birkhäuser Verlag. 363–382.
3. Palozza P, Serini S, Ameruso M, Verdecchia S (2009) Modulation of intracellular signalling pathways by carotenoids. In: Britton G, Liaaen-Jensen S, Pfander H, editors. Carotenoids. Volume 5: Nutrition and Health. Basel: Birkhäuser Verlag. 211 231.
4. Yeum, Aldini G, Russell RM, Krinsky NI (2009) Antioxidant/pro-oxidant actions of carotenoids. In: Britton G, Liaaen-Jensen S, Pfander H, editors. Carotenoids. Volume 5: Nutrition and Health. Basel: Birkhäuser Verlag. 235–268.
5. Burton GW, Ingold KU (1984) ß-Carotene: an unusual type of lipid antioxidant. Science 224: 569–573.
6. The Alpha-Tocopherol Beta Carotene Cancer Prevention Study Group (1994) The effect of vitamin E and beta carotene on the incidence of lung cancer and other cancers in male smokers. New Engl J Med 330: 1029–1035.
7. Hennekens CH, Buring JE, Manson JE, Stampfer M, Rosner B, et al. (1996) Lack of effect of long-term supplementation with beta carotene on the incidence of malignant neoplasms and cardiovascular disease. New Engl J Med 334: 1145–1149.
8. Omenn GS, Goodman GE, Thornquist MD, Balmes J, Cullen MR, et al. (1996) Effects of a combination of beta carotene and vitamin A on lung cancer and cardiovascular disease. New Engl J Med 334: 1150–1155.
9. Sayin VI, Ibrahim MX, Larsson E, Nilsson JA, Lindahl P, et al. (2014) Antioxidants accelerate lung cancer progression in mice. Sci Transl Med 6: 221ra15. DOI: 10.1126/scitranslmed.3007653.
10. Chew BP, Park JS, Wong TS, Kim HW, Weng BB, et al. (2000) Dietary beta-carotene stimulates cell-mediated and humoral immune response in dogs. J Nutr 130: 1910–1913.
11. Blount JD, Metcalfe NB, Birkhead TR, Surai PF (2003) Carotenoid modulation of immune function and sexual attractiveness in zebra finches. Science 300: 125–127.
12. Chew BP, Park JS (2004) Carotenoid action on the immune response. J Nutr 134: 257S–261S.
13. Park JS, Chyun JH, Kim YK, Line LL, Chew BP (2010) Astaxanthin decreased oxidative stress and inflammation and enhanced immune response in humans. Nutr Metab 7: 18–27.
14. Chew BP, Mathison BD, Hayek MG, Massimino S, Reinhart GA, et al. (2011) Dietary astaxanthin enhances immune response in dogs. Vet Immunol Immunopathol 140: 199–206.
15. Britton G (2008) Functions of carotenoid metabolites and breakdown products. In: Britton G, Liaaen-Jensen S, Pfander H, editors. Carotenoids. Volume 4: Natural Functions. Basel: Birkhäuser Verlag. 309–324.
16. Wang X-D (2009) Biological activities of carotenoid metabolites. In: Britton G, Liaaen-Jensen S, Pfander H, editors. Carotenoids. Volume 5: Nutrition and Health. Basel: Birkhäuser Verlag. 383–408.
17. Burton GW, Daroszewski J, Nickerson JG, Johnston JB, Mogg TJ, et al. (2014) ß-Carotene autoxidation: oxygen copolymerization, non-vitamin A products and immunological activity. Can J Chem 92: 305–316. DOI: 10.1139/cjc-2013-0494.
18. Shaw G (1971) The chemistry of sporopollenin. In: Brooks J, Grant PR, Muir M, Gijzel PV, Shaw G, editors. Sporopollenin. London: Academic Press. 305–348.
19. Brooks J, Shaw G (1968) Chemical structure of the exine of pollen walls and a new function for carotenoids in nature. Nature 219: 532–533.
20. Brooks J, Shaw G (1971) Recent developments in the chemistry, biochemistry, geochemistry and post-tetrad ontogeny of sporopollenins derived from pollen and spore exines. In: Heslop-Harrison J, editor. Pollen Development and Physiology. London: Butterworths. 99–114.
21. Brooks J, Shaw G (1978) Sporopollenin: a review of its chemistry, palaeochemistry and geochemistry. Grana 17: 91–97.
22. Hall JA, Grainger JR, Spencer SP, Belkaid Y (2011) The role of retinoic acid in tolerance and immunity. Immunity 35: 13–22.
23. Loudig O, Babichuk C, White J, Abu-Abed S, Mueller C, et al. (2000) Cytochrome P450RAI(CYP26) promoter: a distinct composite retinoic acid response element underlies the complex regulation of retinoic acid metabolism. Mol Endocrinol 14: 1483–1497.
24. Huang Q, Xu X, Mao YL, Huang Y, Rajput IR, et al. (2013) Effects of Bacillus subtilis B10 spores on viability and biological functions of murine macrophages. Anim Sci J 84: 247–252.
25. Perez-Fons L, Steiger S, Khaneja R, Bramley PM, Cutting SM, et al. (2011) Identification and the developmental formation of carotenoid pigments in the yellow/orange Bacillus spore-formers. Biochim Biophys Acta 1811: 177–185.
26. Brooks J (1971) Some chemical and geochemical studies of sporopollenin. In: Brooks J, Grant PR, Muir M, Gijzel PV, Shaw G, editors. Sporopollenin. London: Academic Press. 351–407.

27. Blackwell LJ (2007) Sporopollenin exines as a novel drug delivery system. Ph.D. thesis: University of Hull. Available: https://hydra.hull.ac.uk/resources/hull:7162.

28. Duquette SC, Fischer CD, Feener TD, Muench GP, Morck DW, et al. (2014) Anti-inflammatory benefits of retinoids and carotenoid derivatives: retinoic acid and fully oxidized β-carotene induce caspase-3-dependent apoptosis and promote efferocytosis of bovine neutrophils. Am J Vet Res. 75: in press.

Permissions

All chapters in this book were first published in PLOS ONE, by The Public Library of Science; hereby published with permission under the Creative Commons Attribution License or equivalent. Every chapter published in this book has been scrutinized by our experts. Their significance has been extensively debated. The topics covered herein carry significant findings which will fuel the growth of the discipline. They may even be implemented as practical applications or may be referred to as a beginning point for another development.

The contributors of this book come from diverse backgrounds, making this book a truly international effort. This book will bring forth new frontiers with its revolutionizing research information and detailed analysis of the nascent developments around the world.

We would like to thank all the contributing authors for lending their expertise to make the book truly unique. They have played a crucial role in the development of this book. Without their invaluable contributions this book wouldn't have been possible. They have made vital efforts to compile up to date information on the varied aspects of this subject to make this book a valuable addition to the collection of many professionals and students.

This book was conceptualized with the vision of imparting up-to-date information and advanced data in this field. To ensure the same, a matchless editorial board was set up. Every individual on the board went through rigorous rounds of assessment to prove their worth. After which they invested a large part of their time researching and compiling the most relevant data for our readers.

The editorial board has been involved in producing this book since its inception. They have spent rigorous hours researching and exploring the diverse topics which have resulted in the successful publishing of this book. They have passed on their knowledge of decades through this book. To expedite this challenging task, the publisher supported the team at every step. A small team of assistant editors was also appointed to further simplify the editing procedure and attain best results for the readers.

Apart from the editorial board, the designing team has also invested a significant amount of their time in understanding the subject and creating the most relevant covers. They scrutinized every image to scout for the most suitable representation of the subject and create an appropriate cover for the book.

The publishing team has been an ardent support to the editorial, designing and production team. Their endless efforts to recruit the best for this project, has resulted in the accomplishment of this book. They are a veteran in the field of academics and their pool of knowledge is as vast as their experience in printing. Their expertise and guidance has proved useful at every step. Their uncompromising quality standards have made this book an exceptional effort. Their encouragement from time to time has been an inspiration for everyone.

The publisher and the editorial board hope that this book will prove to be a valuable piece of knowledge for researchers, students, practitioners and scholars across the globe.

List of Contributors

Chris Greening, Jennifer R. Robson and Gregory M. Cook
University of Otago, Department of Microbiology and Immunology, Dunedin, New Zealand

Silas G. Villas-Bôas
University of Auckland, The Centre for Microbial Innovation, Auckland, New Zealand

Michael Berney
University of Otago, Department of Microbiology and Immunology, Dunedin, New Zealand
Albert Einstein College of Medicine, Department of Microbiology and Immunology, Bronx, New York, United States of America

Colin P. McCoy, John F. Cowley, Louise Carson, Áine T. De Baróid,
Greg T. Gdowski, Sean P. Gorman and David S. Jones
Queen's University Belfast, School of Pharmacy, Belfast, United Kingdom

Edward J. ÓNeil
Blue Highway, Inc., Center for Science & Technology, Syracuse University, Syracuse, New York, United States of America

Jun Shigeto, Mariko Nagano, Koki Fujita and Yuji Tsutsumi
Faculty of Agriculture, Kyushu University, Fukuoka, Japan

Yu-Sang Li, Bei-Fan Chen, Xiao-Jun Li, Wei Kevin Zhang and He-Bin Tang
Department of Pharmacology, College of Pharmacy, South-Central University for Nationalities, Wuhan, PR China

Hai-Chao Zhou
Key Laboratory of the Ministry of Education for Coastal and Wetland Ecosystems, Xiamen University, Xiamen, China
Department of Biology and Chemistry, City University of Hong Kong, Hong Kong SAR, China
Futian-CityU Mangrove R&D Centre, City University of Hong Kong Shenzhen Research Institute, Shenzhen, China

Nora Fung-yee Tam
Department of Biology and Chemistry, City University of Hong Kong, Hong Kong SAR, China
Futian-CityU Mangrove R&D Centre, City University of Hong Kong Shenzhen Research Institute, Shenzhen, China

Yi-Ming Lin, Zhen-Hua Ding, Wei-Ming Chai and Shu-Dong Wei
Key Laboratory of the Ministry of Education for Coastal and Wetland Ecosystems, Xiamen University, Xiamen, China

Jianing Wu, Shaoze Yan, Jieliang Zhao and Yuying Ye
Division of Intelligent and Biomechanical Systems, State Key Laboratory of Tribology, Department of Mechanical Engineering, Tsinghua University, Beijing, P. R. China

Camila Hochman-Mendez
Institute of Biomedical Sciences, Federal University of Rio de Janeiro, Rio de Janeiro, Brazil
Institute of Biophysics Carlos Chagas Filho, Federal University of Rio de Janeiro, Rio de Janeiro, Brazil

Marco Cantini and Manuel Salmeron-Sanchez
Division of Biomedical Engineering, School of Engineering, University of Glasgow, Glasgow, United Kingdom

David Moratal
Center for Biomaterials and Tissue Engineering, Universitat Politècnica de València, València, Spain

Tatiana Coelho-Sampaio
Institute of Biophysics Carlos Chagas Filho, Federal University of Rio de Janeiro, Rio de Janeiro, Brazil

Anirban Mitra and Rachel Misquitta
Department of Microbiology and Cell Biology, Indian Institute of Science, Bangalore, India

Valakunja Nagaraja
Department of Microbiology and Cell Biology, Indian Institute of Science, Bangalore, India
Jawaharlal Nehru Centre for Advanced Scientific Research, Bangalore, India

Lea H. Eckhard and Nurit Beyth
Department of Prosthodontics, the Hebrew University – Faculty of Dental Medicine, Jerusalem, Israel

Asaf Sol and Gilad Bachrach
Institute of Dental Science, the Hebrew University - Faculty of Dental Medicine, Jerusalem, Israel

Ester Abtew and Abraham J. Domb
Institute for Drug Research, School of Pharmacology, Faculty of Medicine, the Hebrew University, Jerusalem, Israel

Yechiel Shai
Department of Biological Chemistry, the Weizmann Institute of Science, Rehovot, Israel

Alvaro Díaz-Barrera, Fabiola Martínez, Felipe Guevara Pezoa and Fernando Acevedo
Escuela de Ingeniería Bioquímica, Pontificia Universidad Catolica de Valparaiso, Valparaiso, Chile

Jaqueline N. Silva
Departamento de Engenharia Bioquímica, Escola de Química, Universidade Federal do Rio de Janeiro, Rio de Janeiro, Brazil,

Mateus G. Godoy and Denise M. G. Freire
Departamento de Bioquímica, Instituto de Química, Universidade Federal do Rio de Janeiro, Rio de Janeiro, Brazil

Melissa L. E. Gutarra
Departamento de Engenharia Bioquímica, Escola de Química/Polo Xerém, Universidade Federal do Rio de Janeiro, Rio de Janeiro, Brazil

Daniela Quinteros, Santiago Palma and Daniel Allemandi
Department of Pharmacy, Facultad de Ciencias Químicas, Universidad Nacional de Córdoba, CONICET, Edificio de Ciencias II, Ciudad Universitaria, Córdoba, Argentina

Marta Vicario-de-la-Torre, Vanessa Andrés-Guerrero, Rocío Herrero-Vanrell and Irene T. Molina-Martínez
Department of Pharmacy and Pharmaceutical Technology, Faculty of Pharmacy, Complutense University of Madrid, Pharmaceutical Innovation in Ophthalmology
Research Group, Sanitary Research Institute of the San Carlos Clinical Hospital (IdISSC) and the Ocular Pathology National Net (OFTARED) of the Institute of Health Carlos III, Madrid, Spain

Caroline Louis, Gilles Tinant, Eric Mignolet and Cathy Debier
Institut des Sciences de la Vie, Université catholique de Louvain, Louvain-la-Neuve, Belgium

Jean-Pierre Thomé
Laboratoire d'Ecologie animale et d'Ecotoxicologie, Universitéde Liége, Liège, Belgium

Heejun Choi, Saswata Chakraborty and Runhui Liu
Department of Chemistry, University of Wisconsin-Madison, Madison, Wisconsin, United States of America

Samuel H. Gellman and James C. Weisshaar
Department of Chemistry, University of Wisconsin-Madison, Madison, Wisconsin, United States of America
Molecular Biophysics Program, University of Wisconsin-Madison, Madison, Wisconsin, United States of America

Jun Jia, Zhiliang Wu, Shengqiu Feng, Peng Chen, Pengyan Huang, Leiming Wu and Liangcai Peng
National Key Laboratory of Crop Genetic Improvement and National Centre of Plant Gene Research (Wuhan), Huazhong Agricultural University, Wuhan, P.R. China
Biomass and Bioenergy Research Centre, Huazhong Agricultural University, Wuhan, P.R. China
College of Plant Science and Technology, Huazhong Agricultural University, Wuhan, P.R. China

Bin Yu and Ming Li
National Key Laboratory of Crop Genetic Improvement and National Centre of Plant Gene Research (Wuhan), Huazhong Agricultural University, Wuhan, P.R. China
Biomass and Bioenergy Research Centre, Huazhong Agricultural University, Wuhan, P.R. China
College of Life Science and Technology, Huazhong Agricultural University, Wuhan, P.R. China

Yonglian Zheng
National Key Laboratory of Crop Genetic Improvement and National Centre of Plant Gene Research (Wuhan), Huazhong Agricultural University, Wuhan, P.R. China
College of Life Science and Technology, Huazhong Agricultural University, Wuhan, P.R. China

Hongwu Wang
Institute of Crop Sciences, Chinese Academy of Agricultural Sciences, Beijing, P.R. China

Takanobu Nishizuka, Toshikazu Kurahashi, Tatsuya Hara and Hitoshi Hirata
Department of Hand Surgery, Nagoya University Graduate School of Medicine, Nagoya, Japan

Toshihiro Kasuga
Department of Frontier Materials, Nagoya Institute of Technology, Nagoya, Japan

María Gomariz, Salvador Blaya, Pablo Acebal and Luis Carretero
Departamento de Ciencia de Materiales, Óptica y Tecnología Electrónica, Universidad Miguel Hernández, Elx (Alicante), Spain

Zhengyu Ma
Department of Biomedical Research, Nemours/A.I. duPont Hospital for Children, Wilmington, Delaware, United States of America

David N. LeBard
Department of Chemistry, Yeshiva University, New York, New York, United States of America

Sharon M. Loverde
Department of Chemistry, College of Staten Island, City University of New York, Staten Island, New York, United States of America

Kim A. Sharp
Department of Biochemistry and Biophysics, University of Pennsylvania, Philadelphia, Pennsylvania, United States of America

Michael L. Klein
Institute for Computational Molecular Science and Department of Chemistry, Temple University, Philadelphia, Pennsylvania, United States of America

Dennis E. Discher
Department of Chemical and Biomolecular Engineering, University of Pennsylvania, Philadelphia, Pennsylvania, United States of America

Terri H. Finkel
Department of Pediatrics, Nemours Children's Hospital, Orlando, Florida, United States of America
Department of Biomedical Sciences, University of Central Florida College of Medicine, Orlando, Florida, United States of America

Hans-Peter M. de Hoog, Esther M. Lin JieRong and Sourabh Banerjee
ACM Biolabs Pte Ltd, Research Techno Plaza XF-6, Singapore, Singapore

Fabien M. Décaillot
Singapore Immunology Network, Agency for Science, Technology and Research (A*STAR), Singapore, Singapore

Madhavan Nallani
ACM Biolabs Pte Ltd, Research Techno Plaza XF-6, Singapore, Singapore
Centre for Biomimetic Sensor Science, School of Materials Science and Engineering, Nanyang Technological University, Singapore, Singapore

Yongjun Sun, Huaili Zheng, Jun Zhai, Chun Zhao, Chuanliang Zhao and Yong Liao
Key laboratory of the Three Gorges Reservoir Region's Eco-Environment, State Ministry of Education, Chongqing University, Chongqing, China
National Centre for International Research of Low-carbon and Green Buildings, Chongqing University, Chongqing, China

Houkai Teng
CNOOC Tianjin Chemical Research and Design Institute, Tianjin, China

Marc Bou Zeidan, Giacomo Zara, Ilaria Mannazzu, Marilena Budroni and Severino Zara
Dipartimento di Agraria, University of Sassari, Sassari, Italy

Carlo Viti, Francesca Decorosi and Luciana Giovannetti
Dipartimento di Scienze delle Produzioni Agroalimentari e dell'Ambiente, University of Florence, Firenze, Italy

Sophie S. Abby, Marie Touchon and Eduardo P. C. Rocha
Microbial Evolutionary Genomics, Institut Pasteur, Paris, France UMR3525, CNRS, Paris, France

Bertrand Néron and Hervé Ménager
Centre d'Informatique pour la Biologie, Institut Pasteur, Paris, France

Ping Zhou, Yi Deng and Shicheng Wei
Department of Oral and Maxillofacial Surgery, School and Hospital of Stomatology, Peking University, Beijing, China
Center for Biomedical Materials and Tissue Engineering, Academy for Advanced Interdisciplinary Studies, Peking University, Beijing, China

Beier Lyu and Hongwei Ma
Suzhou Institute of Nano-Tech and Nano-Bionics, Chinese Academy of Sciences, Suzhou, China

Ranran Zhang and Yalin Lyu
Department of Stomatology, Beijing Anzhen Hospital, Capital Medical University, Beijing, China

Hai Zhang
Department of Stomatology, Beijing Anzhen Hospital, Capital Medical University, Beijing, China
Department of Restorative Dentistry, School of Dentistry, University of Washington, Washington, United States of America

James B. Johnston
National Research Council of Canada, Charlottetown, Prince Edward Island, Canada

James G. Nickerson
Avivagen Inc., Charlottetown, Prince Edward Island, Canada

Janusz Daroszewski, Trevor J. Mogg and Graham W. Burton
Avivagen Inc., Ottawa, Ontario, Canada

Index

www.ingramcontent.com/pod-product-compliance
Lightning Source LLC
Chambersburg PA
CBHW080528200326
41458CB00012B/4365
* 9 7 8 1 6 3 2 3 8 5 8 8 8 *

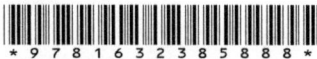